PEARSON

Introduction to Mechanical, Industrial, Manufacturing, and Energy Systems Engineering

Oregon State University

Taken from:
Introduction to Industrial and Systems Engineering, Third Edition
by Wayne C. Turner, Joe H. Mize, Kenneth E. Case, and
John W. Nazemetz

Design Concepts for Engineers, Fourth Edition
by Mark N. Horenstein

Strategies for Creative Problem Solving, Third Edition
by H. Scott Fogler and Steven E. LeBlanc, with Benjamin R. Rizzo

Introduction to Mechanical Engineering
by Robert Rizza

Modern Control Systems, Twelfth Edition
by Richard C. Dorf and Robert H. Bishop

Introduction to Engineering Analysis, Fourth Edition
by Kirk D. Hagen

The Basics of Occupational Safety, Second Edition
by David L. Goetsch

A User's Guide to Engineering
by James N. Jensen

Engineering Economy, Sixteenth Edition
by William G. Sullivan, Elin M. Wicks, and C. Patrick Koelling

Cover Art: Courtesy of Photodisc/Getty Images

Taken from:

Introduction to Industrial and Systems Engineering, Third Edition
by Wayne C. Turner, Joe H. Mize, Kenneth E. Case, and John W. Nazemetz
Copyright © 1993, 1987, 1978 by Pearson Education, Inc.
Published by Prentice Hall
Upper Saddle River, New Jersey 07458

Design Concepts for Engineers, Fourth Edition
by Mark N. Horenstein
Copyright © 2010, 2006, 2002 by Pearson Education, Inc.
Published by Prentice Hall

Strategies for Creative Problem Solving, Third Edition
by H. Scott Fogler and Steven E. LeBlanc, with Benjamin R. Rizzo
Copyright © 2014, 2013 by Pearson Education, Inc.
Published by Prentice Hall

Introduction to Mechanical Engineering
by Robert Rizza
Copyright © 2001 by Pearson Education, Inc.
Published by Prentice Hall

Modern Control Systems, Twelfth Edition
by Richard C. Dorf and Robert H. Bishop
Copyright © 2011, 2008, 2005, 2001 by Pearson Education, Inc.
Published by Prentice Hall

Introduction to Engineering Analysis, Fourth Edition
by Kirk D. Hagen
Copyright © 2014, 2009 by Pearson Education, Inc.
Published by Prentice Hall

The Basics of Occupational Safety, Second Edition
by David L. Goetsch
Copyright © 2015, 2010 by Pearson Education, Inc.
Published by Prentice Hall

A User's Guide to Engineering
by James N. Jensen
Copyright © 2006 by Pearson Education, Inc.
Published by Prentice Hall

Engineering Economy, Sixteenth Edition
by William G. Sullivan, Elin M. Wicks, and C. Patrick Koelling
Copyright © 2015, 2012, 2009, 2006, 2003, 1997 by Pearson Education, Inc.
Published by Prentice Hall

Pearson Learning Solutions, 501 Boylston Street, Suite 900, Boston, MA 02116
A Pearson Education Company
www.pearsoned.com

Printed in the United States of America

1 2 3 4 5 6 7 8 9 10 VOCR 17 16 15 14

000200010271931270

CB /KD

PEARSON

ISBN-10: 1-323-00591-9
ISBN-13: 978-1-323-00591-0

COPYRIGHT ACKNOWLEDGMENTS

Grateful acknowledgment is made to the following sources for permission to reprint material copyrighted or controlled by them:

Chapter 17: "Succeeding in the Engineering Classroom," taken from "Chapter 7: Succeeding in the Classroom," by William C. Oakes, Les L. Leone, and Craig J. Gunn, reprinted from *Engineering Your Future: A Comprehensive Introduction to Engineering, Fifth edition* (2006), by permission of Oxford University Press.

Chapter 4: "Materials Science," taken from "Chapter 5: Materials and Stresses," by Jonathan Wickert and Kemper Lewis, reprinted from *An Introduction to Mechanical Engineering, Third Edition* (2012), Cengage Learning.

BRIEF CONTENTS

CONTENTS

CREDITS

Chapter 1 taken from *Strategies for Creative Problem Solving*, Third Edition by H. Scott Fogler and Steven E. LeBlanc, with Benjamin R. Rizzo.

Chapters 2, 3, and 5 taken from *Introduction to Mechanical Engineering* by Robert Rizza.

Chapters 6 and 7 taken from *Design Concepts for Engineers*, Fourth Edition by Mark N. Horenstein.

Chapters 8, 11 and 12 taken from *Introduction to Industrial and Systems Engineering*, Third Edition by Wayne C. Turner, Joe H. Mize, Kenneth E. Case, and John W. Nazemetz.

Chapter 9 taken from *Modern Control Systems*, Twelfth Edition by Richard C. Dorf and Robert H. Bishop.

Chapter 10 taken from *Engineering Economy*, Sixteenth Edition by William G. Sullivan, Elin M. Wicks, and C. Patrick Koelling.

Chapter 13 taken from *The Basics of Occupational Safety*, Second Edition by David L. Goetsch.

Chapter 14 taken from *Introduction to Engineering Analysis*, Fourth Edition by Kirk D. Hagen.

Chapters 15 and 16 taken from *A User's Guide to Engineering* by James N. Jensen.

Appendix A taken from *Engineering Economy*, Sixteenth Edition by William G. Sullivan, Elin M. Wicks, and C. Patrick Koelling.

Appendix B taken from *Introduction to Engineering Analysis*, Fourth Edition by Kirk D. Hagen.

References taken from *Modern Control Systems*, Twelfth Edition by Richard C. Dorf and Robert H. Bishop.

1 PROBLEM SOLVING

The mere formulation of a problem is far more essential than its solution, which may be merely a matter of mathematical or experimental skill. To raise new questions, new possibilities, to regard old problems from a new angle requires creative imagination and marks real advances in science.

—Albert Einstein

The First Four Steps

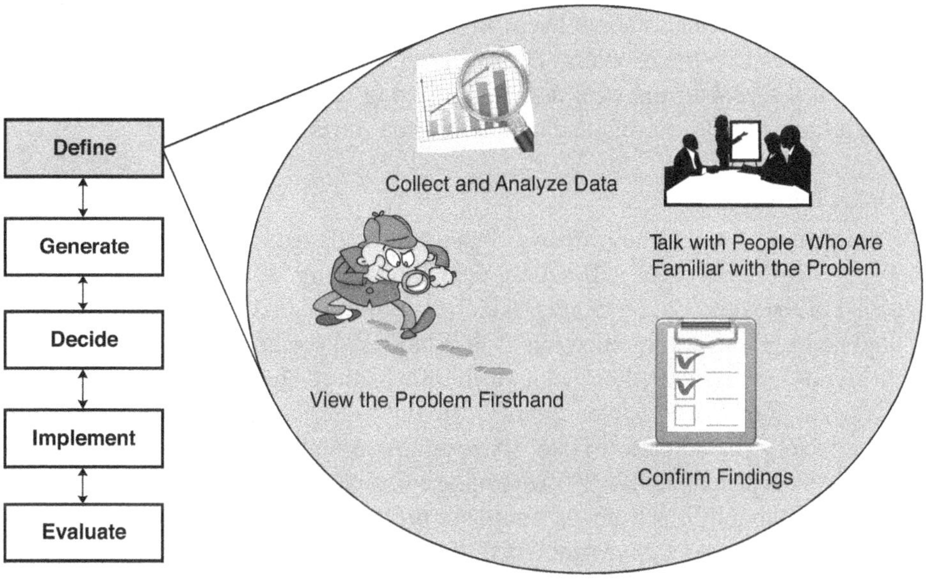

(Images from Shutterstock: iconico, Aleksandr Bryliaev, Nevena Radonja, gnurf)

Often, one of the most difficult aspects of problem solving is understanding and defining the real problem (sometimes referred to as the underlying or root problem). The first task in defining the real problem is to properly gather information about the problem.

THE FIRST FOUR STEPS

The first four steps to understand and define the real problem focus on gathering information.

"Start with an open mind."

"Don't jump to conclusions."

"Look at the big picture."

"Review the obvious."

1. Collect and analyze information and data.
2. Talk with people who are familiar with the problem.
3. If possible, view the problem firsthand.
4. Confirm all findings and continue to gather information.

Step 1: Collect and Analyze Information and Data

Learn as much as you can about the problem. Write down or list everything you can think of to describe the problem. Do an Internet search on all aspects of the problem. Until the problem is well defined, anything might be important. Determine which information is missing and which information is extraneous.

Visualize Information

The information should be properly organized, analyzed, and presented so that it can serve as the basis for subsequent decision making. Make a simple sketch or drawing of the situation. Drawings, sketches, graphs of data, and other illustrations can all be excellent communication tools when used correctly. Analyze the data to show trends, errors, and other meaningful information. Display numerical or quantitative data graphically. Tables can be difficult to interpret and are sometimes misleading (see "Presentation of Data" in Appendix 3 on the Web site). Graphing, by comparison, is an excellent way to organize and analyze large amounts of data. The business research of a private car company provides an interesting example of the use of graphical data to solve problems.

Everyone's Private Driver

A private car company was looking to improve business by reducing the downtime of its drivers by connecting empty private cars to people who want rides. The drivers could make extra money rather than being idle and the passengers get private car service on demand. The company wanted to answer the question, "Where do people want rides in the city?"

The project team began with their own ride history data for their city so they would know where to send idle drivers for the best odds to find customers. They examined their 40 most common pick-up spots in table form but couldn't really make any conclusions.

Street Addresses of the Most Common Pick-up Locations			
104 Maple	8 Mercer	755 Colomy	74 Euclid
556 State	65 George	3673 Perry	56 Rutgers
73 Shyne	466 Durante	745 5th	65 George
45 5th	1233 Perry	754 Harl	7 Grady
832 Evans	25 Nickols	88 9th	74 Kane
7765 Markley	44 Rivington	74 Highland	83 Naper
2343 Ann	23 Charles	99 Almond	890 Madison
56 Rutgers	655 4th	88 Alma	23 Nickols
899 Jane	843 Madison	6443 Burns	34 Broadway
544 Ludlow	6443 Broone	413 Burley	466 Durante

They then moved the data to graphic form by plotting all of the ride locations on a map of the city. As you can see from the graphic, it is much easier to get a sense of where demand for rides is high with the plotted data than a list of ride data.

City Map

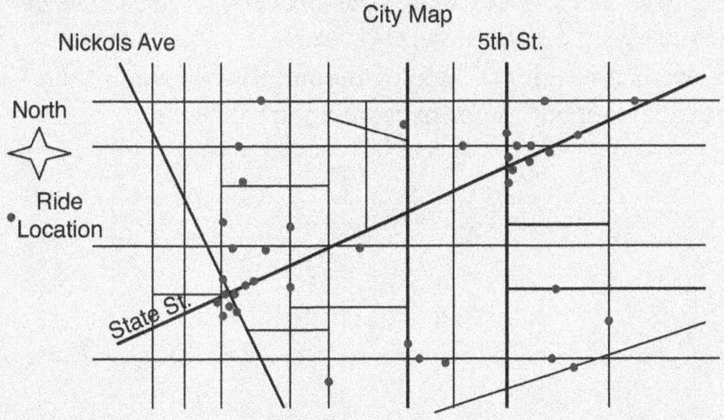

The company began deploying its drivers to the high-density areas, like the intersections of Nickols and 5th with State Street, and business increased immediately.

Order of Magnitude Calculation (Fermi Problems)

When collecting and analyzing data one would like to know what data are needed to calculate the final answer. Then, even before collecting data, one can use a Fermi analysis to obtain an order of magnitude estimate of how large the answer might be. This Fermi number can be used to validate your final answer, by checking that your final answer is within an order of magnitude of what you were expecting. Enrico Fermi was an Italian physicist who received the Nobel Prize for his work on nuclear processes. Fermi was famous for his "back-of-the-envelope order of magnitude calculation," which enabled him to obtain an estimate of the answer to a problem on as little space as the back of an envelope by applying logic and making reasonable assumptions for each number in the calculation of the answer. He used a process to set bounds on the answer by saying it is probably larger than one number and smaller than another, and he arrived at an answer that was within a **factor of 10.** When carrying out a Fermi calculation you should ask the following questions:

- What do you know is relevant?
- What assumptions can you make?
- How plausible are your assumptions?
- Is your chain of reasoning accurate?
- Can you do the problem another way and see if the result is the same?
- In your answer, did you spell out your assumptions, reasoning, solution, and checking procedure clearly?

Fermi calculations are useful when you need to get a reasonable estimate rather than a specific answer. It is important to be cognizant of poor assumptions and fallacies in logic as you estimate numbers, but in this case, the idea is to move forward quickly with a best guess to get your answer. If an exact answer is needed, a more thorough analysis can be done. For example, the National Association of Piano Tuners needs a rough estimate of the number of piano tuners in Chicago, rather than the exact number. A Fermi calculation can be used.

Fermi Example: Piano Tuners in Chicago

The National Association of Piano Tuners (NAPT) is launching a new training series, with its first training session taking place in Chicago. This is the first time such an event will take place, and the NAPT wants to contact all the piano tuners in Chicago to invite them to the training session where they will learn a new technique of piano tuning. The NAPT wants to get a rough idea of about how many piano tuners there are in Chicago so they know how much to budget for mailings and what size room to reserve. The event organizers use a Fermi calculation to get a number, which is outlined in the following.

The city of Chicago has around four million inhabitants. If we estimate that the average number of people living per household is four, then the total number of households in Chicago is around one million. Assuming one-fifth of the households own a piano, then there are 200,000 pianos in Chicago. If an additional assumption is made that a piano needs to be tuned once each year the total of tunes per year would be 200,000 in Chicago. Tuning a piano takes four to five hours, which is two per day. Furthermore, the piano tuners work approximately 250 days per year and will perform 500 tunes per year, since they tune two pianos each day. That brings us to the final calculation of the Fermi Problem calculation of the number of piano tuners in Chicago: 200,000 tunes per year divided by 500 tunes per year for each piano tuner gives us 400 tuners.

Population of Chicago: 4,000,000

Number of people per household: 4

$$\frac{4{,}000{,}000 \textit{ people}}{4 \textit{ people per household}} = 1{,}000{,}000 \textit{ households}$$

Continues

Number of households: 1,000,000

Fraction of households with pianos: 1/5

$$1{,}000{,}000\ \textit{households} \times \frac{1}{5}\ \textit{households own a piano} = 200{,}000\ \textit{pianos in Chicago}$$

Households with pianos: 200,000

Number of tunes each piano needs per year: 1

$$1\frac{tune}{year} \times 200{,}000\ pianos = 200{,}000\frac{tunes}{year}$$

Tunes needed by pianos per year: 200,000

Number of days tuners work per year: 250

Number of tunes each tuner does per day: 2

$$250\frac{workdays}{year} \times 2\frac{tunes}{day} = 500\frac{tunes}{year}$$

Number of tunes by each tuner per year: 500

Number of tuners required for 200,000 tunes per year: 400

$$200{,}000\frac{tunes}{year} \times \frac{1\ tuner}{500\frac{tunes}{year}} = 400\ Tuners$$

Cautions: As you go along in your calculations, always question the strength of your assumptions and fallacies in logic. For example, "Does one out of every five households in Chicago really own a piano?" It's not all the households and it's not one out of 30. It's reasonable to believe that one out of five is pretty close, so we move forward.

The NAPT knew that the Fermi calculation was probably a bit lower than the actual value, because the calculations did not include churches and schools (see Exercise 1.8), so it planned the event accordingly.

Epilogue: The NAPT had 426 piano tuners for its training session in Chicago. The Fermi calculation of 400 allowed the NAPT to budget for mailings and reserve a room months ahead of time, without knowing the exact number.

Source: http://mathforum.org/workshops/sum96/interdisc/sheila2.html.

Step 2: Talk with People Who Are Familiar with the Problem

Ask insightful questions.

Find out who knows about the problem and interview them. Ask penetrating questions using six types of Socratic questions and by doing the following.

- Look past the obvious.
- Ask for clarification when you do not understand something.
- Ask when the problem first occurred.
- Ask how the problem occurred.
- Challenge the basic premise of any explanations given as to the cause of the problem. Ask for *reasons and evidence* to support their premise.
- Ask where the problem is located and where everything is okay.
- Ask what was observed and whether it had been observed previously.
- Ask who else knows about the problem.
- Probe the answers you receive with follow-up questions.
- Listen for what is ***not*** being said as well as what is being said.

The most important thing in communication is to hear what is NOT being said.
—Peter Drucker

Depending on the particular response you receive, additional probing with follow-up questions may prove particularly helpful. Our experience shows that seemingly naive questions (often perceived as "dumb" questions) can produce profound results by challenging established thinking patterns. This act of challenging must be an ongoing process.

You should also describe the problem to other people. Verbalizing the problem to someone else helps clarify in your own mind just what it is you are trying to do. Try to find out who the experts in the field are—and then talk to them.[1]

When interviewing people to collect data and information, try to use as many of the types of critical thinking questions as you can. The following example depicts a situation where Socratic questioning of an individual familiar with the problem is used to find and uncover the real problem.

The Anti-Crime Commissioner

A commissioner ran for office and won on the platform that he would clean up the streets and reduce crime throughout the city. He increased police patrol in high-crime areas and instituted sweeping changes in the judicial system to increase sentence durations; despite these changes, crime rates rose. The commissioner hired a private consulting firm and charged them with the task of solving the following perceived problem: *Find out why more crimes are being committed.*

The consulting firm interviewed the commissioner and some of his officers and asked the following series of Socratic questions to get to the answer.

Questions about the question

Do all crimes get reported in the crime-rate data?

The commissioner says that crimes get reported only when the person thinks the police can help them.

Questions for clarification

Is a higher "reported" crime rate necessarily a bad thing?

The commissioner realizes that a higher reported crime rate does not necessarily mean that more crimes are being committed. Instead, the people are more likely to report committed crimes due to increased confidence in the police department since the addition of more police patrols.

Questions that probe assumptions

Why did the commissioner assume that increased patrols would decrease the crime rate?

The commissioner says that with more police on the streets, the response to a reported crime would be more rapid, and criminals would be more likely to get caught and therefore less likely to commit crimes.

Questions that probe reasons and evidence

What is the commissioner's evidence that there is an increase in the crime rate?

The commissioner says the crime rate is the number of crimes reported in his city during a given time period and this number has increased.

Questions that probe viewpoints and perspectives

Do criminals ever think that with more police they would be more likely to be caught?

The commissioner says that most criminals report that they never thought that they would be caught.

Questions that probe implications and consequences

Have increased patrols and the new changes in the judicial system influenced the public's view of the police?

A town poll shows that the public is now much more confident in the ability of the police force.

Conclusion: The instruction given by the commissioner was to solve the perceived problem: Find out why the number of crimes committed has increased. The real problem was rather to find *why the reported crime rate had increased*. When the firm researched this problem, they found that people had more confidence in the police department so they reported more of the crimes that were committed.

When different individuals interviewed give different interpretations of a situation, it is imperative to be aware that their individual biases can obscure or mask the actual facts. Separate the answers into those that are fact, opinionated fact, and opinion.

<table>
<tr><th colspan="2">Fact</th></tr>
<tr><td>Facts usually come from measurements, reports, tables, figures, firsthand observations, and other data.</td><td>“On August 16th, 45 fish were found dead in the river.”</td></tr>
<tr><th colspan="2">Opinionated Fact</th></tr>
<tr><td>Opinionated facts use phrases to put a bias or spin on the factual data, in an effort to persuade others to adopt a particular point of view.</td><td>The opinionated phrase may denote the significance of the fact (“only 45 fish died”), attach value to the fact (“surprisingly 45 fish died”), suggest generalization (“all 45 fish were killed by…”), or advocate acceptance of the fact (“obviously, 45 dead fish is a serious problem”).</td></tr>
<tr><th colspan="2">Opinion</th></tr>
<tr><td>Opinion is based on years of experience and can be quite useful in analyzing the problem. It can also be used to serve a person’s self-interests.</td><td>“I think you need to look for another reason other than the toxic waste to explain the 45 dead fish.”</td></tr>
</table>

Go Talk to George

When equipment malfunctions, it is a must to talk to the operators because they know the "personality" of the equipment better than anyone. Most organizations have employees who have been around a long time and have a great deal of experience, as illustrated in the following example.

The solution that the company adopted was to replace the printing machines rather than develop different inks. Let's consider a similar situation in which, immediately upon replacement, one of the printing machines began to malfunction. We list, in order, three people to whom we most likely would want to talk.

1. The technician who monitors the printing machine
2. The manufacturer's representative who sold you the printing machine
3. George

Who's George? Every organization has a George. George is the individual who has both years of experience to draw upon and "street smarts." George is an excellent problem solver who always seems to approach the problem from a different viewpoint—one that hasn't been thought of by anyone else. Be sure to tap into this rich source of knowledge when you face a problem. Individuals such as George can often provide a unique perspective on the situation. Strive to become the "George" of your organization.

Search out colleagues who may have useful information and pertinent ideas. Have them play the "What if ...?" game with you: "*What if* the printing machine only printed five bad bills per day out of thousands; would we need to replace it?" or "*What if* the one malfunctioning printing machine was eliminated?" Also have them play the devil's advocate and deliberately challenge your own ideas. This technique stimulates creative interactions. Non-experts are also a rich source of creative solutions, as evidenced by the following example.

Seeking Advice

Joel Weldon, in his tape "Jet Pilots Don't Use Rearview Mirrors," described a problem encountered by a major hotel when it opened several decades ago. The elevator capacity was inadequate for the number of guests, causing a backup on each floor and in the elevator area of the lobby. Even with the installation of mirrors by the elevators, the guests and hotel management found the wait time unacceptable. The manager and assistant manager were lamenting the problem in the lobby one day and were brainstorming about how to increase the elevator capacity. Adding additional elevator shafts would require removal of a number of rooms and a significant loss of income.

The doorman, whose name happened to be George, overheard their conversation and casually mentioned that it was too bad they couldn't just add an elevator on the outside of the building, so as not to disturb things inside. A great idea! It occurred to the doorman because he was outside the building much of the time, and that was his frame of reference. Notice, however, that the doorman's creativity alone was not enough to solve the problem. Knowledge of design techniques was necessary to implement his original idea. A new outside elevator was born, and the rest is history. External elevators have since become quite popular in major hotels. Information, good ideas, and different perspectives on the problem can come from all levels of the organization.

Source: *Chemtech*, 13, 9, p. 517, 1983.

Step 3: View the Problem Firsthand

While it is important to talk to people as a way to understand the problem, you should not rely solely on their interpretations of the situation and problem. If possible, inspect the problem yourself.

Memo Angeles/Shutterstock

> *"You can see a lot just by looking."*
> —Yogi Berra

Viewing the Problem Firsthand

In 2012, for the first time in history, a majority of the people in the world live in cities and it is estimated that as much as 70% of the world's population will live in cities in 2050. All cities are not created equally, however, and urban planners have been researching how to foster creativity and economic growth in metropolises. Initial research showed a direct correlation between population density and innovation, as measured by the number of successful start-up companies, among other metrics. Researchers rationalized that the more people there are in a given area, the more casual interactions there are between diverse ideas, which often leads to innovation and economic growth.

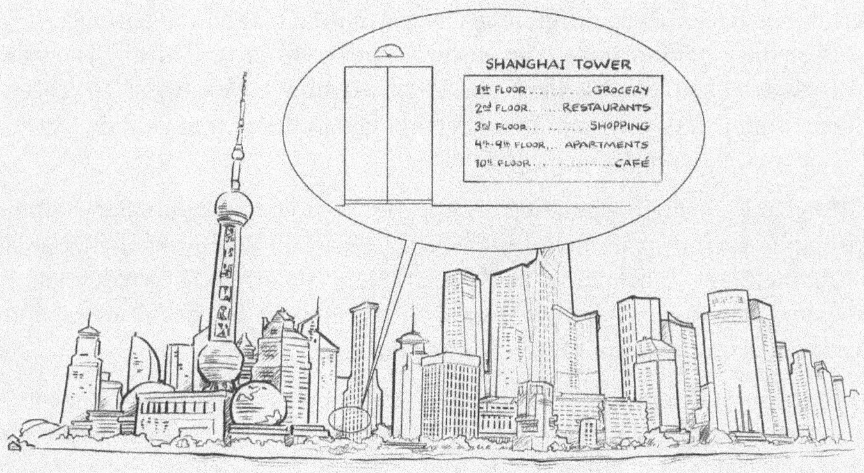

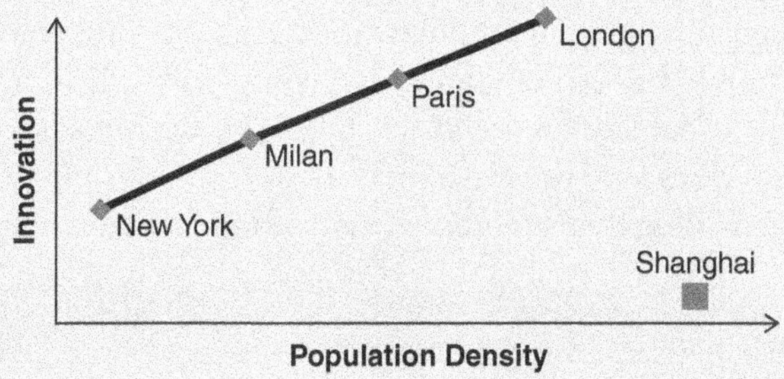

Continues

Cities such as New York, London, Paris, and Milan all fit well with the model of population density equating with innovation, but the urban planners were confounded by Shanghai. With nearly 125,000 people per square mile in some areas, Shanghai should rank highly in innovation according to the population density model; however, this is not the case. Planners went over and over all their maps, data, and criteria and could not figure out why it did not fit; New York, London, Paris, and Milan, all with smaller densities, outpace Shanghai in innovation. The team of urban planners working on the model thought that maybe their survey data were insufficient because Shanghai is such a high-density area. However, before they spent millions of dollars on another more extensive research project, they decided to view the problem firsthand on their own and flew to Shanghai.

The team toured the city extensively and could not see any reason why Shanghai lagged behind in innovation compared to the other cities. On the final day of the trip, the team members were going to have dinner, but it was raining heavily so they asked the hotel reception where they could eat close by. The receptionist said, "Well, you don't need to leave the building, you can eat at one of the seven different restaurants in the mall in the basement, the café on the top floor, or visit the grocery store on the ground level." The team was amazed and asked if most buildings had similar amenities. The receptionist said it was very common for buildings in the area to contain offices, residences, shopping centers, and more.

Shanghai has a high population density, but in order to accommodate all the people in such a small area, lives were compressed into even smaller areas with buildings functioning as self-sufficient suburbs. This reduction in required day-to-day mobility decreased the amount of casual interactions between people that would have led to innovation.

The urban planners followed this idea and found that the most innovative parts of other cities were not the skyscraper-packed streets of high-density population areas, but rather the slightly less population-dense neighborhoods and districts. They mixed multiple purposes, such as combining residential and business, pushing people to move around the city and come into contact with a larger number of ideas than if they simply stayed in a self-sufficient skyscraper. Since the team viewed the problem, in this case their outlying data point, they were able to revise their model to better plan future cities to limit the overcrowding that suppresses innovation and creativity.

Source: http://online.wsj.com/article/SB10000872396390443477104577551133804551396.html?KEYWORDS=for+creative+cities+the+sky+has+its+limit.

Step 4: Confirm All Key Findings

Verify the information you collected. Check and cross-reference data, facts, and figures. Search for biases or misrepresentation of facts. Confirm all important pieces of information, and spot-check others. Distinguish between fact and opinion. Challenge assumptions and assertions.

Confirm All Allegations

The authors of this book were involved in a consulting project for a pulp and paper company we will call Boxright. Several years ago, Boxright had installed a new process for recovering and recycling its chemicals that were used in the paper-making process. Two years after the installation, the process had yet to operate correctly. Tempers flared and accusations flew back and forth between Boxright and Courtland Construction, the supplier of the recycling equipment. Courtland claimed the problem was that Boxright did not know how to operate the process correctly; Boxright contended that Courtland's equipment was designed incorrectly. Boxright finally decided to sue Courtland for breach of equipment performance.

Much data and information were presented by both sides to support their arguments. Courtland presented data and information from an article in the engineering literature that the company claimed proved Boxright was not operating the process correctly. At this point it looked like Courtland had cooked Boxright's goose by presenting such data. Before conceding the case, however, Boxright needed to confirm this claim. When we analyzed this key information in detail, we found in the last few pages of the article a statement that the data would not be expected to apply to industrial-size equipment or processes. Had the allegations been accepted as presented and not confirmed, the suit would have been likely lost in favor of Courtland. Instead, when the article's key information was presented, the lawsuit was settled in favor of the pulp and paper company, Boxright.

FreshPaint/Shutterstock

Throughout the problem-solving process, you should continue to gather as much information as possible by reading texts and literature related to the problem to learn about both the underlying fundamental principles and any peripheral concepts. Literature searches are particularly helpful in this regard. Perhaps a closely related problem has already been solved. George Quarderer of Dow Chemical Company appropriately describes the idea of reinventing the wheel by his statement, "Four to six weeks in the laboratory can save you an hour in the library." The message is clear: Doing a bit of research into the background of the problem may save you hours of time and effort.

The information you gather in the first four steps will be of great help to you as you "define the real problem," which is the topic of this chapter.

SUMMARY

In this chapter we discussed methods for obtaining the information needed to define the problem. Experienced problem solvers use four steps to attack problems:

1. **Collect and analyze information and data.**
   a. Visualize the data
   b. Use Fermi estimates
2. **Talk with people who are familiar with the problem.**
   a. Use Socratic questions
   b. Talk to George
3. **If possible, view the problem firsthand.**
4. **Confirm all findings and continue to gather information.**

WEB-SITE MATERIAL (WWW.UMICH.EDU/~SCPS)

- **Learning Resources**
  Summary Notes
- **Professional Reference Shelf**
  1. Viewing the Problem Firsthand
     – The Off-Spec Plastic
     – A Shocking Installation

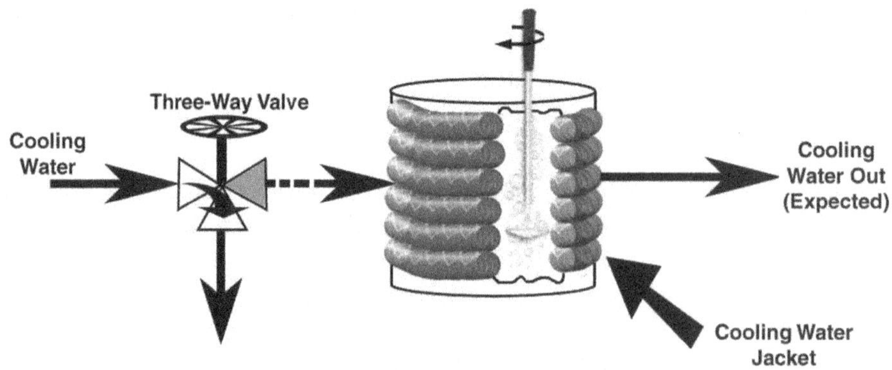

  2. Visualizing Information
     – The Case of the Dead Fish

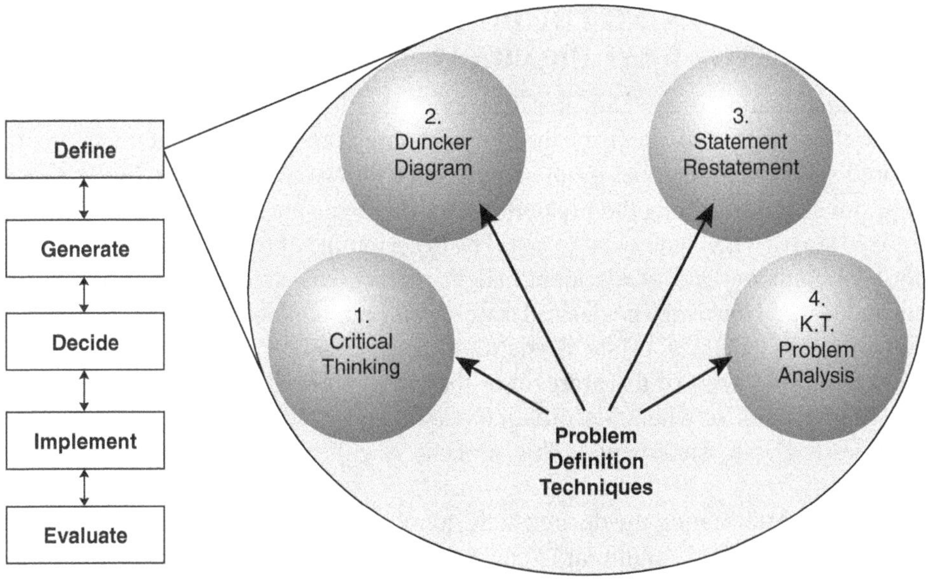

In this chapter, we address the first part of the problem-solving heuristic—problem definition. We present four techniques that will greatly enhance your chances of defining the real problem when combined with the critical thinking skills and information-gathering actions.

The authors of this book conducted a study of the problem-solving methods of experienced engineers in industry. The study revealed some common threads that run through their problem definition techniques. We have classified these common threads into four techniques to help you understand and define the real problem instead of being sidetracked by a perceived problem.

DEFINING THE REAL PROBLEM: FOUR TECHNIQUES

The four steps discussed earlier are all used to gather information about the problem. This information lays the groundwork that allows us to use the four problem definition techniques to help define the real problem instead of the perceived problem.

In this chapter, we add three additional techniques to define the real problem. We will start with one that has been well received when we have presented it in industrial short courses over the last several years, the Duncker diagram.

TECHNIQUE 1: USING THE DUNCKER DIAGRAM TO DEFINE THE REAL PROBLEM

Cleaning up the problem statement.
Luis Santos/Shutterstock

The unique feature of the Duncker diagram is that it contains two major pathways (general solutions) to go from the present state (the problem statement) to the desired state (an acceptable problem solution). Part of the Duncker diagram technique involves describing the **present state** (where you are) and then describing **the desired state** (where you want to go). The descriptions of these states are reworked until all concerns and needs identified in the present state are addressed in the desired state.[1] However, the desired state should not contain solutions to problems that are not in the present state. Sometimes the statements appear to match but they don't get to the heart of the problem or they allow for many solution alternatives. You must recognize when this situation occurs and continue to rework the present state/desired state statements. This process is called *cleaning up the problem statement.*

When determining the desired state, avoid using ambiguous and vague words or phrases like "best," "minimal," "cheapest," "within a reasonable time," or "most efficient," because these words mean different things to different people. Be quantitative where possible. For example, say "The children's playground needs to be completed by July 1st, at a cost less than $50,000" rather than "The playground should be completed in a reasonable time at minimal cost."

There are two sides to a Duncker diagram.[1]

Side 1. General solutions on the left side of the diagram show us how to move from the present state to *achieve the desired state*.

Side 2. General solutions on the right side of the diagram show us how to *make it okay not to achieve the desired state.* (This idea may seem a bit contradictory— but it will be clear in a moment.)

There are two steps involved in each pathway: functional solutions that describe *what you need to do* and then specific solutions that describe *how to do it*. Functional solutions tell us what we could do to move from the present state to either *achieve* or *not achieve* the desired state. After a number of functional solutions are developed, we then generate a number of specific solutions that guide us in implementing *each* of the functional solutions. Specific solutions tell us how to implement the functional solutions.

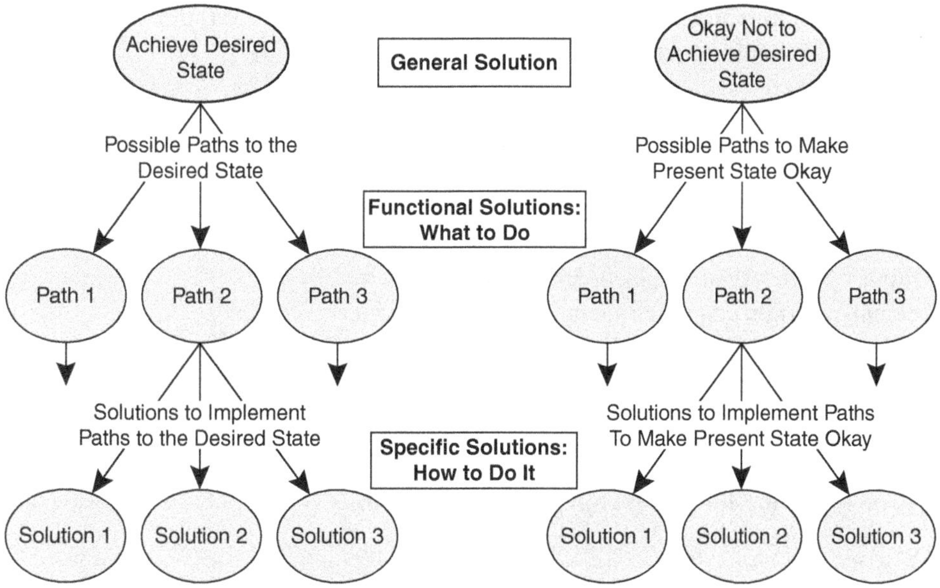

The Duncker Diagram

To help clarify the Duncker diagram, let's consider the following situation. Suppose you are unhappy in your current job at your company (this is the present state—the problem to be solved) and you want to leave your company for a new job (the desired state). The left side of the diagram is the pathway for finding a new job with a different company and therefore achieving the desired state. Possible functional solutions on the left side of the diagram might include retraining for a different type of job or doing the same type of job but with another company. Specific solutions for obtaining the same type of job but with a different company might include networking to learn which companies have openings, searching the Web, and updating your résumé.

The right side of the diagram deals with alternatives that would make it okay to stay in your current job and therefore make it okay *not* to achieve the desired state of a new job. A possible functional solution for the right side of the Duncker diagram might be related to achieving greater job satisfaction. Specific solutions (that is, to have greater job satisfaction) might include greater participation in decision making, more praise, or a salary increase. This side of the diagram helps the solver get to the heart of what is really wrong in the present state and addresses that issue versus proceeding along only one of many possible solution paths to address the perceived problem.

Representing the problem as a Duncker diagram is a creative activity, and as such, there is no right way or wrong way to do it. After completing the Duncker diagram, you should try *to write a new problem statement.* This new problem statement might allow you to identify a compromise solution in which both the present state and the desired state are modified to achieve an acceptable solution. This technique can be a bit confusing at first glance. To see how it works, consider the following example of the present state/desired state technique.

We illustrate the use of the Duncker diagram with an example of a laundry detergent that could be harmful if ingested.

Bitrex

Most laundry detergents and other cleaning agents are poisonous and even fatal if consumed. Sometimes bottles of these cleaners are stored where small children can reach them and possibly drink from them. The instruction given to solve the perceived problem: *Find a way to prevent children from being able to get to the chemicals in the bottle.*

The following ideas were suggested to solve the perceived problem.

1. Print a note to adults to keep the bottle out of the reach of children. However, relying on the parents to follow these instructions is not a good idea because most adults probably will mistakenly never read them.
2. Put child-proof caps on the bottles. Again, there is a chance that the adult will be careless and will not put the cap on correctly. In addition, if the adult has trouble with the cap due to physical problems, he or she might purposely leave the cap only partially attached. The following table shows the Duncker diagram for the perceived problem statement.

General solution	Prevent children from getting to chemicals		Make it okay for children to get to chemicals	
Functional solutions *"What to do"*	Keep bottle away from children	Make bottle impenetrable	Make consumption okay	Remove child's desire to consume detergent
Specific solutions *"How to do it"*	1. Print label directing parents to keep away from children	1. Child proof caps	1. Create a new chemical that isn't poisonous	1. Make it taste bad
	2. Only sell to households without children	2. Individual-use bottles opened by scissors	2. Provide antidote with detergent	2. Educate children on harmfulness of detergent
	3. Offer kids a reward if they don't open the bottle	3. Sell bottles that children aren't strong enough to open	3. Add in large amount of laxative so it leaves the children's system quickly	

The solution that one company arrived at was one much different than previously discussed. Macfarlan Smith Ltd. added a chemical that is harmless to ingest, does not affect the effectiveness of the cleaning solution, and is very bitter to the taste. Only a very small portion of this chemical placed in the solution will make it so bitter that children will not want to drink it after one taste.

New Problem Statement: Determine the exact amount of the bitter-tasting chemical to be added to the cleaning solution.

Source: *Chemical and Engineering News*, January 28, 1991, p. 41 (Matt Gdowski).

Let's consider another real-life example where the Duncker diagram uncovers the real problem and solution as opposed to the perceived problem.

Dead Rats Decaying in the Water

A new water supply system was installed in a western state. Part of the system was a large outdoor holding tank placed in the ground near a farm field. Shortly after installation, unacceptable levels of contamination were found. An investigation revealed that the contamination was caused by dead rats decaying in the holding tank. The problem statement that went out was *find a way to keep the rats out of the tank.*

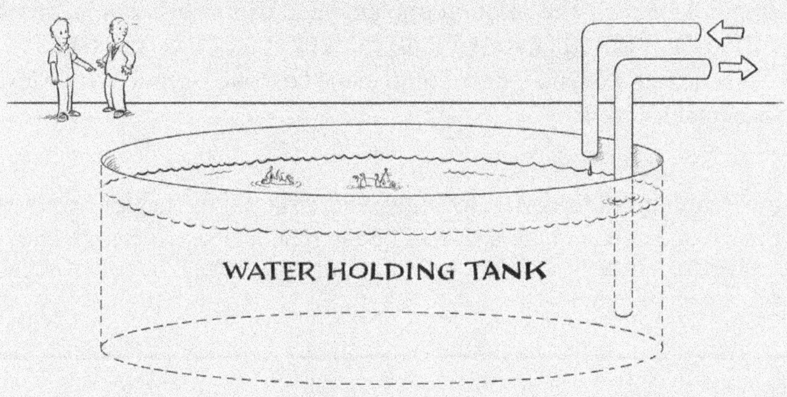

The storage tank was 200 feet in diameter and 30 feet deep, and the water level varied between one and three feet from the top of the tank. The sides of the tank were vertical and smooth and it was not known how the rats got into the tank, but they did. Because the rats are good swimmers, they would survive until exhaustion caused them to drown. Covering the tank would have been very expensive and would make collecting samples for daily testing from different areas of the tank much more difficult, so other ideas were

Continues

considered first. One proposed solution of the perceived problem was to have someone check the tank every hour to see if there are any rats swimming in the tank and then drop a bucket to pull any out. However, this solution would have been time-consuming and costly. Another solution was to always keep the tank full so rats that fell in could get out. Because of the varying water demand, keeping the tank completely full to the top would have been virtually impossible.

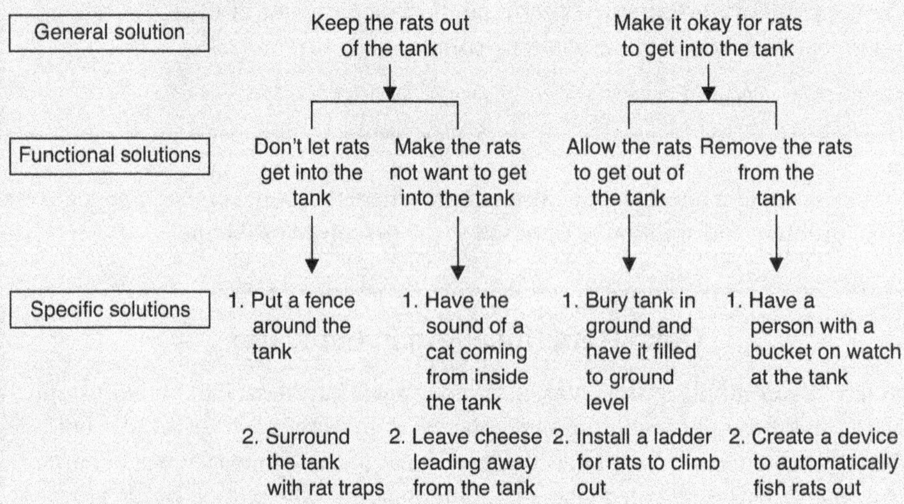

The real problem statement should have been *"Stop rats from dying and decaying in the tank."* The solution implemented was to install a ladder that allowed the rats that fell into the tank to climb out. Once the ladder was installed, dead rats were no longer found in the tank and contamination levels fell to acceptable levels.

A New Problem Statement: How and where should the ladder be placed in the storage tank and what should be the type and size of the ladder?

Source: Adapted from "Drinking Water Comes to a Boil," by Sara Terry, *The New York Times Magazine*, September 26, 1993.

TECHNIQUE 2: USING THE STATEMENT–RESTATEMENT TECHNIQUE

A problem well stated is a problem half solved.
—Charles F. Kettering

The **statement–restatement technique** developed by Parnes[2] is a method to evolve the problem statement to its most accurate representation of the problem. This technique is similar to the present state/desired state technique in that it requires us to rephrase the problem statement.

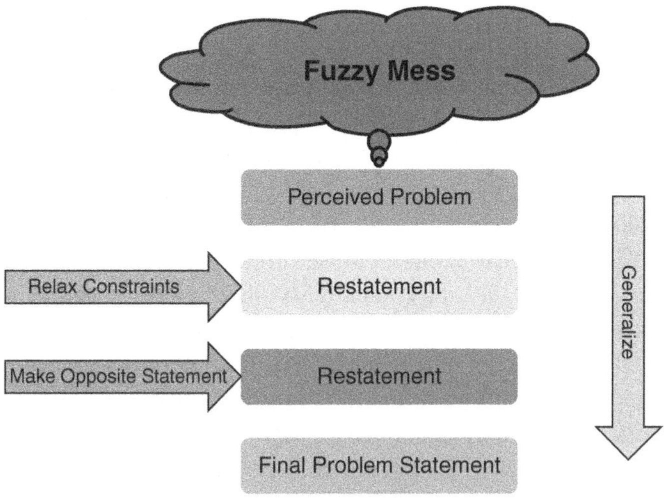

With this technique, we look at the fuzzy or unclear problem situation and write a statement regarding a problem to be addressed. The problem is then restated in different forms a number of times.

When restating the problem, it is important to inject new ideas, rather than simply changing the word order in the restated sentence. Using one or more of the following problem restatement **triggers** should prove helpful in arriving at a definitive problem statement.

Problem Statement Triggers

Trigger 1. Vary the stress pattern—try placing emphasis on different words and phrases.

Trigger 2. Choose a term that has an explicit definition and substitute the explicit definition in each place that the term appears.

Trigger 3. Make an opposite statement, change positives to negatives, and vice versa.

Trigger 4. Change "every" to "some," "always" to "sometimes," "sometimes" to "never," and vice versa.

Trigger 5. Replace "persuasive words" in the problem statement such as "obviously," "clearly," and "certainly" with the argument it is supposed to be replacing.

Trigger 6. Express words in the form of an equation or picture, and vice versa.

Each time the problem is restated, we try to generalize it further to arrive at the broadest form of the problem statement.

As an illustration of the use of these triggers, consider Trigger 3. Instead of asking, "How can my company make the largest profit?" ask, "How can my company lose the most money?" In finding the key activities or pieces of equipment that, when operated inefficiently, will result in the largest loss, we will have identified those pieces that need to be carefully monitored and controlled. This trigger helps us find the **sensitivity** of the system variables and to focus on those variables that dominate the problem.

It is often helpful to relax any constraints placed on the problem, modify the criteria, and idealize the problem when writing the restatement sentence (see trigger 4). Also, does the problem statement change when different time scales are imposed (i.e., are the long-term implications different from the short-term implications)? As we continue to restate and perhaps combine previous restatements, we should also focus on tightening up the problem statement, eliminating ambiguous words, and moving away from a fuzzy, loose, ill-defined statement.

To Market, to Market

Toasty O's was one of the first organic cereals without preservatives when it first came on the market. After several months, however, its sales dropped. The consumer survey department was able to identify that customer dissatisfaction was expressed in terms of a stale taste. The company's management then issued the following instructions to solve the perceived problem: "Streamline the production process to get the cereal on the store shelves faster, thereby ensuring a fresher product."

Unfortunately, there wasn't much slack time that could be removed from the production process to accomplish this goal. Of the steps required to get the product on the shelves (production, packaging, storage, and shipping), production was one of the fastest. Thus the company considered plans for building plants closer to the major markets, as well as plans for adding more trucks to get the cereal to market faster. The addition of either new plants or more trucks would require a major capital investment to solve the problem.

Use the statement–restatement technique to arrive at a new problem statement.

Using the Problem Statement Triggers

Original Problem Statement: The Toasty O's cereal is clearly not getting to market fast enough to maintain freshness.

Trigger 1:
- **Cereal** not getting to market fast enough to maintain freshness. (Do **other products** we have get there faster?)
- Cereal not **getting** to market fast enough to maintain freshness. (Can we make the distance/time shorter?)

- Cereal not getting to **market** fast enough to maintain freshness. (Can we distribute it from a centralized location?)
- Cereal not getting to market fast enough to maintain **freshness.** (How can we keep cereal fresher, longer?)

Trigger 2: • Breakfast food that comes in a box is not getting to the place where it is sold fast enough to keep it from getting stale.
(This restatement makes us think about the box and staleness. How might we change the box to prevent staleness?)

Trigger 3: • How can we find a way to get the cereal to market so slowly that it will never be fresh?
(This restatement makes us think about how long we have to maintain freshness and what controls it.)

Trigger 4: • Cereal is not getting to market fast enough to always maintain freshness.
(This change opens up new avenues of thought. Why isn't our cereal *always* fresh?)

Trigger 5: • The word "clearly" in the problem statement implies that if we could speed up delivery freshness would be maintained. Maybe not! Maybe the store holds the cereal too long. Maybe the cereal is stale before it reaches the store.
(This trigger helps us challenge the implicit assumptions made in the problem statement.)

Trigger 6: • Freshness is inversely proportional to the time since the cereal was baked:

$$(Freshness) = \frac{k}{(Times\ since\ cereal\ baked)}$$

This restatement makes us think about other ways to attack the freshness problem. For example, what does the proportionality constant, k, depend on?

The storage conditions, packaging, type of cereal, and other factors are logical variables to examine. How can we change the value of k?

The total time may be shortened by reducing the time at the factory, the delivery time, or the time to sell the cereal (i.e., shelf time). Once again, this trigger provides us with several alternative approaches to examine to solve the problem: Reduce the time *or* change (increase) k. Of course, the real problem was that the cereal was not staying fresh long enough—not that it wasn't reaching the stores fast enough. Keeping the cereal fresher longer was achieved by improved packaging and the use of additives to slow the rate at which Toasty O's became stale.

The new problem statement is this: "Find how to best improve packaging to keep the cereal fresher longer."

Coating Aspirin: Making an Opposite Statement

To many people, swallowing an aspirin tablet is a foul-tasting experience. A number of years ago, one company that manufactured aspirin decided to do something about it. The manager gave the following instructions to his staff to solve the perceived problem: "Find a way to put a pleasant-tasting coating on aspirin tablets." Spraying the coating on the tablets had been tried in the past, but with very little success. The resulting coating was very non-uniform, which led to an unacceptable product.

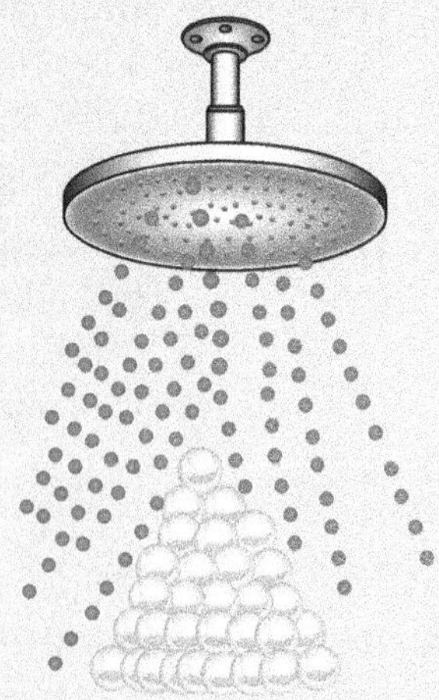

Let's apply a few triggers to this problem.

Trigger 1: Emphasize different parts of the statement.

1. **Put** coating **on** tablet.

Trigger 3: Make an opposite statement.

2. **Take** coating **off** tablet.

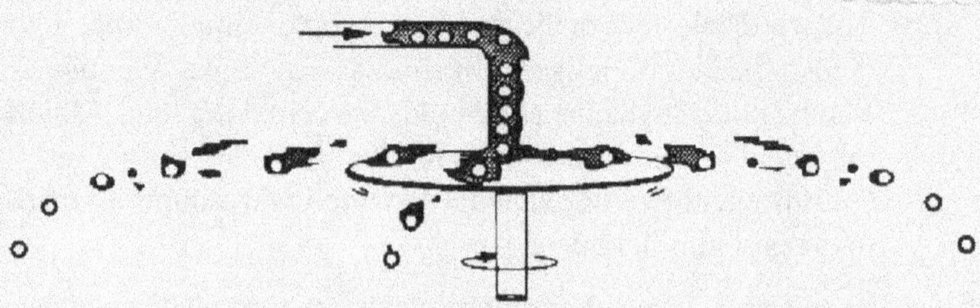

This idea led to one of the newer techniques for coating pills. The pills are immersed in a liquid in which a sweetening agent has been dissolved. The pills and liquid are then passed onto a spinning disk where the centrifugal force causes the fluid and the pills to be thrown off the spinning disk. As they are thrown, some of the liquid separates from the pill and some is held on by surface tension, leaving a nice, even coating around the pill. As the pills and liquid fall by gravity into a collecting device, the liquid evaporates, leaving the sweetening agent on the surface of the pill.

TECHNIQUE 3: KEPNER–TREGOE PROBLEM ANALYSIS

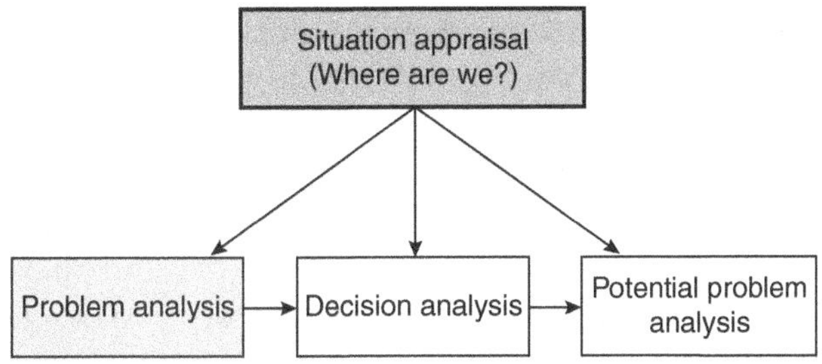

The Kepner–Tregoe[3] (K.T.) problem analysis technique is one of the four components of the K.T. approach, which includes situation appraisal, problem analysis, decision analysis, and potential problem analysis. We will discuss problem analysis in this chapter.

The K.T. problem analysis technique displays information obtained by asking critical thinking questions in a unique way called the "four dimensions of the problem." The four dimensions are what, where, when, and extent. In this technique, distinctions are made in the four dimensions as follows:

- What *is* the problem versus what *is not* the problem?
- Where does the problem occur versus where is everything okay?
- When did the problem first occur versus when was everything okay?
- What is the extent of the problem?

This kind of analysis is most useful in **troubleshooting operations** where the cause of the problem or fault is not known. Problems that lend themselves to K.T. problem analysis are ones in which an undesirable level of performance can be observed and compared with the accepted standard performance.

The basic premise of K.T. problem analysis is that there is something that distinguishes what the problem ***is*** from what it ***is not***. The cause of the problem is usually a change that has taken place and produced undesirable effects. Things were okay in the past and now they're not. The Bureau of Engraving and Printing changed to new printing machines and then found that ink on the new bills smeared when touched. The possible causes of the problem (deviation) were deduced by examining the differences found in the problem. In this case, the only deviation from the previous printing process was the new machines. The most probable cause of the problem is the one that best explains all the observations and facts in the problem statement. It turned out that the new machines were not putting enough pressure on the printing ink to penetrate the new paper, so it wiped off easily. The problem *was* the printing machines, *not* the printing ink.

The real challenge is to identify the distinction between the *is* and the *is not*. Particular care should be taken when filling in the "Distinction" column. Sometimes the distinction statement should be rewritten more than once to sharpen it enough to specify the distinction exactly. For example, in one problem analyzed by the K.T. method, the statement "Two of the filaments were clear (okay) and two were black (not okay)" was sharpened to "Two filaments were clear and two were covered with carbon soot." This **sharpening** of the distinction was instrumental in determining the reason for the black filament. Think in terms of dissimilarities. What distinguishes *this* fact (or category) from *that* fact (or category)? Making such a distinction requires careful analysis, insight, and practice to ferret out the differences between the *is* and the *is not*.

From the possible causes, we try to ascertain the most probable cause. The most probable cause is the one that explains each dimension in the problem specification. The final step is to verify that the most probable cause is the true cause. This step may be accomplished by making the appropriate change to see if the problem disappears.

The problem solver should also separate people's observations from their interpretations of what went wrong. A common mistake is to assume that the most obvious conclusion or the most common is always the correct one. (This point is a good place to start, but not necessarily to stop.) A famous medical school proverb that relates to the diagnosis of disease is this: "When you hear hoof beats, don't think zebras." In other words, look for simple explanations first. Finally, the problem solver should continually reexamine the assumptions and discard them when necessary.

When you hear hoof beats, don't think zebras!

The Four K.T. Dimensions of a Problem

Is, or is not? That is the question, Watson!!!

Dimec/Shutterstock

		Is	Is Not	Distinction	Cause
What	Identify	What is the problem?	What is not the problem?	What is the distinction between the **is** and the **is not?**	What is a possible cause?
Where	Locate	Where is the problem found?	Where is the problem not found?	What is distinctive about the difference in locations?	What is the possible cause?
When	Timing	When does the problem occur?	When does the problem not occur?	What is distinctive about the difference in the timing?	What is a possible cause?
		When was the problem first observed?	When was the problem last observed?	What is the distinction between these observations?	What is a possible cause?

		Is	Is Not	Distinction	Cause
Extent	Magnitude	How far does the problem extend?	How localized is the problem?	What is the distinction?	What is a possible cause?
		How many units are affected?	How many units are not affected?	What is the distinction?	What is a possible cause?
		How much of any one unit is affected?	How much of any one unit is not affected?	What is the distinction?	What is a possible cause?

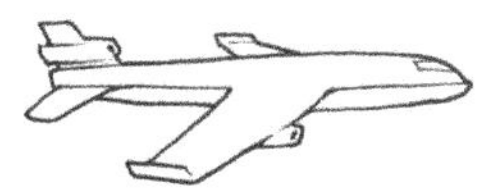

Fear of Flying

A new model of airplane caused an unexpected issue when delivered to a major airline. Immediately after the planes were put into operation, many of the flight attendants developed a red rash on their arms, hands, and faces. The rash did not appear on any other part of the body, and it occurred only on flights that went over water. Fortunately, the rash usually disappeared in 24 hours and caused no additional problems beyond that time. When the attendants flew old planes over the same routes, no ill effects occurred. The rash did not occur on all the attendants of a particular flight, but the same number of attendants contracted the rash on each flight. In addition, those flight attendants who contracted the rash felt no other ill effects.

Continues

The flight attendants' union threatened action because the attendants were upset and worried, and the attendants believed that some malicious force was behind the mysterious rash. Many doctors were called in, but all were in a quandary. Industrial hygienists could not measure anything extraordinary in the cabins

Source: *Chemtech*, 13, 11, p. 655, 1983. Image: idiz/Shutterstock.

Let's apply K.T. problem analysis to this situation to see if we can learn the cause of the problem.

What

The problem appears to be only a rash and not any other illness such as headache or nausea. A rash can be caused by external contact with an allergen such as poison ivy. Consequently, we can complete the first row of the K.T. table as follows:

	Is	Is Not	Distinction
What	Rash	Other illness	External contact

When

The problem occurs only when the new planes are used, but not when the old planes are used. The new planes contain new materials such as lighter composite materials and new fabric, among other things.

	Is	Is Not	Distinction
When	New planes used	Old planes used	New materials

Where

The illness occurs only in flights over water; attendants on flights over land do not develop the rash. Different crew procedures are followed on flights over water than are followed on flights over land.

	Is	Is Not	Distinction
Where	Flights over water	Flights over land	Different crew procedures

Extent

There are two categories of "extent" in this case: (1) how many units are affected and (2) how much of any one unit is affected. In terms of how many units (flight attendants) are affected, only some of the attendants are affected;

not all are affected. The different attendants have different duties. How much of any one unit (attendant) is affected is defined as the fact that the rash appears only on the face, hands, and arms. The culprit must be something contacting the face, hands, and arms.

	Is	Is Not	Distinction
Extent	Only some of the attendants	All of the attendants	Different crew duties
Extent	Face, hands, and arms	Other body parts	Something contacting exposed face, hands, and arms

Let's summarize what we have learned so far:

	Is	Is Not	Distinction
What	Rash	Other illness	External contact
When	New planes used	Old planes used	New materials
Where	Flights over water	Flights over land	Different crew procedures
Extent	Only some of the attendants	All of the attendants	Different crew duties
	Face, hands, and arms	Other body parts	Something contacting exposed face, hands, and arms

When we look at all of the distinctions, we see that (1) something contacting the arms and face could be causing the rash; (2) the rash occurs only on flights over water, and the use of life vests is demonstrated on flights over water; and (3) the life vests on the new planes are made of new materials or of a different brand of materials and usually two flight attendants demonstrate the use of

Continues

these life vests. The new life preservers contained some material that was the rash-causing agent!

A Pre-flight Demo of a Life Vest

Photo: Serg Zastavkin/Fotolia.

Spontaneous Car Starting

Mikhail Bakunovich/ Shutterstock

Molly traded in her old car for a one-year-old, high-end car that had a special keyless entry system. Everything about the car worked perfectly, with one exception: When Molly would park the car, sometimes she would come out to find its motor running. In each instance, she had turned the motor off and taken the key out before leaving the car. This problem most often occurred at the Starbucks on Washtenaw Avenue, a major thoroughfare, but seldom at the Starbucks on Liberty Street, which was a more secluded street. It also occurred a couple of times at Dunkin' Donuts on Stadium Street. Even though the problem mainly occurred on Washtenaw, Molly did not want to stop going there because she enjoyed chatting with the police officers that frequented that location. The Liberty Street Starbucks seemed to have only an occasional police officer stop by. The problem arose during the midmorning hours but not over lunch or dinnertime.

	Is	**Is Not**	**Distinction**
What	Spontaneous starting	Regular starting	Signal to start car

	Is	**Is Not**	**Distinction**
Where	Starbucks on Washtenaw and Dunkin' Donuts on Stadium	Starbucks on Liberty	Police in Washtenaw Starbucks and Dunkin' Donuts

	Is	Is Not	Distinction
When	Midmorning	Lunch	Police taking morning break
	New car	Old car used	New cars have remote entry and ignition system

	Is	Is Not	Distinction
Extent	Only in locations that the police frequent	Locations where there are no police	Police doing something like receiving/sending calls

The newer-model keyless entry system had a remote starting capability when a signal was sent from the key chain. The frequency from the police radios matched the frequency from the key chain that started the engine.

Source: From an idea on the National Public Radio show *Car Talk*.

After the problem is defined, there is more work to be done. Nevertheless, having a well-defined problem puts us well on the way toward finding a solution.

WHICH TECHNIQUE TO CHOOSE?

We do not expect you to apply every problem definition technique to every situation. In fact, when 400 problem solvers were surveyed as to which two techniques presented in this chapter were the most useful to them, their choices were virtually equally divided among those presented in this chapter. In other words, different techniques work better for different individuals and different situations, and your selection of a problem definition technique is a matter of personal choice. The main point is to be organized as well as creative in your approach to problem definition.

DETERMINE WHETHER THE PROBLEM SHOULD BE SOLVED

Establish criteria to judge the solution.

Having defined the real problem, we now need to develop criteria by which to judge the solution to the real problem. One of the first questions experienced engineers ask is this: Should the problem be solved? The figure following this discussion shows how to proceed when answering this question. As you can see from the figure, the first step is to determine whether a solution to an identical problem or a similar problem is available. A literature search may determine whether a solution exists.

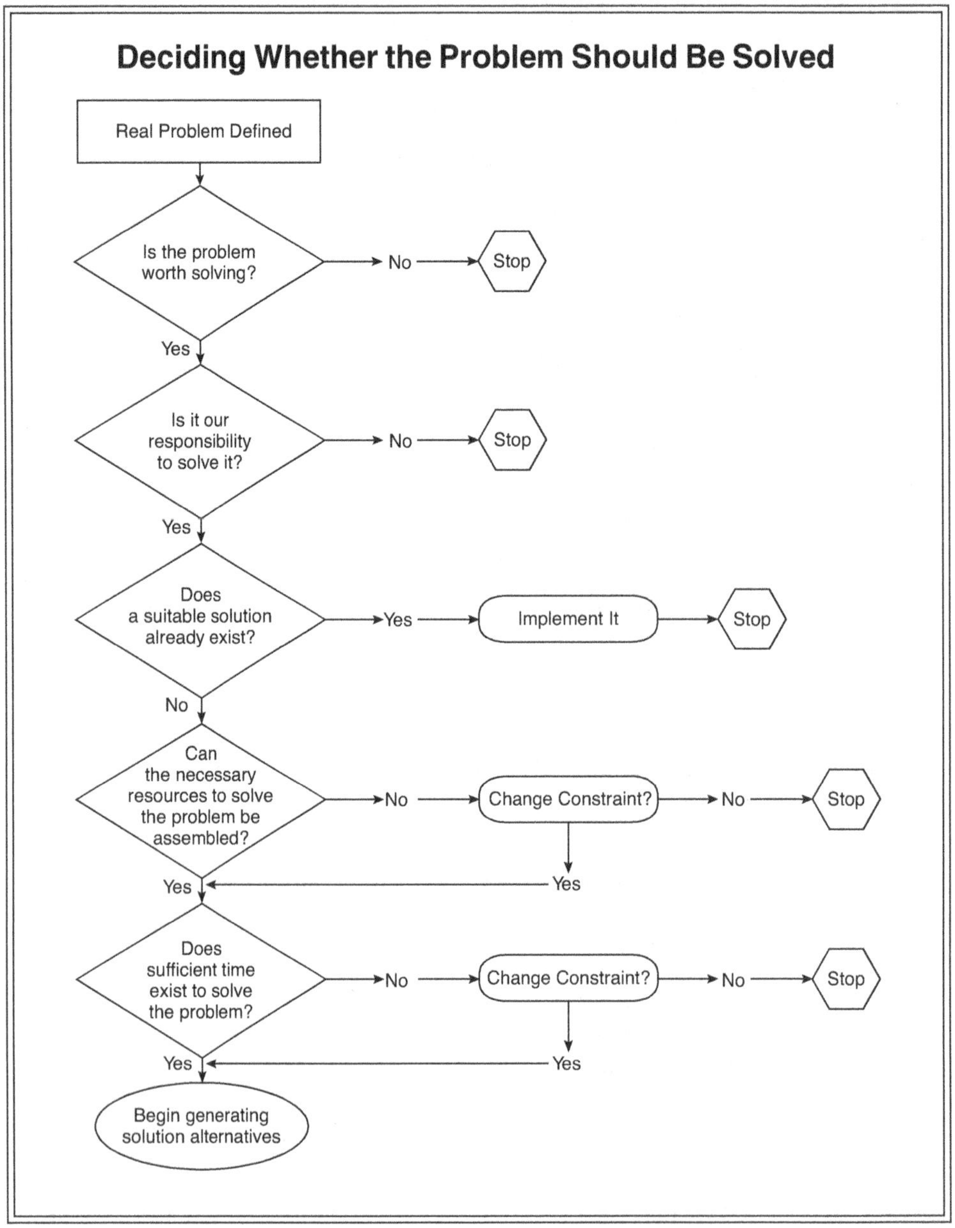

How do experienced engineers go about deciding whether the problem is *worth* solving? Perhaps the problem is just mildly irritating and consequently may be ignored altogether. For instance, suppose the garage door at your plant's warehouse facility is too narrow to allow easy access by some of the delivery vehicles. They can pass through, but the clearance is very tight. This is an annoying problem, but if the fix is quite costly, you could probably "live with it."

Here are some questions you should ask early in the process:

- What are the resources available to solve the problem?
- How many people can you allocate to the problem, and for how long?
- How soon do you need a solution? Today? Tomorrow? Next year?

These are key questions to keep in mind as you take your first steps along the way to finding a problem solution. The quality of your solution is often—but not always—related to the time and money you have to *generate it and carry it through.* In some instances, it may be necessary to extend deadlines to obtain a quality solution.

It may not be possible to completely address the cost issue until we are further along in the solution process. The cost will depend on whether the solution will be a permanent one or a temporary or patchwork solution. Sometimes *two* solutions are required: one to treat short-term symptoms so as to keep the process operating and one to solve the real problem for the long term. Be aware of these two mindsets in the problem-solving process. In some cases the "No" answers in the above figure can be changed to "Yes" by *selling* the project to management so as to get additional resources or extend the deadline. This change can be achieved by showing that the problem is an important one and is relevant to the operation of the company.

"Is the problem worth solving?" The answer is yes, because fungi will continue to affect more fish if we don't do something. Therefore we continue to the next step: "Is it the company's responsibility to solve the problem?" The answer to this question is **no** because the company did not cause the problem and it is the responsibility of the state's Department of Natural Resources (DNR). They should not have the obligation to cure the fish or stop the fungi from spreading just because they were accused of causing the problem. Consequently we should *stop* our problem-solving process.

BRAINSTORM POTENTIAL CAUSES AND SOLUTION ALTERNATIVES

This brings us to the end of the first phase of the creative problem-solving process and to the first step of the second phase of the process: generating solutions to problems.

SUMMARY

In this chapter, we discussed the necessity for defining the real problem. Three problem definition techniques were presented to help you zero in on the true problem definition.

- **Duncker Diagram**
  - Analyze two pathways: (1) a pathway to your desired state and (2) a pathway that makes it okay not to achieve the desired state.

Continues

- **Statement–Restatement**
  - Use the six triggers to restate the problem in a number of different ways.
- **K.T. Problem Analysis**
  - What is? What is not?
  - Where is? Where is not?
  - When is? When is not?
  - Extent is? Extent is not?

WEB-SITE MATERIAL (WWW.UMICH.EDU/~SCPS)

- **Learning Resources**
  Summary Notes
  Thoughts on Problem Solving (access through the home page)
- **Interactive Computer Modules**
  Thoughts on Problem Solving (access through the home page)

Problem Definition 1

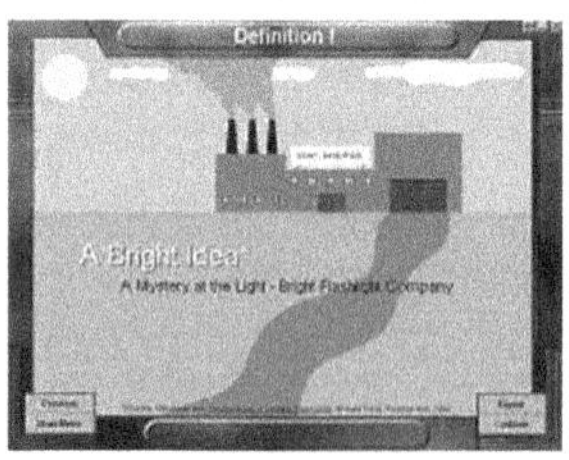

Problem Definition 2

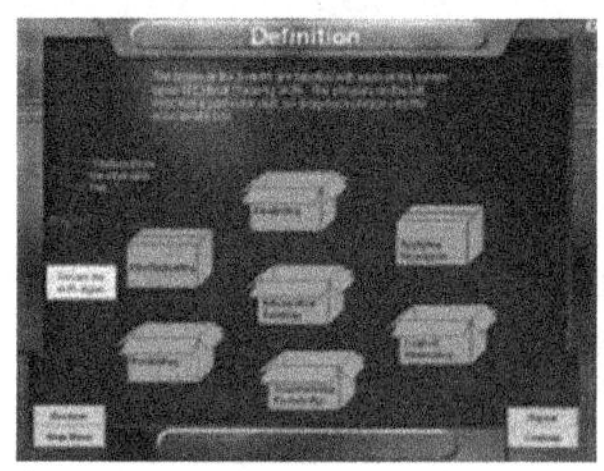

- **Professional Reference Shelf**
  1. Exploring the Problem
     - The Case of the Dead Fish
     - A Real Tooth Ache
     - De-bottlenecking a Process (Heat Exchanger)

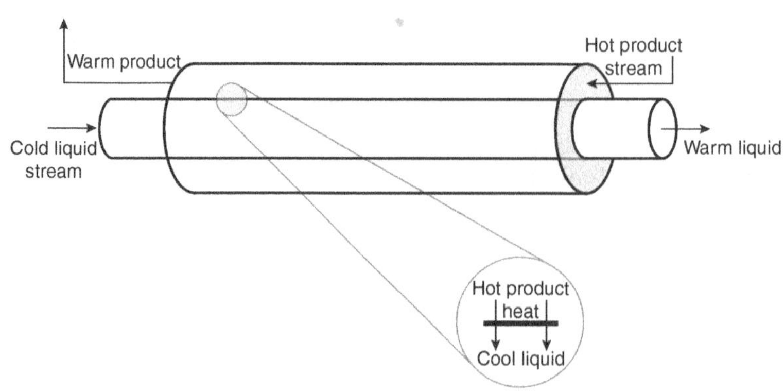

Refrigerator (heat exchanger)

2. Present State/Desired State
   - Missing the Mark
3. Duncker Diagrams
   - Kindergarten Cop
   - To Market, to Market
4. Statement–Restatement Techniques
   - Exotic Materials
   - Arid Land
   - Leaking Flow Meter
   - Wake Up and Smell the Coffee
5. Closed-Ended Algorithm
   - Chemical Reaction Engineering and more

REFERENCES

1. Browne, M. Neil, and Stuart M. Keeley, *Asking the Right Questions: A Guide to Critical Thinking*, Pearson Education, 2010.
2. Higgins, J. S., et al., "Identifying and Solving Problems in Engineering Design," *Studies in Higher Education*, 14, 2, p. 169, 1989.
3. Parnes, S. J., *Creative Behavior Workbook*, Scribner, New York, 1967.
4. Kepner, C. H., and B. B. Tregoe, *The New Rational Manager*, Princeton Research Press, Princeton, NJ, 1981.

EXERCISES

1.1. List the three most important things and the three most interesting things you learned in this chapter.

1.2. Describe a person from your past who has the characteristics of "George."

1.3. Describe a situation from your past in which you encountered a problem or situation and needed to collect information. Which of the four steps did you use?

1.4. Search the Internet to learn more about the Kansas City Hyatt Regency disaster discussed in the Professional Reference Shelf.

1.5. Write out a generic procedure you will use to gather information about a project or problem.

1.6. Load and run "Interactive Computer Module Problem Definition 1" from the SCPS Web site. Record your performance number: _________.

1.7. Load and run "Interactive Computer Module Problem Definition 2" from the SCPS Web site. Record your performance number: _________.

1.8. Identify at least three techniques that you believe will change the ways you think about defining and solving problems. Which problem definition techniques do you find most useful? Prepare a matrix table listing all of the problem definition techniques discussed in this chapter. Identify those attributes that some of the techniques have in common and those attributes that are unique to a given technique.

	Attribute 1	Attribute 2	Attribute 3
Technique A	X		X
Technique B		X	X
Technique C	X	X	

1.9. Write a sentence describing a problem you have. Apply the triggers in the statement–restatement technique to your problem.

Perceived Problem Statement: ______________________________

Restatement 1: ______________________________

Restatement 2: ______________________________

Final Problem Statement: ______________________________

Next apply the Duncker diagram to the same problem.

1.10. Carry out a present state/desired state analysis and prepare a Duncker diagram for the following problem: "I want a summer internship but no one is hiring."

1.11. You have had a very hectic morning, so you leave work a little early to relax a bit before you meet your supervisor, who is flying in to a nearby airport. You have not seen your supervisor from the home office for about a year now. He has written to you saying that he wants to meet with you personally to discuss the last project. Through no fault of yours, everything went wrong: The oil embargo delayed shipment of all the key parts; your project manager had a skiing accident; and your secretary enclosed the key files in a parcel that was sent by mistake to Japan via sea mail. Your supervisor thinks that you have been so careless on this project that you would lock yourself out of your own car.

As you are driving through the pleasant countryside on this chilly late fall afternoon, you realize that you will be an hour early. You spot a rather secluded roadside park about 200 meters away. A quiet stream bubbles through the park, containing trees in all their autumn colors—such an ideal place to just get out and relax. You pull off into the park, absentmindedly get out and lock the car, and stroll by the stream. When you return, you find that you have locked the keys inside the car. The road to the airport is not the usual route; cars pass through about every 10 to 15 minutes. The airport is nine kilometers away; the nearest house (with a telephone) is one kilometer away. The plane is due to arrive in 20 minutes. Your car, which is not a convertible, is such that you cannot get under the hood or into the trunk from the outside. All the windows are up and secured.

Apply the Duncker diagram and one other problem-solving technique to help you decide what to do.

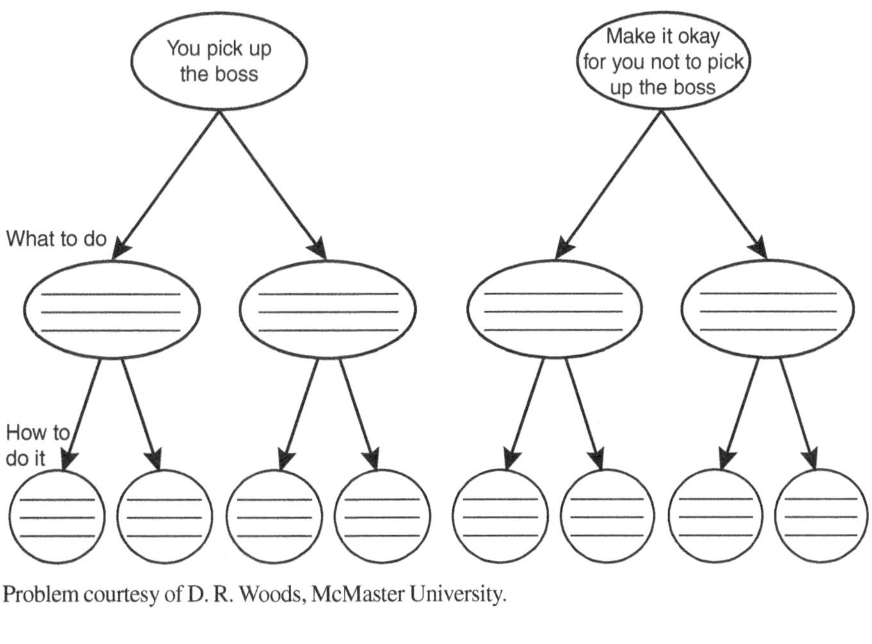

Problem courtesy of D. R. Woods, McMaster University.

1.12. Sodium azide is used as the propellant in an air bag system. This chemical is mixed with an oxidizing agent and pressed into pellets, which are then hermetically sealed into a steel or aluminum can. Upon impact, ignition of the pelletized sodium azide generates nitrogen gas that inflates the air bag. Unfortunately, if the sodium azide contacts acids or heavy metals (e.g., lead, copper, mercury, or their alloys), it forms toxic and sensitive explosives. Consequently, at the end of an automobile's life, a serious problem surfaces when an automobile with an undetonated air bag is sent to the junkyard for compacting and shredding, because it might contact heavy metals during this process. The potential for an explosion during processing represents a serious danger for the people who operate the scrap recycling plant. Apply two or three problem definition techniques to this situation. Source: *Chemtech*, 23, p. 54, 1993.

1.13. Pillsbury, a leader in the manufacture of high-quality baking products, had its origins in the manufacture of flour for the baking industry. At the time Charles Pillsbury purchased his first mill in Minneapolis, the wheat from Minnesota was considered to be substandard compared to the wheat used in the mills in St. Louis, which was then the hub of the milling industry. Part of the problem was that winter wheat, which is commonly used in high-grade flour, could not be grown in Minnesota because of the long and cold winters there. Consequently, the Minnesota mills were forced to use spring wheat, which had a harder shell. At the time, the most commonly used milling machines used a "low grinding" process to separate the wheat from the chaff. The low grinding process refers to the use of stone wheels. A stone wheel rests directly on the bottom wheel, with the wheat to be ground placed between them. With harder wheats, a large amount of heat was generated during the grinding process, discoloring and degrading the product's quality. Thus the flour produced from the Minnesota mills was discolored and inferior, and it had less nutritional value and a shorter shelf life than flour produced in the St. Louis mills. The directions given could have been "Order more river barges to ship winter wheat up the Mississippi River from St. Louis to Minneapolis." Apply two or more problem definition techniques to this situation.

Source: Adapted from "When in Rome" by Jane Ammeson, *Northwestern Airlines World Traveler*, 25, 3, p. 20, 1993.

1.14. Detroit, Michigan, and Windsor, Ontario, are separated by the Detroit River and a distance of approximately one kilometer. Off and on, over the last two years, the residents of Windsor have complained of an annoying noise that sounds like a loud diesel truck idling, metallic grating, or a loud boombox coming from the American side. They have said this hum, that is, noise pollution, is causing illness, whipping dogs into frenzies and disrupting nocturnal baby feedings. The hum is not heard all the time in Canada and has never been heard on the American side of the river. After three months of seismic study, the Canadian government said it was likely the noise was coming from Zug Island in the middle of the Detroit River. A U.S. official said that on over 20 recent trips to the island he never heard the noise. However, hundreds of sleep-deprived locals have complained so frequently that Canada's parliamentary secretary wants the United States to pay for the investigation that will help quiet down the border ruckus. U.S. officials have refused, saying that there have not been any complaints in the Detroit area. Suggest a path forward.

Dan Janisse/*Windsor Star*

Sources: Alistair MacDonald and Paul Vieira, "Canadians Make a Racket over Mysterious 'Windsor Hum': Unexplained Noise Spurs Diplomatic Fracas at Detroit Border; Americans Can't Hear It," *Wall Street Journal*, April 30, 2012; Elizabeth Chuck, "Mysterious Booms in Wisconsin Town," NBC News, March 21, 2012.

1.15. An airline at the Houston Airport tried to please its passengers by always docking its planes at gates within a one- to two-minute walk to the airport entrance and baggage claim and by having all the bags at the baggage claim area within eight to ten minutes. Unfortunately, the airline received many complaints about the time it took to get the bags to the baggage claim area. The airline researched the situation and found that there was virtually no way it could unload the bags to the transport trucks, drive to the unloading zone, and unload the bags any faster. The airline didn't change the baggage unloading procedure, but it did change another component of the arrival process and the complaints disappeared. The airline did not use mirrors to solve the problem (as was the case for the slow elevators).

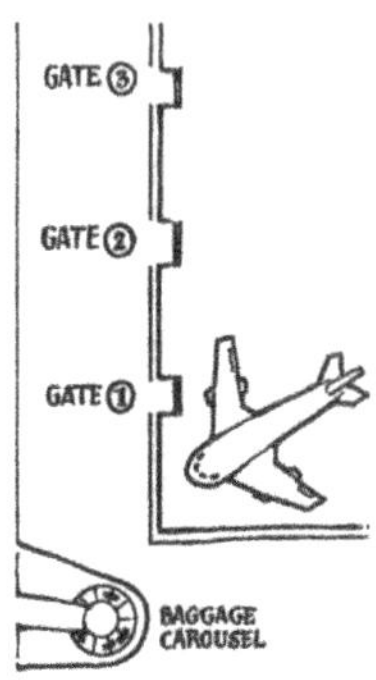

A. What was the real problem?

B. Suggest a number of things that you think the airline might have done to eliminate the complaints. Apply two or more problem-solving techniques.

Source: *Washington Post*, p. A3, December 14, 1992.

NKLRDVC/Shutterstock

1.16. In 1991, 64% of all commercial radio stations in the United States lost money. For a radio station to remain solvent, it must have a significant revenue stream from advertisers. Advertisers, in turn, target the markets they consider desirable (i.e., based on listeners' income, spending, and interests), and for the past several years this target has been the age group from 25 to 54. Along with the revenue loss, the number of radio stations playing the Top 40 songs (i.e., the 40 most popular songs of that week) has decreased by a factor of 2 in the past three years, as did the audience for the Top 40 songs. Many stations tried playing a blend of current hits and hits from 10 to 20 years ago; however, this blend irritated the younger listeners and did not seem to solve the economic problem. Apply two or more problem definition techniques to this situation.

Source: Adapted from *The International Herald Tribune*, p. 7, March 24, 1993.

1.17. FireKing is a small manufacturer of rich-looking fireproof filing cabinets that wants to increase its current market share of 3%. While the company's designs are elegant, the cabinets are also the heaviest ones on the market—and in customers' minds, greater weight means higher quality. However, greater weight also means higher shipping and transportation costs, which makes the cabinets very expensive. FireKing's management asks the following question: "How can we make our product lighter so as to have a competitive price?" However, some executives believed a lighter-weight product might hurt the image of quality. Apply one or more problem definition techniques to this situation.

Problem courtesy of David Turczyn.

1.18. Bug-B-Gone Company has developed a new method for killing roaches that is more effective than any of the other leading products. In fact, no spraying is necessary because the active ingredient is held in a container that is placed on the floor or in corners; the roach problem disappears when the container is activated. This method has the advantage that the product does all the work—the user does not need to search out and spray the live roaches. The product was test-marketed to homemakers in some southern states. Everyone who saw the effectiveness test results agreed that the new product was superior in killing roaches. Despite a massive advertising campaign, however, the standard roach sprays are still far outselling the new product. Apply one or more problem definition techniques to this situation.

Problem courtesy of David Turczyn.

1.19. A major American soap company carried out a massive advertising campaign in an attempt to expand its sales in Poland. The television commercials featured a beautiful woman using the company's soap during her morning shower. Thousands of sample cakes were distributed door to door throughout the country. Despite these massive promotional efforts, the campaign was entirely unsuccessful. Polish television had been used primarily for Communist Party politics, and commercials were relatively rare. What was aired on television was usually party-line politics. Apply one or more problem definition techniques to this situation.

Problem courtesy of Christina Nusbaum.

1.20. Employees of a certain store are allowed to take merchandise out of the department store on approval. The original procedure required each employee to write an approval slip identifying the merchandise taken. Some employees are abusing the system by taking the clothing and destroying the slip, thereby leaving no record of the removed merchandise. Apply one or more problem definition techniques to this situation.

Problem courtesy of Maggie Michael.

1.21. After Crest toothpaste had been on the market for some time, Procter & Gamble, its manufacturer, decided to offer a mint-flavored version in addition to the original, wintergreen-flavored product. In the course of developing the new mint-flavored product, a test batch of mint-flavored product was produced by the same pilot unit used to produce wintergreen-flavored product. The pilot equipment used a tank and impeller device to mix the mint-flavor essence with the rest of the ingredients to form the finished product (which is a very viscous solution). Some of the pilot plant product was packed into the familiar collapsible tubes for further testing. Tubes used in testing the mint flavor were identical to those used for the wintergreen-flavored product. In the packing operation, toothpaste was pumped through lines into the unsealed ends of brand-new tubes. After filling, the open tube ends were heatsealed. The packing operation is illustrated in the figure.

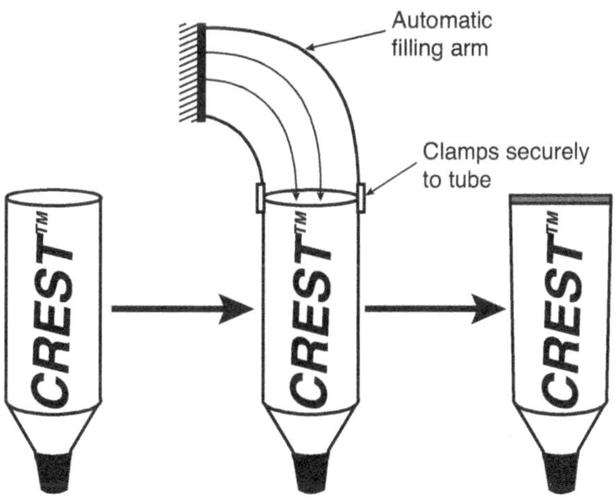

Continuous Movement of Tubes

To assess storage stability, some of the filled tubes were randomly separated into several groups and each group was stored in a constant-temperature room. Storage temperatures varied from 40°F to 120°F.

Early sampling of the mint-flavored product showed nothing unusual. However, several months into the test, a technician preparing to test the product from one of the stored tubes noted that the first 0.25 inch of paste squeezed onto a toothbrush was off-color. The rest of the product in the tube met the color specification. Nothing like this had ever been seen with the original formula. Further testing showed that a person had to squeeze more product out of those tubes that had been stored at higher temperatures and/or stored for longer times before a product that met the color specifications would exit the mouth of the toothpaste tube. Tubes stored for a period of time at 40°F contained no off-color product, whereas tubes stored for the same length of time at higher temperatures produced off-color paste.

The only exception to these results was a single tube, which had been stored at a temperature above 40°F. A leakage of off-color product was found around the base of the cap on this tube, but the product inside the tube met the color specifications.

While other tests showed the off-color product to be safe and effective in cleaning teeth, consumers certainly would not accept a color change in a product expected to have the same color from the first squeeze to the last. Moreover, such a change could have been an early warning of more serious problems to come. This phenomenon had to be understood and eliminated before the new flavor could be marketed. Accordingly, various possible remedies were tested: caps and tubes made of different materials, different mixing methods, and so forth. None of these changes had any effect on the off-color problem. All raw materials, including the new mint-flavor essence, were checked and found to meet specifications. A subsequent batch of the wintergreen product was made and tested for storage stability and, as usual, no off-color problems occurred.

Carry out a K.T. problem analysis to learn the cause of the off-color toothpaste.

1.22. Chocolate butter paste is the primary ingredient used by a number of major bakeries for a wide variety of pastries. The paste is a very viscous liquid that is manufactured by Cocomaker Industries in a major populous city in the Midwest. Cocomaker supplies customers as close as Dolton and as far away as Chicago (which is a long drive from its plant). The paste flows from the production line into five-gallon drums, which are placed immediately into refrigerated trucks for shipment to the respective customers. Until February, all the trucks had been the same size and the drums were stacked in rows three drums wide, four drums high, and eight drums deep. Now two rather small customers, Bell Bakery and Clissold Bakery, each requiring 20 drums per day, have been added in the Chicago area. Supplying these new customers, along with an increased order by the Chicago customer Hoyne Industrial Bakers, necessitated the purchase of a larger truck. The new truck could fit five drums across, four drums high, and eight drums deep. The truck would stop at Bell, then Hoyne, then Clissold in Chicago proper. With the increased market in the Chicago area, Cocomaker's plant is running at close to maximum capacity. Because the ingredients of the paste are mixed by static mixers, the pumps are currently operating at their maximum capacity and the plant is operating 20 hours per day.

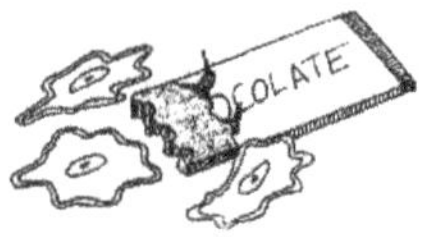

In November, Cocomaker successfully lured two nearby customers, Damon Bakery and Oakley Bakery, away from one of its competitors. By increasing plant operation to 24 hours per day, all orders could be filled. As the Christmas season approached, however, the usual seasonal demand for chocolate butter paste posed a problem for Cocomaker: It needed to meet demands not encountered in previous years. The company decided that if the processing temperature was increased by 20 degrees, the paste would be sufficiently less viscous, and production demands could be met with the current pump limitations. Unfortunately, the increased capacity began to generate problems as Christmas approached. The pumps began failing on a regular basis; a strike at the supplier of the shipping containers caused Cocomaker to buy containers from a new supplier, which claimed to carry only sturdier containers at a 10% increase in price; the safety officer had an emergency appendectomy; and—most troubling—Hoyne called about an unacceptable bacteria count in shipments for the last five days. As a result of the bacteria, people who bought Hoyne's products have been getting ill.

An immediate check of the bacteria levels shows that they are at the same acceptable levels they have always been when leaving Cocomaker's plant. You call Hoyne and tell them that the plant levels show that the paste is within bacteria specifications. Two days later, you receive a call from Hoyne, saying that the bakery had hired an independent firm, which reported that the bacteria levels for the chocolate butter paste are well above an acceptable level. You call the Damon, Bell, Clissold, and Oakley bakeries and ask them to check their bacteria counts; they report back that everything is within the specifications most often reported. A spot check of other customers shows no problems. Now you receive a call from Hoyne, saying the bakery is initiating legal and governmental actions to close your plant down.

Carry out a K.T. problem analysis to learn the cause of the problem.

1.23. "Blue and Green Honey Makes French Beekeepers See Red"

Bee graphic: Pagina/Shutterstock; honeycomp photo: Africa Studio/Shutterstock

France is one of the largest producers of honey within the European Union, producing some 18,330 tons annually, according to a recent audit conducted for the national farm agency FranceAgriMer.

Ribeauville, situated in the middle of Alsace along the scenic wine route southwest of Strasbourg, is best known for its vineyards. The economic situation in the region is excellent, nesting great industries and factories such as Aventis, Brasseries Kronenbourg, Mars Chocolate, and Agrivalor side-by-side with the vineyards. Aventis is a pharmaceutical company manufacturing Ambien and anticancer and other drugs. Kronenbourg is a brewery whose beer is distributed worldwide. Mars Chocolate produces candy such as Mars Bars, Milky Way, Snickers, Skittles, Starburst, M&Ms, Dove Bars, 3 Musketeers, and Twix. Agrivalor is a biogas company that takes waste from these industries and turns it into a gaseous fuel. In addition, living among the winemakers are about 2,400 beekeepers in Alsace who tend some 35,000 colonies and produce about 1,000 tons of honey per year, according to the region's chamber of agriculture.

The continuous success of the region's economy is partially due to the intermittent participation of the local government in creating incentives to develop integration among the local companies. Recently the local authorities have launched a tax program that gives a tax credit for companies that locally integrate their supply chains, including raw materials and waste disposal. As a result Kronenbourg and Mars have been sending their waste to a biogas plant. Each of Mars's products is sent separately in different containers to the plant while Kronenbourg combines all its waste together and ships it in covered barrels.

Recently, bees at a cluster of apiaries, also known as bee yards, in northeastern France have been mysteriously producing honey in shades of blue and green, alarming their keepers. The unsellable honey is a new headache for around a dozen affected beekeepers already dealing with high bee mortality rates and dwindling honey supplies following a harsh winter, said Alain Frieh, president of the beekeepers' union.

Since August, beekeepers around the town of Ribeauville have seen bees returning to their hives carrying unidentified colorful substances that have turned their honey unnatural shades of brown, yellow, blue, and green. Mystified, the beekeepers embarked on an investigation and discovered that the biogas plant was only four kilometers (2.5 miles) away from the apiaries. Because of the unsellable colored honey, the beekeepers decided to call a meeting with the other members of their union. Since the use of pesticides was banned, it was not a reason to be worried. Some of the members of the union brought up the possibility of contamination from the industries in the region. Thus, they perceived that the colored honey was only in the proximities of the biogas plant exclusively. However, the problem only started last August and the biogas plant has been installed there for much longer. The biogas company had to shut down for a couple of weeks in August due to malfunctioning equipment and as a result had to store the open waste containers containing residue from the different products from the Mars factory in an uncovered holding area. The waste from Kronenbourg was stored beside the Mars waste; however, it remained in covered barrels.

Bee numbers have been rapidly declining around the world in the last few years and the French government has banned a widely used pesticide, Cruiser OSR, that one study has linked to high mortality rates.

Carry out a K.T. problem analysis to learn the cause for the colored honey.

Source: Blue and Green Honey Makes French Beekeepers See Red by Patrick Genthon. *Reuters*, Oct. 5, 2012.

1.24. Sparkling mineral water is the primary product of Bubbles, Inc. This firm, which is based in France, serves three major markets—Europe, North America, and Australia. It collects water from a natural spring; the water is then filtered through a parallel array of three filter units, each containing two charcoal filters. The filtration process removes trace amounts of naturally occurring contaminants. The filtered water is stored in separate tank farms, one for each market, until it is transported by tanker truck to one of the three bottling plants that serve the company's markets. When the water arrives at the bottling plant, it is temporarily placed in 3500 m^3 storage tanks until it can be carbonated to provide the effervescence that is the trademark of the producer. Some of the water is also flavored with lemon, cherry, and raspberry additives.

Next, the sparkling water is packaged in a variety of bottle sizes and materials, ranging from 10-ounce glass bottles to 1-liter plastic bottles. The European market receives its shipments directly by truck, usually within three days. Products bound for North America or Australia are shipped first by truck to the waterfront and then by freighters to their overseas destinations.

Business has been good for the last several months, with the North American and European markets demanding as much sparkling water as can be produced. This situation has required that Bubbles contract with additional plastic bottle suppliers to keep up with the increased demand. It has also forced regularly scheduled maintenance for the Australian and North American tank farms to be delayed and rescheduled because of the high demand for the product. There is also, of course, a larger demand placed on the spring that supplies the mineral water for the process.

Unfortunately, the news is not all good for Bubbles. The bottling plant for the Australian market is currently several weeks behind schedule owing to a shipment that was lost at sea. This catastrophe has required that water from the company's reserve springs, which are located many miles from the bottling plant, be used to augment the water supplied by the regular spring so that the bottling plant can operate at an even higher level of production. The availability of water from the reserve springs is hindered by their remote locations, but the water from these springs does not require filtration. In addition, contract negotiations are going badly and it appears there will be a strike at all of the bottling plants. Recent weather forecasts indicate that relief from the ongoing drought, which has already lasted three months, is not likely. Worst of all, customers in the North American and Australian markets are complaining that all shipments of the sparkling water in the last six weeks have contained benzene in unacceptably high concentrations. You know that benzene is often used as an industrial solvent but is also found naturally.

A quick survey of the bottling plant managers shows that the North American–bound products that are currently packaged and awaiting shipment have benzene concentrations in excess of acceptable concentrations. However, the managers of the bottling plants that service the Australian and European markets report that no significant level of benzene was detected in the bottles that are currently stored. Authorities in the North American and Australian markets have already begun recalling the product, with authorities in the European market pressuring Bubbles for a quick solution and threatening to recall products as a precautionary measure.

Carry out a K.T. problem analysis to learn the cause of the problem.

Source: Adapted from *Chemtech*, "When the Bubble Burst," p. 74, February 1992.

1.25. Currently, many drilling platforms in the Gulf of Mexico collect oil from a number of oil wells, which they then pump through a single pipeline from the platform to shore. Most of these wells have always been quite productive. Consequently, the oil flows through the pipeline lying on the ocean floor at a reasonable rate.

When the oil comes out of the wells, it is at a temperature of approximately 145°F; by the time the oil reaches shore, the temperature in most pipelines is approximately 90°F. The temperature of the water on the ocean floor for the majority of the platforms within two miles of shore is approximately 45°F. However, the water depth increases as you move away from shore and the temperature of the water on the ocean floor decreases to 39°F.

Recently, two new platforms (A and B) were erected in the Gulf Coast farther out from shore than the other drilling platforms. About a year and a half after both came on stream, a disaster occurred on Platform A, such that no oil could be pumped to shore through the pipeline from this platform. Platform B continued to operate without any problems. When the crude composition at the well head was analyzed, it was found to have exactly the same chemical composition (e.g., asphaltenes, waxes, gas) as that found in the well heads on all other platforms. The only difference between Platform A and Platform B was that the production rate of Platform A was much lower than that of Platform B. Nevertheless, the production rate from Platform A was still greater than many of the platforms near the shoreline.

Carry out a K.T. problem analysis to learn the reason for the plugging of pipeline A.

1.26. Research or create a statement of a perceived problem and then apply one or more of the following problem definition techniques to that problem statement:

A. Find out where the problem came from

B. Present state/desired state and Duncker diagram

C. Statement–restatement technique

D. Problem analysis

1.27. Dow Chemical Company, a world leader in supplying special chemical products and solutions to all sectors, launched a new product named Dursban, a pesticide specifically for use in Latin American climates. Dow exports significant amounts of pesticide to Latin America where the industrial farms make up 75% of sales and the other 25% are to small individual farmers. Not long after the launch of Dursban, Dow received complaints from some of the large industrial farms that a strong odor was emitted from their fields. They blamed the product Dursban for causing these strong odors because the odor began when they switched to the new pesticide. The management team of engineers at the chemical company decided to install a new multimillion-dollar crystallization separation process at the manufacturing plant in America, in September, to improve the pesticide and to get rid of the odor. The new odor-free Dursban was released later the same year in November. The engineers tested the product thoroughly to make sure the new version was just as effective as the previous version as the initial results showed, and it was. The industrial farms in Latin America praised the new product as their crops were doing just as well with the new product as the old and there was no odor. The company was pleased with its new product and even applied the technology to its North American pesticide to reduce its odors to the applause of Dow's large American customers, which are nearly all large industrial operations. To Dow's surprise, sales of the new product decreased by 25% in Latin America by the beginning of the next year.

Carry out a K.T. problem analysis to find the real problem of why sales of the new product decreased in Latin America.

Source: Steve Pankratz, University of Michigan, BSE Chemical Engineering, Class of 2010.

Fermi Problems

1.28. How many square meters of pizza were eaten by an undergraduate student body population of 20,000 during the fall term of 2015?

1.29. If we could convert all the weight that dieters lose in the United States every year to energy, how many gallons of gasoline would be its equivalent?

1.30. Refine the calculated number of piano tuners in Chicago by including pianos in churches and schools.

1.31. How many exams does a high school geometry teacher give throughout a career?

1.32. How much should you charge to wash all the windows in all the residential homes in Seattle?

1.33. How many vacuum cleaners are made a year?

1.34. Goodcase, a suitcase manufacturer, is launching a new extra large (XL) suitcase, in addition to its current small, medium, and large models. You are a plant engineer at Goodcase, in a meeting where the details of the XL are being rolled out. The additional raw materials for the XL have been ordered and are scheduled to arrive on time. The luggage makers have been trained to make the new model, and the new schedule with an extra shift has been created for the production of 50 XL suitcases per day, seven days a week. Additional trucks to ship the new suitcases are also scheduled to pick up suitcases every Sunday evening.

You are concerned that the rollout team is planning on storing suitcases in the extra supply room, dimensions of 10 ft × 10 ft × 30 ft, while they are waiting to be shipped. Suitcases come in boxes 3 ft by 2 ft by 1.5 ft.

Carry out a Fermi calculation to see if space will be an issue.

FURTHER READING

Copulsky, William. "Vision Innovation." *Chemtech*, 19, p. 279, May 1989. Interesting anecdotes on problem definition and vision related to a number of popular products.

De Bono, Edward. *Serious Creativity*. Harper Business (a division of Harper Collins), New York, 1993. Summary of 20 years of creativity researched by de Bono. Many useful additional problem definition techniques are presented.

Rubenfeld, M., and B. Scheffer. "Critical Thinking: What Is It and How Do We Teach It?" In: J. Dochterman and H. Y. Grace (eds.), *Current Issues in Nursing*. Mosby, St. Louis, MO, 2001, pp. 125–132.

Scheffer, B. K., and M. G. Rubenfeld. "A Consensus Statement on Critical Thinking in Nursing." *Journal of Nursing Education,* 39, pp. 352–359, 2000.

The following books are all published by the Foundation for Critical Thinking (P.O. Box 220, Dillon Beach, CA 94929, www.critical thinking.org):

Linda Elder and Richard Paul. *The Miniature Guide to Analytic Thinking* (2003).

Linda Elder and Richard Paul. *The Miniature Guide to Asking Essential Questions* (2002).

Richard Paul and Linda Elder. *The Miniature Guide to Critical Thinking Concepts and Tools* (2009).

Richard Paul and Linda Elder. *The Thinker's Guide to Critical and Creative Thinking* (2004).

Richard Paul and Linda Elder. *The Thinker's Guide to the Art of Socratic Questioning* (2006).

Richard Paul, Robert Niewoehner, and Linda Elder. *The Miniature Guide to Engineering Reasoning* (2006).

2

Graphing and Units

SECTIONS

OBJECTIVES

After reading this chapter, you should be able to do the following:

- Understand the difference between a unit and a dimension.
- Convert between the different systems of measurement.
- Understand the types of error in a measurement.
- Express large numbers in scientific notation.

2.1 INTRODUCTION

The engineering sciences depend on the quantification of physical phenomena. That is, in order for an engineer to make use of a physical quantity, it must have a numerical value. This numerical value is often obtained through measurement. A unit of measurement is associated with the parameter.

Because measurement instruments are imperfect, there will be some error in the experimentally obtained value. It is important for an engineer to have some indication of the size of this error.

In this chapter, we will examine the systems of measurements commonly used by mechanical engineers. Furthermore, we discuss the types of error that occur whenever a measurement is made.

2.2 DIMENSIONS AND UNITS

In engineering practice, physical quantities are quantified by giving a magnitude and a unit of measurement. For example, the playing surface of a football field is one hundred yards long. In this case, the magnitude is the value one hundred, and the yard is the unit of measurement. A *unit* refers to the arbitrary system of measurement.

Units of measurement are standardized quantities. In the United States, the National Institute of Standards and Technology (NIST) standardizes units of measurement. Standardization of a unit means that within the region where the standardizing organization has jurisdiction, the unit has the same meaning for everyone. For example, the yard in the previous example would imply the same physical quantity of length to a person living in California as it would to a person living in New York. Furthermore, a standard yard would have the same length for both individuals, because the length of a yard is standardized.

Some units are combinations of others. For instance, the unit of force is a combination of the units for mass and acceleration.

A *dimension* is a measurable quantity of a physical parameter. The yard is a unit of length in a particular system of measurement whose dimension is length. *Fundamental dimensions* are independent of one another. For example, a set of fundamental dimensions may be mass, length, time, and temperature.

Let us define the dimension of an engineering quantity by placing a bracket around the quantity. Thus, if a is acceleration, then $[a]$ would imply the dimensions of acceleration. Furthermore, let M, L, T and θ represent the fundamental dimensions of mass, length, time, and temperature, respectively. Then, the dimensions of acceleration would be

$$[a] = \frac{[L]}{[T^2]} = \frac{L}{T^2}.$$

Therefore, the dimensions of acceleration are length per time squared. The dimension of mass is mass. What are the dimensions of force, F?

The law of *Dimensional Homogeneity* requires that all terms in an equation have the same dimensions. Newton's second law says that the force on a system is equal to the mass of the system times the acceleration. This may be written as $F = ma$, where F is the force, m is the mass, and a is the acceleration. Therefore, the dimensions of force must be the same as those of mass times the acceleration in order for Newton's second law to be valid. Then, the dimensions of force are

$$[F] = [ma] = [m][a] = m\frac{L}{T^2}.$$

A nondimensional quantity is a quantity that has dimensions of one. For example, what are the dimensions of $F(ma)$? The answer is

$$\left[\frac{F}{ma}\right] = \frac{\left[\frac{mL}{T^2}\right]}{\left[m\frac{L}{T^2}\right]} = 1.$$

Nondimensional quantities play an important role in fluid mechanics.

PRACTICE!

You may have noticed that when you pull on a spring, there is a force tending to restore the spring to it's unstretched position. If F is this force and x is the amount that the spring was displaced, engineers write the relationship between the force and spring as

$$F = kx,$$

where k is a constant called the spring stiffness. What must the dimensions of k be for this equation to obey the law of dimensional homogeneity?

2.2.1 The British Gravitational System

The units of a quantity depend on the arbitrary system of measurement. One of these systems of measurement is the British Gravitational System (BGS). This particular system uses force, as a fundamental dimension, instead of mass. Because of Newton's second law (i.e., $F = ma$), the dimensions of mass become

$$[m] = \frac{[F]}{[a]} = \frac{FT^2}{L}.$$

In BGS, the unit of force is the pound (lb). Therefore, the unit of mass is related to the unit pound. This unit is called the *slug* and is defined by

$$1 \text{ slug} = 1 \text{ lb} \cdot \text{sec}^2/\text{ft},$$

where ft (foot) is the unit of length.

The weight of an object, W, is the mass times the gravitational acceleration constant, g. This may be written as $W = mg$. In this case, the unit is the pound, as it is a force. Because, on earth, $g = 32.174$ ft/sec^2, a slug weighs

$$\text{Weight of slug} = W = 1 \quad \text{slug} \cdot 32.174 \text{ ft/sec}^2 = 32.174 \text{ lb.} \tag{2-1}$$

Therefore, in BGS, one slug has a weight of 32.174 pounds.

Additional units for BGS are listed in Table 2-1.

TABLE 2-1 Listed in the Table are Some Common Units for Engineering Parameters. The Parameter in Parentheses is the Symbol for That Unit

QUANTITY	BGS	EE	SI
Mass	slug	lbm	kilogram
Length	ft	ft	meter (m)
Time	sec (s)	sec (s)	sec (s)
Area	ft^2	ft^2	m^2
Volume	ft^3	ft^3	m^3
Velocity	Ft/s	Ft/s	M/s
Acceleration	Ft/s^2	Ft/s^2	M/s^2
Density	Slug/ft^3	Lbm/ft^3	Kg/m^3
Force	Lbf	Lbf	newton (N) = kg·m/s^2
Pressure (or Stress)	Lbf/ft^2	Lbf/ft^2	pascal (Pa) = N/m^2
Energy (or Work)	ft·Lbf	ft·Lbf	joule (J) = N·m
Power	ft·Lbf/s	ft·Lbf/s	watt (W) = J/s

2.2.2 The English Engineering System

The English Engineering (EE) system is defined so that 1 unit of mass has a weight of one pound (Lbf) in a standard gravity. The unit of mass in the EE system is the pound-mass (Lbm), and the unit of force is called the pound-force (Lbf). From Newton's second law, we have that 1 Lbf = 1 Lbm·32.174 ft/sec^2. This relation does not have the same units on the right-hand side as the left-hand side of the equation. In order not to violate the law of dimensional homogeneity, the pound-force must be related to the unit of pound mass through the relation:

$$1 \text{ Lbf} = \frac{1 \text{ Lbm} \cdot 32.174 \text{ ft/sec}^2}{32.174 \text{ ft} \cdot \text{Lbm/Lbf} \cdot \text{sec}^2}. \tag{2-2}$$

Thus, in the EE system, when converting between units of mass or force, one must divide or multiply by a conversion factor, g_c, equal to 32.174 ft·Lbm/Lbf·sec^2. Then, in order to have dimensional homogeneity, Newton's second law is properly written as

$$F = \frac{ma}{g_c}. \tag{2-3}$$

EXAMPLE 2.1: What is the force (in Lbf) on a 0.1-Lbm body accelerating at 2 ft/ sec^2?

SOLUTION

The force is given by Equation 2-3. Therefore, we have

$$F = \frac{0.1 \text{ Lbm} \cdot 32.174 \text{ ft/sec}^2}{32.174 \text{ ft} \cdot \text{Lbm/Lbf} \cdot \text{sec}^2} = 0.1 \text{ Lbf}$$

Additional units for the EE system are listed in Table 2-1.

2.2.3 The SI System

The Système International (SI), or metric system, was first established in France in about 1800 and has become the worldwide standard. In the United States, voluntary conversion to the metric system began in 1975, even though the system has been recognized by Congress as early as 1866. In time, it is expected that all units will be in SI form.

The unit of mass in SI is the kilogram (kg). The unit of length is the meter (m) and for volume, the liter (L). The system is set up so that one liter of water has a mass of one kilogram. The unit of weight in the SI system is the newton (N). From newton's second law, one newton is defined as

$$1 \text{ N} = 1 \text{ kg} \cdot \text{m/sec}^2.$$

The weight is given by the expression $W = mg$, where g is 9.81 m/sec^2.

A clear advantage of SI over other systems of measurement is that SI is based on the powers of 10. For example, 1/100th of a meter is a centimeter (cm). To convert from meter to centimeter, all one need do is multiply by 100. This is easily memorized, because "centi" means 1/100th. Therefore, centimeter means 1/100th of a meter. Table 2-2 lists the prefixes defined in the SI system.

TABLE 2-2 Prefixes for the SI System

MULTIPLICATION FACTOR	PREFIX	SI SYMBOL
10^{12}	tera	T
10^{9}	giga	G
10^{6}	mega	M
10^{3}	kilo	k
10^{2}	hecto	h
10^{1}	deka	da
10^{-1}	deci	d
10^{-2}	centi	c
10^{-3}	milli	m

TABLE 2-2 Prefixes for the SI System

MULTIPLICATION FACTOR	PREFIX	SI SYMBOL
10^{-6}	micro	μ
10^{-9}	nano	n
10^{-12}	pico	p
10^{-15}	femto	f
10^{-18}	atto	a

Handling small quantities is very easy in SI. Consider a measurement of 0.03 m. Rather than writing this number in decimal form, multiply it by 100 to get the same measurement in centimeters (i.e., 3 cm). Conversion between SI and the BGS and EE systems is discussed in Section 2.3.

2.3 CONVERSION BETWEEN DIFFERENT UNITS

Table 2-3 contains a table with conversion factors. It is not an exhaustive table, but may be used for the exercises in this book.

TABLE 2-3 The table lists conversion factors that may be used to solve engineering problems. Some values are approximate

DIMENSION	CONVERSION FACTOR
length	0.3048 m = 1 ft
area	1 m^2 = 10.764 ft^2 = 1550 in^2
Mass	1 kg = 2.2046 lbm
Volume	1 m^3 = 264.2 gal = 35.32 ft^3
velocity	1 m/s = 3.2802 ft/s = 2.237 mi/h
Acceleration	1 m/s^2 = 3.2808 ft/s^2
Force	1 N = 0.22481 lbf
Pressure	1 kPa = 0.14504 lbf/in^2
Temperature	$°C = \frac{5}{9}(°F - 32°)$

Except for the definition of mass, the BGS and the EE system have identical units. Conversion between these two units of measurement is straightforward. Because 1 Lbf accelerates 1 Lbm at 32.174 ft/sec^2, 32.174 Lbm would be accelerated at 1 ft/sec^2. Therefore,

$$1 \text{ slug} = 32.174 \text{ Lbm}. \tag{2-4}$$

Expression (2.4) may be used to convert between units of Lbm and slug.

In general, a conversion factor exists between equivalent units. For example, 1 ft = 0.3048 m. When using conversion factors, keep track of the units. In your result, all but the desired unit should drop out from the expression.

EXAMPLE 2.2: A structural component has a length of 2 ft 3 inches. What is the corresponding length in centimeters?

SOLUTION

The conversion between feet and meters is given in Table 2-3 (in particular, 1 ft = 0.3048 m). First, we use the fact that there are 12 inches per foot. Thus, 2 ft 3 inches is equivalent to

$$L = 2\text{ ft} + 3\text{ inch}\frac{1\text{ ft}}{12\text{ inch}} = 2.25\text{ ft},$$

which allows us to convert to meters. Thus,

$$L = 2.25\text{ ft} \times \frac{0.3048\text{ m}}{1\text{ ft}} \times \frac{100\text{ cm}}{1\text{ m}} = 68.58\text{ cm}.$$

Notice how all but the units of centimeter drop out in the expression.

PROFESSIONAL SUCCESS

As we have seen, there are three commonly used systems of measurement in engineering. Although the metric system is gaining in popularity, BGS and EE system are still used. Be very careful when performing calculations with values associated with different systems of measurement.

2.4 ERRORS AND ACCURACY

The engineering sciences are based on measurements of fundamental parameters, such as temperature, velocity, and so on. These measurements are performed for three basic reasons.

The first reason is to obtain information about an unknown process. A classic example is Robert Hooke, who, in the 17th century, measured the extension of wires due to a weight. Hooke was motivated by a desire to understand the deformation of a solid material. In the course of his experiments, he uncovered the relationship between forces acting on a body and the deformation of the body. Today, Hooke's discoveries are contained within a relation known as Hooke's law, which relates stress and strain.

The second reason is to obtain parameters that cannot be obtained by theoretical methods. An example is the forces acting on an aircraft wing. Because it is expensive to numerically obtain these forces, models of the wing or aircraft are placed in a wind tunnel and experimentally measured.

The third reason is to validate theory. A theory is valid only if it explains experimentally obtained observations. A theory is invalid if it contradicts such experimental observations.

Accuracy is defined as the difference between the true value and the measured value of a given parameter. For example, a beaker of water may be measured with a thermometer, and a value of 10.5 °C may be obtained. Suppose the true value of the temperature is 10 °C. There is obviously a difference between the two values, for reasons that will be discussed shortly. The difference between the two values is 0.5 °C. Thus, we would say that the measured temperature is accurate to within 0.5 °C.

For the reasons outlined previously, quantities in engineering depend on one or more measurements. However, all measurements contain error. Obviously, if the error in the measurement is large compared with the measured quantity, then the result is not very useful. Thus, of great importance in the measurement of fundamental engineering parameters is not whether there is error, but is in obtaining an estimate of the error.

The error in a measurement may be divided into two parts. The first part, *bias error*, occurs whenever a particular measurement instrument is used. It is inherent, due to the design and nature of the measurement instrument.

For example, suppose you wanted to cut an 8-foot 2 × 4 piece of lumber to, say, 92 inches. Because the smallest graduation on the tape measure is 1/16 of an inch, the measurement can never be better than half that value. Thus, every time the tape measure is used, the cut will be at least off by 1/32 of an inch.

The second part of the error is called the *precision error* and occurs in a random manner, having a different value whenever the measurement is performed. Precision error is often the result of mistakes made by the technician or external disturbances, such as temperature variations or vibration in a system.

Repeating a measurement several times using different equipment and technicians will randomize the bias error, provided that enough tests are performed. This effectively transforms the bias error into a type of precision error. Statistical analysis of the results of tests will give a mean value for the desired quantity and a level of confidence in the result.

PRACTICE!

In completing the construction of a house, a finish carpenter uses a table saw to rip five pieces from a 4-foot × 8-foot sheet of oak plywood. The desired width of each piece is four inches. After cutting the five pieces, the carpenter accurately measures each piece and obtains the values in Table 2-4. If the bias error is 1/32 of an inch, what is the precision error for each piece?

TABLE 2-4 Width of the Five Pieces

NUMBER	WIDTH (INCHES)
1	4.010
2	3.995
3	4.120
4	3.998
5	4.005

2.5 SIGNIFICANT DIGITS

Engineering calculations are only accurate to a certain number of significant digits, because the calculations are usually based on one or more measurements. The numbers 4.454 and 0.004454 are both given accurate to four significant figures. The absolute error is less than 0.001 for 4.454 and 0.000001 for 0.004454.

For the value 4.454, the true value lies in the range between 4.4535 and 4.4540. However, for the value 0.004454, the true value lies in the range between 0.0044535 and 0.0044540.

Another way to express a number is by the use of scientific notation. Scientific notation is based on the powers of 10. For example, $10^1 = 10$, $10^2 = 100$, $10^3 = 1{,}000$, and so on.

Scientific notation is useful in expressing very large or very small numbers. Consider the average distance between the earth and the moon, which is 240,000 miles. In scientific notation, this may be expressed as $240 \times 1{,}000 = 240 \times 10^3$.

In a number such as 240,000, it is hard to see exactly how many of the digits are significant. Scientific notation may be used to remedy this situation. For example, if five significant figures are required, then 240,000 may be written as 2.40000×10^5. Or, if, say, only three figures are significant, then the number may be written as 240×10^3.

With the advent of calculators, there is a tendency to provide too many figures in a response. Many scientific calculators are available with a nine-digit display. This does not mean that all the figures in the display are significant.

When performing a calculation, remember that the number in the calculation with the least number of significant figures determines the number of significant figures in your answer. For example, consider the calculation:

$$D = 4.25 \text{ ft} \times \frac{0.3048 \text{ m}}{1 \text{ ft}} = 1.30 \text{ m}.$$

In this expression, the number 4.25 has fewer significant figures than 0.3048. Thus, the number of significant figures in 4.25 dictates the number of significant figures in the answer.

The answer has been rounded off. Whenever you round off a number, use the following rule: If the first number discarded is 5 or greater, then increase the preceding number by one.

In the previous example, the result to nine digits was 1.29540000. Because of the number of significant figures in the calculation, this number is rounded off to two decimal places. Because the first number to be discarded is 5, the preceding number (9) is incremented by one. This gives the result 1.30.

PROFESSIONAL SUCCESS

When using scientific notation, be aware of the difference between a number raised to the power of 10 and a number times 10 raised to the power of 10. For example, note that $1.34^{10} \neq 1.34 \times 10^{10}$, because $1.34^{10} = 18.66$ and $1.34 \times 10^{10} = 13{,}400{,}000{,}000$.

KEY TERMS

- accuracy
- British Gravitational System (BGS)
- dimension
- English Engineering (EE) system
- error
- metric system
- scientific notation
- units

2.6 SUMMARY

As we have seen, fundamental quantities in the engineering sciences depend on experimental measurement. These measurements contain error, which is the combination of two types: bias error and precision error. The error in a measurement may be quantified by performing several measurements of the desired quantity. Once the value of the error is quantified, the usefulness of the measurement is known.

Engineering parameters are given in terms of units, which depend on standards codified by governing bodies. A dimension is a measurable quantity of a physical parameter. Dimensions do not depend on the arbitrary system of measurement.

The most widely used system of measurement in the world is SI, or the metric system, although other systems of measurement exist. Once a value of an engineering parameter in known in a particular system of measurement, it may be obtained in another system by using a conversion factor.

2.7 EXERCISES

E1. What is the mass in slugs of an eight-pound bowling ball?

E2. Which has a greater mass, a block of steel weighing 5 pounds or a block of aluminum weighing 10 N?

E3. The speed of light is 299,792,458 m/s. What is the speed of light in km/hr?

E4. Suppose that the steel and aluminum blocks from Problem 2 were each accelerated by a 1-N force. What is the acceleration of the steel block in ft/s^2? What is the acceleration of the aluminum block in m/s^2?

E5. The gravitational acceleration constant on the moon is 1/6 the value on earth. What will be the weight of a man on the moon if the same man weighs 210 pounds on the Earth?

E6. Suppose that the viscosity of an oil is 0.0003 Lbf·s/ft^2. What is the viscosity in N·s/m^2?

E7. What are the dimensions of the unit of pressure Pa (pascal)? (*Hint:* Consider Table 2-1).

E8. The hydrostatic pressure in a fluid is given by

$$p = p_0 + \rho g y,$$

where p_0 is the pressure at the free surface, ρ is the density of the fluid, y is the depth of the fluid, and g is the acceleration due to gravity. Show that this equation obeys the law of dimensional homogeneity.

E9. A Newtonian fluid is governed by the expression

$$\tau = \mu \frac{\partial u}{\partial y},$$

where τ is the shear stress, having dimensions $m/(LT^2)$ and μ is the viscosity of the fluid, with dimensions $m/(LT)$. What must the units of $\partial u/\partial y$ be in order for this equation to obey the law of dimensional homogeneity?

E10. In conduction, heat is transferred in a body between two different temperatures. The amount of heat transferred per unit area, q'', has units of W/m^2 (1W = 1J/s = 1N·m/s) in SI. For steady, one dimensional heat transfer along direction x, q'' is given by the expression

$$q'' = k\frac{\Delta T}{\Delta x}$$

where $\Delta T = T_2 - T_1$ in kelvin, Δx is the change in the dimension x in meters, and k is a constant called the thermal conductivity. Using this relationship and the given information determine the units of the conductivity in SI.

E11. For the equation in Problem E8, what must the units of $\partial u/\partial y$ be if SI is used?

E12. The average distance between the Earth and the Sun is 98 million miles. Write this value in scientific notation with two digits of accuracy.

E13. If the speed of light is 299,792,458 m/s and the distance between the Earth and the Sun is 98 million miles, how long (in minutes) does it take for light to travel from the sun to the Earth?

3

Statics and Dynamics

3.1 INTRODUCTION

This chapter deals with a branch of mechanics that is concerned with bodies at rest or in motion acted upon by forces. The bodies in statics and dynamics are treated as rigid bodies; that is, the bodies undergo no deformation. Thus, statics and dynamics form a branch of mechanics called rigid-body mechanics.

Statics deals with bodies that are in equilibrium with applied forces. Such bodies are either at rest or moving at a constant velocity. On the other hand, dynamics deals with bodies that are accelerating. For an accelerating system, the velocity of the system is not constant.

Rigid-body mechanics is based on the Newtons' laws of motion. These laws were postulated for a particle, which has a mass, but no size or shape. Newton's laws may be extended to rigid bodies by considering the rigid body to be made up of a large number of particles whose relative position from each other do not change. Newtons' laws may be stated as follows:

First law. A particle at rest or in motion with constant velocity along a straight line will remain in its present state unless acted upon by an unbalanced force.

Second law. If an unbalanced force acts upon a particle, then the particle will experience an acceleration that has the same direction as the force. The magnitude of the acceleration is proportional to the magnitude of the force.

Third law. For every force acting on a particle, the particle responds with an equal and opposite reactive force.

Rigid-body mechanics is a foundation of mechanical engineering. Because this subject deals with forces applied to a body, a mechanical engineer must be able to obtain a

SECTIONS

OBJECTIVES

After reading this chapter, you should be able to do the following:

- Understand the concept of a vector.
- Write a vector in component form.
- Understand the concepts of a couple and moment.
- Construct a free-body diagram of a physical system.

representation of the body with the applied forces. This representation is called a free-body diagram. Furthermore, to predict the response of the body under the applied forces, the engineer must be able to mathematically represent the forces. Such forces may be mathematically represented as vectors.

3.2 THE CONCEPT OF A VECTOR

A *scalar* is a quantity having only a magnitude. For example, the temperature of a cup of coffee is a scalar describing a physical quantity of the coffee.

A *vector* is a quantity having both a magnitude and a direction. Forces are vectors. A force vector directed towards the center of the Earth whose magnitude is the weight of the individual may represent the weight of the person.

In this chapter, using bold typeface will indicate a vector. On the other hand, italic typeface will indicate a scalar. For example, **a** is a vector, whereas a is a scalar.

Given two vectors, the vectors will only be equal if both the magnitude and direction of both vectors are equal. A vector that has the same magnitude as another vector, but is in the opposite direction, is a *negative* vector. A vector whose magnitude is unity is called a *unit* vector.

A quantity such as a force has a magnitude, as well as a direction in which it is acting. This implies that vectors may represent forces.

Geometrically, we can represent vectors with an arrow. As shown in Figure 3.1, the head of the arrow points in the direction of the vector, and the length of the arrow indicates the magnitude of the vector.

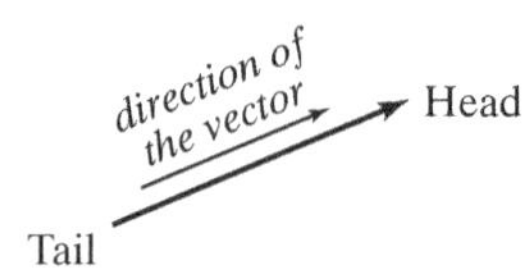

Figure 3.1.

3.2.1 Components of a Vector

If a coordinate system is applied to a physical system, then a vector may be decomposed in vector components that act along each coordinate direction. For example, as shown in Figure 3.2, the vector **F** is the resultant of two components in the two-dimensional Cartesian coordinate system. The vector component $\mathbf{F}_x$ acts along the x direction, while $\mathbf{F}_y$ acts along the y direction. Algebraically, the resultant **F** may be written as

$$\mathbf{F} = \mathbf{F}_x + \mathbf{F}_y. \tag{3-1}$$

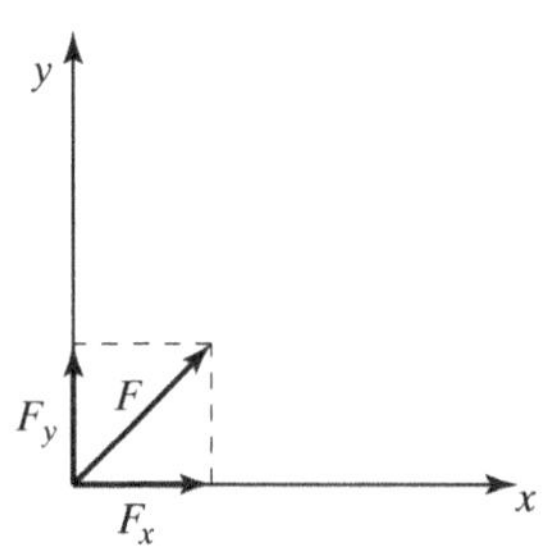

Figure 3.2.

Now, let **i** be a unit vector acting along the x direction and **j** be a unit vector acting along the y direction. The vectors **i** and **j** are sometimes called basis vectors.

Furthermore, suppose that the magnitude of $\mathbf{F}_x$ is F_x and $\mathbf{F}_y$ is F_y, then Equation 3-1 may be written as

$$\mathbf{F} = F_x\mathbf{i} + F_y\mathbf{j}. \tag{3-2}$$

Equation 3-2 treats an expansion for a two-dimensional Cartesian coordinate system. Similar expansions may be obtained for a vector in a non-Cartesian coordinate system.

In general, for a three-dimensional Cartesian coordinate system, a vector **F** may be written as

$$\mathbf{F} = F_x\mathbf{i} + F_y\mathbf{j} + F_z\mathbf{k}, \tag{3-3}$$

where **k** is the unit vector in the z direction.

From geometry (Figure 3.3), it may be shown that the magnitude of a vector is given by

$$\|F\| = \sqrt{F_x^2 + F_y^2 + F_z^2},$$

or the square root of the sum of the squares of the components.

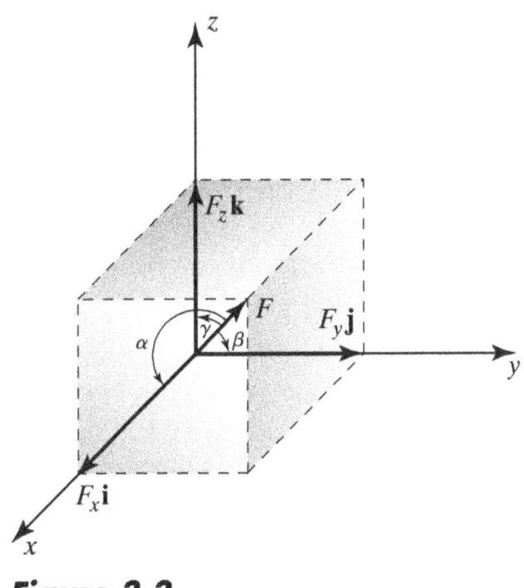

Figure 3.3.

EXAMPLE 3.1: A boy is pulling on a rope that is tied to a barn door, as shown in Figure 3.4. The boy pulls with a force of 22 N along the direction of the rope, which is at 30° to the vertical. The rope lies in a plane. What are the components of this force?

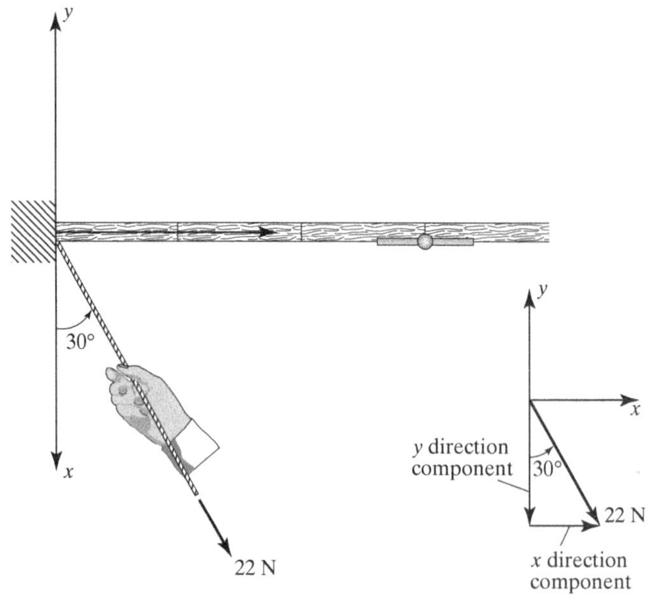

Figure 3.4.

SOLUTION

The problem statement mentions that the 22 N are directed along the direction of the rope. Therefore, we can replace the rope with a force that is located in space in the same manner as the rope. We can place a Cartesian coordinate system where the rope is tied and use geometry to find the components of the force. Because the rope lies in a plane, there are only two components of force, say, F_x and F_y.

The component in the x direction is determined by finding the projection along the x-axis as shown in Figure 3.4. The components of the force along each coordinate axis form a right triangle, as shown in the figure. From trigonometry, the component along the x-axis is

$$F_x = 22\text{ N} \quad \sin 30 = 11\text{ N} \rightarrow,$$

where the arrow indicates the direction of the component.

The component in the y direction is determined by finding the projection along the y-axis, as shown in Figure 3.4. This direction is determined by the cos 30°. Therefore,

$$F_y = 22\text{ N} \quad \cos 30 = 11\sqrt{3}\text{ N}\downarrow.$$

3.2.2 Direction Cosines and Vectors

Another way to write a vector mathematically is to use cosines of the angles that the vector makes with the Cartesian coordinate directions. These are the angles α, β, and γ shown in Figure 3.3. In the figure, notice that the angles are measured between the tail of the vector **F** and the positive direction of the x-, y-, and z-axes. An interesting fact about these angles is that they are never greater than 180°.

From geometry, it can be shown that the cosine of each angle is equal to the projection along the direction defined by the angle divided by the magnitude of the vector. For example, the cosine of α is found as follows:

$$\cos\alpha = \frac{\text{projection of } \mathbf{F} \text{ on the } x\text{-axis}}{\text{magnitude of } \mathbf{F}}. \tag{3-4}$$

Because the projection of the vector **F** on the x-axis is F_x, and the magnitude of **F** is $||F||$ this becomes

$$\text{Cos}\,\alpha = \frac{F_x}{\|F\|}. \tag{3-5}$$

Likewise,

$$\text{Cos}\,\beta = \frac{F_y}{\|F\|} \quad \text{and} \quad \text{Cos}\,\gamma = \frac{F_z}{\|F\|}. \tag{3-6}$$

The direction cosines are such that the sum of their squares is equal to unity. Thus,

$$\cos^2\alpha + \cos^2\beta + \cos^2\gamma = 1. \tag{3-7}$$

Equation 3-7 implies that only two direction cosines need be obtained from the given information. The third direction cosine may be calculated from the equation.

Now let **C** be a vector whose components are the direction cosines; that is, let

$$\mathbf{C} = \cos\alpha\mathbf{i} + \cos\beta\mathbf{j} + \cos\gamma\mathbf{k}. \tag{3-8}$$

Then, by using Equation 3-5 and Equation 3-6, we have

$$\mathbf{C} = \frac{\mathbf{F}}{\|\mathbf{F}\|} = \frac{F_x}{\|\mathbf{F}\|}\mathbf{i} + \frac{F_x}{\|\mathbf{F}\|}\mathbf{j} + \frac{F_x}{\|\mathbf{F}\|}\mathbf{k}. \tag{3-9}$$

Equation 3-9 can be rearranged for the vector **F** in terms of the direction cosines. That is,

$$\begin{aligned} \mathbf{F} &= \|\mathbf{F}\|\mathbf{C} \\ &= F\cos\alpha\mathbf{i} + F\cos\beta\mathbf{j} + F\cos\gamma\mathbf{k} \end{aligned} \tag{3-10}$$

Equation 3-10 implies that if the direction cosines are known, an expression for the vector may be obtained.

EXAMPLE 3.2: Solve Example 3.1 by using direction cosines.

SOLUTION

In this case, there are only two direction cosines that must be found. These are related to angles α and β, which the *tail* of the vector makes with the x- and y-axes, respectively. The angles are shown in Figure 3.5. Their values may be readily obtained from geometry. In fact,

$$\alpha = 90° - 30° = 60°,$$

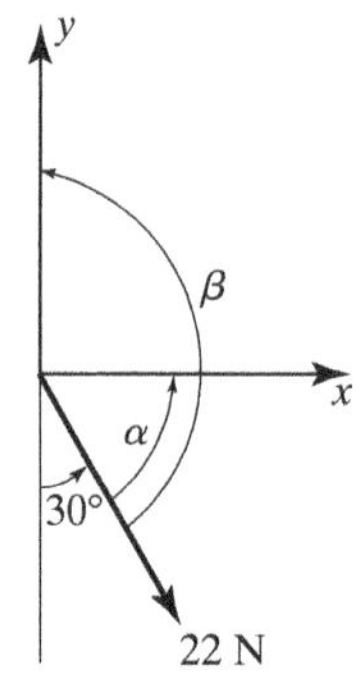

Figure 3.5.

and

$$\beta = \alpha + 90° = 60° + 90° = 150°.$$

Then, by using Equation 3-5 and Equation 3-6, the components are obtained as

$$F_x = F \,\mathrm{Cos}\,\alpha = 22 \text{ N} \cdot \mathrm{Cos}\,60° = 11 \text{ N} \rightarrow$$

$$F_y = F \,\mathrm{Cos}\,\beta = 22 \text{ N} \cdot \mathrm{Cos}\,150° = 11\sqrt{3} \text{ N}\downarrow$$

These are the same results obtained in Example 3.1.

PRACTICE!

A car is climbing a hill that has a slope of 20°, as shown in Figure 3.6. The car weighs 1,800 lb. The weight of the car may be assumed to act as shown in the figure. What are the components of this force in the x and y directions for the coordinate system shown in the figure?

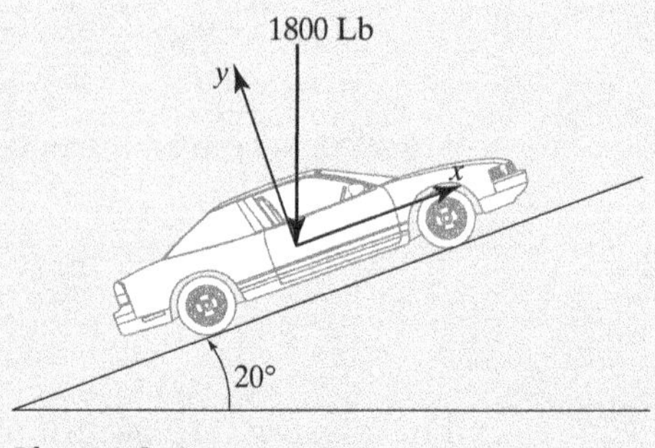

Figure 3.6.

PROFESSIONAL SUCCESS

As you may have noticed from Example 3.1 and Example 3.2, solving for the components of vectors requires extensive use of geometry and trigonometry. These topics are typically studied in high school. If they are not fresh in your mind, then this might be the time for a review. One way to do this is to take a refresher course in precalculus mathematics. Most universities offer such a course.

3.2.3 Addition of Vectors

Because many forces may act on a rigid body and since forces are vectors, to solve problems involving vectors, one must be able to find the resultant of the applied forces by adding the vectors. This may be done by using a geometric technique or by summing the components of the force vector.

Vectors may be added geometrically "by joining the vectors head to tail" or "tip to tail" and making use of geometric trigonometric relations. When using this approach, use the following sequence of steps:

1. Construct a sketch by rearranging the vectors so that the head of one vector touches the tail of the other. Be sure to maintain the directions of the vectors.
2. Draw a vector from the tail of the first vector to the head of the last vector. This is the resultant vector.
3. Use geometry or trigonometry to obtain the magnitude and direction of the resultant vector.

If the vectors are drawn to scale, then the direction and the magnitude of the resultant vector may be obtained graphically through direct measurement on the drawing. If direct measurement is not possible, then the unknown quantities may be obtained by using the sine or the cosine law from geometry. If the vectors form a 90° triangle, then trigonometry may be used.

The geometric method works well for two coplanar vectors, but for systems with more than two vectors, extensive calculations are required. In such cases, it is better to add the components of the vector. Of course, this requires that the vectors first be written in terms of their components, which is the reason for discussing the material in Section 3.2.1 and Section 3.2.2. When using the component summation approach, use the following steps:

1. Obtain the components of each vector. This may be done by finding the projection of the vector along the desired direction, as discussed in Section 3.2.1, or by finding the direction cosines and using Equation 3-10 as discussed in Section 3.2.2.
2. Calculate the components of the resultant vector by adding the components of the vectors along the appropriate direction.

EXAMPLE 3.3: The magnitude of a vector is the sum of its components. Show that the solutions of Example 3.1 and Example 3.2 are correct by adding the components.

SOLUTION

We add the components by placing the two vectors head to tail as shown in Figure 3.7. Let **A** be the sum of these two vectors, which makes an angle α with respect to the horizontal.

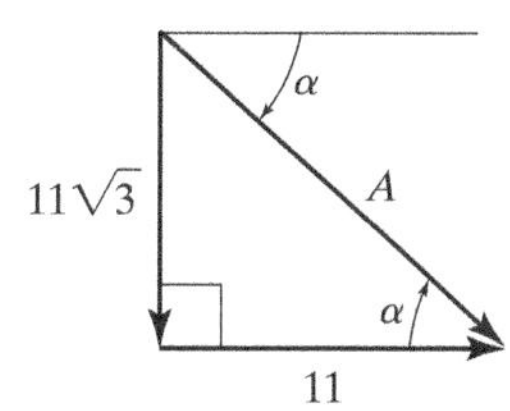

Figure 3.7.

The magnitude of **A**, $||\mathbf{A}||$, is obtained by using the Pythagorean theorem. That is,

$$\|\mathbf{A}\| = \sqrt{(11)^2 + (11\sqrt{3})^2} = 22,$$

which is the magnitude of the initial vector.

The angle this vector makes with the horizontal may be obtained from geometry. Since

$$\tan\alpha = \frac{11\sqrt{3}}{11},$$

it follows that

$$\alpha = \tan^{-1}\left(\frac{11\sqrt{3}}{11}\right) = 60°.$$

This means that the vector makes an angle of 30° with the vertical, as defined in the original problem statement to Example 3.1.

PRACTICE!

For the two vectors shown in Figure 3.8, find the resultant vector and the angle it makes with the x-axis.

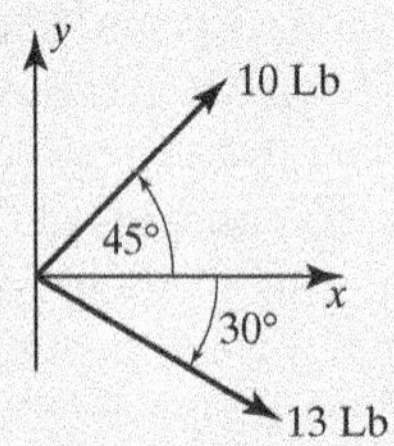

Figure 3.8.

3.3 FORCES, COUPLES, AND MOMENTS

According to Newton's first law, a force is any action that causes a body to move, stop, or accelerate. A force may be represented as a vector.

Now suppose that two parallel forces, F, are applied to a rigid body as shown in Figure 3.9. The forces are a distance d apart and are coplanar; that is, they lie in the same plane. Furthermore, the forces have the same magnitude, but act in opposite directions. The action of such a system of forces causes the body to rotate. The forces may be replaced with a moment, M_c, which is equal to the magnitude of the forces times the distance between them. Mathematically,

$$M_c = Fd. \tag{3-11}$$

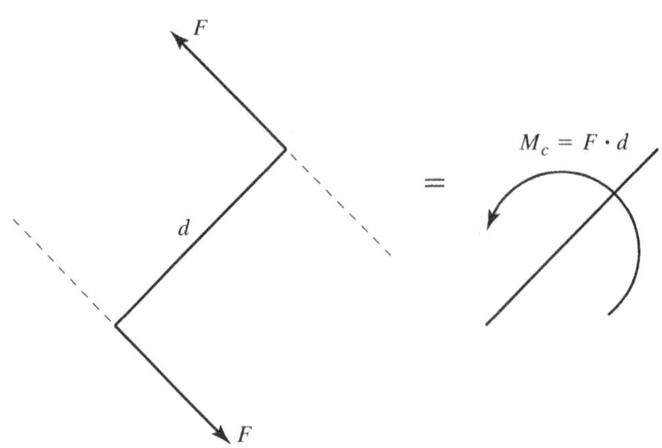

Figure 3.9.

Equation 3-11 gives the magnitude of a couple. The direction of the couple is given by the right-hand rule, with the thumb indicating the direction of the moment and the figures in the direction of the forces. Therefore, a positive couple generates rotation in a counterclockwise sense.

Couples are free vectors; that is, they may be applied at any point *P* on a body and added in a manner similar as that used to add vectors.

A moment is a measure of rotation about an axis due to a force. The magnitude of a moment is given by Equation 3-11. Again, the right-hand rule is used to indicate a positive moment.

EXAMPLE 3.4: A man is pushing a 3-foot-high dresser across a room. If he applies a horizontal force of 100 lbf, find the equivalent moment of the force that would cause rotation about the point o in Figure 3.10.

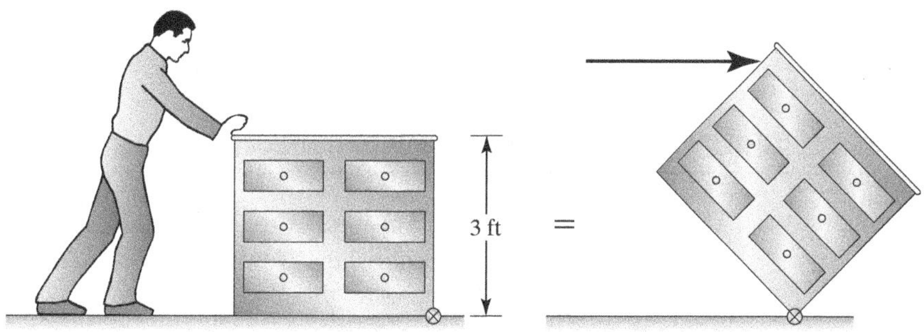

Figure 3.10.

SOLUTION

The equivalent couple is equal to the force times the moment arm. The moment arm is the perpendicular distance between point *o* and the direction of the force. This moment arm is equal to 3 ft. Thus, the moment is

$$M = F \cdot d = 100lbf \cdot 3ft = 300lbf \cdot ft.$$

This moment acts in a clockwise fashion about an axis through *o*.

PRACTICE!

Water exits from the arms of a sprinkler with a force equal to 2 N, as shown in Figure 3.11. This causes the sprinkler to spin. What is the equivalent moment (in N·m) acting on the sprinkler? Does the sprinkler spin clockwise or counterclockwise?

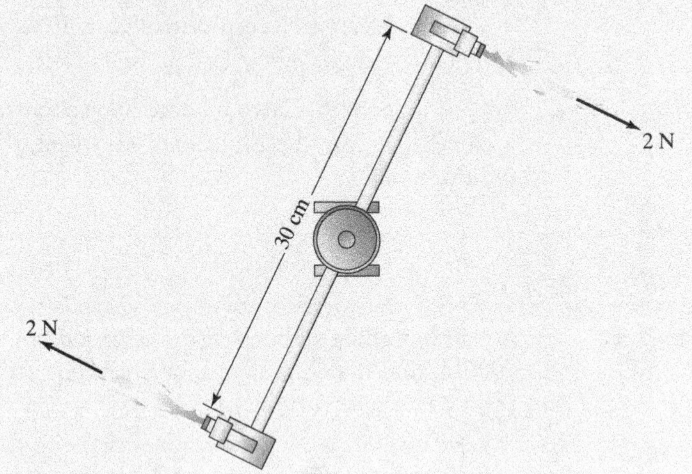

Figure 3.11.

3.4 EQUILIBRIUM AND FREE-BODY DIAGRAMS

A rigid body is in static equilibrium if the resultant force acting on the body is zero. For a system in static equilibrium, Newton's second law implies that the acceleration is zero. Notice this does not mean that the velocity of the body is zero; rather, it means that the velocity is a constant. If **F** and **M** are forces and moments on a system, then for a system in static equilibrium, Newton's second law may be written as follows:

$$\Sigma \mathbf{F} = 0, \qquad \Sigma \mathbf{M} = 0. \tag{3-12}$$

The symbol Σ is used to indicate a sum. In particular, $\Sigma \mathbf{F}$ indicates the sum of all the forces acting on the rigid body.

Because Equation 3-12 is in terms of vectors, the equations are often written in terms of the components of the forces and moments. The solution of Equation 3-12 for different types of loading forms the crux of statics. To solve Equation 3-12, the forces and moments acting on a body must be represented within the context of what is happening physically to the body. This is accomplished by constructing a free-body diagram.

A free-body diagram is a sketch showing the loads acting on a rigid body. These loads may be forces and moments. The loads may be internally or externally applied. In solving problems involving rigid-body mechanics, it is imperative that an accurate free-body diagram be constructed.

Free-body diagrams represent the physical state of the rigid body. That is, the loads on the rigid body must be accurately represented so that the physical state of the body is represented.

Students often have difficulty constructing a proper free-body diagram. Because free-body diagrams are so important in statics and dynamics, such students will have difficulty with the courses.

When constructing free-body diagrams, use the following steps:

1. Isolate the body from its physical surroundings, and draw a simplified representation of the body.
2. Pick an appropriate coordinate system, and sketch on your drawing.
3. Add the appropriate loads to the sketch, bearing in mind what is happening physically to the body. Do not forget to add internal loads if needed.
4. Label known quantities with the magnitude and direction.

3.5 FRICTIONAL FORCES

In problems involving the contact of two bodies, if the contact is perfectly smooth, then there will only be a reaction normal to the contact surfaces. On the other hand, if the contact is not smooth, a reaction will occur along the line of contact. This reaction is a force of resistance called the friction. Frictional forces inhibit or prevent slipping.

Frictional forces occur in fluids and solid bodies. In fluids, friction occurs between a fluid and a solid surface, such as between air and a turbine blade. Friction also occurs between different phases of fluids. In solid bodies, friction occurs at the contact surface and is called *coulomb*, or *dry*, friction.

Provided that there is no slipping at the contact surface and that the body is not accelerating, experimental studies have shown that the frictional force is related to the normal contact force by the equation

$$f_s = \mu_s N, \tag{3-13}$$

where f_s is the static frictional force and N is the normal contact force. The constant μ_s is called the coefficient of static friction. Values for μ_s have been experimentally measured for various bodies in contact. However, the value varies due to several factors, including surface finish, humidity, and temperature. For example, for the contact between metal and ice, the value of coefficient of static friction has been obtained in the range 0.03 to 0.05.

If the body is accelerating, but not slipping, then the frictional force has a value less than the static value. This frictional force, f_s, is called the kinetic frictional force and is related to the normal force as

$$f_k = \mu_k N, \tag{3-14}$$

where μ_k is the coefficient of kinetic friction. Values of μ_k are as much as 25% smaller than values for μ_s.

In cases where slipping is occurring at the contact surface, then the frictional force may be less than the value given by Equation 3-13 and Equation 3-14. In solving problems involving slippage, the normal force must be obtained independently of the frictional force.

EXAMPLE 3.5: What force is required to begin pushing the 15-kg crate shown in Figure 3.12 up the side of a hill? The hill has a 30° slope, and the coefficient of static friction is 0.30.

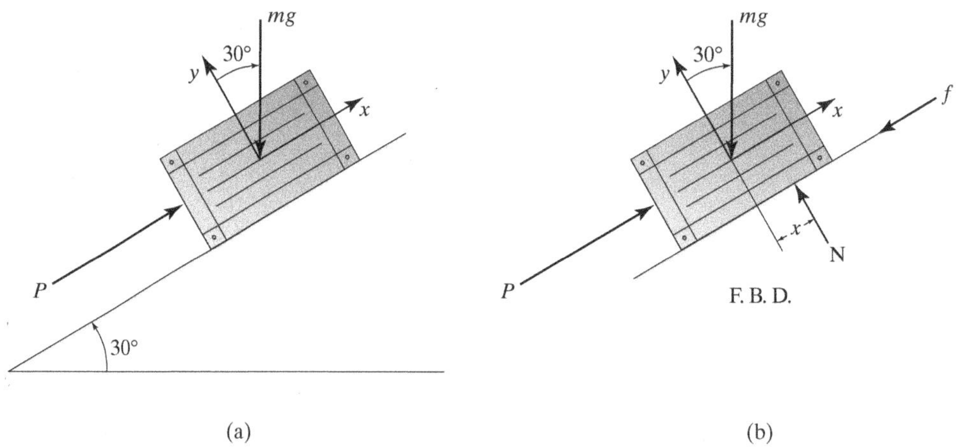

Figure 3.12.

SOLUTION

Let the unknown force be P. Because of the position of the force, it is possible that the crate will tip, rather than slide.

Whether the crate will tip depends on the position of the normal load. Let us call this position x, as shown in the figure. We will assume that the crate is on the verge of sliding and calculate the value of x. If x is less than 0.3 m, then our assumption is correct.

If we assume sliding, then

$$f = \mu_s N = 0.3N. \tag{1}$$

Now, we apply Newton's law, Equation 3-12, along each coordinate direction. Summing the forces in the x direction gives

$$\overset{+}{\longrightarrow} \Sigma F_x = ma_x \Rightarrow P - mg\sin 30 - f = 0. \tag{2}$$

Summing the forces in the y direction gives

$$+\uparrow \Sigma F_y = ma_y \Rightarrow N - mg\cos 30° = 0. \tag{3}$$

Taking the sum of the moments about o gives:

$$+\ \Sigma M_o = ma_y \Rightarrow -f(0.1) + N(x) = 0. \tag{4}$$

Substituting Equation (1) into Equation (4) and solving for x, we obtain

$$-f(0.1) + N(x) = -0.3N(0.1) + N(x)$$

$$\Rightarrow x = 0.03.$$

Because $0.03 < 0.3$, the crate is sliding, not tipping, and the frictional force is related to the normal force by Equation 3-13, as we assumed.

The value of the normal force is obtained from (3). Thus,

$$N = mg\cos 30^\circ = 15\text{ kg}\cdot 9.81\frac{\text{m}}{\text{s}^2}\cos 30^\circ = 127.4\text{ N}.$$

The value of the force P may be obtained from (2), that is,

$$P = mg\sin 30 + f = 15\text{ kg }?9.81\frac{\text{m}}{\text{s}^2}\sin 30 + 0.3?(127.4\text{ N})$$

$$\text{P} = 111.8\text{N}$$

EXAMPLE 3.6: Suppose that the crate from Example 3.5 is not moving. What is the value of the frictional force to prevent the crate from sliding?

SOLUTION

Because the crate is not moving, the frictional force is not given by Equation 3-13. Applying Newton's Laws to the free body diagram shown in Figure 3.13, we obtain

$$+\blacklozenge F_y = ma_y\,?\,N - mg\cos 30^\circ = 0$$

for the y direction and for the x direction:

$$??^{+}\blacklozenge\ F_x = ma_x\,?f - mg\sin 30 = 0$$

These two equations may be solved for the normal force and the frictional force. Thus,

$$N = mg\cos 30^\circ = 15\text{ kg }?9.81\frac{\text{m}}{\text{s}^2}\cos 30^\circ = 127.4\text{ N}$$

and

$$f = mg\sin 30 = 15\text{ kg }?9.81\frac{\text{m}}{\text{s}^2}\sin 30 = 73.6?\text{N}$$

Notice that the frictional force is not equal to the value given by Equation 3-13, that is, $(0.3)\,N = (0.3)(127.4) = 38.2N$.

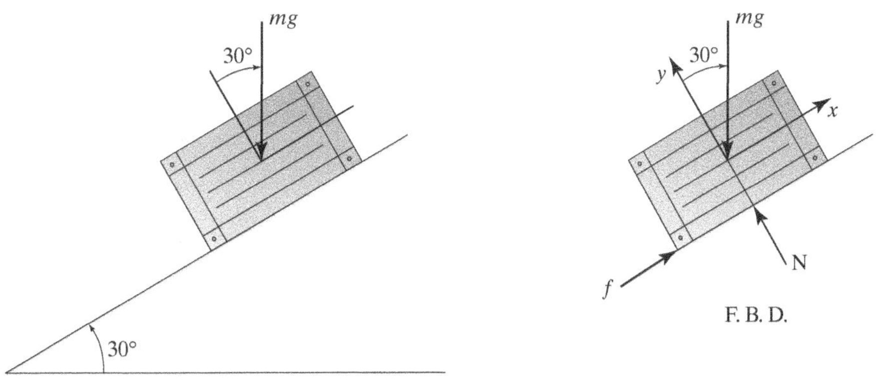

(a) (b)

Figure 3.13.

3.6 MOTION OF A RIGID BODY

If the loads on a body are unbalanced, then the body will accelerate according to Newton's second law. The study of the acceleration of rigid bodies is undertaken in dynamics.

Under the action of unbalanced forces and moments, Equation 3-12 becomes

$$\Sigma\mathbf{F} = m\mathbf{a},$$
$$\Sigma\mathbf{M} = I\alpha, \quad (3\text{-}15)$$

where m is the mass of the rigid body, $\mathbf{a}$ (in m/s^2) is the acceleration of the body's center of mass, I is called the mass moment of inertia (in kg·m^2 or slug·ft^2), and α is the angular acceleration of the center of mass. The angular acceleration is analogous to linear acceleration and has units of rad/s^2.

The branch of rigid-body mechanics dealing with dynamic loads is divided into two fields of study. The first of these fields, kinematics, deals with determining quantities such as position, velocity, and acceleration. The second field of study, kinetics, deals with determining forces and moments.

Acceleration $\mathbf{a}$, is defined as the rate of change of velocity. The average acceleration, $\mathbf{a}_{avg}$, is the change in velocity divided by the change in the time interval.

PROFESSIONAL SUCCESS

When constructing a free-body diagram for an accelerating body, the procedure is as follows:

1. Isolate the body from its physical surroundings, and draw a simplified representation of the body.
2. Pick an appropriate coordinate system, and sketch it on your drawing.
3. Add the appropriate loads to the sketch, bearing in mind what is happening physically to the body. Do not forget to add internal loads if needed.
4. Label known quantities with the magnitude and direction.

However, the acceleration and angular acceleration must be indicated on the diagram. Place these terms at the center of the mass of the rigid body.

EXAMPLE 3.7: A crate (see Figure 3.14a) accelerates as it slides down an incline, due to its weight. The incline is 30° to the horizontal. Draw the free-body diagram for the crate. You may neglect the air resistance, but do not neglect friction between the crate and the incline.

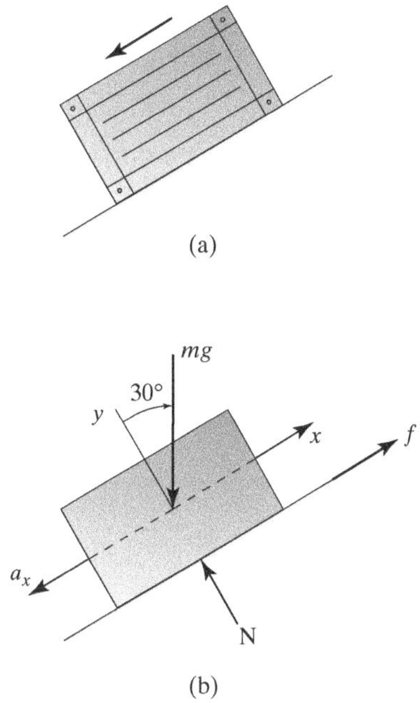

Figure 3.14.

SOLUTION

The forces on the crate include a frictional force, which acts in the direction opposite to the direction of motion of the crate. Let the force, which acts normal to the incline surface, be indicated by the letter N. By placing a coordinate system as shown in Figure 3.14b, we see that there is no motion in the y direction and hence no acceleration in the y direction.

The free-body diagram is shown in Figure 3.14b. Notice that the acceleration is placed so that it acts through the center of mass.

EXAMPLE 3.8: Assuming that the coefficient of friction is equal to 0.3, calculate the acceleration of the crate from Example 3.7.

SOLUTION

To calculate the acceleration, we apply Equation 3-15 along each coordinate direction. Thus, the sum of the forces in the y direction gives

$$+\uparrow \Sigma F_y = ma_y \Rightarrow N - mg\cos 30° = 0. \quad (1)$$

Therefore, the normal force has a value of

$$N = mg\cos 30. \quad (2)$$

Summing the forces in the x direction gives

$$\xrightarrow{+} \quad \Sigma F_x = ma_x \Rightarrow -mg\sin 30 + f = ma_x. \tag{3}$$

Assuming sliding without slip, we can relate the frictional force, f, directly to the normal force N. Thus, we have

$$f = \mu N = 0.3N = 0.3(mg\cos 30°), \tag{4}$$

where we have used Equation (2). Equation (4) may be substituted into Equation (3). This gives a relationship for the acceleration:

$$a_x = -g\sin 30° + (0.3)g\cos 30°$$

$$= g(0.3\cos 30° - \sin 30°).$$

Using $g = 9.81\ \text{m/s}^2$ and the values of cos 30° and sin 30° gives

$$a_x = 9.81\frac{\text{m}}{\text{s}^2}(0.3\cos 30 - \sin 30)$$

$$a_x = 2.36\frac{\text{m}}{\text{s}} \leftarrow . \tag{5}$$

PRACTICE!

A 30-lb, 1-ft-diameter wheel is rolling along the ground due to a 30-lb · ft couple moment. The moment of inertia of this wheel, I, is 0.750 slug · ft². Draw the free-body diagram for this wheel. What is the linear and angular acceleration of the wheel if the static and kinetic coefficients of friction between the wheel and the ground are 0.3 and 0.25, respectively?

3.7 SUMMARY

Rigid-body mechanics, which includes statics and dynamics, is a branch of science that deals with forces and motion of bodies that do not deform under the applied loads. Rigid-body mechanics is one of the fundamental building blocks upon which subsequent courses in mechanical engineering are developed. Thus, the importance in understanding the material in this subject cannot be over emphasized.

A diagram called a free-body diagram may be established for a physical system. In such a diagram, the body under consideration is isolated from its surroundings, and loads acting on the body are shown. The direction and magnitudes of the loads must be properly indicated or the analysis will fail.

Writing a force as a vector, which has magnitude, as well as a direction, allows one to solve problems in rigid-body mechanics involving forces. The projection of a vector along a coordinate direction is called a component of the vector. The components of the vectors may be obtained through geometry. Vectors may be added or subtracted by using a geometric approach. If the vectors are known in terms of their components, then by adding the corresponding component from each vector, the vectors may be summed.

KEY TERMS

components of a vector	direction cosines	free-body diagram
couple	equilibrium force	Newton's laws vector

3.8 EXERCISES

E1. Find the resultant of the vectors shown in Figure 3.15 and the angle the resultant makes with the x-axis.

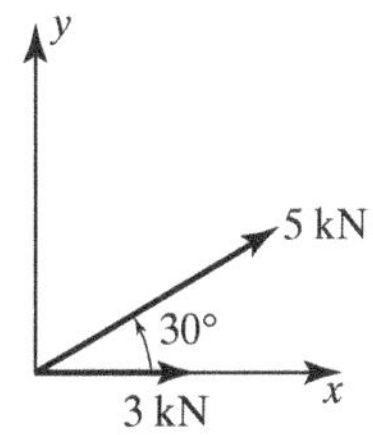

Figure 3.15.

E2. Find the resultant vector, as well as the angle that the vectors make with the x-axis, for the vectors shown in Figure 3.16.

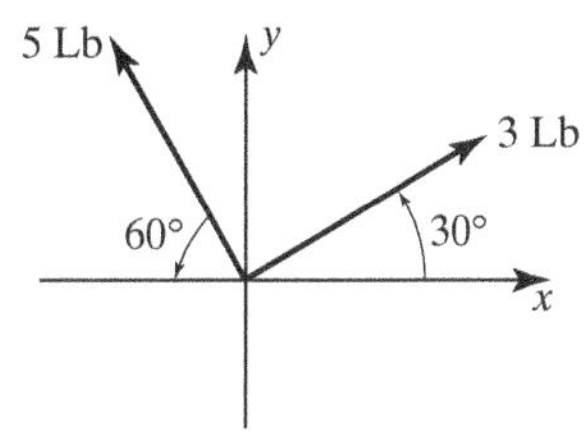

Figure 3.16.

E3. Considering Figure 3.17, find the value of the resultant force, **F**, so that the forces are in equilibrium.

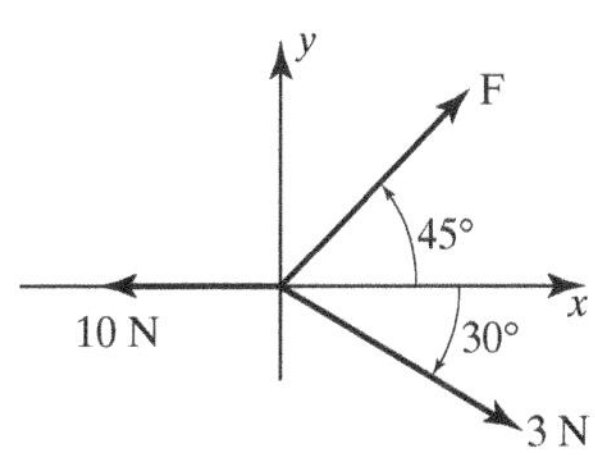

Figure 3.17.

E4. For the system of forces shown in Figure 3.18, find the value of the resultant force, **F**, so that the system is in equilibrium.

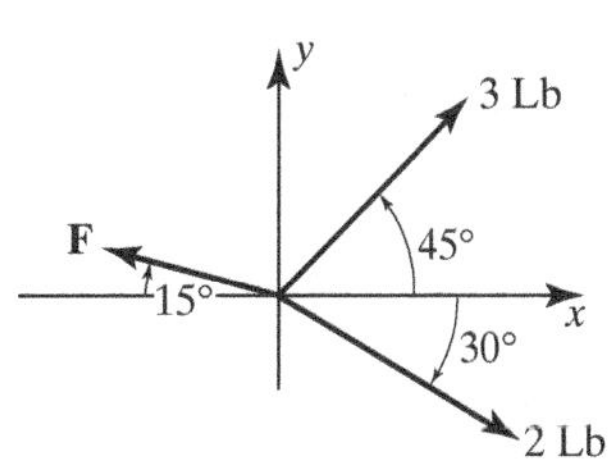

Figure 3.18.

E5. Find $\mathbf{F}_1$ and $\mathbf{F}_2$ shown in Figure 3.19, so that the system of forces is in equilibrium.

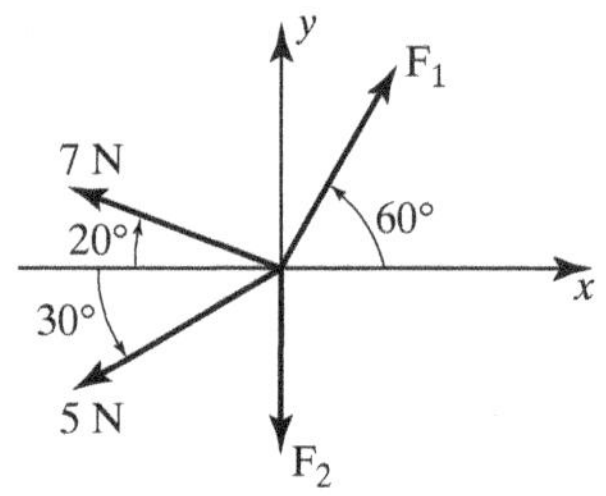

Figure 3.19.

E6. A 10-foot ladder is leaning against the side of a building, making a 30° angle with the ground, as shown in Figure 3.20. Suppose that the weight of the ladder is taken at its midpoint, draw the free-body diagram of the ladder.

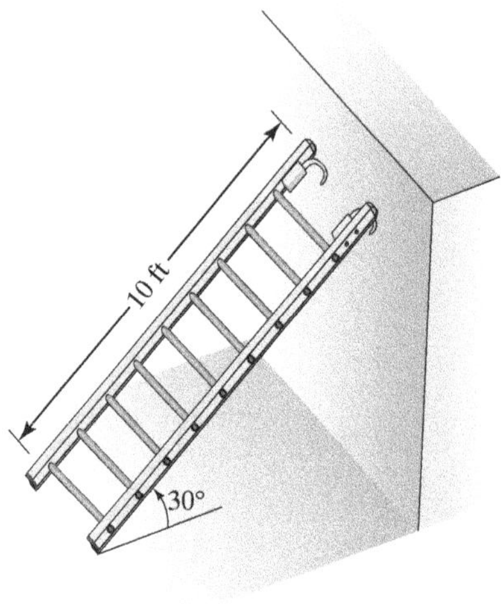

Figure 3.20.

E7. A rocket is launched vertically as illustrated in Figure 3.21. Draw the free-body diagram of the rocket.

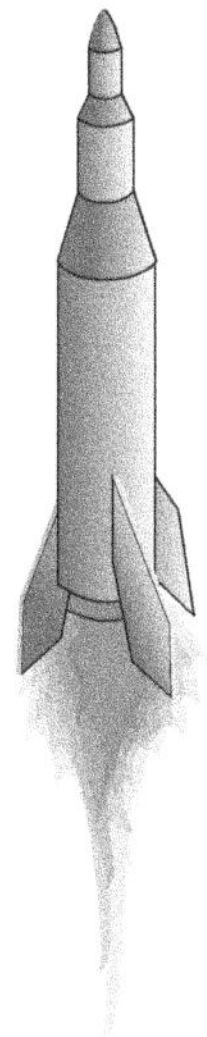

Figure 3.21.

E8. A cannon ball is shot out of a cannon as shown in Figure 3.22. Draw the free-body diagram of the ball if air resistance is neglected.

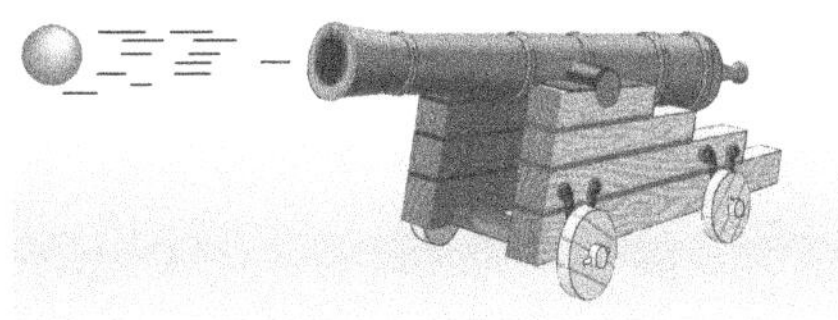

Figure 3.22.

E9. An aircraft is in steady flight when the forces due to drag (D), lift (L), thrust (T) and the weight of the aircraft are in equilibrium, as shown in Figure 3.23. Suppose the weight of the aircraft is 20,000 lb and the two engines each provide 50,000 pounds of thrust. Calculate the values of the lift and drag forces.

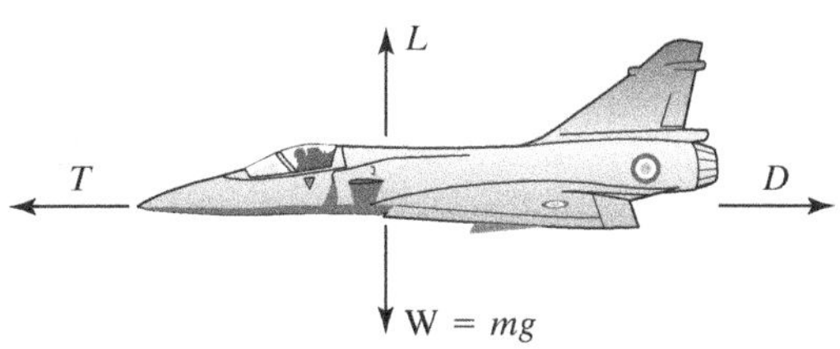

Figure 3.23.

E10. The spring and block shown in Figure 3.24 model the suspension of a certain automobile. Suppose the block has an equivalent mass of 3-kg. A force of 30 N is applied to press the mass downward, compressing the spring as shown in the figure. The spring resists this action with an equal an opposite force, what is the magnitude of this force? Do not neglect the weight of the block.

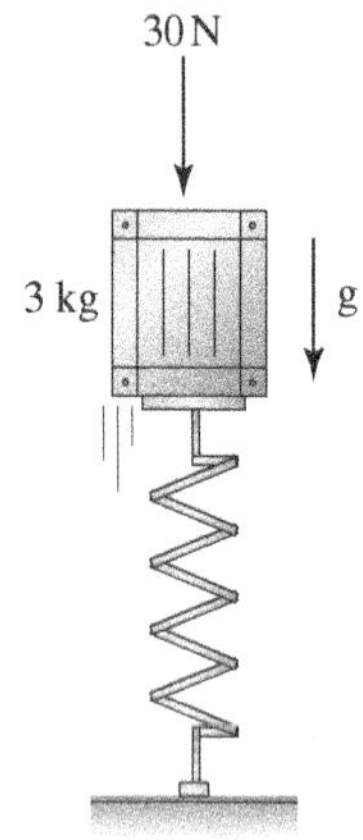

Figure 3.24.

E11. The car in Figure 3.25 is accelerating up along a 20° incline. What force **F** must be applied so that the 1.5-ton car achieves an acceleration of 15 ft/s^2 in the x direction?

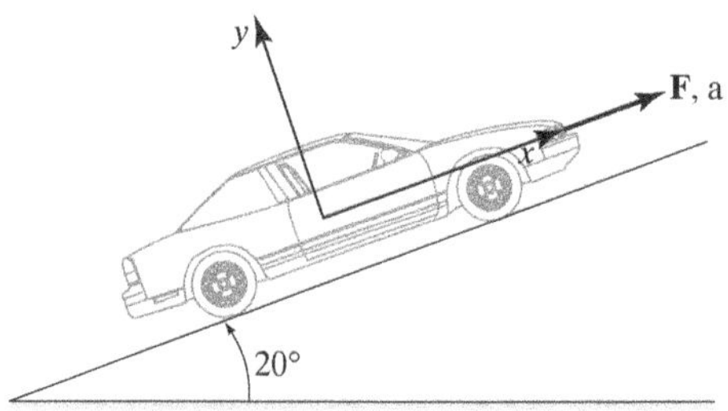

Figure 3.25.

4

Materials Science

4.1 OVERVIEW

As one of their responsibilities, mechanical engineers design hardware so that it shouldn't break when being used and so that it can carry the forces acting on it reliably and safely. As an example, consider Boeing's 767 commercial airliner, which weighs up to 350,000 lb when fully loaded. When the airplane is parked on the ground, its weight is supported by the landing gear and wheels. During flight, the aircraft's wings create an upward lift force that exactly balances the weight. Each wing, therefore, carries a force that is equal to half of the airplane's weight, which in this case is equivalent to some 70 family-sized automobile sedans. Subjected to the lift force, the wings bend upward, and, if the flight happens to encounter rough weather, the wings will bend up and down by an additional considerable amount as the plane is buffeted by turbulence. When engineers selected the aircraft's materials, they took into account the facts that an aircraft's wings are subjected to large forces, that they sag under their self-weight, and that they bend upward in response to the lift forces. The wings are designed to be strong, safe, and reliable while being no heavier than necessary to meet the design requirements.

By applying the properties of force to systems, you have seen how to calculate the magnitudes and directions of forces that act on certain structures and machines. Knowing those forces alone, however, is not enough information to decide whether a certain

SECTIONS

OBJECTIVES

- Identify circumstances in which a mechanical component is loaded in tension, compression, or shear, and calculate the stress present.
- Sketch a stress-strain curve, and use it to describe how a material responds to the loads applied to it.
- Explain the meaning of the material properties known as the elastic modulus and yield strength.
- Understand the differences between elastic and plastic responses of materials and between their ductile and brittle behaviors.
- Discuss some of the properties and uses for metals and their alloys, ceramics, polymers, and composite materials.
- Apply the concept of a factor of safety to the design of mechanical components that are subjected to tension or shear stress.

piece of hardware will be strong enough and not fail in its task. By "fail" or "failure," we not only mean that the hardware will not break, but also that it will not stretch or bend so much as to become significantly distorted. A 5-kN force, for instance, might be large enough to break a small bolt or to bend a shaft so much that it would wobble and not spin smoothly. A larger-diameter shaft or one that is made from a higher-grade material, on the other hand, might very well be able to support that force without incurring any damage.

With those thoughts in mind, you can see that the circumstances for a mechanical component to break, stretch, or bend depend not only on the forces applied to it, but also on its dimensions and the properties of the material from which it is made. Those considerations give rise to the concept of *stress* as a measure of the intensity of a force applied over a certain area. Conversely, the *strength* of a material describes its ability to support and withstand the stress applied to it. Engineers compare the stress present in a component to the strength of its material in order to determine whether the design is satisfactory. As an example, the broken crankshaft shown in Figure 4.1 was removed from a single-cylinder internal-combustion engine. This failure was accelerated by the presence of sharp corners in the shaft's rectangular keyway, which is used to transfer torque between the shaft and a gear or pulley. The spiral shape of the fracture surface indicates that the shaft had been overloaded by a high torque prior to breaking. Engineers are able to combine their knowledge of forces, materials, and dimensions in order to learn from past failures and to improve and evolve the design of new hardware.

Stress Strength

In this chapter, we will discuss some of the properties of engineering *materials* and examine the *stresses* that can develop within them. Within these topics lie the discipline known as solid mechanics, and they fit into the hierarchy of mechanical engineering topics shown in Figure 4.2. Tension, compression, and shear stress are quantities that engineers calculate when they relate the dimensions of a mechanical component to the forces acting on it. Those stresses are then compared to the material's physical properties to

Element 4: Materials and stresses

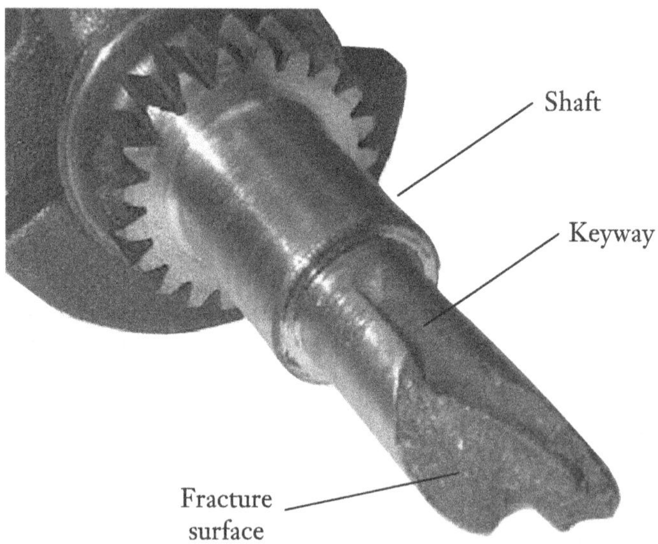

Figure 4.1. *A broken crankshaft that was removed from a single-cylinder internal-combustion engine.*

Image courtesy of the authors.

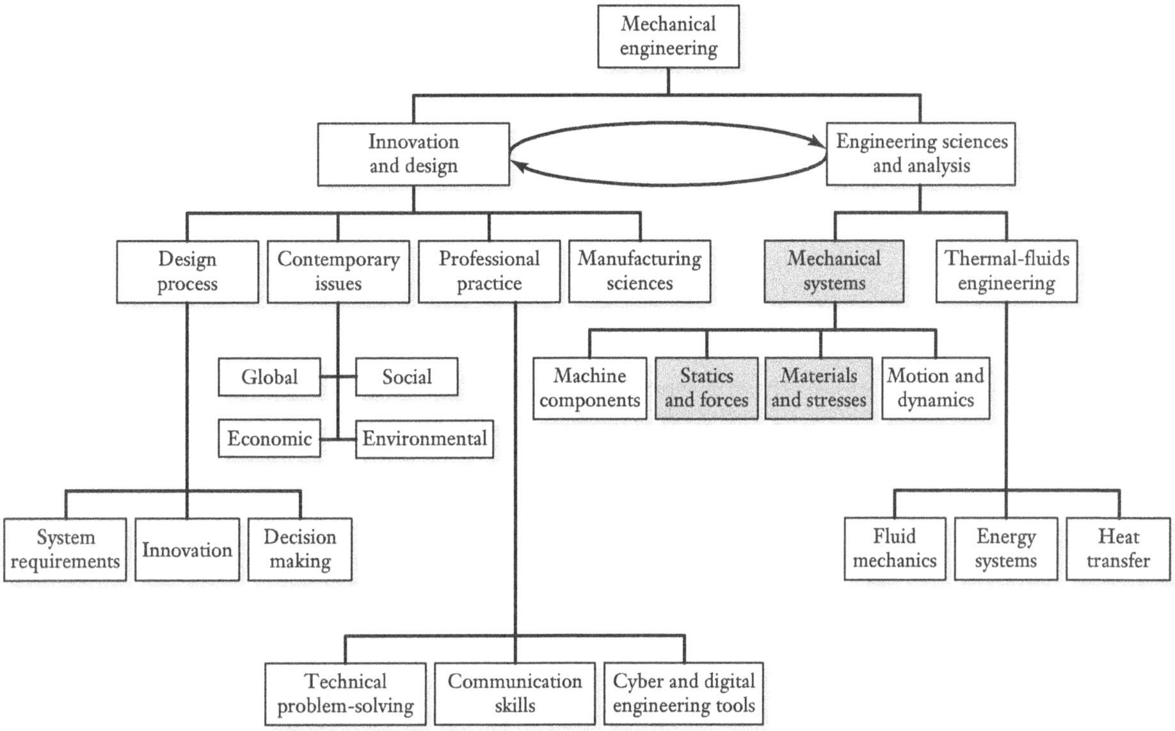

Figure 4.2. *Relationship of the topics emphasized in this chapter (shaded boxes) relative to an overall program of study in mechanical engineering.*

determine whether failure is expected to occur. When the strength exceeds the stress, we expect that the structure or machine component will be able to carry the forces without incurring damage. Engineers conduct these types of force, stress, materials, and failure analyses while they design products.

4.2 TENSION AND COMPRESSION

The type of stress that is most readily visualized and useful for you to develop intuition about materials and stresses is called tension and compression. Figure 4.3 (see on page 84) shows a built-in round rod that is held fixed at its left end and then later placed in tension by the force F that pulls on the right end. Before the force is applied, the rod has original length L, diameter d, and cross-sectional area.

$$A = \pi \frac{d^2}{4} \tag{4.1}$$

Engineers usually calculate the cross-sectional area of round rods, bolts, and shafts in terms of their diameter rather than their radius r ($A = \pi r^2$) since it's more practical to measure the diameter of a shaft using a caliper gauge. As the force F is gradually applied, the rod stretches along its length by the amount ΔL shown in Figure 4.3(b). In addition,

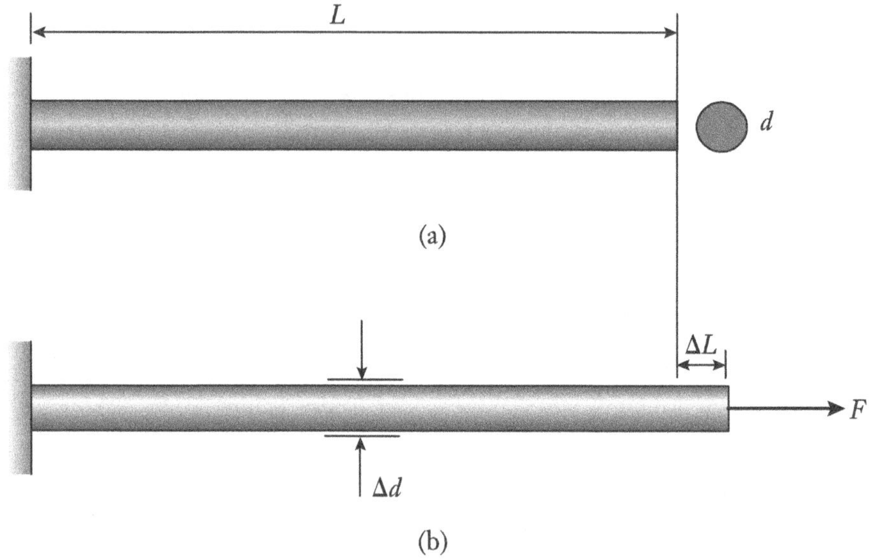

Figure 4.3. *A straight rod is stretched and placed in tension.*

Poisson's contraction

the diameter of the rod shrinks by a small amount due to an effect known as *Poisson's contraction*, which is a topic described in the next section. In any event, the change in diameter Δd is smaller and usually less noticeable than the lengthwise stretch ΔL. To gauge the relative amounts of ΔL and Δd, try stretching a rubber band to notice how its length, width, and thickness change.

Elastic behavior

Plastic behavior

If the force is not too great, the rod will return to its original diameter and length (just like a spring) when F is removed. If the rod is not permanently deformed after F has been applied, the stretching is said to occur *elastically*. Alternatively, the force could have been great enough that the rod would be *plastically* deformed, meaning that, when the force was applied and then removed, the rod would be longer than it was initially. You can experiment with a desktop paper clip to see firsthand the difference between the elastic and plastic behavior of materials. Bend one end of the paper clip a small amount—perhaps just a millimeter or two—and notice how it springs back to the original shape when you release it. On the other hand, you could unwrap the paper clip into a nearly straight piece of wire. In that case, the paper clip does not spring back. The force was large enough to permanently change the material's shape through plastic deformation.

Internal force

Although the force is applied at only one end of the rod in our example, its influence is felt at each cross section along the rod's length. As shown in Figure 4.4, imagine slicing through the rod at some interior point. The segment that is isolated in the free body diagram of Figure 4.4(b) shows F applied to the rod's right-hand end and an equivalent *internal force* on the segment's left- hand end that balances F. Such must be the case; otherwise the segment shown in the free body diagram would not be in equilibrium. The location of our hypothetical slice through the rod is arbitrary, and we conclude that a force of magnitude F must be carried by the rod at each of its cross sections.

Since the rod is formed of a continuous solid material, we do not realistically expect that the internal force will be concentrated at a point as depicted by the force vector arrow in Figure 4.4(b). Instead, the influence of the force will be spread out and smeared

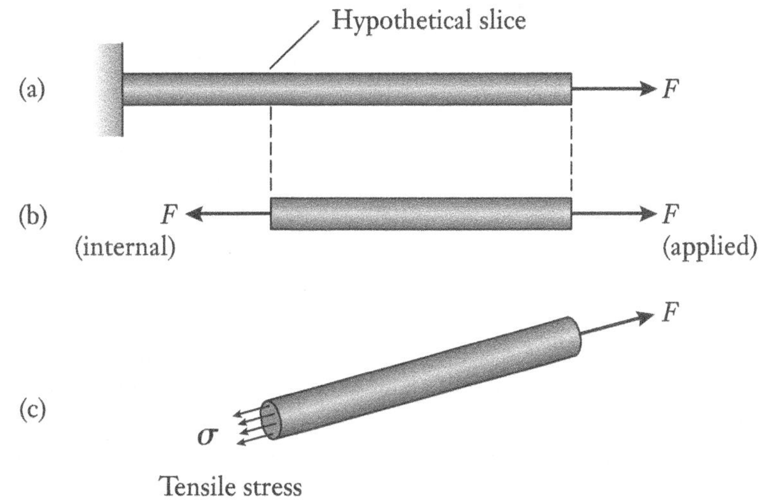

Figure 4.4. *(a) A rod that has been stretched, (b) A section that is sliced from the rod to expose the internal force, (c) The tensile stress that is distributed over the rod's cross section.*

over the rod's cross section; this process is the basic idea behind stresses in mechanical components. *Stress* is essentially an internal force that has been distributed over the area of the rod's cross section [Figure 4.4(c)], and it is defined by the equation

Stress

$$\sigma = \frac{F}{A} \tag{4.2}$$

Like the force F, the stress σ (the lowercase Greek character sigma) is oriented perpendicularly to a hypothetical slice made through the cross section. When the stress tends to lengthen the rod, it is called *tension*, and $\sigma > 0$. On the other hand, when the rod is shortened, the stress is called *compression*. In that case, the direction of σ in Figure 4.4 reverses to point inward, and $\sigma < 0$. The orientations of forces in tension and compression are shown in Figure 4.5.

Tension

Compression

Similar to the pressure within a liquid or gas, stress is also interpreted as a force that has been distributed over an area. Therefore, stress and pressure have the same dimensions. In the SI, the derived unit of stress is the pascal (1 Pa = 1 N/m^2), and the dimension pound-per-square-inch (1 psi = 1 lb/in^2) is used in the USCS. Because large numerical quantities frequently arise in calculations involving stresses and material properties, the prefixes "kilo" (k), "mega" (M), and "giga" (G) are applied to represent factors of 10^3, 10^6, and 10^9, respectively. Therefore,

$$1\,\text{kPa} = 10^3\,\text{Pa} \quad 1\,\text{MPa} = 10^6\,\text{Pa} \quad 1\,\text{GPa} = 10^9\,\text{Pa}$$

Despite mixing formats with the SI, it is also conventional to apply the prefixes "kilo" and "mega" when representing large numerical values for stress in the USCS. Mechanical engineers abbreviate 1000 psi as 1 ksi (without the "p"), and 1,000,000 psi as 1 Mpsi (with the "p"):

$$1\,\text{ksi} = 10^3\,\text{psi} \quad 1\,\text{Mpsi} = 10^6\,\text{psi}$$

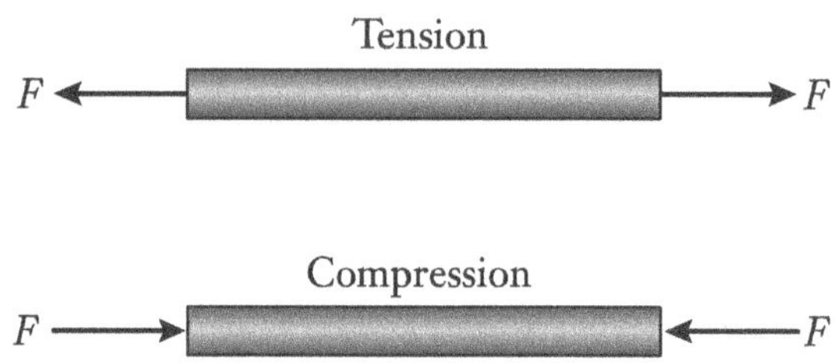

Figure 4.5. *Directions of tensile and compressive forces.*

One billion psi is an unrealistically large value for USCS calculations involving materials and stresses in mechanical engineering applications, and so it is not used. Numerical values for stresses can be converted between the USCS and SI using the factors listed in Table 4.1. Referring to the table's first row, for instance, we see that

$$1\,\text{psi} = 10^{-3}\,\text{ksi} = 6.895 \times 10^{3}\,\text{Pa} = 6.895\,\text{kPa} = 6.895 \times 10^{-3}\,\text{MPa}$$

Elongation

While stress is related to the intensity of a force's application, the engineering quantity called strain measures the amount that the rod stretches. The *elongation* ΔL in Figure 4.3 is one way to describe how the rod lengthens when F is applied, but it is not the only way, nor is it necessarily the best. If a second rod has the same cross-sectional area but is only half as long, then according to Equation (4.2), the stress within it would be the same as in the first rod. However, we expect intuitively that the shorter rod would stretch by a smaller amount. To convince yourself of this principle, hang a weight from two different lengths of rubber band, and notice how the longer band stretches more. Just as stress is a measure of force per unit area, the quantity called *strain* is defined as the amount of elongation that occurs per unit of the rod's original length. Strain ε (the lowercase Greek character epsilon) is calculated from the expression

Strain

$$\varepsilon = \frac{\Delta L}{L} \tag{4.3}$$

Because the length dimensions cancel out in the numerator and denominator, ε is a dimensionless quantity. Strain is generally very small, and it can be expressed either as a decimal quantity (for instance, $\varepsilon = 0.005$) or as a percentage ($\varepsilon = 0.5\%$).

TABLE 4.1 Conversion Factors between USCS and SI Units for Stress

psi	ksi	Pa	kPa	MPa
1	10^{-3}	6.895×10^{3}	6.895	6.895×10^{-3}
10^{3}	1	6.895×10^{6}	6.895×10^{3}	6.895
1.450×10^{-4}	1.450×10^{-7}	1	10^{-3}	10^{-6}
0.1450	1.450×10^{-4}	10^{3}	1	10^{-3}
145.0	0.1450	10^{6}	10^{3}	1

EXAMPLE 4.1: Hanger Rod in the Kansas City Hyatt Regency Hotel's Skyway

During the investigation of the collapse involving the hotel's elevated walkways, the connection between the hanger rod and the upper walkway was found to have broken at a load of approximately 20,500 lb. In the dimensions of ksi, determine the stress in the 1.25-in.-diameter hanger rod when the tension was 20,500 lb.

Approach

The hanger rod carries a tension force in the manner shown in Figure 4.3 with $F = 2.05 \times 10^4$ lb and $d = 1.25$ in. We are to calculate the stress by applying Equation (4.2).

SOLUTION

By using Equation (4.1), the cross-sectional area of the hanger rod is

$$A = \frac{\pi (1.25\,\text{in.})^2}{4} \qquad \leftarrow \left[A = \frac{\pi d^2}{4} \right]$$
$$= 1.227\ \text{in}^2$$

The tensile stress is

$$\sigma = \frac{2.05 \times 10^4\ \text{lb}}{1.227\ \text{in}^2} \qquad \leftarrow \left[\sigma = \frac{F}{A} \right]$$
$$= 1.670 \times 10^4 \frac{\text{lb}}{\text{in}^2}$$
$$= 1.670 \times 10^4\,\text{psi}$$

where we have substituted the definition of the derived unit psi (pound per square inch) for stress from Table 2.5. It is conventional to apply the abbreviation "ksi" to represent the factor of 1000 psi in a more compact form:

$$\sigma = \left(1.670 \times 10^4\ \text{psi}\right)\left(10^{-3} \frac{\text{ksi}}{\text{psi}}\right)$$
$$= 16.70 \left(\cancel{\text{psi}}\right)\left(\frac{\text{ksi}}{\cancel{\text{psi}}}\right)$$
$$= 16.70\ \text{ksi}$$

Discussion

As we'll see in the following section, this level of stress is not particularly high when compared to the strength of steel materials. Although the tension may not have been sufficient to break the hanger rod, it was great enough to destroy the connection between the rod and the pedestrian walkway. As was the case in the skyways' collapse, it is not an uncommon situation in mechanical engineering for the connections between components to be weaker than the components themselves.

$$\sigma = 16.70\ \text{ksi}$$

EXAMPLE 4.2: U-Bolt Clamp

The U-bolt is used to attach the body (formed with I-beam construction) of a commercial moving van to its chassis (formed from hollow box channel). (See Figure 4.6.) The U-bolt is made from a 10-mm-diameter rod, and the nuts on it are tightened until the tension in each straight section of the U-bolt is 4 kN. (a) Show how forces are transferred through this assembly by drawing free body diagrams of the U-bolt and its nuts, the body and chassis stack, and the clamping plate. (b) In the units of MPa, calculate the tensile stress in a straight section of the U-bolt.

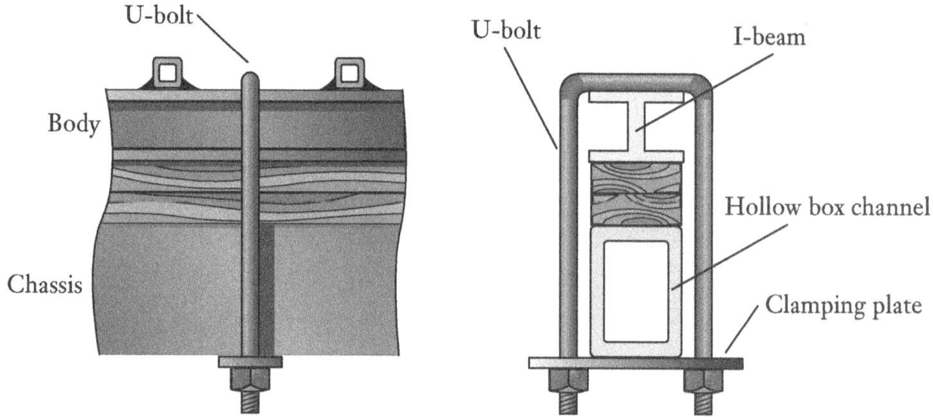

Figure 4.6.

Approach

In part (a), we will isolate three free bodies, and each must be in equilibrium: the U-bolt and its nuts, the body and chassis stack, and the clamping plate. The forces acting within the assembly will be equal in magnitude but oppositely directed on adjacent components. The straight sections of the U-bolt are placed in tension in the manner shown in Figure 4.3 with $F = 4$ kN and $d = 10$ mm. We will calculate the stress by using Equation (4.2).

SOLUTION

(a) Because each of the two straight sections of the U-bolt carries 4 kN of tension, their 8-kN resultant is transferred as compression to the body and chassis stack. (See Figure 4.7.) The 8-kN load is likewise applied by the box channel to the clamping plate. The equal and oppositely directed 4-kN forces act between the clamping plate and the nuts threaded onto the U-bolt.

(b) The cross-sectional area of the bolt is

$$A = \frac{\pi (10\,\text{mm})^2}{4} \qquad \leftarrow \left[A = \frac{\pi d^2}{4}\right]$$

$$= 78.54\,\text{mm}^2$$

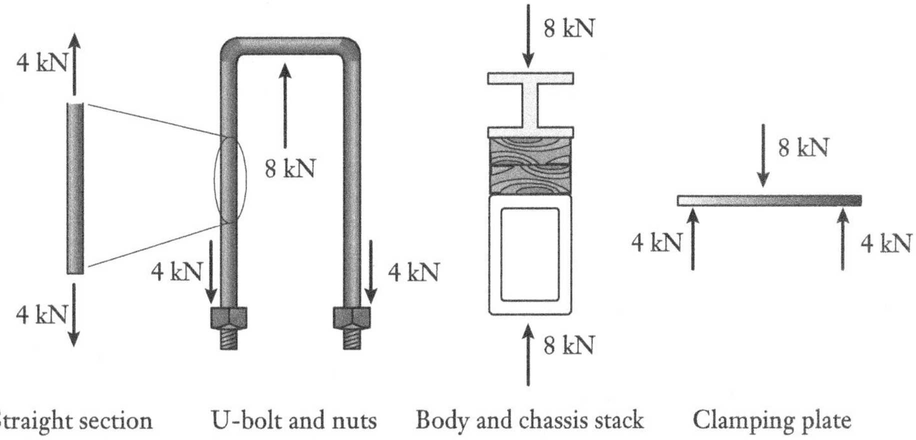

Figure 4.7.

which must be converted to dimensionally consistent units for stress in the SI:

$$\begin{aligned} A &= (78.54\,\text{mm}^2)\left(10^{-3}\,\frac{\text{m}}{\text{mm}}\right)^2 \\ &= 7.854\times10^{-5}\,(\text{mm}^2)\left(\frac{\text{m}^2}{\text{mm}^2}\right) \\ &= 7.854\times10^{-5}\,\text{m}^2 \end{aligned}$$

The tensile stress becomes

$$\begin{aligned} \sigma &= \frac{4000\,\text{N}}{7.854\times10^{-5}\,\text{m}^2} \qquad \leftarrow\left[\sigma=\frac{F}{A}\right] \\ &= 5.093\times10^{7}\,\frac{\text{N}}{\text{m}^2} \\ &= 5.093\times10^{7}\,\text{Pa} \end{aligned}$$

where we have applied the definition of the derived unit pascal. The SI prefix "mega" (Table 4.1) condenses the large power-of-ten exponent in order to write the result in a more conventional form:

$$\begin{aligned} \sigma &= (5.093\times10^{7}\,\text{Pa})\left(10^{-6}\,\frac{\text{MPa}}{\text{Pa}}\right) \\ &= 50.93\,(\text{Pa})\left(\frac{\text{MPa}}{\text{Pa}}\right) \\ &= 50.93\,\text{MPa} \end{aligned}$$

Discussion

This is a significant tensile stress in the rod's straight sections; this makes sense because the applied load is large. In the upper portion of the U-bolt where the rod has 90° corners and is in contact with the I-beam, a more complicated state of stress is present, which would require a different analysis.

$$\sigma = 50.93 \text{ MPa}$$

Focus On MASDAR CITY

Masdar City has been designed to be one of the world's most sustainable urban environments (Figure 4.8.). Located 17 km outside Abu Dhabi, Masdar City is designed to be carbon neutral with no carbon emissions of any kind. Critical to the city's design is an infrastructure of streets, walkways, and buildings that are naturally cooled by the innovative structural design, solar shades, and wind towers. There will be no automobiles in the city. Rather, the residents will use an underground personal rapid transit system powered by solar-powered batteries. During the design of the city, engineers had to account for a wide range of loading conditions, forces, stresses, and material behaviors. As part of this design process, engineers had to select the best structural configurations and materials for each force-bearing element, considering cross-sectional shapes, lengths, connection types, safety factors, and potential failure modes. By developing innovative solutions for solar energy use and sustainable urban infrastructures in the design of Masdar City, engineers are directly addressing the NAE grand challenges.

Figure 4.8. *An artist's depiction of Masdar City.*

Courtesy of Iain Masterton/Alamy.

4.3 MATERIAL RESPONSE

The definitions of stress and strain, in contrast to force and elongation, are useful because they are scaled with respect to the rod's size. Imagine conducting a sequence of experiments with a collection of rods made of identical material but having various diameters and lengths. As each rod is pulled in tension, the force and elongation would be measured. In general, for a given level of force, each rod would stretch by a different amount because of the variations in diameter and length.

For each individual rod, however, the applied force and elongation would be proportional to one another, following

$$F = k\Delta L \tag{4.4}$$

Stiffness

Hooke's law

where the parameter k is called the *stiffness*. This observation is the basis of the concept known as *Hooke's law*. In fact, British scientist Robert Hooke wrote in 1678 that

> *the power of any spring is in the same proportion with the tension thereof; that is, if one power stretch or bend it one space, two will bend it two, and three will bend it three, and so forward.*

Note that Hooke used the term "power" for what we today call "force." In this respect, any structural component that stretches or bends can be viewed as a spring having stiffness k, even though the component itself might not necessarily look like a "spring" in the sense of being a coiled wire helix.

Continuing with our hypothetical experiment, we next imagine constructing a graph of F versus ΔL for each of the different rods. As indicated in Figure 4.9(a), the lines on these graphs would have different slopes (or stiffnesses) depending on the values of d and L. For a given force, longer rods and ones with smaller cross sections stretch more than the other rods. Conversely, shorter rods and ones with larger cross-sectional areas stretch less. Our graph would show a number of straight lines, each having a different slope. Despite the fact that the rods are made from the same material, they would appear to be quite different from one another in the context of the F-versus-ΔL graph.

On the other hand, the rods would behave in an identical manner when their stretching is instead described by stress and strain. As shown in Figure 4.9(b), each of the F-versus-ΔL lines would collapse onto a single line in the stress-strain diagram. Our conclusion from such experiments is that while the stiffness depends on the rod's dimensions, the stress-strain relationship is a property of the material alone and is independent of the test specimen's size.

Stress-strain curve

Elastic and plastic regions

Proportional limit

Figure 4.10 shows an idealized *stress-strain curve* for a typical structural-quality steel. The stress-strain diagram is broken down into two regions: the low-strain *elastic region* (where no permanent set remains after the force has been applied and removed) and the high-strain *plastic region* (where the force is large enough that, upon removal, the material has permanently elongated). For strains below the *proportional limit* (point A), you can see from the diagram that stress and strain are proportional to one another and that they therefore satisfy the relationship

$$\sigma = E\varepsilon \tag{4.5}$$

Elastic modulus

The quantity E is called the *elastic modulus,* or Young's modulus, and it has the dimensions of force per unit area. In the SI, the units GPa are typically used for the elastic modulus, and the dimension Mpsi is used in the USCS. The elastic modulus is a physical material property, and it is simply the slope of the stress-strain curve for low strains. By combining Equations (4.2) and (4.3), the rod's elongation when it is loaded below the proportional limit is

$$\Delta L = \frac{FL}{EA} \tag{4.6}$$

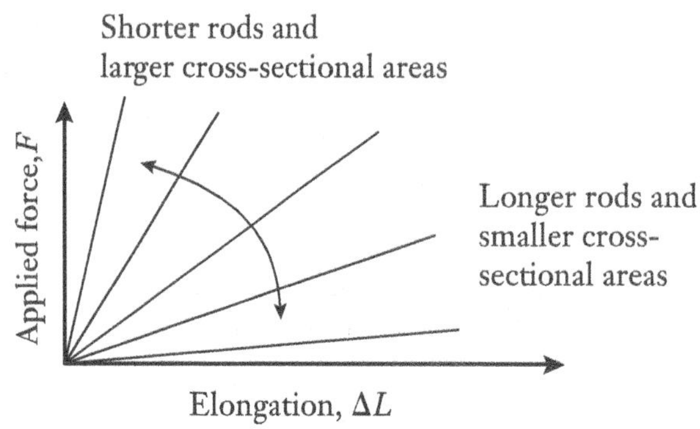

(a)

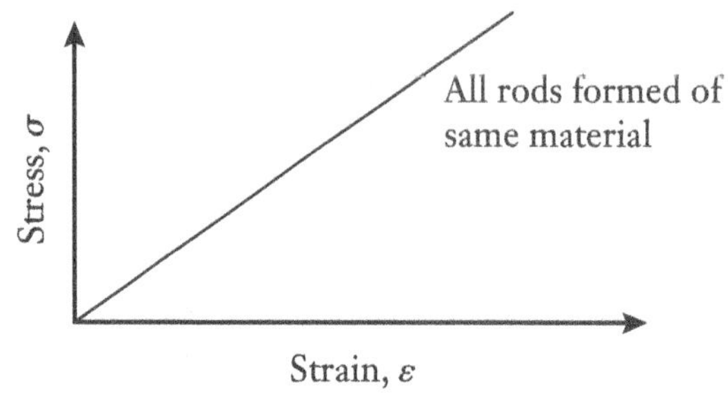

(b)

Figure 4.9. *(a) Force–elongation behaviors of rods having various crosssectional areas and lengths, (b) Each rod has similar stress-strain behavior.*

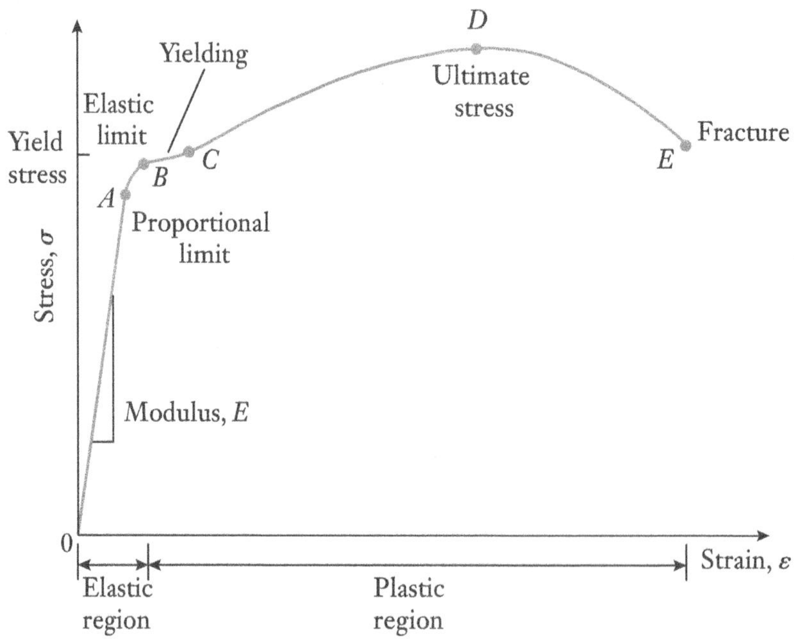

Figure 4.10. *Idealized stress-strain curve for structural steel.*

with the stiffness in Equation (4.4) being

$$k = \frac{EA}{L} \tag{4.7}$$

In as much as metals and most other engineering materials are combinations of various chemical elements, the elastic modulus of a material is related to the strength of its interatomic bonds. As one example, steel alloys contain different fractions of such elements as carbon, molybdenum, manganese, chromium, and nickel. A common material known as 1020 grade steel contains 0.18–0.23% carbon (by weight), 0.30–0.60% manganese, and a maximum of 0.04% phosphorus and 0.05% sulfur. This alloy, like all other compositions of steel, is comprised primarily of iron, and so the value of the elastic modulus for one alloy of steel is not much different from that for another. A similar situation exists for various alloys of aluminum. The following numerical values for the elastic modulus of steel and aluminum are sufficiently accurate for most engineering calculations and design purposes:

$$E_{\text{steel}} \approx 210\,\text{GPa} \approx 30\,\text{Mpsi}$$
$$E_{\text{aluminum}} \approx 70\,\text{GPa} \approx 10\,\text{Mpsi}$$

You can see that the elastic modulus of aluminum is lower than that of steel by a factor of three. In light of Equation (4.6), one implication of that difference is that, with the same dimensions and applied force, an aluminum rod would stretch three times as much as a steel rod. The elastic modulus for a specific sample could differ from these values, and so, in critical applications, the material properties should be measured whenever practical.

As we described in the preceding section, after a rod is stretched, its diameter will also contract by a slight amount. Conversely, the diameter would enlarge slightly if a compressive force is applied. This cross-sectional effect is known as Poisson's contraction or expansion, and it represents a dimensional change that occurs perpendicular to the direction of the applied force. When a soft material (such as a rubber band) is stretched, those dimensional changes can often be observed without any special equipment. For metals and other materials in engineering applications, the changes are very small and must be measured using precision instrumentation.

Poisson's ratio

The material property that quantifies the contraction or expansion of a cross section is *Poisson's ratio*, represented by v (the lowercase Greek character nu). It is defined by changes in the rod's diameter Δd and length ΔL:

$$\Delta d = -vd\frac{\Delta L}{L} \tag{4.8}$$

The negative sign in this equation sets the sign convention that tension (with $\Delta L > 0$) causes the diameter to contract ($\Delta d < 0$), and compression causes the diameter to expand. For many metals, $v \approx 0.3$, with numerical values generally falling in the range 0.25–0.35.

Elastic limit

Yielding

Returning to our discussion of the stress-strain diagram of Figure 4.10, point B is called the *elastic limit.* For loading between points A and B, the material continues to behave elastically, and it will spring back upon removal of the force, but the stress and strain are no longer proportional. As the load increases beyond B, the material begins to show a permanent set. *Yielding* starts to occur in the region between B and C; that is, even for small changes in stress, the rod experiences a large change in strain. In the yielding region, the rod stretches substantially even as the force grows only slightly because of the shallow slope in the stress-strain curve. For that reason, the onset of yielding is often

Yield strength

Ultimate strength

Materials testing machine

taken by engineers as an indication of failure. The value of stress in region B–C defines the material property called the *yield strength,* S_y. As the load is even further increased beyond point C, the stress grows to the *ultimate strength,* S_u at point D. That value represents the largest stress that the material is capable of sustaining. As the test continues, the stress in the figure actually decreases, owing to a reduction in the rod's cross-sectional area, until the sample eventually fractures at point E.

Stress-strain curves are measured on a device called a *materials testing machine.* Figure 4.11 shows an example of equipment in which a computer both controls the test stand and records the experimental data. During tension testing, a specimen such as a rod of steel is clamped between two jaws that are gradually pulled apart, placing the specimen in tension. A load cell that measures force is attached to one of the jaws, and a second sensor (called an extensometer) measures how much the sample is stretched. A computer records the force and elongation data during the experiment, and those values are then converted to stress and strain by using Equations (4.2) and (4.3). When the stress-strain data is drawn on a graph, the slope in the low-strain region is measured to determine E, and the value of the stress S_y at the yield point is read off the curve.

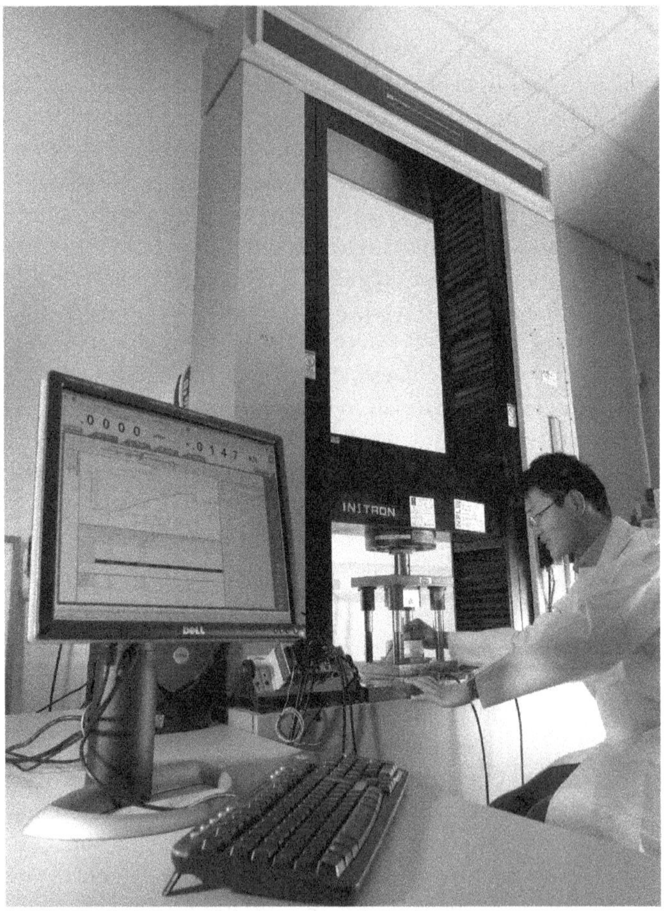

Figure 4.11. *The engineer is using a materials testing machine to bend a metal bar between two supports. The computer controls the experiment and records the force and deflection data.*

Courtesy of Andrew Brookes, National Physical Laboratory/Science Source.

A stress-strain diagram for a sample of structural steel is depicted in Figure 4.12. The stress is shown in the USCS dimensions of ksi, and the dimensionless strain is shown as a percentage. We can use this diagram to determine the elastic modulus and yield strength of this sample. The relation between σ and ε is nearly straight for small strains (up to about 0.2%), and the modulus E is determined from the slope of the line in that region. The strain was zero when no stress was applied, and, as read off the axes of the

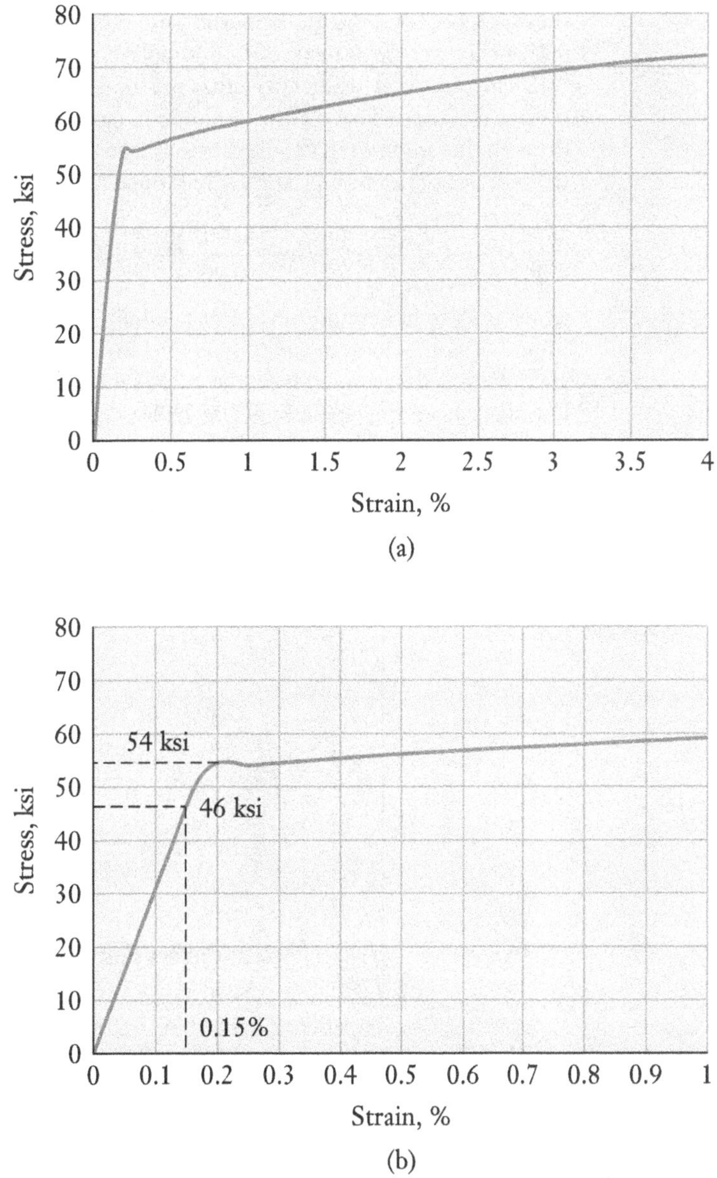

Figure 4.12. *A measured stress- strain curve for a sample of structural steel (a) shown over a wide range of strain, and (b) magnified to highlight the low- strain proportional and yielding regions.*

graph, the specimen was subjected to 46 ksi at a strain of 0.15%. Following Equation (4.5), the elastic modulus is

$$E = \frac{46 \times 10^3 \text{ psi}}{0.0015} = 3.06 \times 10^7 \text{ psi} = 30.6 \text{ Mpsi}$$

which is quite close to the rule-of-thumb value for steel (30 Mpsi). The yield point is also evident in Figure 4.12(b), and we measure $S_y = 54$ ksi directly from the graph's ordinate.

For aluminum and other nonferrous metals, the sharp corner appearing at the yield point in the stress-strain diagram in Figure 4.10 and the narrow yielding region *B–C* are generally not seen. Instead, such materials tend to exhibit a smoother and more gradual transition between the elastic and plastic regions. For such cases, a technique called the *0.2% offset method* is instead used to define the yield strength. As an illustration, Figure 4.13 shows a stress-strain curve that was measured for an aluminum alloy on a materials testing machine. The elastic modulus is again determined by drawing a straight line through the origin to match the stress-strain curve in the proportional region. When the stress is 15 ksi, the strain is 0.14%, and the elastic modulus therefore becomes

$$E = \frac{15 \times 10^3 \text{ psi}}{0.0014} = 1.1 \times 10^7 \text{ ksi} = 11 \text{ Mpsi}$$

which is close to aluminum's nominal value (10 Mpsi). However, unlike the steel alloy of Figure 4.12, the onset of yielding is not sharp and evident. In the 0.2% offset method, the yield stress is determined from the curve's intersection with a line that is drawn at slope E but offset from the origin by 0.2%. In Figure 4.13, the straight construction line drawn with slope 11 Mpsi from the offset point intersects the stress-strain diagram at the yield point, and $S_y = 22$ ksi. That value is taken as the level of stress where the material begins to yield appreciably and becomes unacceptable for further use.

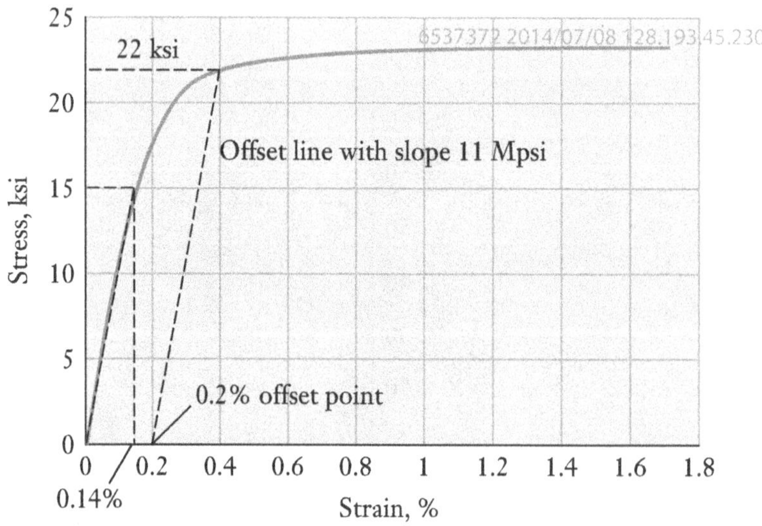

Figure 4.13. *A measured stress-strain curve of an aluminum sample. The elastic modulus is determined from the slope of the curve at low strain, and the yeild stress is found by using the 0.2% offset yeild method.*

TABLE 4.2 Elastic Modulus and Weight Density of Selected Metals*

MATERIAL	ELASTIC MODULUS, *E* Mpsi	ELASTIC MODULUS, *E* GPa	POISSON'S RATIO, ν	WEIGHT DENSITY, $\rho\omega$ lb/ft³	WEIGHT DENSITY, $\rho\omega$ kN/m³
Aluminum alloys	10	72	0.32	172	27
Copper alloys	16	110	0.33	536	84
Steel alloys	30	207	0.30	483	76
Stainless steels	28	190	0.30	483	76
Titanium alloys	16	114	0.33	276	43

*The numerical values given are representative, and values for specific materials could vary with composition and processing.

Tables 4.2 and 4.3 list material property data for several metals, including the elastic modulus, Poisson's ratio, weight density, and ultimate and yield strengths. The properties of any given metallic sample, however, could differ from those listed in the table. Whenever possible, particularly for applications where failure could result in an injury, the material properties should be measured directly, or the material's supplier should be contacted. Common uses for the metals and alloys listed in Table 4.3 are discussed in Section 4.5.

TABLE 4.3 Ultimate and Yield Strengths of Selected Metals*

MATERIAL		ULTIMATE STRENGTH, S_u ksi	ULTIMATE STRENGTH, S_u MPa	YIELD STRENGTH, S_Y ksi	YIELD STRENGTH, S_Y MPa
Aluminum alloys	3003-A	16	110	6	41
	6061-A	18	124	8	55
	6061-T6	45	310	40	276
Copper alloys	Naval brass-A	54	376	17	117
	Cartridge brass-CR	76	524	63	434
Steel alloys	1020-HR	66	455	42	290
	1045-HR	92	638	60	414
	4340-HR	151	1041	132	910
Stainless steels	303-A	87	600	35	241
	316-A	84	579	42	290
	440C-A	110	759	70	483
Titanium alloy	Commercial	80	551	70	482

*The numerical values given are representative, and values for specific materials could vary with composition and processing. A = annealed, HR = hot-rolled, CR = cold-rolled, and T = tempered.

EXAMPLE 4.3: U-Bolt's Dimensional Changes

For the 10-mm-diameter steel U-bolt in Example 4.2, determine the (a) strain, (b) change in length, and (c) change in diameter of the bolt's 325-mm-long straight section. Use the rule-of-thumb value $E = 210$ GPa for the elastic modulus, and take the Poisson's ratio as $\nu = 0.3$. (See Figure 4.14.)

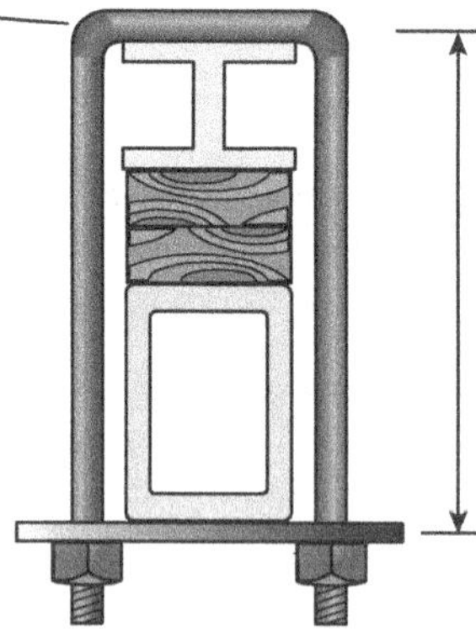

Figure 4.14.

Approach

The stress in the U-bolt's straight section was previously found in Example 4.2 to be $\sigma = 5.093 \times 10^7$ Pa. We will calculate the strain, change in length, and change in diameter by applying Equations (4.5), (4.3), and (4.8) to parts (a), (b), and (c), respectively.

SOLUTION

(a) The strain in the straight section is

$$\varepsilon = \frac{5.093 \times 10^7 \text{ Pa}}{210 \times 10^9 \text{ Pa}} \quad \leftarrow \left[\varepsilon = \frac{\sigma}{E}\right]$$
$$= 2.425 \times 10^{-4} \frac{\cancel{\text{Pa}}}{\cancel{\text{Pa}}}$$
$$= 2.425 \times 10^{-4}$$

Because this value is a small dimensionless number, we will write it as the percentage $\varepsilon = 0.02425\%$.

(b) The change in the U-bolt's length (the elongation) is

$$\Delta L = (2.425 \times 10^{-4})(0.325 \text{ m}) \quad \leftarrow [\Delta L = \varepsilon L]$$
$$= 7.882 \times 10^{-5} \text{ m}$$

We convert this numerical value to the derived SI unit of microns following the definition in Table 3.2:

$$\Delta L = (7.882 \times 10^{-5}\,\text{m})\left(10^{6}\,\frac{\mu\text{m}}{\text{m}}\right)$$
$$= 78.82(\cancel{\text{m}})\left(\frac{\mu\text{m}}{\cancel{\text{m}}}\right)$$
$$= 78.82\,\mu\text{m}$$

Here the SI prefix "micro" represents the factor of one-millionth.

(c) The diameter change is even smaller than the U-bolt's elongation:

$$\Delta d = -(0.3)(0.01\,\text{m})\left(\frac{7.882 \times 10^{-5}\,\text{m}}{0.325\,\text{m}}\right) \qquad \leftarrow \left[\Delta d = -\nu d\frac{\Delta L}{L}\right]$$
$$= -7.276 \times 10^{-7}\,\text{m}\left(\frac{\cancel{\text{m}}}{\cancel{\text{m}}}\right)$$
$$= -7.276 \times 10^{-7}\,\text{m}$$

To suppress the large negative power-of-ten exponent, we apply the "nano" prefix:

$$\Delta d = (-7.276 \times 10^{-7}\,\text{m})\left(10^{9}\,\frac{\text{nm}}{\text{m}}\right)$$
$$= -727.6(\cancel{\text{m}})\left(\frac{\text{nm}}{\cancel{\text{m}}}\right)$$
$$= -727.6\,\text{nm}$$

Discussion

The U-bolt's elongation is small indeed; this is expected because the U-bolt is made of steel. The elongation is approximately the same as the diameter of a human hair or slightly larger than the 632.8 nm wavelength of light in a helium-neon laser. The bolt's diameter therefore contracts by an amount only somewhat more than one wavelength of light. Even though the bolt carries a load of 4 kN (approximately 900 lb), the changes in its dimensions are imperceptible to the eye and would require specialized equipment to measure.

$$\varepsilon = 0.02425\%$$
$$\Delta L = 78.82\,\mu\text{m}$$
$$\Delta d = -727.6\,\text{nm}$$

EXAMPLE 4.4: Rod Stretching

A round rod is made from the steel alloy having the stress-strain characteristics shown in Figure 4.12. (See Figure 4.15.) When it is subjected to 3500 lb of tension (approximately equal to the weight of an automobile sedan), calculate (a) the stress and strain in the rod, (b) the amount that it stretches, (c) its change in diameter, and (d) its stiffness, (e) If the force was only 1000 lb, by what amount would the rod have stretched? Use the value $\upsilon = 0.3$ for the Poisson's ratio.

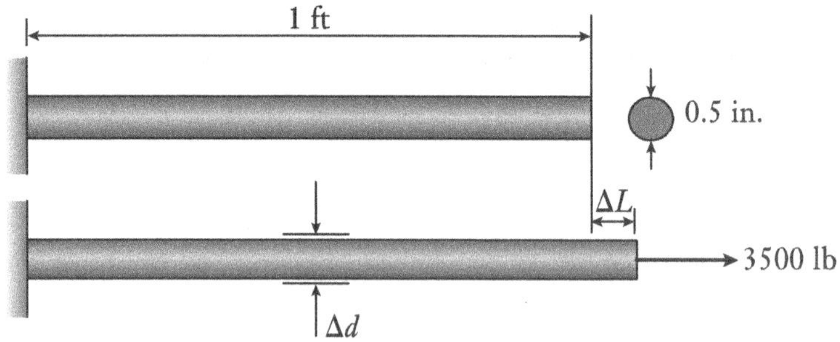

Figure 4.15.

Approach

We will first calculate the stress by applying Equation (4.2). Then, we will calculate the strain, change in length, change in diameter, and stiffness by applying Equations (4.5), (4.3), (4.8), and (4.7) respectively.

SOLUTION

(a) The rod's cross-sectional area is

$$A = \frac{\pi (0.5 \text{ in.})^2}{4} \qquad \leftarrow \left[A = \frac{\pi d^2}{4} \right]$$
$$= 0.1963 \, \text{in}^2$$

and the tensile stress is

$$\sigma = \frac{3500 \text{ lb}}{0.1963 \, \text{in}^2} \qquad \leftarrow \left[\sigma = \frac{F}{A} \right]$$
$$= 1.783 \times 10^4 \, \frac{\text{lb}}{\text{in}^2}$$
$$= 1.783 \times 10^4 \text{ psi}$$

The USCS dimension ksi is next used to represent the factor of 1000 psi:

$$\sigma = \left(1.783 \times 10^4 \text{ psi}\right)\left(10^{-3} \, \frac{\text{ksi}}{\text{psi}}\right)$$
$$= 17.83 \left(\cancel{\text{psi}}\right)\left(\frac{\text{ksi}}{\cancel{\text{psi}}}\right)$$
$$= 17.83 \, \text{ksi}$$

The strain within the rod is given by

$$\begin{aligned} \varepsilon &= \frac{1.783\times10^{4}\,\text{psi}}{30.6\times10^{6}\,\text{psi}} \qquad \leftarrow\left[\varepsilon=\frac{\sigma}{E}\right] \\ &= 5.825\times10^{-4}\left(\frac{\cancel{\text{psi}}}{\cancel{\text{psi}}}\right) \\ &= 5.825\times10^{-4} \end{aligned}$$

or 0.05825%. Because stress and elastic modulus have the same units, their dimensions cancel when strain is calculated by using Equation (4.5).

(b) The 3500-lb load elongates the rod by

$$\begin{aligned} \Delta L &= \left(5.825\times10^{-4}\right)(12\,\text{in.}) \qquad \leftarrow[\Delta L=\varepsilon L] \\ &= 6.990\times10^{-3}\,\text{in.} \end{aligned}$$

(c) Using $\upsilon=0.3$ for steel, the diameter changes by the amount

$$\begin{aligned} \Delta d &= -(0.3)(0.5\,\text{in.})\left(\frac{6.990\times10^{-3}\,\text{in.}}{12\,\text{in.}}\right) \qquad \leftarrow\left[\Delta d=-\nu d\frac{\Delta L}{L}\right] \\ &= -8.738\times10^{-5}\,(\text{in.})\left(\frac{\cancel{\text{in.}}}{\cancel{\text{in.}}}\right) \\ &= -8.738\times10^{-5}\,\text{in.} \end{aligned}$$

and the negative sign convention indicates that the diameter contracts.

(d) The rod's stiffness is determined from the material's elastic modulus, cross-sectional area, and length:

$$\begin{aligned} k &= \frac{\left(30.6\times10^{6}\,\text{psi}\right)\left(0.196\,\text{in}^{2}\right)}{12\,\text{in.}} \qquad \leftarrow\left[k=\frac{EA}{L}\right] \\ &= 5.007\times10^{5}\left(\frac{\text{lb}}{\cancel{\text{in}}^{2}}\right)\left(\cancel{\text{in}}^{2}\right)\left(\frac{1}{\text{in.}}\right) \\ &= 5.007\times10^{5}\,\frac{\text{lb}}{\text{in.}} \end{aligned}$$

In canceling the dimensions, we have expanded the stress unit psi as lb/in^2.

(e) With a force of only 1000 lb, the rod would stretch by only

$$\begin{aligned} \Delta L &= \frac{1000\,\text{lb}}{5.007\times10^{5}\,\text{lb / in.}} \qquad \leftarrow\left[\Delta L=\frac{F}{k}\right] \\ &= 1.997\times10^{-3}\left(\cancel{\text{lb}}\right)\left(\frac{\text{in.}}{\cancel{\text{lb}}}\right) \\ &= 1.997\times10^{-3}\,\text{in.} \end{aligned}$$

Discussion

In the USCS, the derived unit mil is equivalent to one-thousandth of an inch, and it is convenient for representing small length and diameter changes. A sheet of writing paper is only 3–4 mils thick, and so the rod stretches by about the thickness of two sheets of standard paper, which is expected with the given load and choice of steel for the rod. The diameter contracts by a far smaller amount, less than 0.10 mil. To place that small change in perspective, if the rod's initial diameter had been measured to five significant digits as 0.50000 in., the diameter after extension would be0.49991 in. Indeed, measuring a change at the fifth decimal place requires sensitive and well calibrated instrumentation.

$$\sigma = 17.83\,\text{ksi}$$
$$\varepsilon = 0.05825\%$$
$$\Delta L\ (\text{at } 3500\,\text{lb}) = 6.990\,\text{mils}$$
$$\Delta d = -0.08738\,\text{mil}$$
$$k = -5.007 \times 10^5\ \frac{\text{lb}}{\text{in.}}$$
$$\Delta L\ (\text{at } 1000\,\text{lb}) = 1.997\,\text{mils}$$

4.4 SHEAR

In Figure 4.4, the tensile stress σ acts along the length of the rod, and it is also oriented perpendicular to the rod's cross section. Roughly speaking, a tensile stress will stretch a mechanical component and tend to pull it apart. However, excessive forces can damage hardware in other ways as well. One example is a shear stress, which develops when a force tends to slice or cut through a structure or machine component.

Shear differs from tension and compression in that the stress is oriented in the same plane as the rod's cross section. That is, a shear stress is associated with a force that acts parallel to the surface of the cross section. Consider the block of elastic material in Figure 4.16 that is being pressed downward and pushed between the two rigid supports. As the force F is applied, the material tends to be sliced, sheared, or cut along the two edges that are marked as *shear planes* in the figure. A free body diagram of the block is shown in Figure 4.16(b), and equilibrium in the vertical direction requires that $V = F/2$. The two forces V are called *shear forces*, and you can see that they lie in the shear planes and are parallel to them.

Shear plane

Shear force

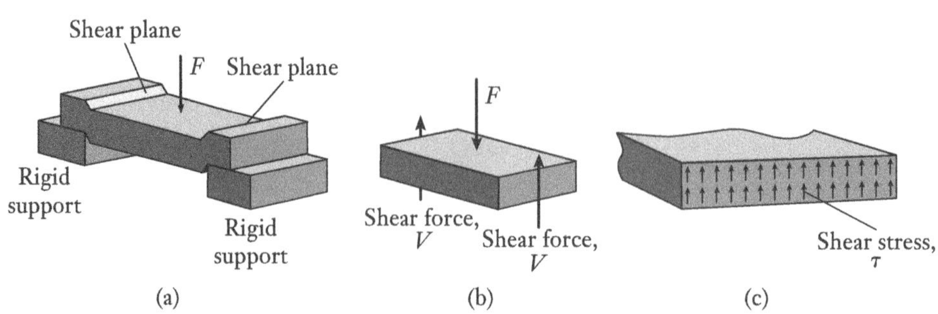

Figure 4.16. *Shear forces and stresses act on material that is being pressed between two rigid supports.*

Like tension and compression, shear forces are also spread continuously within material over any cross section that we can imagine making. The shear force V results from the combination of the shear stresses in Figure 4.16(c) acting over the entire exposed area. In this case, the shear stress τ (the lowercase Greek character tau) is defined

$$\tau = \frac{V}{A} \tag{4.9}$$

where A is the exposed area of the cross section.

Shear stress is often associated with connections that are made between components in a structure or machine, including bolts, pins, rivets, welds, and adhesives. Two types of attachments that are seen in practice are known as single and double shear. The terminology refers to the manner in which the shear forces are transmitted between the two objects joined together. Figure 4.17 shows these configurations for the illustrative case of adhesively bonded lap joints. By using free body diagrams, we imagine disassembling the pieces to expose the shear forces within the adhesive layers.

Single shear
Double shear

For the case of *single shear* in Figure 4.17(a), the full load is carried by just one layer of adhesive, and $V = F$. *Double shear* is illustrated in Figure 4.17(b). Because two surfaces share the load, the shear stress is halved and $V = F/2$. Double shear connections transfer shear forces along two planes simultaneously.

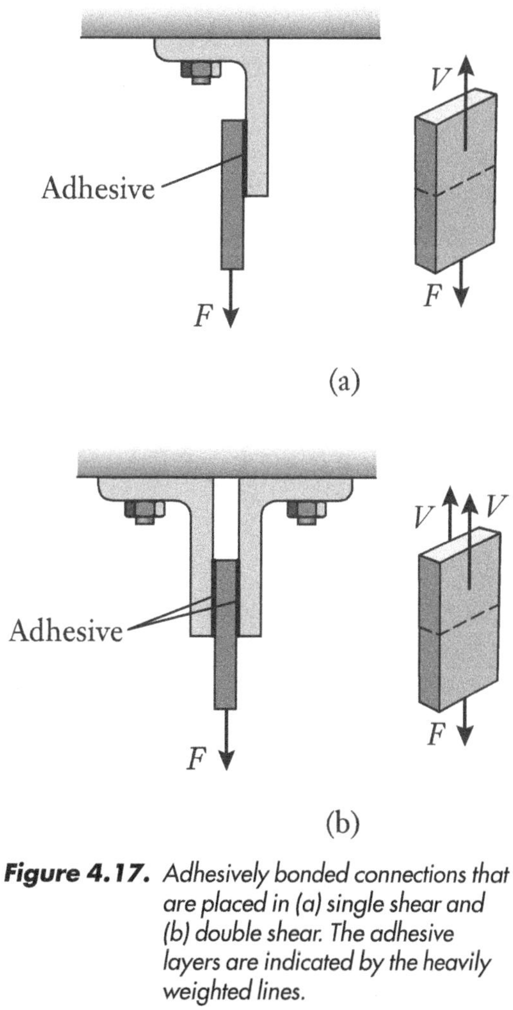

Figure 4.17. *Adhesively bonded connections that are placed in (a) single shear and (b) double shear. The adhesive layers are indicated by the heavily weighted lines.*

EXAMPLE 4.5: Wire Cutter's Hinge Connection

A wire cutter's hinge pin B must support a 385-N force when the handles are pressed together. If the diameter of the hinge pin is 8 mm, determine the pin's shear stress in the SI dimensions of MPa. (See Figure 4.18.)

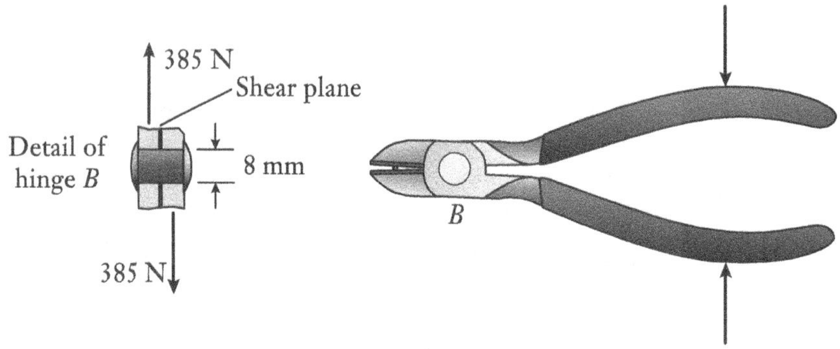

Figure 4.18.

Approach

The force transmitted by pin B between one combined jaw/handle, and the other was found by applying the requirements of static equilibrium. We now extend that analysis to examine how intensely the hinge pin's material is being loaded. Since the 385-N force is transmitted from one jaw/handle to the other across only one shear plane, the hinge pin is loaded in single shear and V = 385 N. We calculate the shear stress by applying Equation (4.9).

SOLUTION

The pin's cross-sectional area is found from Equation (4.1):

$$A = \frac{\pi (0.008\,\text{m})^2}{4} \quad \leftarrow \left[A = \frac{\pi d^2}{4} \right]$$
$$= 5.027 \times 10^{-5}\,\text{m}^2$$

The shear stress is

$$\tau = \frac{385\,\text{N}}{5.027 \times 10^{-5}\,\text{m}^2} \quad \leftarrow \left[\tau = \frac{V}{A} \right]$$
$$= 7.659 \times 10^6 \frac{\text{N}}{\text{m}^2}$$
$$= 7.659 \times 10^6\,\text{Pa}$$

where in the last step we used the definition of the derived unit pascal. We convert the numerical value to the conventional stress dimensions of MPa as follows:

$$\tau = \left(7.659 \times 10^{6}\,\text{Pa}\right)\left(10^{-6}\,\frac{\text{MPa}}{\text{Pa}}\right)$$
$$= 7.659\,(\cancel{\text{Pa}})\left(\frac{\text{MPa}}{\cancel{\text{Pa}}}\right)$$
$$= 7.659\,\text{MPa}$$

Discussion

As the 385-N force is transmitted from one jaw/handle to the other, the action of that force tends to slice pin *B* into two pieces. The hinge pin is loaded in single shear, across one plane, and at the intensity of 7.659 MPa, which is much less than the yield strength of steel, a common material for hinge pins.

$$\tau = 7.659\,\text{MPa}$$

EXAMPLE 4.6: Clevis Joint

The threaded rod is subjected to 350 lb of tension, and that force is transferred through the clevis joint to the fixed base. In the USCS dimensions of ksi, determine the shear stress acting within the $\frac{3}{8}$-in.-diameter hinge pin. (See Figure 4.19.)

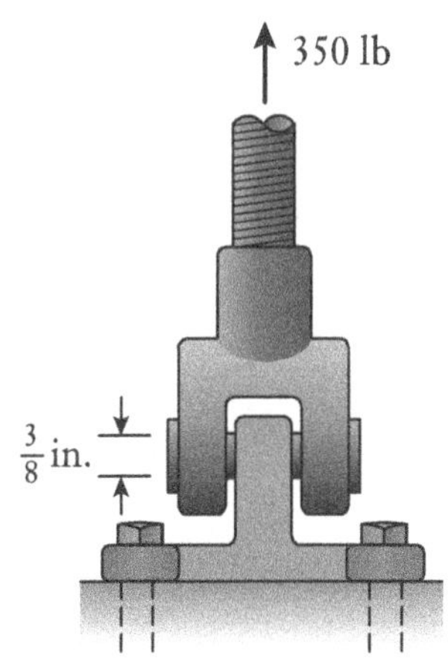

Figure 4.19.

Approach

The threaded rod and the bolts attached to the fixed base are loaded along their length in tension. The clevis joint's hinge pin carries forces perpendicular to its length, and it is loaded in shear. To determine the magnitude of the shear force, we will draw free body diagrams of the clevis joint to show how forces are carried within it. (See Figure 4.20.)

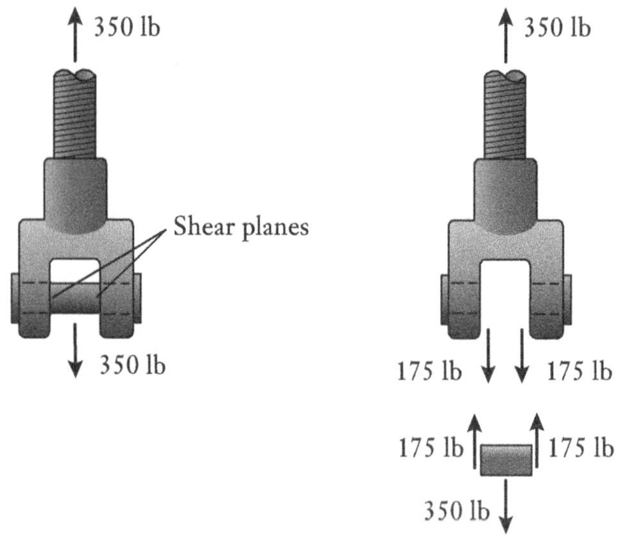

Figure 4.20.

SOLUTION

The 350-lb force is transmitted from the pin to the base through two shear planes, and $V = 175$ lb. The pin's cross-sectional area is

$$A = \frac{\pi (0.375 \text{ in.})^2}{4} \quad \leftarrow \left[A = \frac{\pi d^2}{4} \right]$$
$$= 0.1104 \text{ in}^2$$

The shear stress is

$$\tau = \frac{175 \text{ lb}}{0.1104 \text{ in}^2} \quad \leftarrow \left[\tau = \frac{V}{A} \right]$$
$$= 1.584 \times 10^3 \frac{\text{lb}}{\text{in}^2}$$
$$= 1.584 \times 10^3 \text{ psi}$$

where we have used the abbreviation psi to denote lb/in^2. To place the numerical value in conventional form, we convert the stress to the dimension ksi by using Table 4.1:

$$\tau = (1.584 \times 10^3 \text{ psi}) \left(10^{-3} \frac{\text{ksi}}{\text{psi}} \right)$$
$$= 1.584(\cancel{\text{psi}}) \left(\frac{\text{ksi}}{\cancel{\text{psi}}} \right)$$
$$= 1.584 \text{ ksi}$$

Discussion

The forces acting on the clevis joint tend to cleave or cut through the pin at two locations. The connection is loaded in double shear at the intensity of 1.584 ksi, which again is much less than the yield strength of steel, a common material for hinge pins.

$$\tau = 1.584 \text{ ksi}$$

4.5 ENGINEERING MATERIALS

At this point, we have discussed some of the fundamental characteristics of engineering materials with respect to how they respond when subjected to stress. The next step involves deciding what type of material should be used in a particular design application. A wide variety of materials is available for engineering products, and choosing the correct ones is an important aspect of the design process. Mechanical engineers select materials in the context of both the product's purpose and the processes that will be used during its manufacture. The main classes of materials encountered in mechanical engineering are:

- Metals and their alloys
- Ceramics
- Polymers
- Composite materials

Electronic materials comprise another class that includes the semiconductors that are used widely in electronic, computer, and telecommunication systems. Devices such as microprocessors and memory chips use metal materials, such as electrical conductors and ceramic materials as insulators.

Engineers select materials based on their performance, cost, availability, and past track record in similar applications. Because the production of engineering materials involves the consumption of natural resources and energy, environmental concerns are also factors involved in the selection process. The more manufacturing steps that are required to produce the material and form it into a final product, the greater the material's cost both economically and environmentally. The fall life cycle of a material involves the use of natural resources, for instance in the form of ores; processing raw materials; fabricating and manufacturing products; using the product; and then either disposing of the product or recycling its materials.

In selecting the materials to be used in a product, an engineer first needs to decide on the class of materials to use. Once the class has been chosen (for instance, metals and their alloys), an engineer next determines which material within the class is most appropriate (for instance, steel or aluminum). Many products are designed to use combinations of different classes of materials, each one best suited for a specific task. Automobiles, for instance, contain about 50–60% steel in the frame, engine, and drivetrain components; 5–10% aluminum for engine and body components; and 10–20% plastics used for trim and interior components. The remaining traction includes glass tor the windows, lead for the battery, rubber for the tires, and other materials.

4.5.1 Metals and Their Alloys

Metals are relatively stiff and heavy materials; in other words, from a technical standpoint, they generally have large values for their elastic modulus and density. The strength of metals can be increased by mechanical and heat treatments and by *alloying*, which is the process of adding small amounts of other carefully chosen elements to a base metal. From a design standpoint, metals are a good choice to use in structures and machines that must carry large forces. On the negative side, metals are susceptible to corrosion, and, as a result, they can deteriorate and weaken over time. Another attractive feature of metals is that many methods exist to make them, shape them, and attach them. Metals are versatile materials because they can be manufactured by casting, extrusion, forging, rolling, cutting, drilling, and grinding.

Ductility

Some metals, by virtue of their processing and alloys, have high degrees of *ductility*, that is, the ability of a material to withstand a significant amount of stretching before it fractures. In the stress-strain curve of Figure 4.10, a ductile material has a wide region over which plastic deformation occurs; the steel that is used in paper clips is a good example of a ductile metal. A brittle material like glass, on the other hand, exhibits essentially no plastic deformation. For obvious reasons, ductile metals are well suited for use in structures and machines because, when they become overloaded, the materials give prior warning by noticeably stretching or bending before they break.

Metals include a number of important alloys such as aluminum, copper, steel, and titanium (Table 4.3):

- The 3003 grade aluminum alloy is often produced in the form of wide flat sheets, and it can be easily bent and shaped to form boxes and covers for electronic equipment, among other products. Used for machined mechanical components that are subjected to moderate forces, the 6061 alloy is available either annealed (A) or tempered (T6). Annealing and tempering are processing steps that involve heat treatment to improve the material's strength.

- Copper alloys include brasses (which are yellowish alloys of copper and zinc) and bronzes (which are brownish alloys of copper and tin). These materials do not have particularly high strengths, but they are resistant to corrosion and can be easily joined by soldering. Copper alloys are used in gears, bearings, and the tubing in condensers and heat exchangers.

- Grade 1020 steel is a common, easily worked, and relatively inexpensive medium-grade alloy of steel. The 4340 grade is a higher-strength and more expensive material than grade 1020. Although all steels are comprised primarily of iron and a small fraction of carbon, they are differentiated based on mechanical and heat treatments and on the presence of additional alloying metals, including carbon, manganese, nickel, chromium, and molybdenum.

- For stainless steels, the 316 alloy is used for corrosion-resistant nuts, bolts, and pipe fittings, and the higher-strength material 440C is used to form the raceways of rolling element bearings.

- Titanium alloys are strong, lightweight, and corrosion resistant. On the other hand, they are also far more expensive and difficult to machine than other metals. Titanium is used in chemical industrial pipes, gas turbine blades, high-performance aircraft structures, submarines, and other demanding material applications.

4.5.2 Ceramics

When you think of ceramics, images of coffee mugs, dinner plates, and artwork probably come to mind. Engineering ceramics, on the other hand, are used in the automotive, aerospace, electronics, telecommunications, computer, and medical industries for applications encompassing high temperatures, corrosion, electrical insulation, and wear resistance. Ceramics are produced by heating naturally occurring minerals and chemically treated powders in a furnace to form a rigid mechanical component.

Ceramics are hard, brittle, crystalline materials that can comprise metals and nonmetals. Ceramics have large values of the elastic modulus, but, because they are brittle and tend to break suddenly when overloaded, ceramics are not appropriate for supporting large tensile forces. Mechanical components made from ceramics become significantly weakened by the presence of small defects, cracks, holes, bolted connections, and so forth.

An important characteristic of ceramics is that they can withstand extreme temperatures and insulate other mechanical components from heat. Ceramics are used as thermal barrier coatings to protect turbine blades from the high temperatures developed in jet engines. The space shuttle used tens of thousands of lightweight ceramic tiles to insulate the spacecraft's structural frame from temperatures that reached 2300°F during reentry.

Some examples of ceramics are the compositions silicon nitride (Si_3N_4), alumina (Al_2O_3), and titanium carbide (TiC). Alumina is sometimes formed into a honeycomb-like support structure that is used in an automobile's exhaust system and catalytic converter. Because of its mechanical, electrical, and thermal characteristics, the advanced ceramic AlTiC (64% Al_2O_3 and 36% TiC) is used in computer hard disk drives to support the recording heads above the surface of the rotating disks.

Engineers and physicians are finding an increasing range of applications for ceramic materials in medical applications, including the repair or replacement of human hips, knees, fingers, teeth, and diseased heart valves. Ceramics are one of the few materials that can withstand the corrosive environment of the human body's interior for an extended length of time. Ceramic implants and coatings of ceramics on metallic joint replacements have been found to stimulate bone growth and to protect the metallic portions of an implant from the immune system.

4.5.3 Polymers

Plastics and elastomers are two types of polymers. The root of "polymer" is a Greek word meaning "of many parts," and it emphasizes the fact that polymers are giant molecules formed as long chains of smaller, building-block molecules. These polymer macromolecules have enormous molecular weights, and they can contain hundreds of thousands of atoms. Each macromolecule is made up of a large number of simpler units that are joined together in a regular repeating pattern. Polymers are organic compounds; that is, their chemical formulation is based on the properties of the element carbon. Carbon atoms are able to attach themselves to one another more than other elements can, and other atoms (such as oxygen, hydrogen, nitrogen, and chlorine) are attached to those carbon chains. From a chemical standpoint, therefore, engineering polymers are formed from large-chained molecules having a regular pattern and based on carbon.

Rubber and silk are two naturally occurring macromolecules, but chemists and chemical engineers have developed hundreds of useful macromolecular materials.

Synthetic polymers are classified into two groups: *plastics* (which can be extruded into sheets and pipes or molded to form a wide range of products) and *elastomers* (which are compliant in a manner characteristic of rubber). Unlike the first two classes of materials—metals and their alloys and ceramics—plastics and elastomers are relatively soft materials. They typically have an elastic modulus that is many times smaller than metals. In addition, their properties also change significantly with temperature. At room temperature, polymers may stretch and behave elastically, but, as the temperature is lowered, they become brittle. These materials are not well suited for applications where strength is required or for operation at elevated temperatures. Nevertheless, plastics and elastomers are widely used and remarkable engineering materials. They are relatively inexpensive, lightweight, good insulators against heat and electricity, and easy to shape and mold into complex parts.

Plastics are one of the most utilized engineering materials in any industry, and some of the most common forms are polyethylene, polystyrene, epoxy, polycarbonate, polyester, and nylon. Elastomers, the second category of polymers, are the synthetic rubber-like macromolecules that are elastic and stretchable in a manner that is characteristic of rubber. Elastomers can be greatly deformed and still return to their original shape after being released.

In one of their largest applications, elastomers are used to make tires for vehicles ranging from mountain bikes to aircraft. Other elastomers include the polyurethane foam that is used to insulate buildings, silicone sealant and adhesive, and neoprene, which is resistant to chemicals and oils. Elastomers are also used to make supports and mounting blocks that can reduce the vibration produced by a machine. Vibration isolation mounts that incorporate elastomers are used to attach an automobile's engine to its chassis and to isolate the hard disk drive in a laptop or tablet computer from shock and vibration in the event the computer is accidentally dropped.

4.5.4 Composite Materials

As their name implies, composites are mixtures of several materials, and their formulation can be customized and tailored for specific applications. Composite materials are generally comprised of two components: the matrix and the reinforcement. The matrix is a relatively ductile material that holds and binds together the strong reinforcing particles or fibers embedded in it. Some composite materials comprise a polymer matrix (usually epoxy or polyester) that is reinforced by many small-diameter fibers of glass, carbon, or Kevlar®.

Composites are not well suited for high temperatures because, like plastics and elastomers, the polymer matrix softens as the temperature increases. The main idea behind fiber-reinforced composites is that the strong fibers carry most of the applied force. Other examples of composite materials are concrete that has been reinforced with steel rods, automobile tires that include steel reinforcing belts in an elastomer matrix, and power transmission belts that use fiber or wire cords to carry the belt's tension.

Composite materials are an example of the adage that "the whole is greater than the sum of its parts" in that their mechanical properties are superior to those of the constituent materials alone. The primary advantages of composites are that they can be made very stiff, strong, and lightweight. However, the additional processing steps necessary to produce these materials raise their cost.

Figure 4.21. *This aircraft is fabricated from aluminum, titanium, composites, and other advanced materials in order to meet the competing performance requirements of low weight and high strength.*
Courtesy of VanderWolf Images/Fotolia.

The widespread usage of fiber-reinforced composite materials began in the aerospace industry (Figure 4.21) where weight is at a premium. A substantial amount of an aircraft's weight can be reduced by incorporating composite materials into the airframe, horizontal and vertical stabilizers, flaps, and wing skins. Approximately 30% of the external surface area of the Boeing 767 commercial airliner is formed from composite materials.

As the technology of composites has matured and costs have decreased, these materials have been adopted in automobiles, spacecraft, boats, architectural structures, bicycles, skis, tennis rackets, and other consumer products.

Focus On THE DESIGN OF NEW MATERIALS

Engineers and scientists are continually developing new materials that can be used in the design of innovative products. Combining over twice the yield strength of titanium and the characteristics of plastics, a new class of metal alloys known as Liquidmetal® was developed by a team of researchers from the California Institute of Technology, NASA, and the U.S. Department of Energy. Liquidmetal® alloys consist of a mixture of zirconium, titanium, copper, nickel, and beryllium, and they possess unique amorphous atomic structures that, by contrast to the crystalline structures exhibited by standard metals, show no discernable patterns in their atomic structure. As a result properties superior to the limits of conventional metals can be achieved. For example, material tests have demonstrated yield strengths well over 250 ksi, twice that of conventional titanium alloys. Moreover, Liquidmetal® alloys have a profound ability of retaining their original shape after being exposed to intense loads and stresses, and they are highly resistant to wear and corrosion. As a result, Liquidmetal® alloys are becoming effective design choices in a diverse set of products and systems including flash drives, MP3 players, mobile phones, sports equipment, protective coatings for industrial machinery, and medical, military and space-related equipment.

Other research teams are taking inspiration from the design of biological systems. Researchers at the Royal Institute of Technology in Sweden have developed so-called nanopaper by tightly weaving together nano-sized cellulose threads, found in the cell walls of plants and algae. The resulting structure is stronger and tougher than cast iron and can be manufactured from renewable materials at relatively low temperature and pressure. Another form of nanopaper is made from metal nanowires. With its ability to soak up to 20 times its weight in oil, the nanopaper is being used to help clean oil spills and other environmental toxins. Researchers at the Massachusetts Institute of Technology have developed a new type of self-assembling photovoltaic cell. These cells mimic the repair mechanisms found in plants and are expected to be 40% more efficient at energy conversion than current photovoltaic cells. These new cells could potentially prolong the currently limited life span of solar panels indefinitely.

This discovery might someday lead a revolution in materials science and engineering. Imagine being able to work with materials that can repair themselves over time. You might design a product with glass capable of repairing cracks or with metals that can renew their protective coating to prevent corrosion. These are some of the material technologies that could soon become realistic resources for engineers in the design of products, systems, and machines.

EXAMPLE 4.7: Selecting Materials to Minimize Weight

In designing a truss structure, an engineer finds that the metallic material for a rod should be chosen subject to three constraints. (See Figure 4.22.) The rod must:

- Support a force of magnitude F
- Have length L
- Elongate by less than ΔL

The rod's cross-sectional area and the material from which it will be made remain to be selected by the designer. Among steel, aluminum, and titanium, which material would you recommend to minimize the rod's weight?

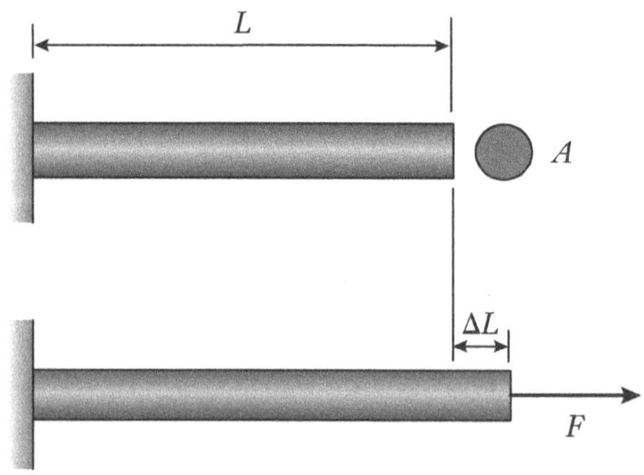

Figure 4.22.

Approach

To determine the best material, we must consider two material properties— density and elastic modulus— in the calculations involving the rod's weight and elongation. The rod's weight is $w = \rho_\omega AL$, where ρ_ω is the metal's weight per unit volume listed in Table 4.2. The length and cross-sectional area, the elastic modulus of the rod's material, and the applied force are related by Equation (4.6).

SOLUTION

Although the rod's area is not known, it can be expressed in terms of the given quantities (F, L, and ΔL) and the elastic modulus E:

$$A = \frac{FL}{E\Delta L} \quad \leftarrow \left[\Delta L = \frac{FL}{EA}\right]$$

By substituting into the expression for the rod's weight and making some algebraic manipulations, we have

$$\begin{aligned} w &= (\rho_\omega L)\left(\frac{FL}{E\Delta L}\right) \quad \leftarrow [\omega = \rho_\omega AL] \\ &= \left(\frac{FL^2}{\Delta L}\right)\left(\frac{\rho_\omega}{E}\right) \\ &= \frac{FL^2 / \Delta L}{E / \rho_\omega} \end{aligned}$$

The numerator in this expression is a grouping of variables that are given in the problem's statement: the applied force, the rod's length, and the allowable elongation. Thus, the numerical value of the numerator is fixed. On the other hand, the denominator E/ρ_ω is characteristic of the material alone. To minimize the rod's weight, we should choose a material having as large a value of E/ρ_ω as possible. The rod's cross-sectional area can then be determined subsequently from the expression $A = FL/E\Delta L$. By using the elastic modulus and weight-density values that are listed in Table 4.2, the ratio's value for steel is

$$\begin{aligned} \frac{E}{\rho_\omega} &= \frac{207 \times 10^9 \text{ Pa}}{76 \times 10^3 \text{N/m}^3} \\ &= 2.724 \times 10^6 \left(\frac{\text{N}}{\text{m}^2}\right)\left(\frac{\text{m} \cdot \text{m}^2}{\text{N}}\right) \\ &= 2.724 \times 10^6 \text{ m} \end{aligned}$$

and in a similar fashion, we find the values for aluminum (2.667×10^6 m) and titanium (2.651×10^6 m). By tabulating these numbers, we can see that steel offers a slight advantage over aluminum and titanium, but the difference amounts to only a few percent.

TABLE 4.4

MATERIAL	E/ρ_ω (m)
Steel	2.724×10^6
Aluminum	2.667×10^6
Titanium	2.651×10^6

Discussion

The parameter E/ρ_ω is a measure of a material's stiffness per unit weight. For an application involving tension, we should choose a material that maximizes E/ρ_ω for the structure to have high stiffness and low weight. As it turns out, steel has a slight advantage, but the E/ρ_ω ratio is nearly the same for aluminum and titanium. Once the material's cost is factored in, however, steel would clearly be the preferable choice since aluminum is somewhat more expensive than steel, and titanium is significantly more expensive. Of course, our analysis has not taken into account corrosion, strength, ductility, and machinability. Those factors could be important for the task at hand and influence the final recommendation. You can see how selecting a material involves making trade-offs between cost and performance.

Steel has a slight advantage.

4.6 FACTOR OF SAFETY

Mechanical engineers determine the shape and dimensions and select materials for a wide range of products and types of hardware. The analysis that supports their design decisions takes into account the tensile, compressive, and shear stresses that are present, the material properties, and other factors that you will encounter later in your study of mechanical engineering. Designers are aware that a mechanical component can break or otherwise be rendered useless by a variety of causes. For instance, it could yield and be permanently deformed, fracture suddenly into many pieces because the material is brittle, or become damaged through corrosion. In this section, we introduce a simple model that engineers use to determine whether a mechanical component is expected to yield because of either tension or shear stress. This analysis predicts the onset of yielding in ductile materials, and it is a useful tool that enables mechanical engineers to prevent the material in a piece of hardware from being loaded to, or above, its yield stress. Yielding is only one of many possible failure mechanisms, however, and so our analysis in this section obviously will not offer predictions regarding other kinds of failure.

From a practical standpoint, engineers recognize that, despite their analysis, experiments, experience, and design codes, nothing can be built to perfection. In addition, no matter how carefully they might try to estimate the forces that are going to act on a structure or machine, the mechanical component could be accidentally overloaded or misused. For those reasons, a catchall parameter called the factor of safety is generally introduced to account for unexpected effects, imprecision, uncertainty, potential assembly flaws, and material degradation. The *factor of safety* is defined as the ratio of the stress

Factor of safety

at failure to the stress during ordinary use. The factor of safety to guard against ductile yielding is often chosen to fall in the range of 1.5–4.0; that is, the design is intended to be between 150% and 400% as strong as it needs to be for ordinary use. For engineering materials with better-than-average reliability or for well controlled operating conditions, the lower values for the factor of safety are appropriate. When new or untried materials will be used or other uncertainties are present, the larger factors will result in a safer design.

When a straight rod is subjected to tension as in Figure 4.3, the possibility of it yielding is assessed by comparing the stress σ to the material's yield strength S_y. Failure due to ductile yielding is predicted if $\sigma > S_y$. Engineers define the tensile-stress factor of safety as

$$n_{\text{tension}} = \frac{S_y}{\sigma} \tag{4.10}$$

If the factor of safety is greater than one, this viewpoint predicts that the component will not yield, and if it is less than one, failure is expected to occur. For components that are instead loaded in pure shear, the shear stress τ is compared to the yield strength in shear, which is denoted by S_{sy}. As developed in more advanced treatments of stress analysis, one perspective on the failure of materials relates the yield strength in shear to the value in tension according to the expression

$$S_{sy} = \frac{S_y}{2} \tag{4.11}$$

The shear yield strength can therefore be determined from the strength values obtained from standard tensile testing, as listed in Table 4.3. To evaluate the possibility of a ductile material yielding in shear, we compare stress and strength in terms of the shear factor of safety

$$n_{\text{shear}} = \frac{S_{sy}}{\tau} \tag{4.12}$$

The numerical value for the factor of safety that an engineer chooses for a particular design will depend on many parameters, including the designer's own background, experience with components similar to the one being analyzed, the amount of testing that will be done, the material's reliability, the consequences of failure, maintenance and inspection procedures, and cost. Certain spacecraft components might be designed with the factor of safety being only slightly greater than one in order to reduce weight, which is at a premium in aerospace applications. To counterbalance that seemingly small margin for error, those components will be extensively analyzed and tested, and they will be developed and reviewed by a team of engineers having a great deal of collective experience. When forces and load conditions are not known with certainty or when the consequences of a component's failure would be significant or endanger life, large values for the factor of safety are appropriate. Engineering handbooks and design codes generally recommend ranges for safety factors, and those references should be used whenever possible. In part, design codes set safety standards for many mechanical products.

EXAMPLE 4.8: Designing a Gear-to-Shaft Connection

The spur gear is used in a transmission, and it is attached to the 1-in.-diameter shaft by a key having a 0.25 × 0.25-in. cross section and length of 1.75 in. (See Figure 4.23.) The key is made from grade 1045 steel, and it fits into matching slots that are machined in the shaft and gear. When the gear is driven on its tooth by the 1500-lb force, determine (a) the shear stress in the key and (b) the factor of safety against yielding.

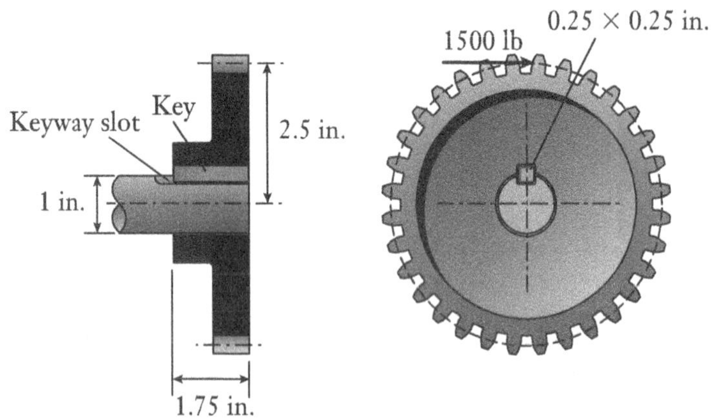

Figure 4.23.

Approach

Torque is transmitted between the gear and the shaft through the key, which tends to be sliced along a single shear plane by the force transmitted between the gear and the shaft. (See Figure 4.24.) We determine the magnitude of the shear force by applying the condition of rotational equilibrium to the gear. Although the 1500-lb force tends to rotate the gear in the clockwise direction, the shear force between the gear and key balances that torque in the counterclockwise direction.

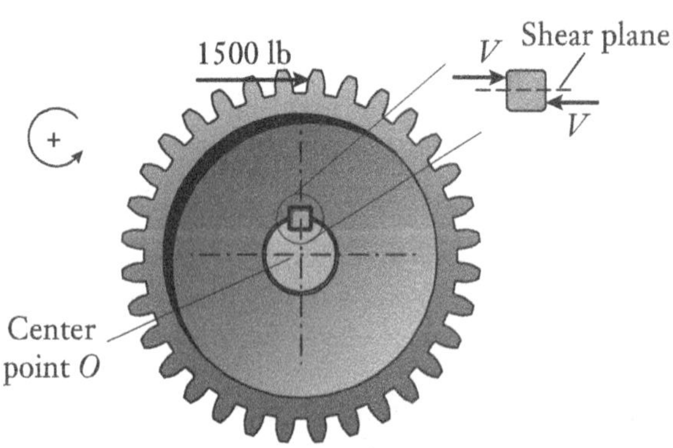

Figure 4.24.

SOLUTION

(a) The shear force on the key is determined from the balance between its torque and that of the 1500-lb tooth force about the shaft's center:

$$-(1500\,\text{lb})(2.5\,\text{in.})+V(0.5\,\text{in.})=0 \quad \leftarrow\left[\sum_{i=1}^{N} M_{O,i}=0\right]$$

and $V=7500$ lb. This force is distributed over the shear plane of the key having cross-sectional area

$$A=(1.75\,\text{in.})(0.25\,\text{in.})=0.4375\,\text{in}^2$$

The shear stress is

$$\begin{aligned}\tau &= \frac{7500\ \text{lb}}{0.4375\ \text{in}^2} \quad \leftarrow\left[\tau=\frac{V}{A}\right]\\ &=1.714\times10^4\ \frac{\text{lb}}{\text{in}^2}\\ &=1.714\times10^4\ \text{psi}\end{aligned}$$

To place the numerical value in more conventional form, we convert to the dimension ksi:

$$\begin{aligned}\tau &= (1.714\times10^4\,\text{psi})\left(10^{-3}\,\frac{\text{ksi}}{\text{psi}}\right)\\ &=17.14(\cancel{\text{psi}})\left(\frac{\text{ksi}}{\cancel{\text{psi}}}\right)\\ &=17.14\,\text{ksi}\end{aligned}$$

(b) The yield strength $S_y = 60$ ksi is listed in Table 4.3, but this is the strength for tensile loading. Following Equation (4.11), the shear yield strength of this alloy is

$$\begin{aligned}S_{sy} &= 60\frac{\text{ksi}}{2} \quad \leftarrow\left[S_{sy}=\frac{S_y}{2}\right]\\ &=30\,\text{ksi}\end{aligned}$$

The factor of safety for the key against ductile yielding is

$$\begin{aligned}n_{\text{shear}} &= \frac{30\,\text{ksi}}{17.14\,\text{ksi}} \quad \leftarrow\left[n_{\text{shear}}=\frac{S_{sy}}{\tau}\right]\\ &=1.750\frac{\cancel{\text{ksi}}}{\cancel{\text{ksi}}}\\ &=1.750\end{aligned}$$

which is a dimensionless number.

Discussion

Since the factor of safety is greater than one, our calculation indicates that the connection between the gear and shaft is strong enough to prevent yielding by an additional margin of 75%. If the force on the gear's tooth would grow by more than that amount, then the key would be expected to yield, and the design would be unsatisfactory. In that event, the key's cross-sectional area would need to be increased, or a higher strength material would need to be selected.

$$\tau = 17.14 \text{ ksi}$$

$$n_{\text{shear}} = 1.750$$

4.7 SUMMARY

One of the primary functions of mechanical engineers is to design structures and machine components so that they are reliable and do not break. Engineers analyze stress, strain, and strength to determine whether a component will be safe or risks being overloaded to the point that it might deform excessively or fracture. The important quantities introduced in this chapter, common symbols for them, and their units are summarized in Table 4.5, and the key equations are reviewed in Table 4.6.

Generally, engineers conduct a stress analysis during the design process, and the results of those calculations are used to guide the choice of materials and dimensions. When the shape of a component or the loading circumstances are particularly complicated, engineers will use computer-aided design tools to calculate the stresses and deformations. Figure 4.25 illustrates the results of one such computer-aided stress analyses for a clevis connection (as described in Example 4.6).

TABLE 4.5 Quantities, Symbols, and Units that Arise when Analyzing Stresses and Material Properties

QUANTITY	CONVENTIONAL SYMBOLS	CONVENTIONAL UNITS	
		USCS	SI
Tensile stress	σ	psi, ksi, Mpsi	Pa, kPa, MPa
Shear stress	τ	psi, ksi, Mpsi	Pa, kPa, MPa
Elastic modulus	E	Mpsi	GPa
Yield strength			
Tension	S_y	ksi	MPa
Shear	S_{sy}	ksi	MPa
Ultimate strength	S_u	ksi	MPa
Strain	ε	—	—
Poisson's ratio	v	—	—
Factor of safety	$\eta_{\text{tension}}, \eta_{\text{shear}}$	—	—
Stiffness	k	lb/in.	N/m

TABLE 4.6 Key Equations that Arise when Analyzing Materials and Stresses

Tension and compression	
Stress	$\sigma = \dfrac{F}{A}$
Strain	$\varepsilon = \dfrac{\Delta L}{L}$
Material response	$\sigma = E\varepsilon$
Rod deformation	
Elongation	$\Delta L = \dfrac{FL}{EA}$
Diameter change	$\Delta d = -\nu d \dfrac{\Delta L}{L}$
Hooke's law	$F = k\Delta L$
Stiffness	$k = \dfrac{EA}{L}$
Shear	
Stress	$\tau = \dfrac{V}{A}$
Yield strength	$S_{sy} = \dfrac{S_y}{2}$
Factor of safety	
Tension	$n_{tension} = \dfrac{S_y}{\sigma}$
Shear	$n_{shear} = \dfrac{S_{sy}}{\tau}$

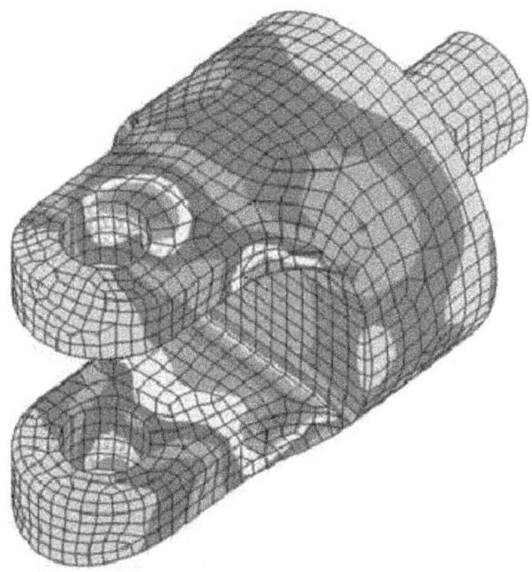

Figure 4.25. *Computer-aided stress analysis of a clevis connection. The shading levels indicate the magnitude of stress.*

In this chapter, we discussed the loading conditions known as tension, compression, and shear, as well as a failure mechanism called ductile yielding. The following concepts are central to selecting materials and setting dimensions in the design of structures and machines:

- *Stress* is the intensity of a force distributed over an exposed area of material. Depending on the direction in which the stress acts, it can be tension or shear.
- *Strain* is the change in length per unit of original length. Because of its definition as a ratio of two lengths, strain is a dimensionless quantity, and it is often expressed as a decimal percentage. At a strain of 0.1 %, a rod that was 1 m long will have stretched by a factor of 0.001, or 1 mm.
- *Strength* captures the ability of a material to withstand the stresses acting on it. Mechanical engineers compare stresses to the strength of materials to assess whether yielding will occur.

An important part of mechanical engineering involves choosing the materials that will be used in the design of a structure or machine component. This choice is an important aspect of the design process and mechanical engineers must consider performance, economic, environmental, and manufacturing issues when making this decision. In this chapter, we introduced some of the characteristics of the main classes of materials encountered in mechanical engineering: metals and their alloys, ceramics, polymers, and composite materials. Each class of materials has its own advantages, special characteristics, and preferred applications.

Once the forces acting on a mechanical component have been determined, the dimensions of the component have been assigned, and the materials to be used in production have been identified, the reliability of the design is assessed. The factor of safety is the ratio of the stress at failure to the stress during ordinary use. It has been said that there are *knowns* (things that we know and that we can account for in a design), *known unknowns* (things that we don't know but that we at least know we don't know them), and *unknown unknowns* (an unknown that might surprise us because we are unaware of it and that might unexpectedly compromise a design). The factor of safety is intended to improve a design's reliability and to account for known and unknown unknowns in the forms of uncertainty in usage, materials, and assembly.

4.6.1 Self-Study and Review

4.1. How are stress and strain defined for a rod that is loaded in tension?

4.2. In the SI and USCS, what are the conventional dimensions for stress and strain?

4.3. Sketch a stress-strain diagram, and label some of its important features.

4.4. What is the difference between the elastic and plastic behavior of materials?

4.5. Define the following terms: elastic modulus, proportional limit, elastic limit, yield point, and ultimate point.

4.6. What are the approximate numerical values for the elastic modulus of steel and aluminum?

4.7. How is the yield strength found using the 0.2% offset method?

4.8. What is Poisson's ratio?

4.9. In what ways do tensile and shear stresses differ?

4.10. Discuss some of the characteristics and applications for metals and their alloys, ceramics, polymers, and composite materials.

4.11. How is the shear yield strength S_{sy} related to the yield strength S_y that is obtained from a tensile test?

4.12. What is the factor of safety? When is the factor of safety too small? Can it be too large?

4.13. Discuss some of the trade-offs that an engineer would consider when deciding whether a design's factor of safety is too large or too small.

4.8 PROBLEMS

Problem P4.1

Find a real physical example of a mechanical structure or machine that has tensile stress present.

(a) Make a clear, labeled drawing of the situation.
(b) Estimate the dimensions of the structure or machine and the magnitudes and directions of the forces that act on it. Show these on the drawing. Briefly explain why you estimate the dimensions and forces to have the numerical values that you assigned.
(c) Calculate the magnitude of the stress.

Problem P4.2

A 1-Mg container hangs from a 15-mm-diameter steel cable. What is the stress in the cable?

Problem P4.3

Grade 1020 steel has a yield strength of 42 ksi and an elastic modulus of 30 Mpsi. Another grade of steel has a yield strength of 132 ksi. What is its elastic modulus?

Problem P4.4

A steel cable of diameter $\frac{3}{16}$ in. is attached to an eyebolt and tensioned to 500 lb (Figure P4.4, see on page 122). Calculate the stress in the cable, and express it in the dimensions psi, ksi, Pa, kPa, and MPa.

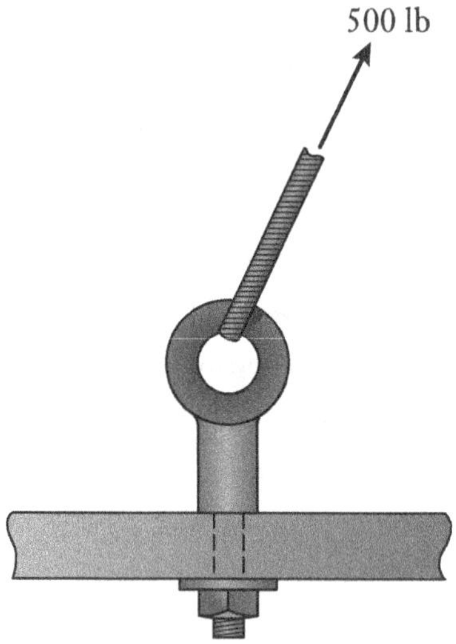

Figure P4.4.

Problem P4.5

When a 120-lb woman stands on a snow-covered trail, she sinks slightly into the snow because the compressive stress between her ski boots and the snow is larger than the snow can support without crumbling. Her cross-country skis are 65 in. long and $1\frac{7}{8}$ in. wide. After estimating the dimensions of the boot's sole, calculate the percentage reduction in stress applied to the snow when she is wearing skis instead of boots.

Problem P4.6

As a machinist presses the handles of the compound-action bolt cutters, link *AB* carries a 7.5-kN force (Figure P4.6). If the link has a 14×4-mm rectangular cross section, calculate the tensile stress within it.

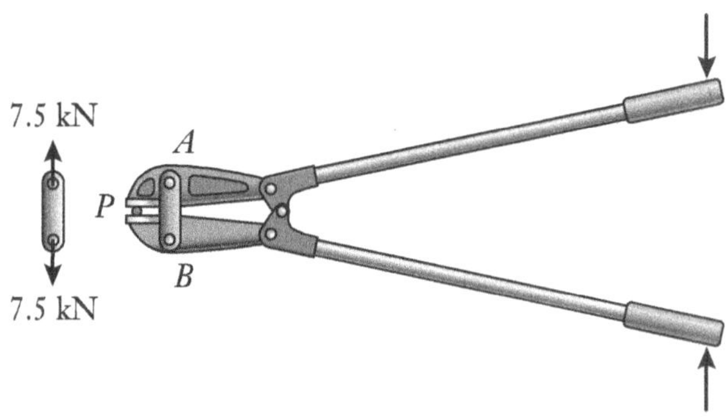

Figure P4.6.

Problem P4.7

See Figure P4.7.

(a) By using either the vector algebra or polygon methods for finding a resultant, determine the magnitude of F that will cause the net effect of the three forces to act vertically.

(b) For that value of F, determine the stress in the bolt's $\frac{3}{8}$-in.-diameter straight shank.

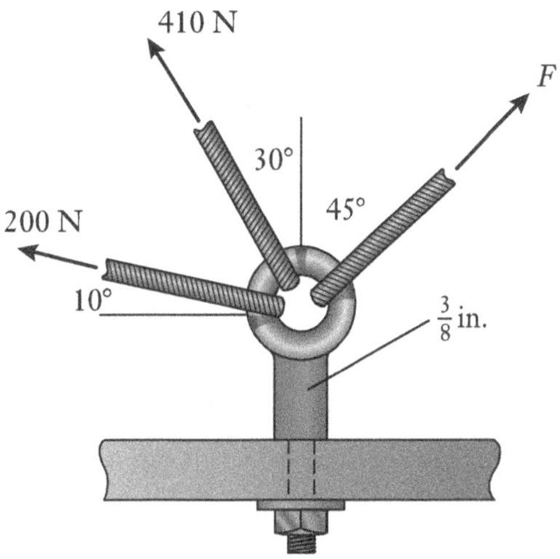

Figure P4.7.

Problem P4.8

In a machine shop, the band saw blade is cutting through a work piece that is fed between the two guide blocks B (Figure P4.8). To what tension P should the blade be set if its stress is to be 5 ksi during use? Neglect the small size of the teeth relative to the blade's width.

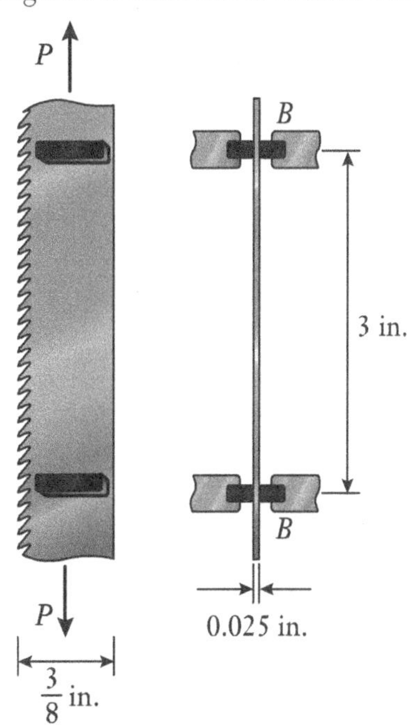

Figure P4.8.

Problem P4.9

An 8-mm-diameter aluminum rod has lines scribed on it that are exactly 10 cm apart (Figure P4.9). After a 2.11-kN force has been applied, the separation between the lines has increased to 10.006 cm.

(a) Calculate the stress and strain in the rod.
(b) To what total length has the rod stretched?

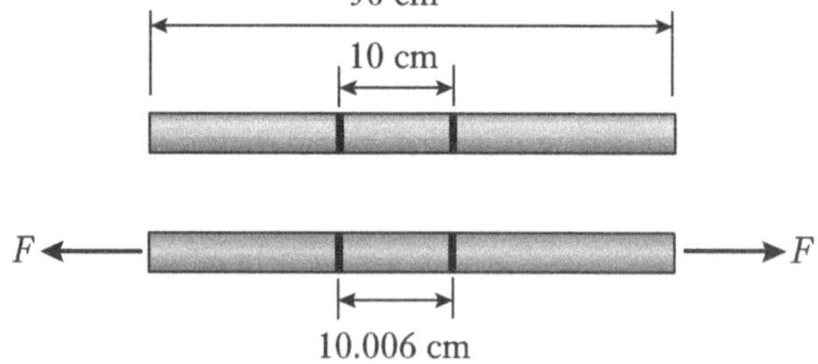

Figure P4.9.

Problem P4.10

The tires of the 4555-lb sedan are 6.5 in. wide (Figure P4.10). Each tire contacts the ground over a distance of 4.25 in. as measured along the vehicle's length. Calculate the compressive stress between each tire and the road. The locations of the vehicle's mass center and the wheelbase dimensions are shown.

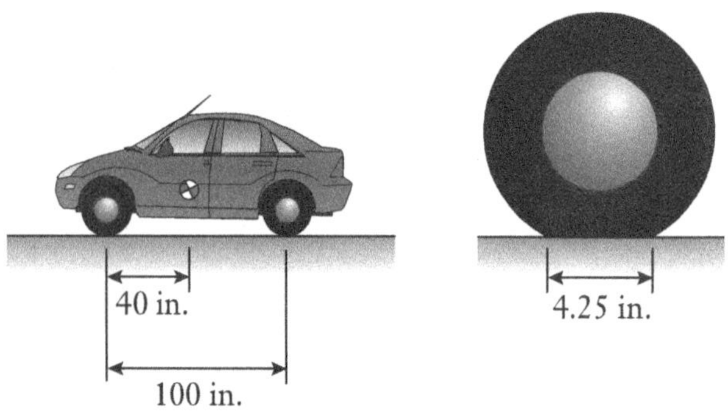

Figure P4.10.

Problem P4.11

Determine the elastic modulus and the yield strength for the material having the stress-strain curve (Figure P4.11, see on page 125). Use the 0.2% offset method.

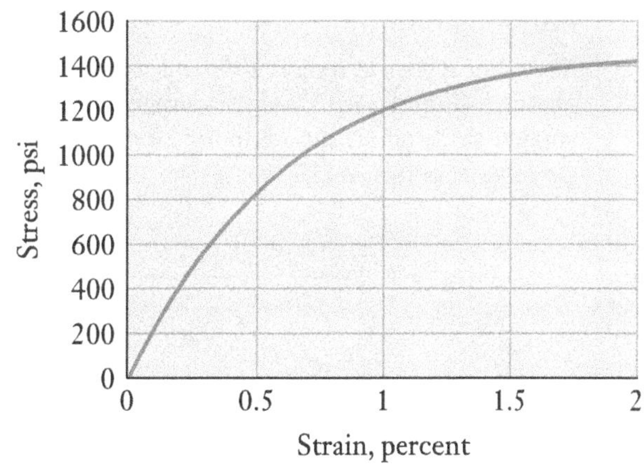

Figure P4.11.

Problem P4.12

A 1-ft-long rod is made from the material of Problem P4.11. By what amount must the rod be stretched from its original length for it to begin yielding?

Problem P4.13

The steel bolt and anchor assembly is used to reinforce the roof of a passageway in an underground coal mine (Figure P4.13). In installation, the bolt is tensioned to 5000 lb. Calculate the stress, strain, and extension of the bolt if it is formed from 1045 steel alloy.

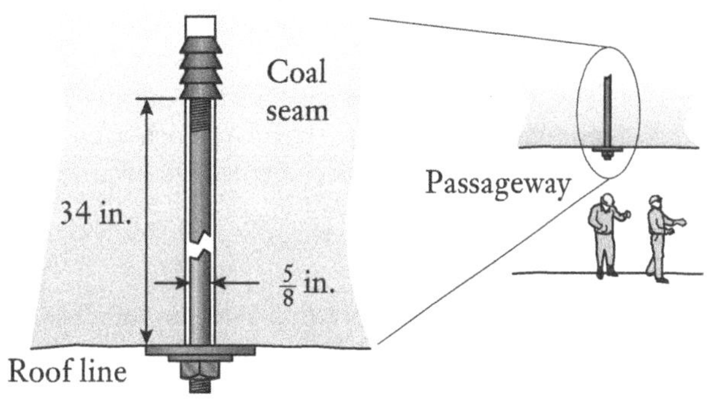

Figure P4.13.

Problem P4.14

A circular rod of length 25 cm and diameter 8 mm is made of 1045 grade steel.

(a) Calculate the stress and strain in the rod, and its extension, when it is subjected to 5 kN of tension.
(b) At what force would the rod begin to yield?
(c) By what amount would the rod have to be stretched beyond its original length in order to yield?

Problem P4.15

A two-tier system to repair bridge infrastructures is supported by two steel cables for each platform (Figure P4.15). All the support cables have a diameter of 0.50 in. Assume the applied loads are at the midpoint of each platform. Determine how much each cable stretches at points *A*–*D*.

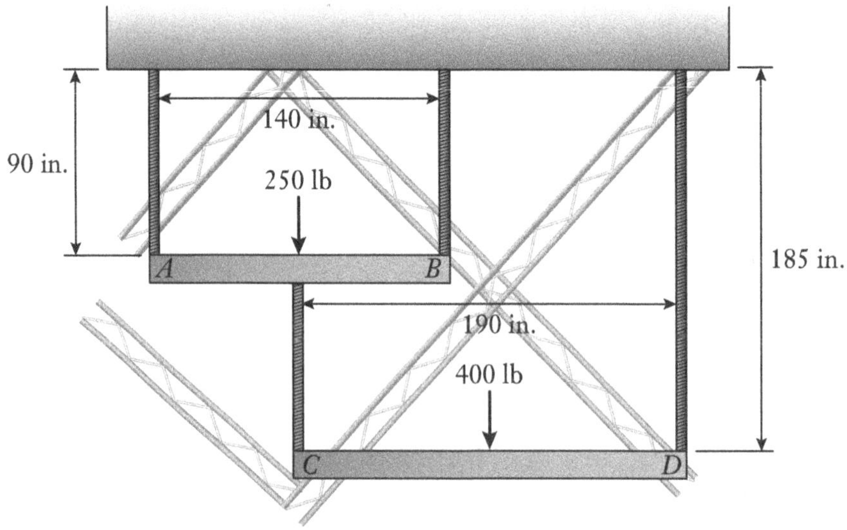

Figure P4.15.

Problem P4.16

For the system in Problem P4.15, assume that two 180-lb workers are going to be standing on the system to repair a bridge, one on each platform. The following design requirement must be met: The stretch of each cable must be less than 0.01 in. at the worst-case loading condition for each cable. Determine the minimum diameter required for each cable.

Problem P4.17

An engineer determines that a 40-cm-long rod of 1020 grade steel will be subjected to a tension of 20 kN. The following two design requirements must be met: The stress must remain below 145 MPa, and the rod must stretch less than 0.125 mm. Determine an appropriate value for the rod's diameter to meet these two requirements. Round up to the nearest millimeter when reporting your answer.

Problem P4.18

Find a real physical example of a mechanical structure or machine that has shear stress present.

(a) Make a clear, labeled drawing of the situation.
(b) Estimate the dimensions of the structure or machine and the magnitudes and directions of the forces that act on it. Show these on the drawing. Briefly explain why you estimated the dimensions and forces to have the numerical values that you assigned.
(c) Calculate the magnitude of the stress.

Problem P4.19

Adhesive tape is capable of supporting relatively large shear stress, but it is not able to support significant tensile stress. In this problem, you will measure the shear strength of a piece of tape. Refer to Figure P4.19.

(a) Cut about a dozen segments of tape having identical length L and width b. The exact length isn't important, but the segments should be easily handled.
(b) Develop a means to apply and measure the pull force F on the tape. Use, for instance, dead weights (cans of soda or exercise weights) or a small fishing scale.
(c) Attach a segment of tape to the edge of a table, with only a portion of tape adhering to the surface. In your tests, consider attachment lengths ranging between a fraction of an inch and several inches.
(d) Being careful to apply the pull force straight along the tape, measure the value F necessary to cause the adhesive layer to slide or shear off the table. Tabulate pull-force data for a half dozen different lengths a.
(e) Make graphs of pull force and shear stress versus a. From the data, estimate the value of the shear stress above which the tape will slide and come loose from the table.
(f) At what length a did the tape break before it sheared off the table?
(g) Repeat the tests for the orientation in which F is applied perpendicular to the surface, tending to peel the tape instead of shearing it. Compare the tape's strengths for shear and peeling.

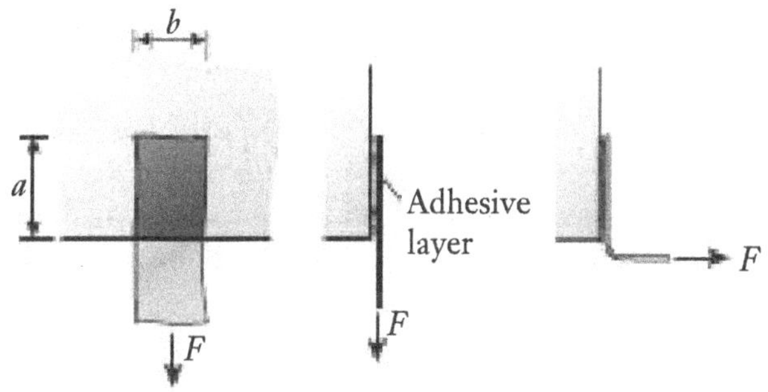

Figure P4.19.

Problem P4.20

The small steel plate is connected to the right angle bracket by a 10-mm-diameter bolt (Figure P4.20, see on page 128). Determine the tensile stress at point A in the plate and the shear stress in the bolt.

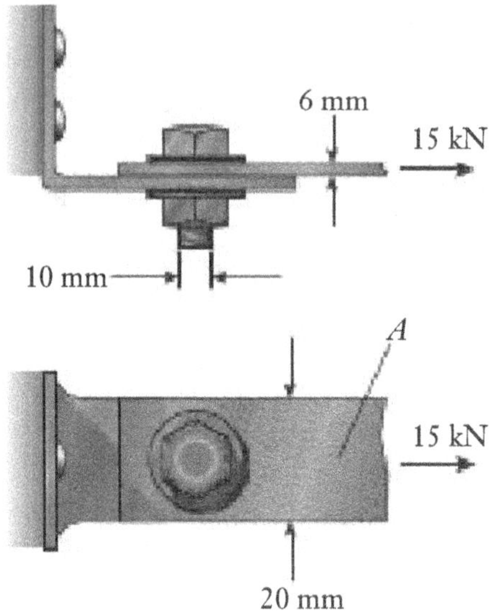

Figure P4.20.

Problem P4.21

A 600-lb force acts on the vertical plate, which in turn is connected to the horizontal truss by five $\frac{3}{16}$-in.-diameter rivets (Figure P4.21).

(a) If the rivets share the load equally, determine the shear stress in them.
(b) In a worst-case scenario, four of the rivets have corroded, and the load is carried by only one rivet. What is the shear stress in this case?

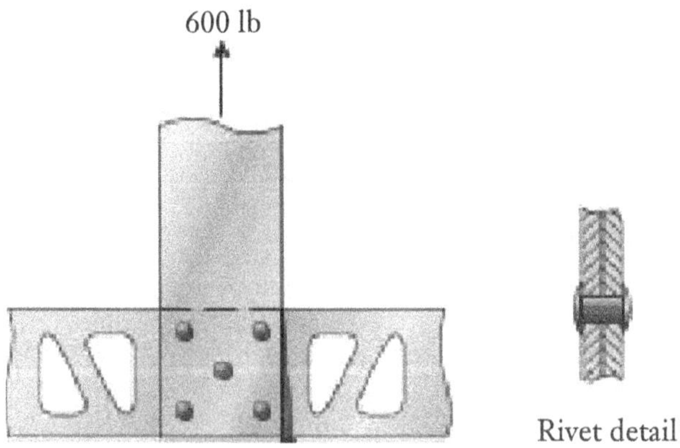

Figure P4.21.

Problem P4.22

A detailed view is shown for the connection B in the concrete trough (Figure P4.22). Determine the shear stress acting in the shackle's $\frac{3}{8}$-in. bolt.

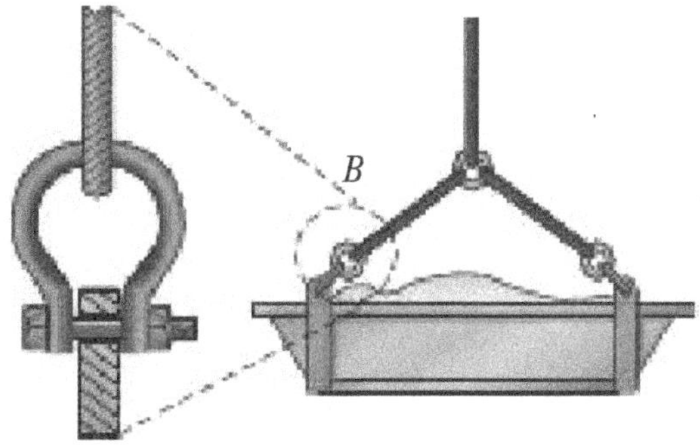

Figure P4.22.

Problem P4.23

The tie-down mount in Figure P4.23 is bolted to the deck of a cargo truck and restrains the steel cable under 1.2 kN of tension. Determine the shear stress in the 6-mm-diameter bolt. Neglect the small horizontal and vertical offset distances between the bolt and the cable.

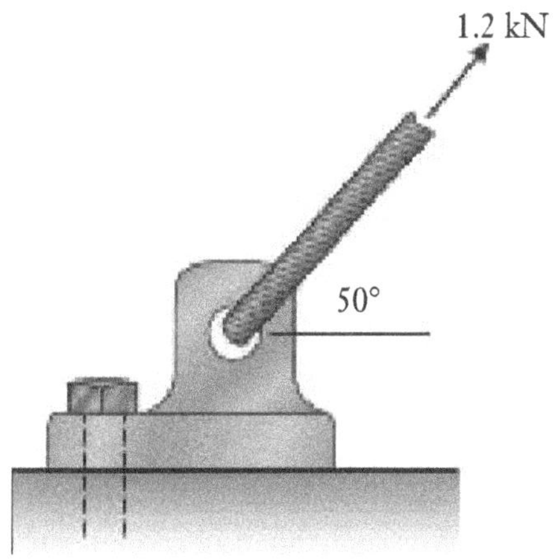

Figure P4.23.

Problem P4.24

A spur gear transmits 35 N · m of torque to the 20-mm-diameter driveshaft. The 5-mm-diameter setscrew threads onto the gear's hub, and it is received in a small hole that is machined in the shaft. Determine the shear stress in the screw along the shear plane B–B (Figure P4.24, see on page 130).

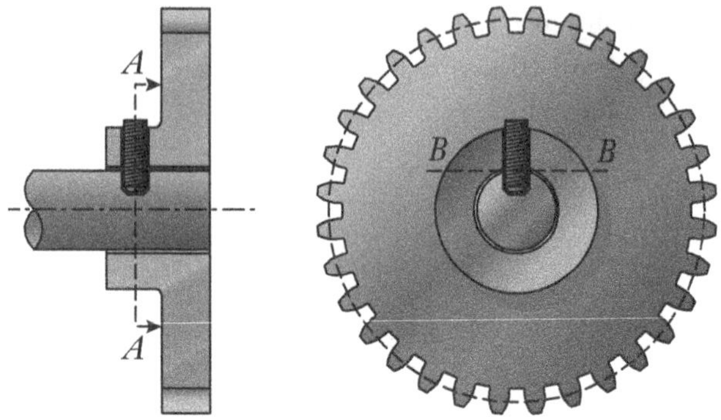

Figure P4.24.

Problem P4.25

A plastic pipe carries deionized water in a microelectronics clean room, and one end of it is capped (Figure P4.25). The water pressure is $p_0 = 50$ psi, and the cap is attached to the end of the pipe by an adhesive. Calculate the shear stress τ present in the adhesive.

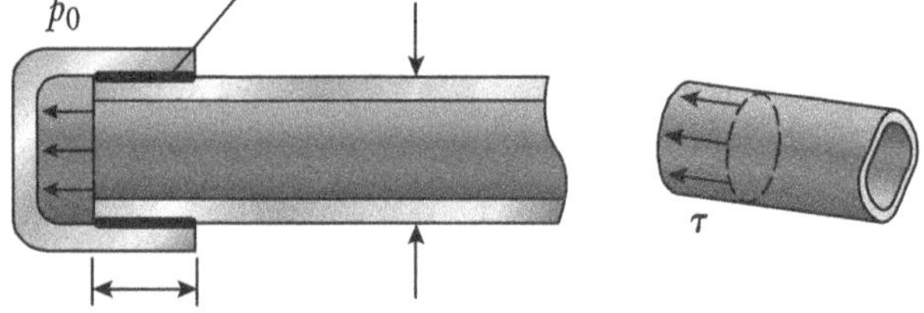

Figure P4.25.

Problem P4.26

Structural steel and stainless steel are both primarily made of iron. Why doesn't stainless steel rust? Research the issue and prepare an approximately 250-word report describing the reason. Cite at least three references for your information.

Problem P4.27

Recent technological advances have allowed for the production of graphene, which is a sheet of carbon atoms one atom thick, densely packed into a honeycomb structure. This material has excellent electrical, thermal, and optical properties, making it an ideal candidate for the design of electronic components. These characteristics are further complemented by graphene's ability to kill bacteria and by its breaking strength, which initial tests demonstrate is 200 times greater than steel. Research this material and prepare an approximately 250-word report on the current engineering applications for graphene. Cite at least three references for your information.

Problem P4.28

A small stepladder has vertical rails and horizontal steps formed from a C-section aluminum channel (Figure P4.28). Two rivets, one in front and one in back, secure the ends of each step. The rivets attach the steps to the left- and right-hand rails. A 200-lb person stands in the center of a step. If the rivets are formed of 6061-T6 aluminum, what should be the diameter d of the rivets? Use a factor of safety of 6, and round your answer to the nearest $\frac{1}{16}$ of an in.

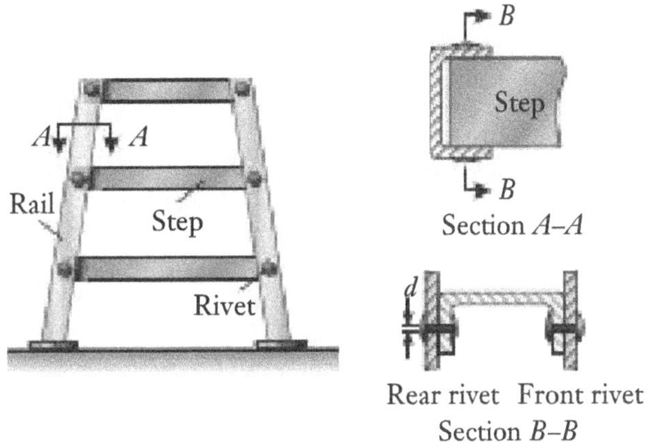

Figure P4.28.

Problem P4.29

For the exercise of Problem P4.28, and for the most conservative design, at what location on the step should you specify in your calculation that the 200-lb person stand? Determine the rivet diameter for that loading condition.

Problem P4.30

A $\frac{3}{8}$-in.-diameter bolt connects the marine propeller to a boat's $1\frac{1}{4}$-in. driveshaft (Figure P4.30). To protect the engine and transmission if the propeller accidentally strikes an underwater obstacle, the bolt is designed to be cut when the shear stress acting on it reaches 25 ksi. Determine the contact force between the blade and obstacle that will cause the bolt to be sheared, assuming a 4-in. effective radius between the point of contact on the blade and the center of the driveshaft.

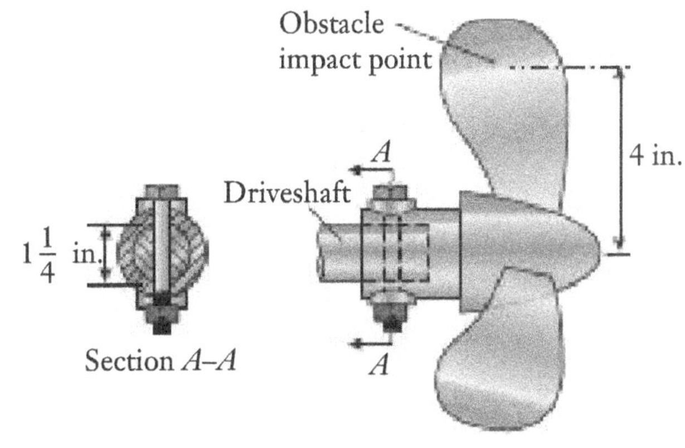

Figure P4.30.

Problem P4.31

A machinist squeezes the handle of visegrip pliers while loosening a frozen bolt (Figure P4.31). The connection at point A, which is shown in a magnified cross-sectional view, supports a 4.1-kN force.

(a) Determine the shear stress in the 6-mm-diameter rivet at A.
(b) Determine the factor of safety against yielding if the rivet is formed from 4340 steel alloy.

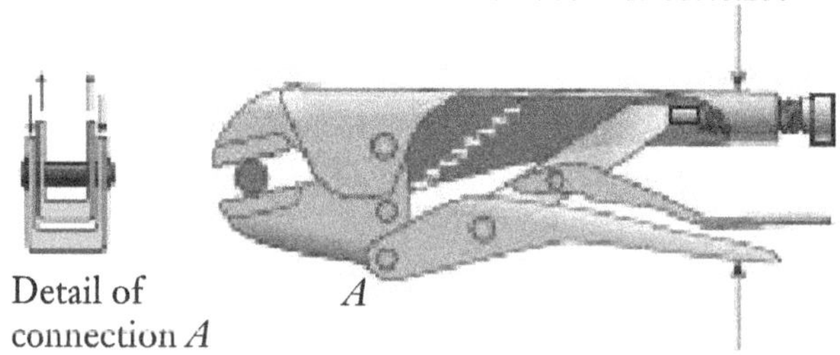

Figure P4.31.

Problem P4.32

Compound lever shears cut through a piece of wire at A (Figure P4.32).
(a) By using the free body diagram of handle CD, determine the magnitude of the force at rivet D.
(b) Referring to the magnified cross-sectional drawing of the connection at D, determine the shear stress in the rivet.
(c) If the rivet is formed from 4340 steel alloy, what is the factor of safety?

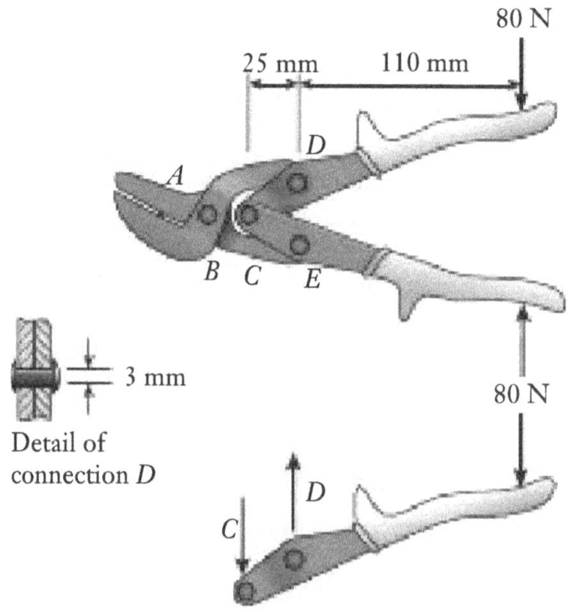

Figure P4.32.

Problem P4.33

Plates and rods are frequently used to help rehabilitate broken bones (Figure P4.33). Calculate the shear stress in the lower 5-mm-diameter biomedical bolt if it is supporting a 1300-N force from the bone.

Figure P4.33.

Problem P4.34

Using the bridge repair system in Problem P4.15, develop two additional two-dimensional configurations that utilize different suspension arrangements. The following design requirements must be met: Two-level access to the bridge must be provided with the top platform at 90 in. below the road and the lower level 185 in. below the road; each level must accommodate no more than two 200-lb workers; standard structural steel cables ($S_y = 250$ Mpa) of 0.50-in. diameter with varying lengths are to be used. Evaluate the three designs (your two and the one from Problem P4.15) with respect to their factor of safety in worst-case loading conditions and identify which configuration has the highest factor of safety.

References

Ashby, M. F., *Materials Selection in Mechanical Design.* Butterworth-Heinemann, 1999.

Askeland, D. R., and Phulé, P. P., *The Science and Engineering of Materials,* 4th ed. Thomson-Brooks/Cole, 2003.

Dujardin, E., Ebbesen, T. W., Krishan, A., Yianilos, P. N. and Treacy, M. M. J., "Young's Modulus of Single-Walled Nanotubes," *Physical Review B,* **58**(20), 1998, pp. 14013–14019.

Gere, J. M., and Timoshenko, S. P., *Mechanics of Materials,* 4th ed. PWS Publishing, 1997.

Hooke, R., *De Potentia Restitutiva.* London, 1678, p. 23.

Iijima, S. "Helical Microtubules of Graphitic Carbon," *Nature,* **354**, 1991, pp. 56–58.

Yu, M. F., Files, B. S., Arepalli, S., and Ruoff, R. S., "Tensile Loading of Ropes of Single Wall Carbon Nanotubes and Their Mechanical Properties," *Physical Review Letters,* **84**, 2000, pp. 5552–5555.

5

Thermal-Fluid Sciences

5.1 INTRODUCTION

A fluid is a substance that deforms continuously under an applied shear stress. Fluids include liquids, such as water, and gases, such as air. For a Newtonian fluid, proportionality exists between the shear stress and the rate of deformation. For a fluid, the deformation is equal to the velocity gradient (i.e., the rate at which the velocity changes in the fluid). Thus, for a Newtonian fluid, we may write

$$\tau = \mu \frac{\partial u}{\partial y}, \tag{5-1}$$

where u is the velocity of the fluid, y is a coordinate direction, τ is the shear stress, and μ is a constant called the *viscosity* (in $N \cdot s/m^2$). The viscosity is a property of the fluid, though it does vary with temperature. If the viscosity is zero, then the fluid is said to be *inviscid*. Inviscid fluids do not actually exist. However, there are many problems where the assumption of an inviscid fluid might be made to simplify the analysis and still yield meaningful results.

Sometimes, the viscosity is given in terms of the density. Let ρ be the density of the fluid. Then, the *kinematic viscosity* ν is defined as

$$\nu = \frac{\mu}{\rho}. \tag{5-2}$$

Fluid mechanics is concerned with Newtonian fluids, which are at rest or moving. Because fluids often interact with a structure (e.g., a river pushing against a dam), the determination of forces characterizing the interaction are often the focus of the analysis.

SECTIONS

OBJECTIVES

After reading this chapter, you should be able to a do the following:

- Understand the difference between a fluid in motion and a fluid at rest.
- Find the pressure in a hydrostatic fluid.
- Use Newton's law and calculate the force on a surface due to a fluid.
- Understand the concept of temperature and convert a temperature from one scale to another.
- Discuss the different modes of heat transfer.
- Understand that the design or analysis of a machine operating in a thermal cycle is governed by the first and second laws of thermodynamics.

5.2 FLUIDS AT REST

Many engineering applications involve fluids in a state of rest; that is, the fluid is not moving. A branch of fluid mechanics called hydrostatics addresses such problems. An example of a fluid at rest is coffee in a coffee cup. The coffee is a liquid that is not moving, yet it exerts a force on the walls of the cup. The force on each wall is related to the pressure in the fluid.

Anyone who has ever dived into a swimming pool clearly knows that the pressure is highest at the bottom of the pool. This is because the weight of each element of the water is pressing down. At the bottom of the pool, the entire volume of the water is bearing its weight on the diver.

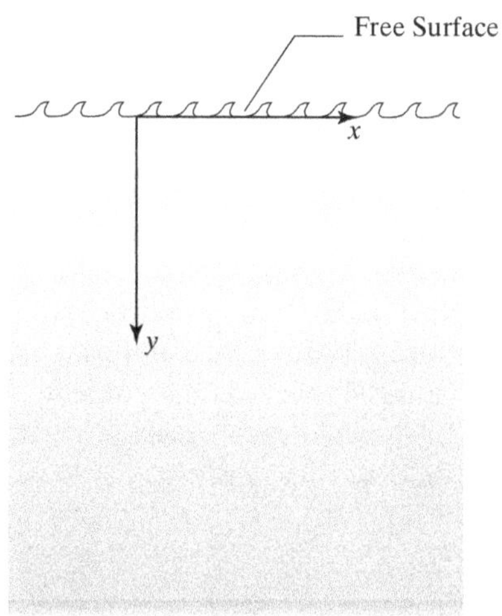

Figure 5.1.

To formulate an expression for this phenomenon, let us take a fluid and place a coordinate system at the top of the fluid with positive y acting in the direction of increasing depth, as shown in Figure 5.1. The pressure increases as y increases and is affected by the weight of the fluid per unit volume. The weight of the fluid is ρg, where ρ is the density of the fluid. Also, suppose that the pressure at the surface of the fluid, the free surface, has some value p_0. Thus, we have

$$p = p_0 + \rho g y. \tag{5-3}$$

The force acting on the bottom of the cup is equal to this pressure times the area of the bottom of the cup.

This equation only works for a surface that is uniformly submerged below a fluid. For surfaces inclined with respect to the free surface, a differential form of Equation 5-3 must be obtained.

Buoyancy, or Archimedes' principle, is a restating of Equation 5-3. This principle states that a submerged body is subject to an upward force F_B equal to the weight of the fluid displaced. The principle may be expressed as:

$$F_B = \rho g \times \text{volume}. \tag{5-4}$$

EXAMPLE 5.1: What is the pressure at the bottom of a three-meter-deep swimming pool?

SOLUTION

The pressure is given by Equation 5-3, where P_0, the pressure at the free surface, is equal to one atmosphere of pressure (101 kPa). The density of water is 1000 kg/m^3. Thus, Equation 5-3 yields

$$p = p_0 + \rho g y,$$

$$p = 101 \times 10^3\,pa + 1000\,\frac{\text{kg}}{\text{m}^3} \cdot 9.81 \cdot \frac{\text{m}}{s^2}\,3\,\text{m},$$

$$p = 101.81\ \text{kPa}.$$

PRACTICE!

A hot-air balloon rises off the ground because the density of the air in the ballon is less than the density of the atmosphere. Suppose that a spherical balloon has a diameter of 10 feet and that the density of the air is 0.071 lbm/ft^3. How much weight can the balloon lift and remain aloft? For this analysis, neglect the cooling of the air in the balloon.

PROFESSIONAL SUCCESS

Hydrostatic pressures variations are neglected in hydraulic systems, because they are quite small compared with the pressure of the fluid. For example, automobile hydraulic brakes can develop pressures as high as 10 MPa (1500 psi) and aircraft and machinery hydraulic systems have pressures up to 30 MPa (4500 psi).

5.3 FLUIDS IN MOTION

For a fluid at rest, the velocity is zero. However, for a fluid in motion, the velocity becomes an unknown quantity and may vary from point to point in the fluid. At points where the fluid interacts with another surface, the fluid will have the same velocity as the surface. This is called the no-slip condition. For example, the velocity at the bottom of the river is zero, because the water is in contact with the ground, which is not moving.

In solving problems involving fluids in motion, two approaches may be taken. The first of these is to track every particle of the fluid as it moves. However, because fluids deform continuously, this would require tracking the particle continuously. Another approach is to pick a region in the fluid and watch the fluid as it enters and leaves the region. This second approach is called the *control volume* approach and is more practical for solving engineering problems involving the flow of a fluid. In using the control volume approach to solve problems, conservation laws such as conservation of mass and the laws of Newton are applied.

EXAMPLE 5.2: Suppose we are designing a water-piping system for a building with two apartments and wish to supply water to two faucets, which are connected by a tee configuration, as shown in Figure 5.2. We want the velocity of the water to be same when it leaves both faucets. What should be the velocity of the water source?

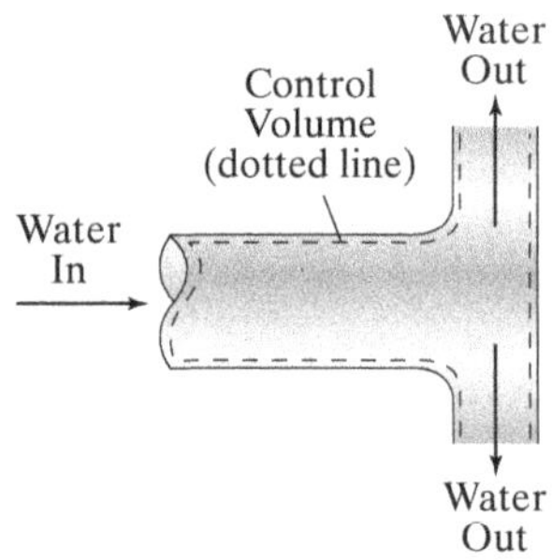

Figure 5.2.

SOLUTION

We pick a control volume, as shown in Figure 5.2, and apply mass conservation. In this problem, it is practical to assume that the mass of the fluid per unit time entering the control volume is equal to the amount leaving. This will be true if the source of the water remains constant and if fluid is not trapped within the control volume. Let us label the inlet as region 1, with a corresponding velocity v_1, and the two outlets as regions 2 and 3. The mass per unit time is called the mass flowrate and is equal to the density times velocity and the cross-sectional area of the pipe. Then, conservation of mass gives

$$\dot{m}_{\text{in}} = \dot{m}_{\text{out}},$$

$$\dot{m}_1 = \dot{m}_2 + \dot{m}_3,$$

$$\rho v_1 A_1 = \rho v_2 A_2 + \rho v_3 A_2,$$

where A_1, A_2, and A_3 are the cross-sectional areas of the pipe at Section 5.1, Section 5.2, and Section 5.3, respectively. If the density of the water is the same throughout, then the density term ρ will drop out of every term in the equation. Furthermore, if the diameter of the pipe is the same everywhere, then the cross-sectional areas are the same everywhere. Then, we have

$$v_1 = v_2 + v_3.$$

However, we want $v_1 = v_2$. Thus,

$$v_1 = 2v_2 = 2v_3.$$

Therefore, in order for the velocity to be the same at each outlet, v_1 must be twice the value at Section 5.1 or Section 5.2.

In Example 5.2, we assumed that the water was steady (i.e., that the water was not trapped in control volume). Had water accumulated in the control volume, the amount of water leaving would not be equal to the amount entering the control volume.

A fluid with constant density is called an incompressible fluid. Compressible fluid flow comes into consideration in fluids that are moving at high speed, such as the area around an aircraft.

If the speed of a fluid exceeds the speed of sound, the flow is called supersonic. Qualitatively, new phenomena, such as shock waves, occur at supersonic speeds. The phenomenon of shock waves are very complex and beyond the scope of this book.

Nevertheless, it is important to note that shock waves need to be taken into consideration when designing supersonic aircraft and projectiles.

Moving fluids pose some practical engineering problems that have only recently been understood. For example, aircraft flight only became possible in the last century because investigators did not fully understand the mechanics of fluid flow. It is the interaction of air and the wing of an aircraft that allows the aircraft to become airborne.

The section of an aircraft wing is called an airfoil. The upper and lower surfaces are curved. The upper surface has more curvature than the lower, as shown in Figure 5.3. Because the upper surface is more curved, the air must speed up in order to flow over the surface. For a moving fluid, an increase in velocity corresponds to a decrease in pressure. Thus, a smaller pressure exists on the top surface of the wing than on lower surface. The higher pressure essentially pushes the wing up.

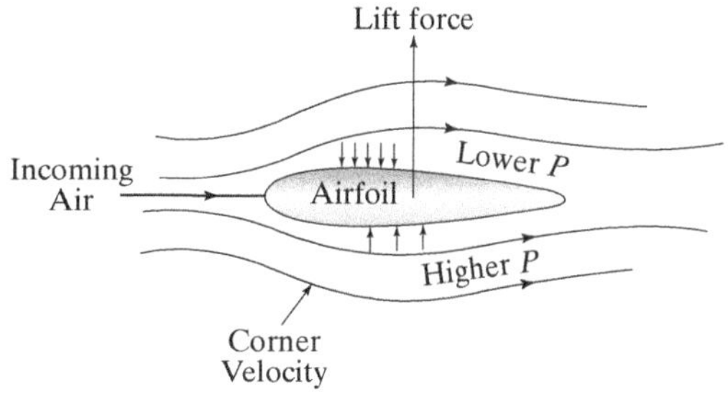

Figure 5.3.

Engineers called aerodynamicists can calculate this force, which results due to this imbalance in pressure. This force is called the lift.

It is hard for a fluid particle to flow at high speed on a very curved path. Thus, if the airfoil is too thick or if the angle that the airfoil makes with the horizontal is too large, the fluid will not be able to remain attached to the upper surface of the airfoil. As shown in Figure 5.4, if the fluid ceases to move across the top of the wing, then the wing is stalled, and the imbalance in the pressure no longer exists. Stall may lead to a catastrophic failure of the aircraft.

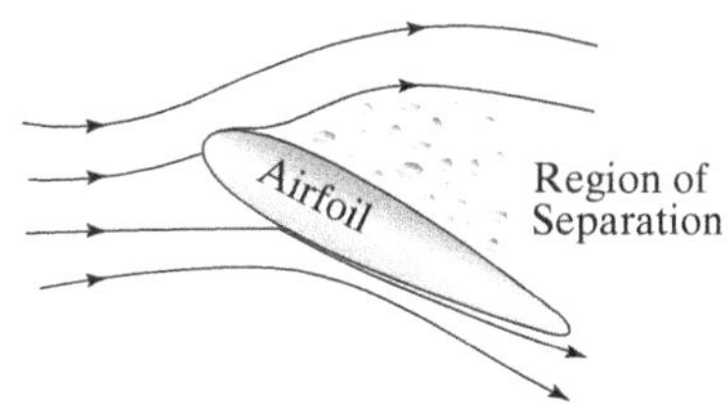

Figure 5.4.

As previously stated, if the momentum of a fluid is known, then forces may be obtained by using Newton's second law. In general, formulating an expression suitable for solving engineering problems requires the use of calculus. However, if we assume steady one-dimensional flow, then we may write that the force in the kth direction, F_k, as

$$F_k = \dot{m}(V_{2k} - V_{1k}), \tag{5-5}$$

where $\dot{m}$ is the mass flowrate and $(V_{2k} - V_{1k})$ is the difference in the velocity of the fluid from station 1 to station 2 in the kth direction.

EXAMPLE 5.3: Water with a velocity of 10 m/s strikes a turbine used for power generation and is rotated 60° from the horizontal by the blade, as shown in Figure 5.5. The cross section of the inlet water is 0.003 m². What is the force on a turbine blade shown in the figure?

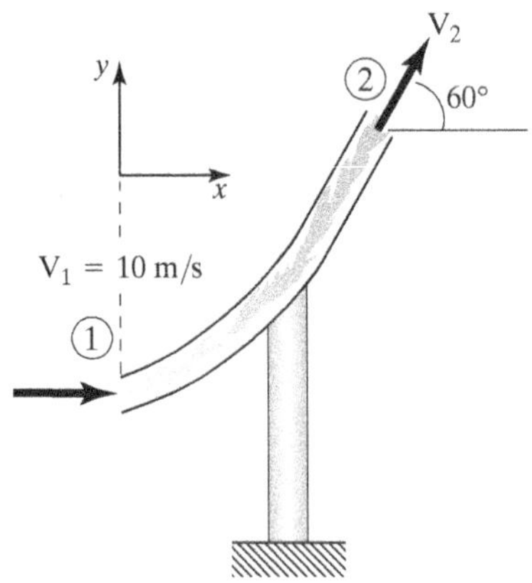

Figure 5.5.

SOLUTION

In solving this problem, we employ Equation 5-5 and the concept of the conservation of mass. We may assume incompressible flow, so that $\rho_1 = \rho_2$. For this problem, we also assume that $A_1 = A_2$. The fact that the water is initially horizontal means that $V_{1y} = 0$.

Applying the idea of conservation of mass indicates that

$$\dot{m}_1 = \dot{m}_2 = \rho_1 V_1 A_1 = \rho_2 V_2 A_2 = \dot{m}$$

$$\dot{m} = 1000 \frac{\text{kg}}{\text{m}^3} \cdot 10 \frac{\text{m}}{\text{s}^2} \cdot 0.003\,\text{m}^2$$

$$\dot{m} = 30 \frac{\text{kg}}{\text{s}^2}.$$

Notice that because $A_1 = A_2$, then $V_1 = V_2$.

Applying Equation 5-5 to the x direction yields

$$F_x = \dot{m}(V_{2x} - V_{1x}) = \dot{m}(V_2 \cos 60° - V_1) = \dot{m}V_1(\cos 60° - 1).$$

Substituting the appropriate numbers into this expression gives

$$F_x = 30 \frac{\text{kg}}{\text{s}} \cdot 10 \frac{\text{m}}{\text{s}} \cdot (\cos 60 - 1) = 150\ \text{N} \leftarrow .$$

The arrow indicates the direction of the x component of the force.

Likewise, in the y direction, we have

$$F_y = \dot{m}(V_{2y} - V_{1y}) = \dot{m}(V_2 \sin 60° - 0) = \dot{m}V_1 \sin 60°$$

$$F_y = 30\,\frac{\text{kg}}{\text{s}} \cdot 10\,\frac{\text{m}}{\text{s}} \sin 60° = 259.8 \text{ N}\uparrow.$$

PRACTICE!

What would be the angle of the turbine in Example 5.3 if the vertical force (F_y) was zero? What would be the value of the horizontal force (F_x)?

PROFESSIONAL SUCCESS

To understand how a fluid reacts around complicated machinery, fluid mechanists use experimental methods and visualization techniques. One method, which works well with air, involves injecting a stream of smoke into the air stream. A model of the object, such as an aircraft, is placed in a wind tunnel and the smoke injected. As the smoke flows past the aircraft, it creates a series of lines called streaklines. Each streakline represents all the fluid particles passing through that point in the fluid.

5.4 SUMMARY

In engineering problems involving fluid mechanics, the fluid may be moving or may be at rest. If the fluid is at rest, then the velocity of the fluid is everywhere zero. Solution of such problems requires obtaining the value of the pressure in the fluid. The pressure will linearly increase from the free surface.

On the other hand, if a fluid is moving, the velocity of the fluid becomes a variable and may vary from point to point in the fluid. One approach to solving engineering problems with moving fluids is to pick a region of the fluid called a control volume and apply conservation laws such as conservation of mass and momentum to the volume. Newton's second law may be used to obtain forces, because this law relates force to the change in momentum.

Control volumes must be properly chosen. In problems involving the interaction of a fluid with a structure, the control volume is chosen so as to include the structure or the reaction of the structure to the fluid.

KEY TERMS

control volume
density
hydrostatic pressure
viscosity

5.5 EXERCISES

For problems E1–E4, use a handbook to find the viscosity for each fluid at the given temperature.

E1. SAE 30 oil at 20° C.

E2. SAE oil at 100° C.

E3. Kerosine at 20° C.

E4. Kerosine at 100° C.

E5. What is the pressure and force acting on the rectangular underwater gate shown in Figure 5.6? The gate is 1-m wide by 1-m deep.

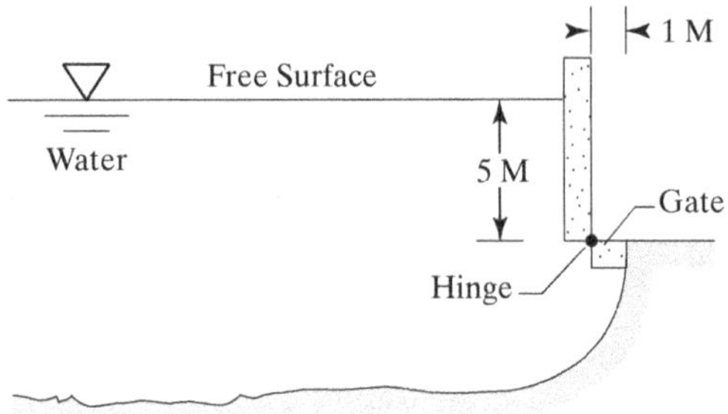

Figure 5.6.

E6. A tank, as illustrated in Figure 5.7, contains two layers of water, each of which is at a different temperature and hence a different density. The densities of each fluid are such that $\rho_1 = 1000$ kg/m^3 and $\rho_2 = 1500$ kg/m^3. What is the hydrostatic pressure at the bottom of the tank?

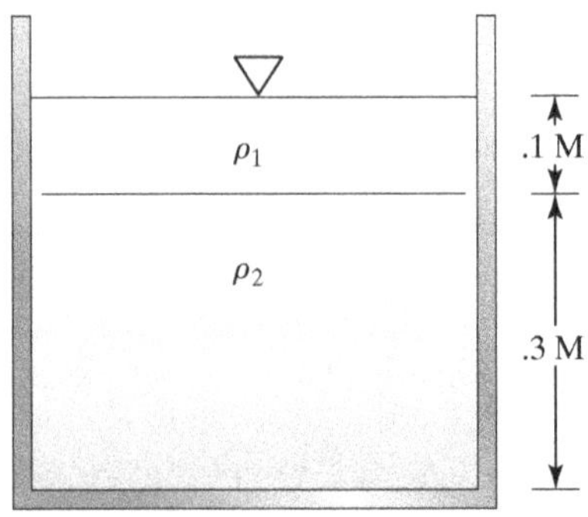

Figure 5.7.

E7. In Figure 5.8, water enters a converging two-dimensional channel with a velocity of 5 m/s. If the inlet is 0.5-m wide, the exit is 0.2-m, and the channel is 0.5-m wide perpendicular to the plane of the paper, what is the velocity of the fluid at the exit?

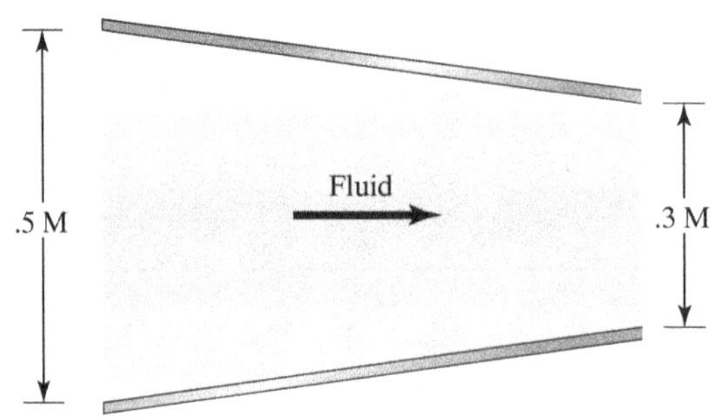

Figure 5.8.

E8. Air flows through a ventilation duct. The section is a rectangular box shown in Figure 5.9. If the fluid enters at stations 1 and 2 and leaves at station 3, what is the velocity of the fluid at station 3? The areas at the stations are $A_1 = 0.1\ \text{m}^2$, $A_2 = 0.3\ \text{m}^2$, and $A_3 = 0.15\ \text{m}^2$. The velocity at stations 1 and 2 are known to be $V_1 = 1$ m/s and $V_2 = 5$ m/s, respectively.

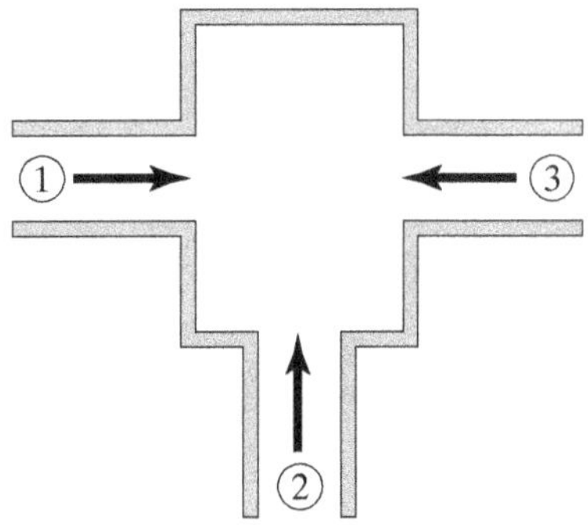

Figure 5.9.

E9. In Figure 5.10, water strikes a hemispherical cup and is rotated 45°. If the inlet velocity of the water is 5 m/s, what is the force on the cup? The jet of water has an initial diameter of 100 mm and 50 mm at each outlet.

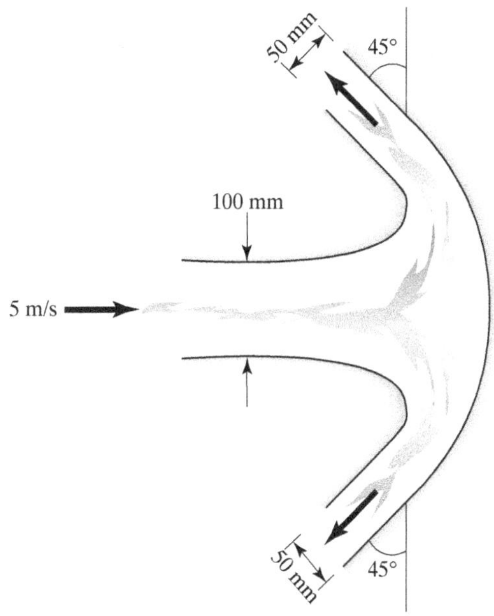

Figure 5.10.

E10. What would be the force on the cup if the cup from Exercise E9 had a 50-mm central hole, as shown in Figure 5.11?

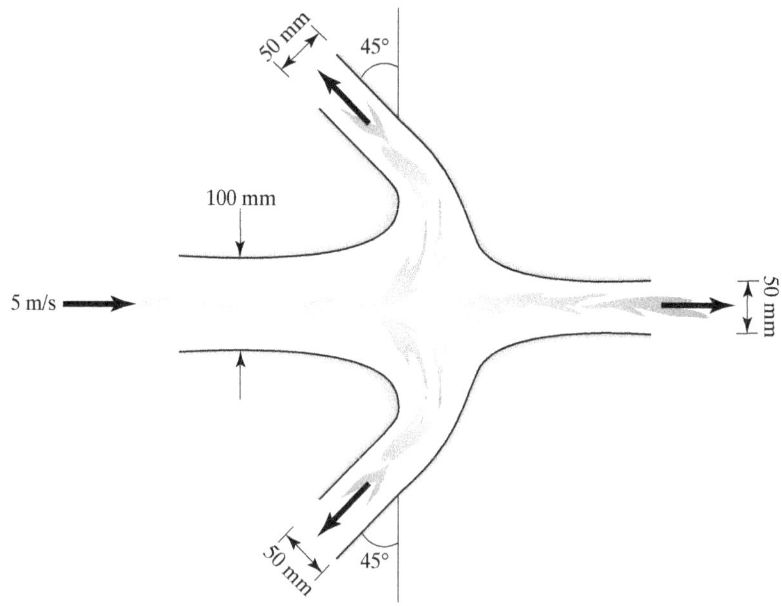

Figure 5.11.

5.6 INTRODUCTION

Thermal science is an area of scientific thought encompassing thermodynamics and heat transfer. Thermodynamics is a science that deals with heat and work and the properties of substances related to heat and work. Heat transfer is a science that deals with the transfer of thermal energy.

Mechanical engineers are often involved in the design and analysis of machinery wherein the working fluid or substance undergoes a thermal cycle. In undergoing a thermal cycle, the substance passes through a succession of thermodynamic states. Some systems produce work (as in power systems), while others require work input to produce other effects, as in refrigeration.

Examples of such systems include internal combustion engines, refrigerators, heating and air-conditioning systems, and fossil-fuel power plants. The basic objective of these power systems is to take a fuel source and transform the chemical energy contained within that fuel source into work. Often, the net work output is used to propel a vehicle or to produce electric power.

Engineers define the efficiency η of a power system as the net work output divided by the net heat input (i.e., the useful effect divided by the required input), or

$$\eta = \frac{\text{net work out}}{\text{net heat in}}. \tag{5-6}$$

The efficiency is often given as a percent. This is obtained by multiplying the ratio from Equation 5-6 by 100.

A power system's efficiency increases as the net work output increases or as the net heat input decreases. Thus, a highly efficient system will produce very much work output, while requiring little heat input.

Air-conditioning systems remove heat from the air in an enclosed environment such as a house. To do this, a refrigerant must be compressed, which requires work input in the form of electrical energy.

In a way similar to that of efficiency, a coefficient of performance (COP) is defined for a refrigeration system as the heat removed by the working fluid from the refrigerated space divided by the required work input:

$$\text{COP} = \frac{\text{heat removed}}{\text{net work in}}.$$

In this case, the useful effect is the heat removed.

In the past 20 years, the cost of fossil fuels has risen considerably because petroleum and natural gas prices have risen. Inefficient systems use more fuel than efficient systems to produce the same effect; therefore, an objective of a mechanical engineer working in designing machinery is to increase the efficiency. To accomplish this goal, the engineer must have a clear understanding of the science involved in the system, as well as the mechanical limitations of the machinery.

5.7 THE CONCEPTS OF TEMPERATURE AND HEAT TRANSFER

What is temperature? We may say it is 10 below zero outside, but what does that mean?

Temperature is a measurement of the energy of molecules due to their rapid movement. Temperature is measured indirectly using a transducer. A typical transducer, the thermometer, measures the temperature by the expansion and contraction of a liquid (usually mercury) in a glass tube. In the case of a mercury thermometer, if the temperature rises, the mercury expands and hence rises in the tube. If the temperature drops, the opposite occurs.

Various scales have been defined to quantify this phenomenon. These scales are somewhat arbitrary. For example, the Fahrenheit scale, named after Daniel Fahrenheit (1686–1736), places the freezing point of water at 32 degrees and the boiling point of water at 212 degrees. The Celsius scale, named after Anders Celsius (1701–1744), places the freezing point at zero degrees and the boiling point at 100 degrees. These values may be used to obtain the following relationship between the Celsius scale and the Fahrenheit scale:

$$°\text{F} = 1.8(°\text{C}) + 32°; \tag{5-7a}$$

$$°\text{C} = \frac{5}{9}(°\text{F} - 32°). \tag{5-7b}$$

The Kelvin scale, named after Physicist Lord Kelvin, is an absolute scale, with zero Kelvin defined as when the motion of all molecules ceases. This point occurs at $-273\ °\text{C}$; thus, there is a difference of 273 degrees between the Celsius and Kelvin scales. In particular,

$$\text{K} = °\text{C} + 273. \tag{5-8}$$

EXAMPLE 5.4: What is the temperature in °C and Kelvin for 70 °F?

SOLUTION

Equation 5-7 and Equation 5-8 give the conversion to each temperature scale. In Celsius, using Equation 5-7, we see that the temperature would be

$$70°\text{F} = 1.8(°\text{C}) + 32°,$$

$$°\text{C} = \frac{70 - 32}{1.8} = 21\ °\text{C}.$$

In Kelvin, using Equation 5-8, we see that the temperature would have the value

$$\text{K} = 21°\text{C} + 273 = 294\text{K}.$$

PROFESSIONAL SUCCESS

Notice that when dealing with temperature differences, there is no need to convert from degrees Celsius to Kelvin. This is because a change in one degree in the Celsius scale is equivalent to a change of one unit in the Kelvin scale.

Heat transfer is a thermodynamic process. Heat transfer will occur whenever there is a temperature difference. Heat transfer is quantified in terms of the heat Q (in Joules, J), heat rate $\dot{Q}$ (in Watts, W), or heat per unit area $q''(\text{W/m}^2)$, called the heat flux.

There are three different modes of heat transfer, as shown in Figure 5.12. These modes are conduction, convection, and radiation.

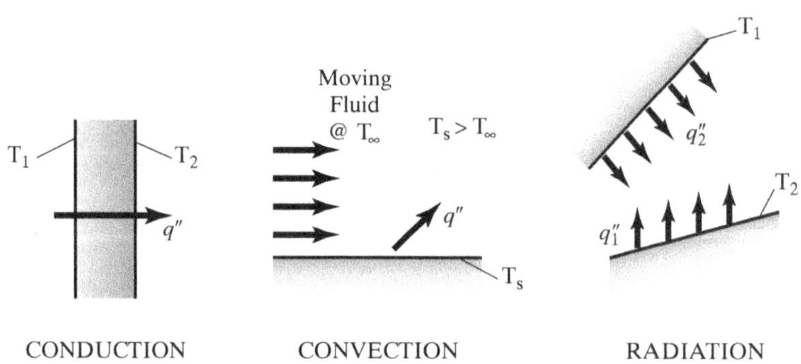

Figure 5.12.

In conduction, the heat is transferred between warmer and colder regions of the same body. In this case, the heat flux is from the high temperature to the low temperature and is caused by the temperature gradient within the body. A law by Fourier relates the heat flux to the temperature gradient as

$$q'' = -k\nabla T,$$

where T is the temperature, k is a proportionality constant called the thermal conductivity, and ∇ is the del operator, or gradient. The term ∇T describes how the temperature varies as a function of position.

The thermal conductivity has units of W/(M·K) in SI and is a property of a material. Insulating materials will have a small thermal conductivity.

If the thermal conductivity is constant throughout the body and the temperature varies with respect to one dimension only, say, x, then Fourier's law can be reduced to

$$q'' = -k\frac{\Delta T}{\Delta x}, \tag{5-9}$$

where ΔT is the change in the temperature and Δx is the change in the x dimension.

In convection, the heat is transferred from a surface to a moving fluid when a temperature difference exists between the surface and the fluid. Heat will be transferred more readily if the fluid is moving rapidly across the surface. An example of this is the wind chill effect. On a cold day, an individual's heat is transferred to the air because a temperature difference exists between the individual and the air. If the wind blows, much more heat is transferred, by convection. The harder the wind blows, the greater the heat transfer. Forced convection is used to describe convection wherein the fluid is propelled by an external means such as a fan, wind or pump.

The heat flux transferred by convection is given by:

$$q'' = k\Delta T \tag{5-10}$$

where h is called the *heat transfer coefficient*. Equation 5-10 is called Newton's law of heat convection.

The heat transfer coefficient has the same units of W/m^2 in the SI system. The larger the value of the heat transfer coefficient, the larger the amount of heat transfer for a given temperature difference. Typically, the heat transfer coefficient for free convection is in the range $2 \leq h \leq 25$ W/(m^2 · K), whereas for a gas under forced convection, the heat transfer coefficient will have a value in the range $25 \leq h \leq 250$ W/(m^2 · K). The heat transfer coefficient can be on the order of 10^3 or higher for forced convection involving liquids.

Heat may be transferred between two surfaces without an intervening medium if the surfaces are at different temperatures. This form of heat transfer is called radiation and it propagates in the form of electromagnetic waves. Energy is emitted in radiation by all matter that has a finite temperature because of the changes in the atoms composing the matter. The Stefan-Boltzmann law, Equation 5-11, gives the maximum amount of energy that may be transmitted.

$$q'' = \sigma T_s^4, \tag{5-11}$$

where $\sigma = 5.67 \times 10^{-8}\text{W}/(\text{m}^2 \cdot \text{K}^4)$ is a constant called the Stefan–Boltzmann constant and T_s is the temperature of the radiating surface. A surface that radiates energy according to the Stefan–Boltzmann law is called an *ideal radiator*, or *blackbody*. Real surfaces emit radiation at a lower value. An expression for a real radiator may be obtained by multiplying the right-hand side of Equation 5-11 by a factor ε called the emissivity, which has a value between zero and one. Also, suppose that the radiation is occurring to the surroundings, which has a temperature T_{sur}. Then, Equation 5-11 becomes

$$q'' = \varepsilon\sigma(T_s^4 - T_{sur}^4). \tag{5-12}$$

In many practical applications, heat transfer occurs in one or more of these modes. Often, one or more mode may be predominant, which allows one to neglect the remaining mode(s). This greatly simplifies the model of the heat transfer.

EXAMPLE 5.5:

On a cold day, the temperature outside of a house reaches − 30 °F, whereas the temperature inside the house is 70 °F. What is the heat transferred through a wall of the house with thickness of 4.5 inches if the wall is made of concrete with a thermal conductivity of 1 W/(m·K)? Assume that heat transfer by radiation and convection is negligible.

SOLUTION

The heat transfer is given by Equation 5-9. First, the temperature values are converted to Celsius. Using the methods shown in Example 5.4, we find that − 30 °F = − 34.4 °C and 70 °F = 21 °C. Also, 4.5 inches = 0.1143 m. Then, by substitution into Equation 5-9, we get

$$q'' = k\Delta T = 1\frac{\text{W}}{\text{m}\cdot{}^\circ\text{C}}\frac{(-34.4 - 21)^\circ\text{C}}{0.1143\text{m}}(-34.4 - 21)^\circ\text{C} = 484.7\frac{\text{W}}{\text{m}^2}.$$

PRACTICE!

A wall in a house has a window with the same thickness as the wall. The heat loss through this window is being considered. Suppose that the wall has a thermal conductivity of 0.135 W/(m·K) and that the glass window has a thermal conductivity of 0.60 W/(m·K). Compare the heat flux lost through the window with that through the wall, and suggest a way to reduce the heat loss through the window.

PROFESSIONAL SUCCESS

When solving problems in the thermal sciences, use the following format:

1. Draw a schematic of the system under consideration. Show the boundaries of this system. Identify relevant heat and work transfers.
2. List all assumptions that are going to made in solving the problem.
3. List the property values used in the analysis. Include tables or diagrams where the values were obtained.
4. Apply the laws of thermodynamics in a manner consistent with your assumptions.
5. Consider your results, and draw the appropriate conclusions. Review your assumptions.

5.8 THE FIRST LAW OF THERMODYNAMICS: ENERGY CONSERVATION

A system is a quantity of matter of fixed mass and identity. This system will have a thermodynamic state defined by properties, including temperature, pressure, volume, and mass. This system will undergo a change in thermodynamic state if any of these properties are changed. Energy may be exchanged with the surroundings as work or heat.

This principle of energy conservation, which is the first law of thermodynamics, includes the energy transferred to work and heat. The work may be done by or on the system, while heat may be added to or removed from the system.

In solving engineering problems, a mathematical expression for the first law is needed. This may be obtained by letting E be the change in energy in going from thermodynamic state 1 to thermodynamic state 2, Q the heat transfer in, and W the work done. Then, the first law may be written as

$$\Delta E = Q - W. \tag{5-13}$$

In the special case where two bodies of different temperature are brought in contact, there may be no work done, but heat is transferred. Then, energy conservation requires that the change in energy be equal to the heat transfer, or that

$$\Delta E = Q.$$

The energy of the system E is composed of three separate parts: internal energy U, kinetic energy KE, and potential energy PE. That is,

$$E = U + \text{KE} + \text{PE}. \tag{5-14}$$

Internal energy is defined by the thermodynamic state of the system. A system will have kinetic energy if it is moving. In many applications, the kinetic energy will be

$$\text{KE} = \frac{1}{2}mv^2, \tag{5-15}$$

where m and v are the mass and velocity of the system, respectively.

If the potential energy is due to a reference to a gravitational potential, then the potential energy may be written as

$$\text{PE} = mgz, \tag{5-16}$$

where z is the height above the gravitational reference and g is the gravitational acceleration.

Employing Equation 5-14, Equation 5-15, and Equation 5-16, we find that the first law becomes

$$\Delta U + \frac{m\Delta(V^2)}{2} + mg\Delta z = Q - W, \tag{5-17}$$

where ΔU, $\Delta(V^2/2)$, and Δz are the change in the internal energy, kinetic energy, and height of the system, respectively, from state 1 to state 2.

The work done in compressing or expanding a substance is called pv work, because it is characterized by a change in pressure and volume. The work done by a system during a process between an initial volume v_1 and final volume v_2 is

$$W = \int_{v_1}^{v_2} p\,dv, \tag{5-18}$$

where p_2 and p_1 are the final and initial pressures, respectively, and ν_1 and ν_2 are the volumes at the initial and final states. Equation 5-18 indicates that the variation of the pressure as a function of the volume v is needed in order to evaluate the work.

EXAMPLE 5.6: Air in a 0.01-m^3 cylinder is heated as shown in Figure 5.13. The cylinder has a frictionless piston at the top of the cylinder on which several small weights are placed. During the process the volume of the gas doubles. If the initial pressure is 100 kPa, what is the work done in expanding the air?

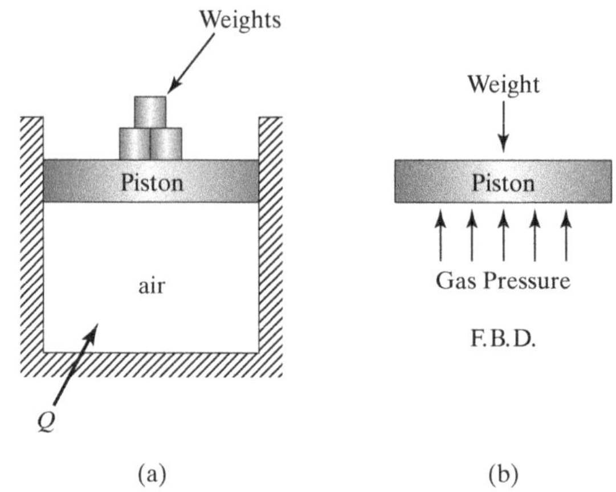

Figure 5.13.

SOLUTION

We recognize that the air undergoes pv work. This is a quasiequilibrium process, wherein deviation from thermodynamic equilibrium is small. Thus, the gas may be considered to be in equilibrium.

As shown in Figure 5.13b, a free-body diagram indicates that the force due to the mass of the weights and the pressure are in equilibrium. Therefore, the pressure of the gas is constant while it expands.

The value of the work is given by Equation 5-18. This work is done on the piston by the gas. By integration of Equation 5-18, we get

$$W = \int_{v_1}^{v_2} p dv = p(v_2 - v_1)$$

$$= 100 \text{ kPa}(0.005 \text{ m}^3 - 0.01 \text{ m}^3) = -0.500 \text{ kJ}.$$

EXAMPLE 5.7: Water flowing over a paddle wheel has been used from ancient times as a source of power. Suppose that water from a river is moving at 3 m/s before falling over a paddle wheel. After striking the paddle, the river is moving at 1 m/s. If the difference in height of the river is 5 m, what is the amount of work per unit mass of the water done by the river in striking the paddle wheel?

SOLUTION

We can assume that the water has the same thermodynamic state before and after striking the paddle wheel. Also, this is an adiabatic process; that is, there is no heat transfer. Then, the only nonzero terms in the first law of thermodynamics [Equation 5-17] are the terms relating to the height and kinetic energy differential. Thus, the work done per unit mass is

$$\frac{W}{m} = \frac{\Delta(V^2)}{2} + g\Delta z = \frac{1}{2}(1^2 - 3^2)\frac{\text{m}^2}{\text{s}^2} + 9.81\frac{\text{m}}{\text{s}^2}(5\text{m}) = 45.1\frac{\text{J}}{\text{kg}}.$$

PRACTICE!

According to the first law of thermodynamics, it is possible for the efficiency of a power system, as defined in the introduction, to have a value of 100 %. Describe how this may be possible.

5.9 THE SECOND LAW OF THERMODYNAMICS

To understand the second law of thermodynamics, let us consider a simple example. Suppose that you have a hot cup of coffee. Because the coffee is warmer than its surroundings, heat is transferred from the coffee to the surroundings according to the first law of thermodynamics. Given enough time, the cup of coffee will reach thermodynamic equilibrium and will not revert to its initial temperature by itself.

The second law of thermodynamics defines the direction of a process. In the case of the cup of coffee, the direction of the process was such that the coffee cooled by heat transfer to the surroundings. Certainly, after the coffee has cooled, it is possible to warm the coffee.

In the case of the cooling of the coffee, heat was transferred from a high temperature to a cooler temperature. If we wanted to transfer heat from a cooler temperature to a warmer temperature, then work input is required. This is how a refrigerator works.

In addition to the direction of a process, the second law of thermodynamics specifies the limits for the efficiency of thermal devices. A device that has a higher efficiency violates the second law and is called a perpetual-motion machine.

PRACTICE!

While it is possible for the efficiency of a power system to be 100% according to the first law of thermodynamics, in practice, the efficiency will be less than this value. Describe how the second law of thermodynamics places an upper bound on the efficiency.

PROFESSIONAL SUCCESS

Mechanical engineers designing a new machine or analyzing existing ones always check for violation of the first and second laws of thermodynamics.

5.10 SUMMARY

Mechanical engineers involved in the design and analysis of machinery must be aware of the change in the thermodynamic state during the cycle of the machinery. Many practical systems take a fuel source and convert the chemical energy of the fuel source to work, heat, or both. In doing so, the total energy must be conserved, as stated by the first law of thermodynamics. Real processes always obey the first law of thermodynamics.

Natural heat transfer is from a body at a high temperature to a body at a low temperature. The potential of heat to go naturally from high to low temperatures can be utilized to produce work through a heat engine. However, not all the available heat can be converted to work, since some heat is always lost to the surroundings.

KEY TERMS

Celsius	Fahrenheit	temperature
energy	heat	laws of thermodynamics

5.11 EXERCISES

E1. What is the temperature of air in Fahrenheit if it has a temperature of 13 °C?

E2. If water is at 70 °F, what is its corresponding temperature in Celsius?

E3. What is the temperature in Kelvin of water if its temperature in Fahrenheit is 50 °F?

E4. What is the temperature in Fahrenheit for air if the air has a temperature of 515 K?

E5. Water in a tank is stirred. If 6000 kJ of work are done to the fluid and 1000 kJ of heat are dissipated, what is the change in the internal energy of the fluid?

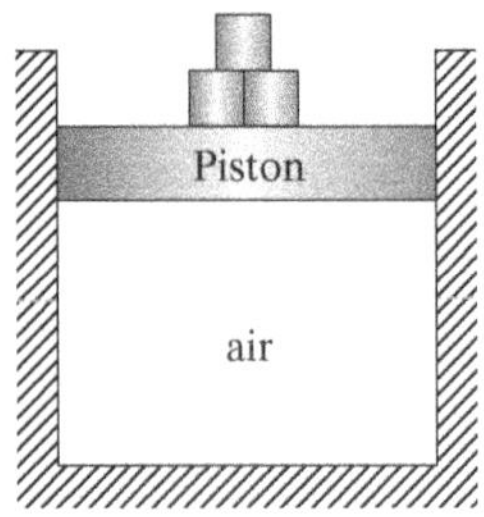

Figure 5.14.

E6. Air is contained in a cylinder as shown in Figure 5.14. The piston at the top of the cylinder contains small weights. Initially, the gas is 150 kPa and has a volume of 0.02 m^3. Suppose that the weights are slowly removed so that the gas expands according to $pv^{1.2}$ = constant until the volume is 0.1 m^3. What is the work done in raising the weights?

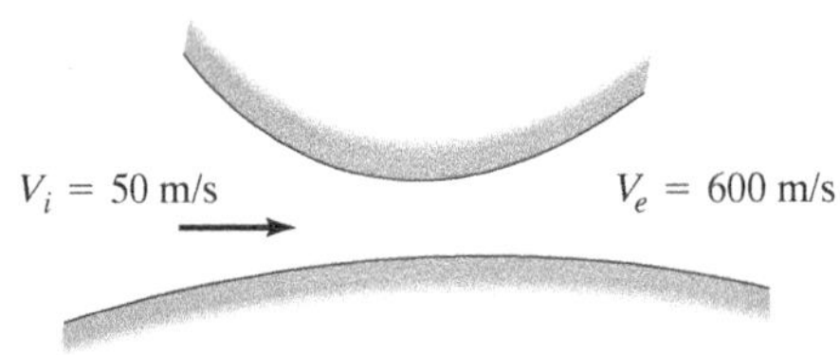

Figure 5.15.

E7. Air enters an insulated nozzle as shown in Figure 5.15 with a velocity of 50 m/s and leaves with a velocity of 600 m/s. What is the change in internal energy of the substance?

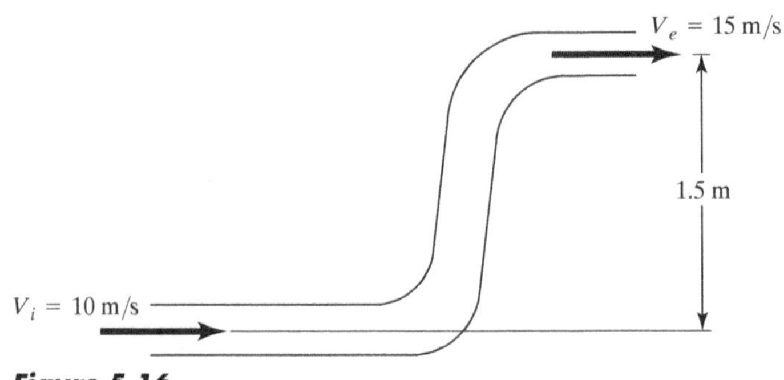

Figure 5.16.

E8. 5000 kJ of work are done to elevate water 1.5 m as shown in Figure 5.16. The process is adiabatic. What is the change in internal energy of the fluid?

E9. The temperature of the air in a hot-air balloon is much warmer than the surrounding atmosphere. Describe the cooling of the air in the balloon as dictated by the first and second laws of thermodynamics. (See Figure 5.17.) What must be done in order to keep the balloon aloft?

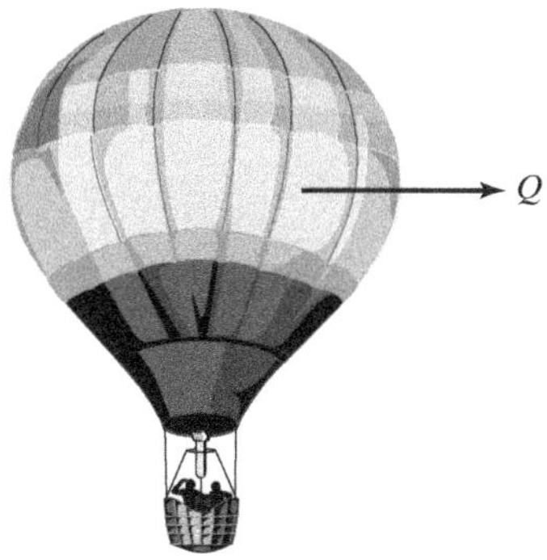

Figure 5.17.

E10. An automobile engine works by combustion of gasoline. This produces an explosion that drives a piston and in turn a camshaft, as in Figure 5.18. The turning of the shaft returns the piston to its initial position, and the process is repeated. Qualitatively describe how the first and second laws of thermodynamics dictate the heat transfer, the work, and the efficiency of this cycle.

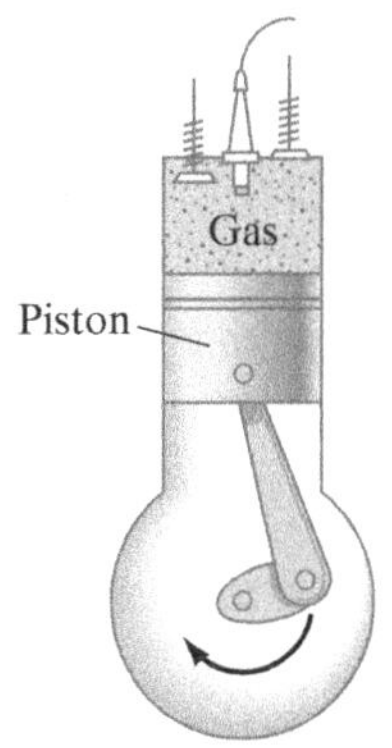

Figure 5.18.

CHAPTER 6

Project Management

Objectives

In this chapter, you will learn about:

- The importance of project management in ensuring design success
- Teamwork as an essential element of engineering design
- How engineers determine the tasks that will lead to a finished product
- Scheduling and managing design tasks
- The role of the project manager
- How members of a design team interact
- Documentation and its vital role in the design process
- Legal issues and their relevance to the design engineer

This chapter is devoted to the subject of project management. As a student of engineering, you may ask, "Why is this chapter necessary? Why is the topic important?" The answer lies in the nature of design problems. Design problems are inherently open ended; the more complex the task, the greater the need for a management infrastructure to ensure that elements of the project ultimately fit together. Accomplishing this goal requires that all teams and team members work together within a common strategy. Sometimes, managing a project requires only the most basic application of common sense: get things done on time, hold regular meetings to discuss progress, write things down so you don't forget them. At other times, a more formal structure is required; this formal structure often relies on time-tested methods of project management. A good engineer must know how to work within the framework of a team, keep a project on track, maintain good documentation, address legal issues, and work within a well-defined management plan. This chapter introduces several project management skills that are essential elements of engineering design.

6.1 WORKING IN TEAMS

The spirit of rugged individualism is pervasive in society, persisting as a theme in books, movies, and television. The image of a lone hero striving for truth and justice against insurmountable odds appeals to our sense of adventure and daring. Our pioneering spirit dreams of becoming the sole entrepreneur who endures economic hardship to change the face of technology, or perhaps the head of a startup company that single-handedly bests Microsoft or Intel. Yet, in real life, engineers seldom work alone. Most engineering problems are interdisciplinary, and true progress requires cooperative teamwork and the contributions of many individuals. This concept is easy to understand in the context of designing large structures, such as bridges, airports, skyscrapers, or oil refineries. It would be unimaginable to even approach such large tasks without a well-defined hierarchy of managers, job directors, engineers, and craftspeople. Similarly, the great engineering accomplishments in space exploration such as the Apollo moon landing, the International Space Station, the Hubble Space Telescope, and the Mars Exploration Rover required hundreds (in some cases thousands) of engineers working with teams of physicists, chemists, astronomers, material scientists, medical specialists, mathematicians, and project managers. Teamwork is also critical to the design of small but complicated devices such as medical prostheses, copy machines, children's mechanical toys, kitchen appliances, and cellular telephones. These devices (not an exhaustive list) cannot be designed by one person alone.

Teamwork is an important skill, and you must master it if you are to be a good engineer. Working in a team, such as the one depicted in Figure 6.1, requires that you speak clearly, write effectively, and have the ability to assimilate another person's point of view. All members of a team must understand how their tasks relate to the responsibilities of the team as a whole.

When addressing a design problem, dealing with the technical issues is only half the story. Determining how people and their time will be managed is just as important. By properly assigning tasks to the right people and ensuring that they work together cooperatively, the success of the project can be ensured. In the sections that follow, we address some of the major issues related to people management.

6.1.1 Building an Effective Team

An effective team is one that works well together. It functions at its maximum potential and thrives on a symbiosis of the special capabilities of its individual members. One key characteristic of an effective team is a good supportive attitude

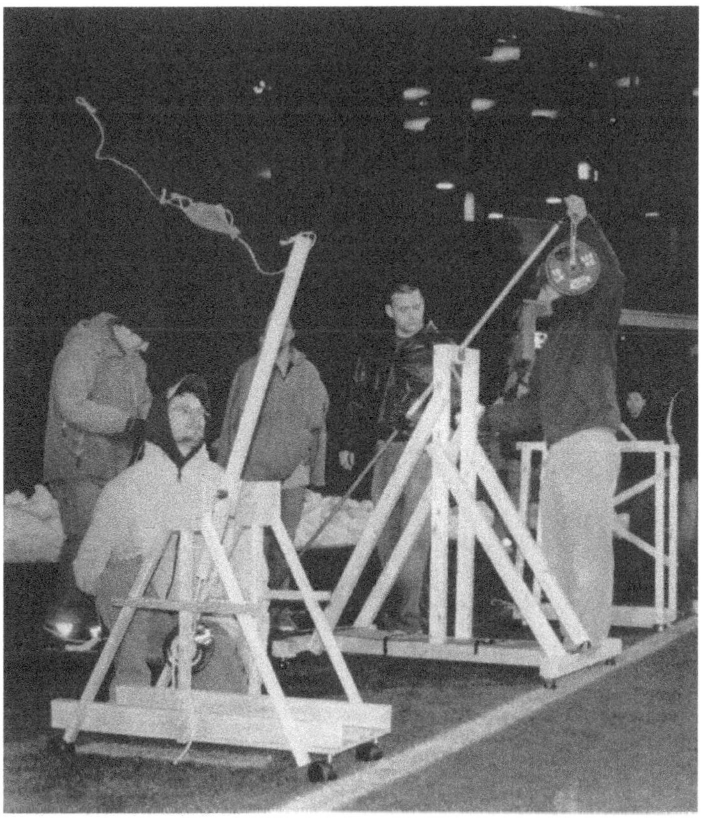

Figure 6.1
Design team works on the prototype of a Trebuchet catapult. (Photo courtesy of the Daily Free Press.)

among fellow teammates. Team morale and a sense of professionalism can be enhanced if team members agree on some basic rules of behavior. The following set of guidelines illustrates one possible approach to building an effective design team.

1. *Agree Upon Goals.* The members of your team should agree upon the goals of the project. This consensus is not as easily achieved as you might think. One teammate may want to solve the problem using a traditional, time-tested approach, while another may want to attempt a far-out, esoteric path to success. Define a realistic set of goals at the outset. If the approach you choose brings surprises, you can always redefine your goals midway through the project.
2. *Define Clear Roles.* Each team member should be assigned a specific function within the team. The responsibilities of each individual should be defined *before* project work begins. Individual roles need not be mutually exclusive, but they should be defined so that all aspects of the design problem fall within the jurisdiction of at least one person. In that way, no task will "fall between the cracks" during the design effort.
3. *Define Procedures.* You and your teammates should agree on a set of procedures for getting things done. Everything should follow a predetermined procedure, including documentation; the ordering of parts; prototype construction; and communicating with professors, clients, and customers. In that way, misunderstandings about conduct can be greatly reduced.
4. *Develop Good Interpersonal Relationships.* You must learn to work with everyone on your team, even with those individuals whom you may personally dislike. In the real world, a client will not care about any conflicts that occur behind the scenes. Engineering professionalism demands that you rise above personality clashes and concentrate on the job at hand. Be nice. Be professional. Forbid name calling, accusations, or assigning fault to team members.
5. *Define Leadership Roles.* Some teams work best when a single person emerges as the chosen leader. Others work better by consensus using distributed leadership or even no leadership at all. Regardless of your team's style, make sure that leadership hierarchy is clearly defined and agreed upon at the start of the project.

Professional Success: You're the Team Leader, One Teammate has Disappeared, and You're Doing All the Work

It's impossible to get along with all people all the time. When you work closely with other individuals, personal conflicts are inevitable. At times, these disagreements occur because one team member has failed to meet assigned responsibilities. At other times, the conflict arises from fundamental differences in personal outlook or priorities. However complicated your team relationships may become, remember that your customer does not care about them. Your customer is interested in receiving a well-designed product that reflects your best engineering abilities. It's up to you to resolve team conflicts internally. This resolution may mean that some team members will do more work than others, even if they will not be rewarded for their extra efforts. A good leader understands this trade-off and devises a plan to work around an errant teammate. Such situations may seem frustrating and unfair, but they happen in the real world all the time. Learning how to deal with them as a student is part of your engineering education.

Practice!

Define the roles, goals, and procedures that might apply to the following.

1. The design and construction of an eight-lane highway system under a major metropolitan city (*www.masspike.com/bigdig*).
2. A team of software engineers developing an auction website (*www.ebay.com*).
3. A team of electrical and mechanical engineers developing a radio-controlled robot (*www.usfirst.org*).
4. A team of biomedical engineers developing an artificial heart for mechanical implantation inside human subjects.
5. A software system for scheduling production in a "just in time" textile factory that fills customer orders as they arrive (*www.crohmiq.com*).
6. The design and construction of the Mars Exploration Rover (*origin.mars5.jpl.nasa.gov*).
7. An oil pipeline from Prudhoe Bay to Valdez, Alaska (*www.alyeska-pipe.com*).
8. A package tracking system for an air/ground shipping company (*www.fedex.com*).
9. A system for compiling, printing, and distributing a daily national newspaper (*www.usatoday.com*).
10. The design of a mineral extraction plant that produces copper, uranium, gold, and silver (*www.wmc.com*).
11. An oil refinery that must produce a variety of products, including gasoline, diesel, and home heating oil.
12. A machine to selectively apply pesticides to an agricultural area only where parasites have been detected by distributed sensors.
13. A system for monitoring the energy flow and power production in a multiplant electric power generation system.
14. A system for allowing an expert surgeon to conduct an operation via remote control from a distant location.
15. The design and construction of a portable air conditioning system for roadside tent camping.
16. The design of a command center for communicating with football players wearing individual earpieces.
17. The design of a food court in a shopping mall.
18. The design of a transportation system consisting of autonomous (driver-free) subway trains.
19. The development of an order a system for filling orders at a pizza restaurant.
20. The design of an agricultural management system for a large farm collective.

6.1.2 Organizational Chart

When engineers work on a team-oriented project, they often establish some hierarchy among individuals. It would be nice if an engineering team could always function as a simple group of colleagues, but inevitably some team members will be

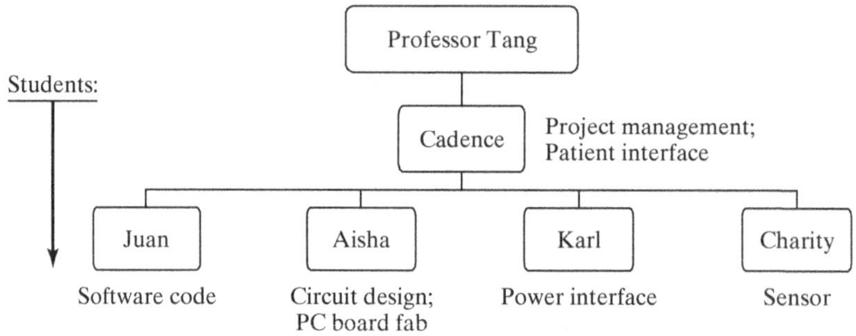

Figure 6.2
Team organizational chart for heart defibrillator project.

burdened more than others even if everyone's responsibilities are clearly spelled out. Likewise, it will be easy for some tasks to fall between the cracks if roles and responsibilities are not understood by all. One vehicle for specifying the management structure of a team is called the *organizational chart.* An organizational chart indicates who is responsible for each aspect of an engineering project. It also describes the hierarchy and reporting structure of the team. Figure 6.2 illustrates a simple organizational chart that might be used by students working on a design project for an engineering class. The objective of this particular project is to design a heart rate monitor for a professor doing research in cardiac defibrillation. Cadence acts as the team leader, and she, in turn, reports to the professor in charge for leadership and guidance. In the corporate world, where the structuring of workers and bosses can become complex, organizational charts are essential because each employee must understand the entire responsibility chain beginning with upper management.

6.1.3 The Job Description

When you work for an engineering company (or any company, for that matter), your responsibilities are spelled out in a document called your *job description.* A job description can be as informal as a simple letter of employment, or it can be a detailed, multipage document. In general, the larger the company and the more complex the organizational chart, the more detailed the job description for each person. Smaller companies may have more loosely defined job descriptions because fewer people are available to handle all the tasks required to keep the company running. A job description is tied to the work to be performed and not to any specific individual who may hold the job at any particular time. Job descriptions become most prominent when a job is vacant and is subsequently advertised.

A typical formal job description might contain the following information:

- Job title
- Grade level (defines the salary range of the job)
- Immediate supervisor (specified as another job position, not a named individual)
- List of responsibilities

Note that job descriptions often form the basis of employment postings as well. One example of a specific engineering job description might read as follows:

Structural Engineer, Zilmore Civil Engineering, SG 74
Tracking Code 1743/G274

Job Description Develop and manage construction of various large scale project such as the food court renovation in the Peach Tree Mall. Coordinate new equipment purchases, installation, and commissioning for food service equipment. Collaborate with Zilmore staff and partner companies to create processes and manage workflow at construction site. Train interns, hourly staff, and personnel from partner companies to use equipment and supervise its use. Responsible for the maintenance of the job site: order supplies, establish and enforce policies and procedures, and maintain and calibrate equipment.

Required Skills Master of science or equivalent in civil or architectural engineering and four to six years of experience with onsite construction project management. Excellent interpersonal skills with the ability to work with a variety of individuals. Professional registration required.

6.1.4 Team Contact List

As obvious as it may seem, it's important to maintain a contact list for all personnel working on a particular project. One team member (or a staff person if one exists) should have the responsibility of maintaining the list. It should contain e-mail addresses and telephone numbers for everyone on the team, as well as home or cell phone numbers and instructions for when it's permissible to call outside of working hours. If resources permit, it's a good idea to maintain the list as a web page so that it is easily updated and constrained to a *single* accurate list. Password protection may be needed in some situations.

6.1.5 Team Meetings

It's important for an engineering design team to meet regularly to review project status, design issues, and problem resolution. Frequency is entirely up to the team, but it is usually decided by the team leader. Meeting once per week seems to be the norm in many engineering organizations, including college design courses. Others may choose to meet only when deemed necessary (called meeting "on demand"), while other projects may require daily meetings. However often your team meets, it will be important to impose some type of structure so that meetings do not degrade into free-for-all chaos. Although a preset agenda is always helpful, setting the tone for the meeting is the responsibility of the team leader. This task can be particularly difficult in a course situation where the team is composed of otherwise equal peers. Even in company settings, personality differences between employees can get in the way of team cooperation and progress. Learning to deal with these situations in college will help you prepare for the job market. It is an important component of your engineering education.

6.1.6 Working with Other Teams in the Organization

More often than not, a complex design effort will involve more than one team. Assigning multiple teams to a project is common practice in companies that manufacture products for sale. The classic division of labor, still used in many companies, divides responsibilities into the broad areas of research, development, manufacturing, and marketing. The *research team* conducts fundamental studies that seek to probe new forefronts of knowledge relating to the company's core expertise. The *development team*, often composed of different individuals, is assigned the task of transforming fundamental discoveries into product concepts that culminate in working prototypes. The *manufacturing team* must take the prototype and develop a method for mass producing the product in a cost-effective and profitable way. The *marketing team*, which often contains engineers who understand how the product works, has the job of seeking and securing customers.

Traditional thinking in project management asks these four teams to interact serially, with each one passing its work "over the wall" to the next group down the chain of command. In such a scenario, Research thinks up the ideas, Development reduces them to practice, Manufacturing determines how to make the product in quantity, and Marketing seeks out new customers. Traditionally, the only feedback in the loop might be from Marketing back to Research, wherein the former communicates customers' needs to the latter. For long-term reference, teams typically operate as separate entities, with each holding on to its own set of priorities and rivalries. Often these rivalries bog down the design process and greatly increase the time span from concept to finished product.

More modern project management philosophies recognize the importance of continuous interaction between the various components of a company's organizational chart. The most successful companies are those in which the needs and priorities of each team are considered from the start and given appropriate weight when decisions are made. In such an interactive team environment, for example, Development might not be given sole authority to specify materials and dimensions of a product. Rather, manufacturing will ask that options be limited to pre-existing processes that can be enacted without requiring expensive new tooling. Marketing may insist that the design incorporate ergonomic features that will allow the product to fare well against a competitor's product. Research may seek the assistance of Development and Manufacturing to decide which projects to address, and Development may seek the help of Marketing to assess customer needs. Manufacturing may collaborate with Development to adopt rapid-prototyping methods.

In some companies, the classic team divisions described above may not have distinct boundaries at all. Research, Development, and Manufacturing may be the responsibilities of one blended group of individuals. This approach has found success, for example, in the Saturn Corporation, a subsidiary of General Motors. Their team philosophy and work unit structure are summarized in the abridged diagram of Figure 6.3. The structure of Saturn is based on "work units" and "modules." Work units are responsible for the primary tasks related to automobile production. Modules are groups of work units that perform similar types of labor. Both work units

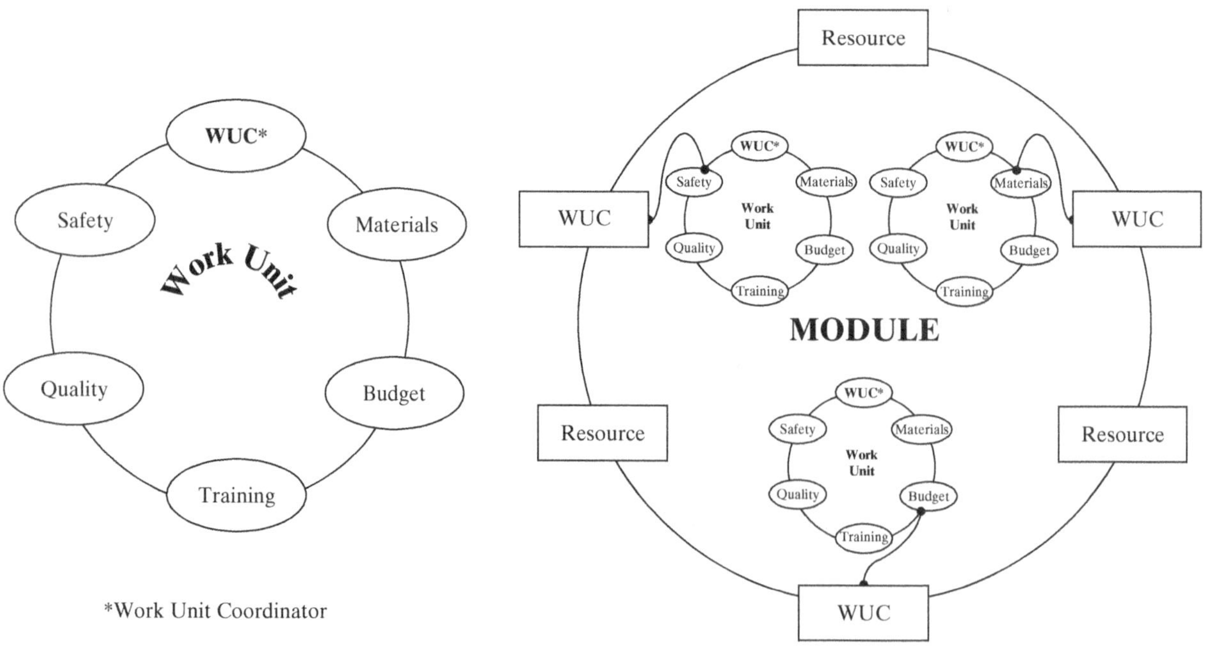

Figure 6.3
Work unit and module organizational structure for the Saturn automobile manufacturing company.

and modules are responsible for all decisions made within the circle; each circle has an elected representative on the next highest order circle. When first introduced, this self-directed, team-integrated approach to project management was unique to the U.S. auto industry and still stands out today as a refreshing way to approach large-scale endeavors.

When confronted with the classic compartmentalized approach to product design, it's important for engineers to avoid rivalries and "turf wars." Remember that ultimate success requires the symbiotic cooperation of all teams involved in the design process. While the various team structures may occupy distinct branches of the company's organizational chart, as in Figure 6.4, it's the unseen bridges between them that will keep the company strong and lead to rapid and successful product development.

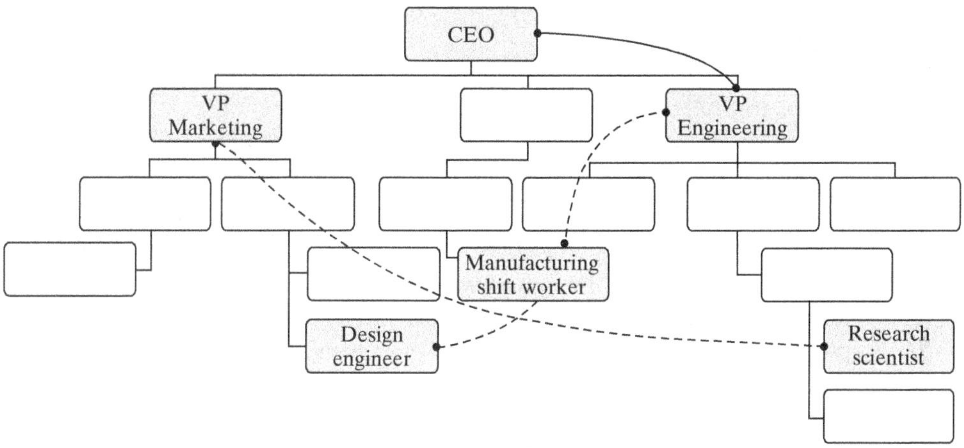

Figure 6.4
The unseen bridges between branches of an organizational chart keep the company strong and lead to rapid and successful product development.

Practice!

1. Devise an organizational chart and time line for building stage scenery for a drama production at your college or university.
2. A large, nonprofit organization is planning a walk-a-thon in which 3,000 people will walk 60 miles in 3 days and will sleep in tent camps each night. Develop an organizational chart for all persons involved in planning and implementing this event.
3. A student organization is planning to enter the Mini Baja vehicle competition for students sponsored by the Society of Automotive Engineers (*students.sae.org*). Develop an organizational chart that will guide the team through the design and construction of a competition vehicle.
4. Imagine that you work for a company that makes wind turbines for electric power generation. Define an organizational chart for such a company from design and fabrication through installation and contracted maintenance.

5. Even sports teams have organizational charts. Choose your favorite sports team and draw its organizational chart.
6. Your college or university has an administrative organizational structure. Learn about this structure and draw an organizational chart, from the president down to the level of department heads.
7. Draw the organizational chart for a major petroleum company. Include all aspects of exploration, recovery, refinery, and distribution.
8. Write the job description for an engineer assigned to a team designing a plastics processing plant.
9. Write the job description for someone who works for on online hardware store.
10. Write the job description for an engineer who has just joined a company that makes underwater hydrophones. The company has a total of ten employees.
11. Write the job description for an engineer who works for a large defense contractor that makes military aircraft.
12. Write the job description for an engineer who works for a company that makes pipeline valves.
13. Write the job description for an engineer who works for a company and helps them optimize factory floor operations for a cosmetics manufacturer.
14. Suppose you want to hire an industrial engineer for a company that makes large kitchen appliances. Write an advertisement that you might put on a nationally circulated job site (e.g., *www.monster.com*).
15. Suppose that you want to hire a nuclear engineer to oversee the operations of an electric power plant construction project. Write the advertisement that you might submit to a national trade magazine.

6.2 MANAGING TASKS: KEEPING THE PROJECT ON TRACK

Time management is critical to the success of any engineering project. In a perfect world, engineers would have as much time as needed to work on all aspects of a project; in the real world, however, deadlines have a nasty habit of creating pressure to get the product out the door. Demands for demonstrations of progress, prototype tests, something for "sales and marketing to show," and the pressures of corporate life require that an engineer develop a sense of how much time will be needed for each aspect of product development.

Even the simplest design project requires time management. A systematic approach to design tasks is always preferable to a random, hit-or-miss approach. While the subject of time management can (and does) occupy the contents of entire books, several time management tools form a basic set that should be understood by all engineers.

6.2.1 Checklist

A simple checklist enables the monitoring of engineering tasks related to a specific design project. It differs very little from the "to-do" list that you might compile for a set of weekend chores. The tasks enumerated on an engineering checklist need not be completed in any particular order. An example of a simple checklist that might apply to the design of a wheelchair ramp is shown in Figure 6.5.

Wheel Chair Ramp To-Do List

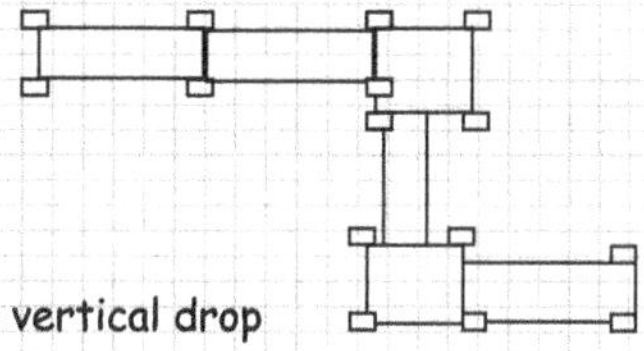

- Measure vertical drop
- Calculate needed ramp length based on 1:12 incline (ADA requirement)
- Spreadsheet of lumber needs and costs
- Spreadsheet of bolts, deck and screws, etc.
- Search for vendor of lumber/hardware
- Sink support posts
- Screw together framework
- Cut and nail down deck boards.

Figure 6.5
Checklist for the design of a wheelchair ramp.

6.2.2 Timeline

A *timeline* is a valuable tool for keeping a project on schedule. A timeline is similar to a checklist, but its various tasks are expected to be completed sequentially by specified milestone dates. The timeline is very appropriate when the entire team works in unison, with the differentiation of tasks occurring serially in time. If a given task is in danger of not being completed before its designated milestone date, it's the job of the project manager to allocate more time, and overtime if necessary, so that the task can be completed on schedule. A typical timeline (in this case, one prepared by students designing a vehicle for a class design project) is shown in Figure 6.6.

6.2.3 Gantt Chart

When a project is complex and involves many people, a simple timeline may be inadequate for managing all aspects of the project. Similarly, if the project's various tasks are interdependent, so that the completion of one phase depends on the success of several others, the *Gantt Chart* of Figure 6.7 may be a more appropriate time-management tool. The Gantt chart is simply a two-dimensional plot in which the

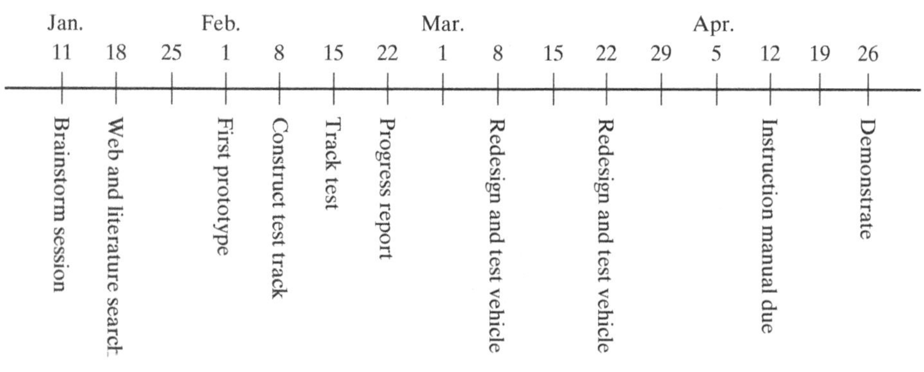

Figure 6.6
Timeline for tasks pertaining to a vehicle design competition.

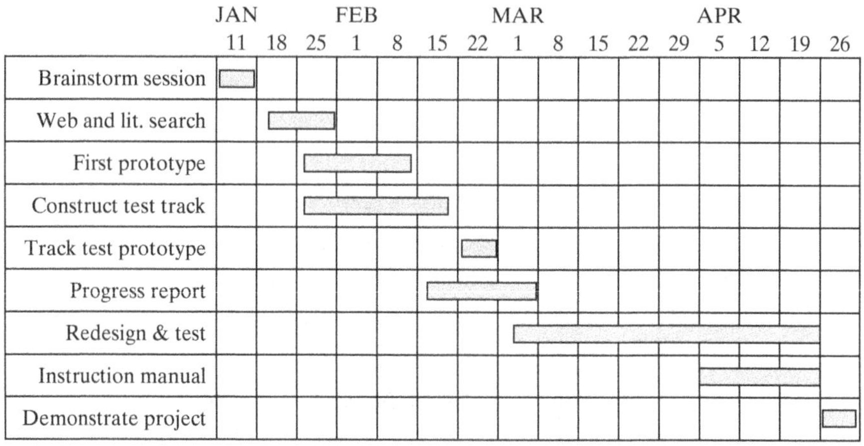

Figure 6.7
A Gantt chart provides a more comprehensive, two-dimensional method of scheduling the tasks shown in Figure 6.6.

horizontal axis reflects time measured in days, weeks, or months, and the vertical axis represents either the tasks to be completed or the individuals responsible for those tasks. Unlike the checklist, which simply enumerates the tasks to be completed, and the one-dimensional timeline, which merely displays sequential time allocations for each phase of the project, the Gantt chart shows how much time is allotted to multiple tasks performed in parallel. It also provides for overlapping time periods that help indicate the interdependency between the various aspects of the project. When a particular task has been completed, it can be shaded on the Gantt chart, so that the status of the project can be determined at a glance.

6.2.4 PERT Chart

The Project Evaluation and Review Technique (PERT) was first proposed by the United States Navy and developed by the consulting firm of Booz, Allen, and Hamilton in 1958. Its purpose at the time was to coordinate the activities of over 10,000 separate subcontractors involved in the Polaris missile development program. The PERT method is very similar to the *critical path method* (CPM), and these terms are essentially interchangeable. The PERT technique is fundamentally a method for prioritizing and scheduling complex interrelated activities. The PERT chart helps to identify the most time-critical events in the design process.

The essence of the technique is embodied in a graphical network called the *PERT chart*, depicted in generic form in Figure 6.8. The PERT chart consists of

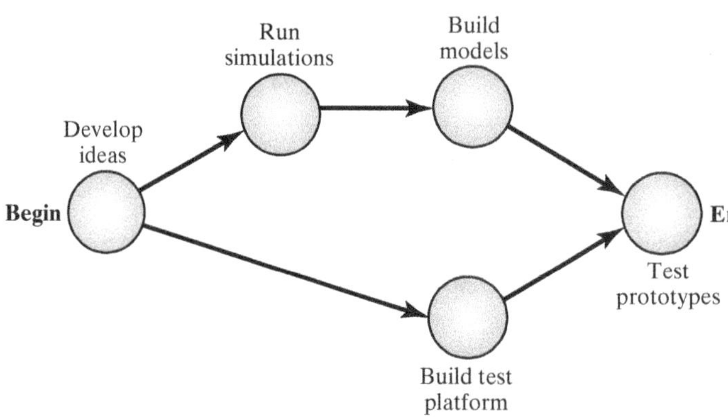

Figure 6.8
The generic form of a PERT chart. The circles represent milestone points, and each arrow represents a task leading up to a node. Each task is allocated a completion time.

numbered milestone circles and pathway branches. Each branch is labeled by a time interval allocated to the completion of the task to which the branch leads. Branches are labeled in appropriate time units (e.g., days, weeks, or months). Like the Gantt chart, the PERT chart summarizes the time allocated for each task and notes task completion milestones. Unlike the Gantt chart, however, the PERT chart also shows the way in which the tasks depend on one another. This interdependency is indicated by the branch lines that connect the task-completion circles of the chart.

A PERT chart must have a starting point, or *node*, and a single ending milestone (the last node to which all pathways must lead). As in the Gantt chart, time progresses from left to right. The time allocated for the pathway between any two milestone nodes will be equal to the sum of the times allocated for each task in the pathway. In Figure 6.9, for example, the sequence of tasks leading from milestone A to milestone C is allocated a total of $3 + 5 = 8$ days.

As each task depicted on a PERT chart is completed, the project manager checks it off on the chart. The manager can thus monitor the progress of the entire project and be alerted to any possible path delays. Some project managers prefer the PERT chart over the Gantt chart because it clearly illustrates task dependencies. A PERT chart, however, can be much more difficult to interpret, especially on complex projects. Alternatively, some project managers may choose to use both techniques.

When compiling a PERT chart for project management, it's possible (and often preferable) for the sum of branch times over one pathway to be less than the sum of times over a parallel pathway. The excess, called *slack time,* can be used to compensate for the inevitable delays that will accompany the design process. If the delay experienced along the way does not exceed the slack time of the pathway, the entire project will still be on track. Of most interest, therefore, are so-called *critical pathways*. The latter are those branches containing zero slack time. Any delays in a

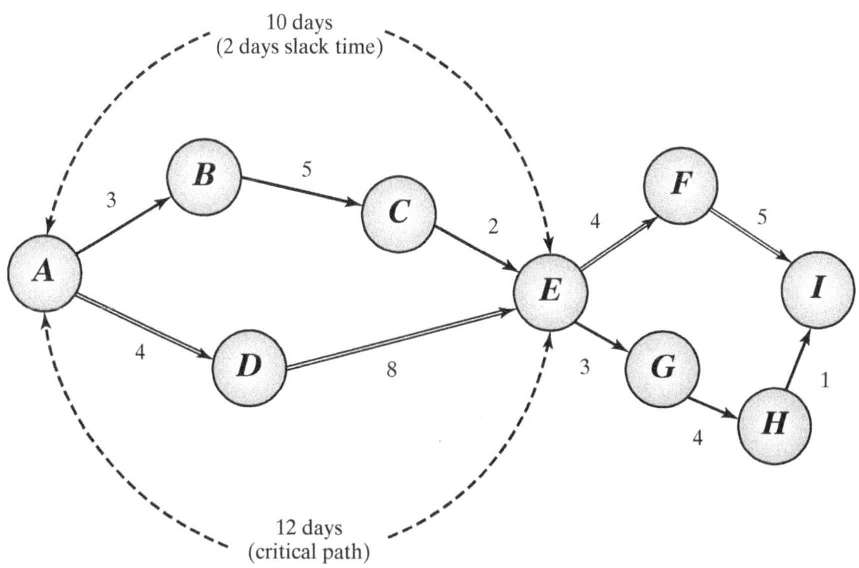

Figure 6.9
A PERT chart showing milestone nodes and task time allocations in days. On this chart, critical pathways are shown as double lines.

critical pathway can jeopardize the time flow of the entire project and should be monitored carefully. The pathway from node A to node E (via node D) in Figure 6.9, for example, is a critical pathway.

EXAMPLE 6.1

Managing a Building Construction Project

The major steps required to erect a framed house (somewhat simplified in this case) are summarized by the PERT chart of Figure 6.10. Such a chart might typically be used by a general contractor who must hire many different subcontractors to complete various aspects of the project. Node 0 denotes the starting point of construction. Before anything else can transpire, the building site must be excavated so that the foundation can be framed, as well as molded and poured. In Figure 6.10, the former is allocated 4 days, and the latter 2 days. After the foundation has been poured and cured, the external skeleton or frame of the building can be erected. This frame will form the support structure for all the exterior walls of the building, as well as any interior walls specified by the architect. This task has been allocated a total of 5 days.

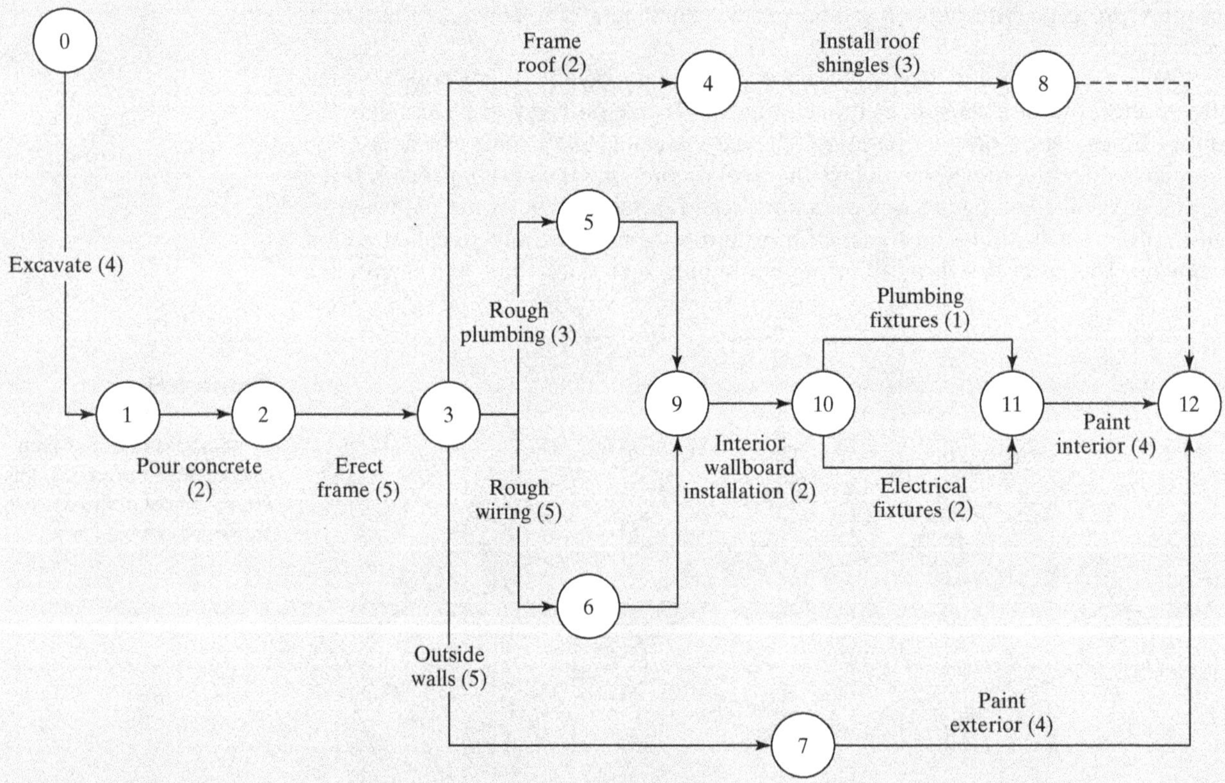

Figure 6.10
PERT chart that describes the principal steps involved in building a house. Numbered circles (nodes) designate task completion points. Each pathway is labeled with a time allocated to that task. The numbering of the nodes is arbitrary.

After the building frame has been completed, the major tasks continue over parallel pathways. The framing for the roof must be completed, followed by the weather protection (in this case, standard roofing shingles). These tasks have been allocated 2 and 3 days, respectively, leaving considerable slack time compared to the other paths. Similarly, following arrival at node 3, work begins in parallel on the installation of the building's exterior walls. The rough plumbing and electrical tasks, to be described in the next paragraph, require that the interior walls remain open, but not the exterior walls.

In parallel with the finishing of the roof and outside walls, the installation of the rough plumbing is now initiated. Rough plumbing refers to the piping—both supply and waste—that must lie within the walls of the structure. Obviously, this task must be completed before the interior walls themselves can be installed. Yet another parallel pathway begins after node 3, consisting of the rough electrical wiring. The latter must also be completed before the interior walls can go up. The rough plumbing and wiring tasks are allocated 3 and 5 days, respectively.

Following the completion of the rough work, the interior walls (e.g., standard gypsum wallboard; see *www.nationalgypsum.com*) can now be installed (2 days). Once the interior walls have been completed, the plumbing and electrical crews can return to install fixtures; afterwards, the interior walls may be painted. Usually, this task is performed after the electrical and plumbing work are finished, because the latter tasks might cause workers to mar the finished paintwork. Some contractors, however, prefer to paint before fixture installation, because the resulting paint job is much cleaner around the edges of the plumbing and electrical fixtures.

The completion of the roof, interior paint, and exterior paint on the PERT chart all arrive at the last node on the diagram (node 12), signifying the completion of the job. The house may now be occupied. The critical pathway in this PERT chart consists of the following tasks, beginning at node 3: rough electrical wiring (5 days), wallboard installation (2 days), wiring completion (2 days), and interior painting (4 days). These tasks require a total of $5 + 2 + 2 + 4 = 13$ days. Any delays over this critical pathway will lead to a delay in occupancy of the house.

EXAMPLE 6.2

Managing a Software Project

Figure 6.11 shows a PERT chart used by a development team tasked with the design of a cell phone running on firmware (machine-level binary software). The unit, once manufactured and ready for sale, is to be shipped with a user manual and online help files. Once the specifications have been set, the PERT chart divides into three distinct pathways—one related to hardware, one to the firmware, and one to user documentation. The critical pathway in this project runs from node 1 to node 2, and then via nodes 3, 5, 6, 8, and 7 to the final node 13, for a total time allocation of $2 + 4 + 2 + 3 + 4 + 1 + 4 = 20$ weeks.

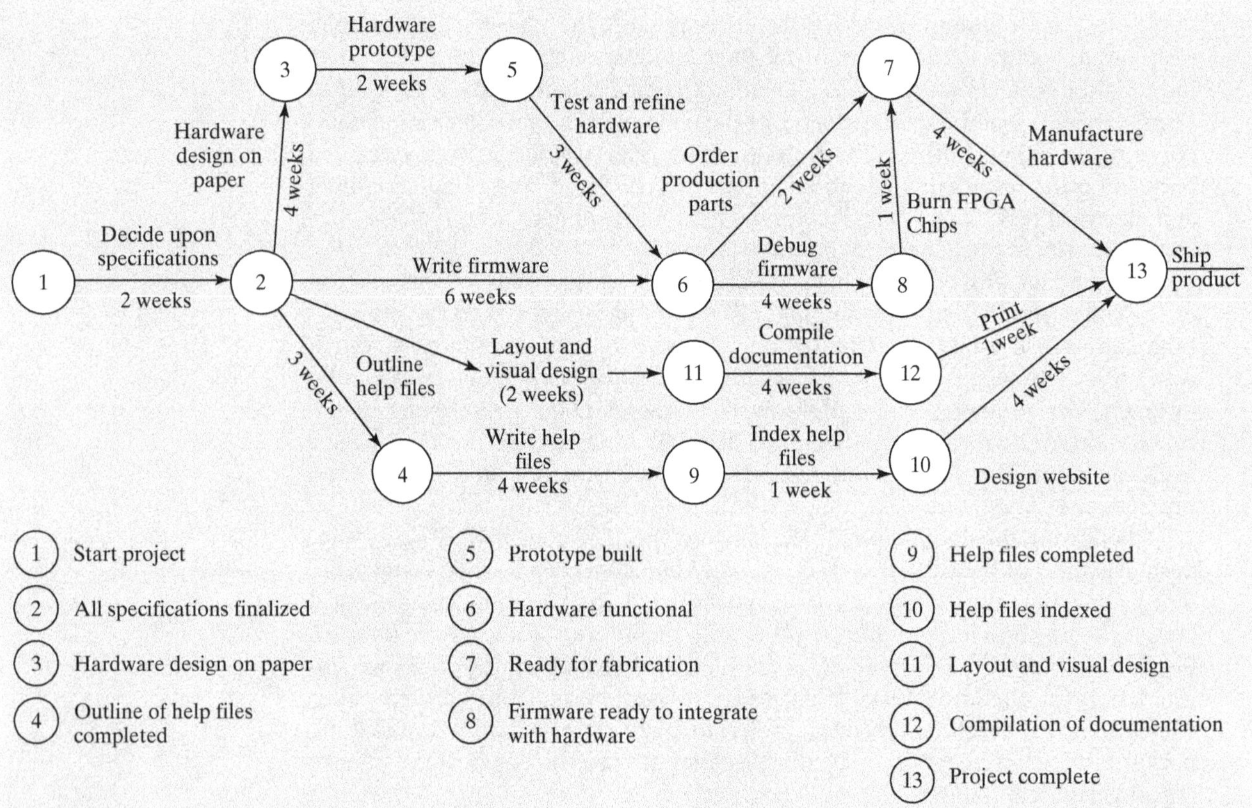

Figure 6.11
PERT chart that describes the development of a cell phone running on firmware.

Professional Success: The Laws of Time Estimation

How long will it take to perform a particular design task? How much time should you allocate to each segment of a PERT chart or Gantt chart? The following Laws of Time Estimation will help you to determine the time required for a given design task:

1. Everything takes longer than expected.
2. If you've done the same task before on a different project, estimate the amount of time required to finish the task. The actual amount of time required will be about twice as long. Something will always be different in the current project.
3. If you've worked on something *similar* before, but not exactly the same task, estimate the amount of time required to finish the task. The actual amount of time required will be about four times as long.

4. If you've *never* worked on something similar before, estimate the amount of time required to finish the task. The actual amount of time required will be the next highest time unit. For example, something estimated to take an hour will really take a day; something estimated to take a day will take a week, etc.

6.3 DOCUMENTATION: THE KEY TO PROJECT SUCCESS

Engineering design is never performed in isolation. Even the simplest of projects involves a designer and an end user. More often, a design effort involves considerably more individuals concerned with various aspects of the product. Additionally, the uses of the product may involve an entire segment of the population. The design of an automobile, for example, encompasses the work of mechanical, electrical, industrial, materials, and safety engineers, and the entire driving public constitutes the group of end users. A consumer products company involved in the production of cellular telephones will bring together computer, systems, electrical, mechanical, and manufacturing engineers in a multidisciplinary team that may include people from sales and marketing. A public works management corporation, tasked with designing a waste water treatment facility, might solicit the help of civil, chemical, mechanical, and environmental engineers, along with the city planners and systems engineers. Complex engineering projects are successful only if everyone on the design team communicates with everyone else at all phases of the design effort.

One way in which engineers communicate with each other is through careful recordkeeping. Good documentation is essential when you work as a member of a design team. As a practicing engineer, it will be your responsibility to maintain a comprehensive collection of design concepts, sketches, detailed drawings, test results, redesigns, reports, and schematics—whatever records are pertinent to the project. This *documentation trail* will serve as a tool for passing information on to team members who may need to repeat or verify your work. The documentation trail is critical to those who will manufacture your product from a prototype, apply for patents based on your inventions, or even take over your job if you are promoted or move to another company. Written records are also a good way to communicate with yourself. Many an engineer has been unable to reproduce design accomplishments due to sloppy recordkeeping. Indeed, one of the marks of a professional engineer is the discipline needed to keep organized, neat, up-to-date records. Documentation should never be performed as an afterthought. If a project is dropped by one team member, the state of documentation should always be such that another team member can resume the project without delay. As a student of engineering, you should learn the art of recordkeeping and develop good documentation habits early in your career. Most companies, laboratories, and other technical firms require their employees to keep records that document the results of engineering efforts.

6.3.1 Paper versus Electronic Documentation

Today, just about every piece of engineering documentation, with the exception of the engineer's logbook (described in the next section), is produced electronically. This situation was not true only a decade ago. Examples of documents destined for preservation include text documents, spreadsheets, computer codes, schematics, drawings, design layouts, and simulated test results. Most engineers prefer to store information electronically so that it can be viewed onscreen and printed only as needed. Some still prefer to preserve documentation by printing everything on paper and storing the documents in a physical file cabinet. Whichever method you choose, take heed of the following important guidelines:

- *Organize your information.* It's important to store documentation in an organized and logical manner. If the project is small, its documentation should be stored in a single folder (paper or electronic). Larger projects may require a group of folders, each relating to different aspects of the project. The folders should be labeled and dated with informative titles (such as "Propulsion System for XYZ Project") and kept in a place that will be easy to find in the future.
- *Back up your information.* It's equally important to store a duplicate copy of all documentation. This guideline applies to written as well as electronic information. Fire, flood, theft, misplacement, the all-too-common disk crash, and the unfortunate havoc wreaked by so-called "malware" can lead to the loss of a project's entire documentation trail. Archival storage of records in a different physical location will help to keep a project on track should one of these catastrophes occur. Companies exist that will provide secure offsite storage of critical data.

6.3.2 The Engineer's Logbook (Notebook)

One important vehicle for recordkeeping is the *engineer's logbook,* sometimes called the *engineer's notebook.* A well-maintained logbook serves as a permanent record that includes all ideas, calculations, innovations, and test results that emerge from the design effort. When engineers work in a team, each member keeps a separate logbook reflecting that individual's assigned tasks. When the project is brought to completion, the logbooks of all team members are placed in an archive and remain the property of the company. An engineering notebook thus serves as an archival record of new ideas and engineering research achievements *whether or not they lead to commercial use.*

A complete logbook also serves as evidence of inventorship and establishes the date of conception and "reduction to practice" of a new idea. It shows that the inventor (e.g., you) has used diligence in advancing the invention to completion. In this respect, the engineer's logbook is more than just a simple lab notebook. It serves as a valuable document that has legal implications. When you work as an engineer, you have a professional responsibility to your employer, your colleagues, and to the integrity of your job to keep a good logbook.

The notebook shown in Figure 6.12 is typical of many used in industry, government labs, and research institutions. The company, laboratory, or project name is printed at the top, and the notebook is assigned a unique number by the user. In some large institutions, a central office may assign notebook numbers to its employees when the notebook is issued.

The techniques for logbook use differ from those used in most science and (unfortunately) some introductory engineering classes where instructors encourage students to write things down first on loose scratch paper and then recopy relevant items into a neat notebook. This procedure is bad practice for the design engineer. Although notebooks prepared in this way are easier for instructors to grade, the finished notebook seldom resembles a running record of what *actually* occurred in the laboratory; it is not especially useful for engineering design projects. Design is as much a *process* as it is a final product, and the act of recording ideas as they emerge

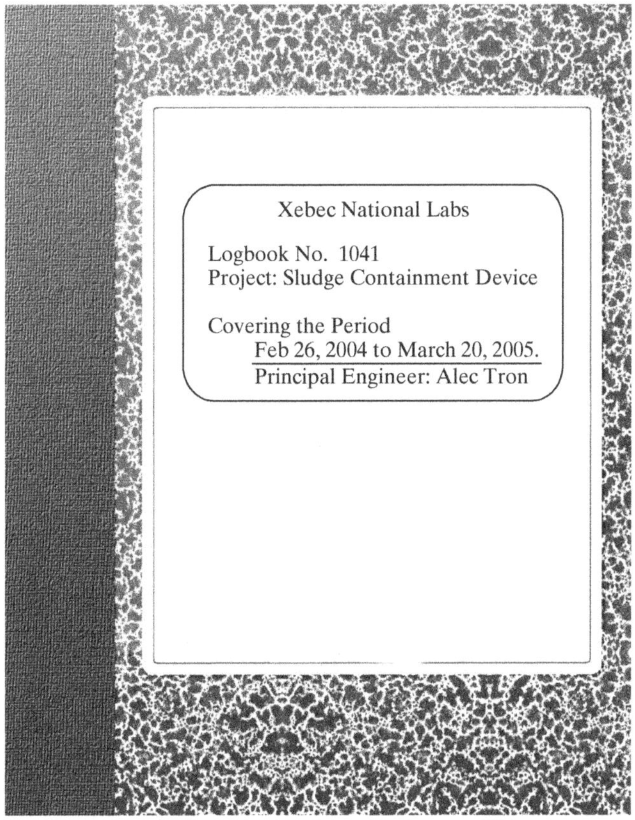

Figure 6.12
Cover and label of a typical engineer's logbook.

and events as they happen helps engineers to think and be creative. Also, keeping a record of what did *not* work is just as important as recording what did work; this practice ensures that mistakes will not be repeated in the future.

6.3.3 Logbook Format

An engineering logbook should be used as a design tool. Enter everything into your logbook, no matter how seemingly irrelevant. Write down ideas as you think of them, even if you have no immediate plans to pursue them. Keep an ongoing record of successes and failures. Record the results of every test—mechanical, structural, electrical, system, flight, flow, pressure, performance—even if the results may not be used in the final design. Stop to write things down! This habit—sometimes hard, but important to learn—will require discipline but will always be worth the effort. Important information, including some you might otherwise have forgotten, will be in your logbook and will be preserved for a time when you may need it.

Any logbook format that meets your needs and those of your team is suitable, as long as it forms a permanent record of your contributions to the design project. Ban loose paper from the laboratory. It is easily lost, misfiled, or damaged. Resist the temptation to reach for loose paper when you need to do a calculation. Instead of grabbing that pad to draw a sketch or discuss an idea, take the time to open your logbook. You'll be glad you did when those numbers and sketches you need are readily available. Unbound paper used for anything other than doodling has no place in an engineering laboratory.

6.3.4 Using Your Engineer's Logbook

As chief author of your logbook, you have the freedom to set your own objectives for its use. The following guidelines, however, are typical of those used by many engineers and design teams:

1. Each person working on a project should keep a separate logbook specifically for that project. When the logbook is full, it should be stored in a safe place specifically designated for logbook storage. In that way, everyone will know where to find the logbook when it's needed.
2. All ideas, calculations, experiments, tests, mechanical sketches, flowcharts, circuit diagrams, etc., related to the project should be entered into the logbook. Entries should be dated and written in ink (pencil has a nasty habit of smudging). It's no crime to cross out errors made in pen. Any computer-generated plots, graphics, schematics, or photos printed on loose paper should be affixed to bound logbook pages. This procedure will help prevent loss of important data.
3. Logbook entries should outline the problem addressed, tests performed, and so forth, but subjective conclusions about the success of the tests (e.g., "I believe") should be avoided. The facts should speak for themselves.
4. The voice of the logbook may be written in the first person (*I tested the widget ...*) , but it should speak to another reader. Assume that your logbook will be read by teammates, your boss, or perhaps someone from marketing.
5. In settings where intellectual property is at stake, the concluding page of each session should be dated and, where appropriate, signed. This practice eliminates all ambiguity with regard to dates of invention and disclosure.
6. Logbook pages should not be left blank. If a portion of a page would otherwise be left blank, a vertical or slanted line should be drawn through it. Pages should be numbered consecutively and not be torn out. Do not make changes using correction fluid. Cross out instead. These procedures will prevent you from creating obscure or questionable entries should your logbook be entered as legal evidence in patent or liability actions. Although this precaution probably won't be relevant to logbooks you keep for college design courses, it's a good idea to begin following it now so that the procedure becomes a career habit.

EXAMPLE 6.3

An Engineer's Logbook

The following example illustrates proper use of an engineering logbook. Imagine that the logbook pages shown describe your team's design for a telescoping flag for a motorized wheelchair. The intent is to increase visibility when the user operates the wheelchair in city streets. The first page, Figure 6.13, shows your preliminary sketch of a basic concept based on the use of a commercially available telescoping car radio antenna. The second page, Figure 6.14, contains some calculations that estimate the battery drain as the flag is raised and lowered. The entries on the third page, Figure 6.15, show a list of parts and materials that your team has decided to purchase. These parts will enable you to build a prototype and test your system's ability to raise and lower the flag on demand.

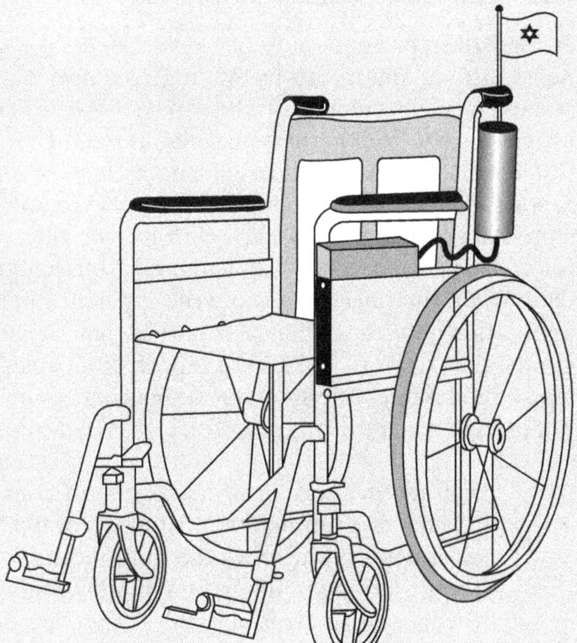

Figure 6.13
Preliminary logbook sketch for a telescoping wheelchair flag.

Battery Energy Requirements *10/1/09*

Measured motor current: 0.45 A $\approx \frac{1}{2}$A

Rod Speed: (estimated) abt. 10 cm/sec

$\rightarrow$ *Motor* ON *time = Dist/Rate =* $\dfrac{121 \text{ cm}}{10 \text{ cm/sec}} = 12.1$ *sec*

Here 12 seconds = $\dfrac{1}{5}$ *minute =* $\dfrac{1}{300}$ *hour*

. . . so each extension requires about. . .

$0.45 \text{ A} \times \dfrac{1}{300} \text{ hour} \approx 1.5 \times 10^{-3} \text{ A-h}$

A 0.3 amp-hour battery can provide about

$0.3 \text{ A-h} \;\; 1.5 \times 10^{-3} \text{ A-h} \approx 200$ *extensions of flag.*

Note to Mike: Good for a start

Figure 6.14
Calculations to show estimates of battery drain.

6.3.5 Technical Reports and Memoranda

A logbook represents only one component of a good documentation trail. Engineers also communicate by writing technical reports at significant project milestones. A technical report describes a particular accomplishment and perhaps provides some background material as well. The report may contain theory, data, test results, calculations, design parameters, or fabrication dimensions. Technical reports form the backbone of a company's technical database. Reports are typically stored in archival format, each with its own title and catalog number. Information for technical reports is easily gathered from logbooks that are accurate and up to date. When the time comes to write a patent disclosure, journal paper, or product application note, the technical report becomes an indispensable reference tool. Taking the time to write a technical report about a negative result or design failure can save considerable time should a design concept be revisited by engineers who were not present when the original project was undertaken.

> Yet another way in which engineers communicate their ideas to other engineers is through the writing of formal journal articles. A journal article is appropriate when the work represents new knowledge unknown to those working in the field. While often used to report experimental and theoretical findings, journal articles are also frequently used to report design innovations. The structure of a typical journal paper is discussed in more detail in other areas of this book.
>
> *Peer review* is the process by which a paper is evaluated and critiqued, then either accepted or rejected for publication. The editor of the journal will typically send a submitted paper to one or more knowledgeable reviewers who will return a list of comments and questions for the author. It is common for a submitted paper to be returned at least one time to the author for revision. It's most unusual for a paper to be accepted unchanged upon its first submission. Conversely, some papers may be rejected as unsuitable for publication. This peer-review method ensures that papers appearing in the best journals are accurate, relevant, and up to date.

6.3.6 Software Documentation and the Role of the Engineering Notebook

Of all design endeavors, software writing is the one most prone to poor documentation. The revision loop of a software design cycle can be extremely rapid, because the typical software development tool enables a programmer to make small changes and test their effects immediately. This rapid-fire method of development is good for prototyping but invites poor documentation habits. Seldom does the software engineer find a good time to stop and document the flow of a program, because most pauses are short and change is frequent. As a result, the documentation for many software programs is added after the fact, if at all.

If you find yourself writing software, get into the habit of including documentation in your program as you go along. All software development tools provide a means for embedding comments in the program code. Add them frequently to explain why you've taken a certain approach or written a particular section of the code. Explain the meaning of object names and program variables. Your program documentation should enable other engineers on your team to completely understand and take over your sections of the code simply by reading the comment lines. Good documentation will also be invaluable should you need to modify your code at a later time. It's amazing how quickly a programmer can forget the internal logic of a program after setting it aside for only a short time.

If your program is destined for commercial sale, then good internal documentation and truly helpful "help" files are essential. Documentation included inside the program on a regular basis will easily translate into help files when the need arises. One trick used by top-notch software developers is to write the help and instruction files as the program code is developed, rather than as an afterthought. Changes to the help files can be made at the same time that changes are made in the program code. The abundance of commercial software packages with pathetic or poorly written help files is testimony to generations of software engineers who have perpetuated a tendency toward poor documentation habits. If you master the skill of documenting software, your software products will be better utilized and more successful than those having poor documentation.

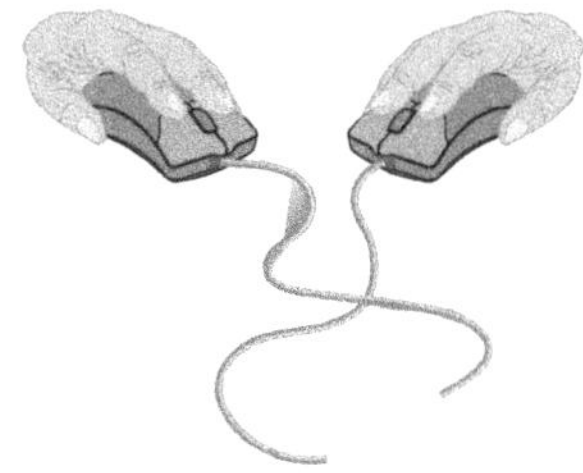

Although keeping engineering logbooks is less relevant to software development than to other types of engineering, logbooks still play a role. On the pages of your notebook, you can outline the overall structure of the program (e.g., using flowcharts) and the interconnections between its various modules. You can enter sketches of graphical user interfaces without writing actual computer code. You can similarly draw block diagrams of relational databases and make lists of the variables to be used in the software.

Professional Success: How to Keep Good Records All the Time

If you want to keep a good documentation trail, carry some sort of logbook with you at all times. It will then be available whenever you have an idea that needs recording. Buy a medium-size notebook that can fit easily into your backpack. Clip a pen right inside the front cover. Be sure to write your name and contact information on the cover in case you lose it! A tiny 3" × 5" bound notebook will do nicely. Although writing space will be limited because of the smaller size, you'll be more likely to carry it if it's not overly large. You can then tape its smaller pages into those of your central, full-sized logbook.

Practice!

1. Refer to the logbook calculations in Figure 6.14. Revise the estimate of the number of extensions if the motor requires 2.1 A. Rewrite the logbook page.
2. Refer to the logbook calculations in Figure 6.14. Revise the estimate of the number of extensions if the motor requires 300 mA, and a 200 A-h battery (similar to a lead-acid car battery) is used. Rewrite the logbook page.
3. Do some Internet research and identify sources and vendors for the telescoping antenna shown in Figure 6.13. Prepare the resulting logbook page.
4. For the flag system outlined in Figure 6.13, what are the advantages and disadvantages of using a separate battery for the flagpole (as

opposed to tapping the primary storage battery that powers the wheelchair.) Write a logbook page that records your ideas.

5. On a simulated logbook page, outline some ideas for marketing your telescoping flag invention. Include cost estimates for its manufacture.
6. Imagine that you are one of the Wright brothers developing the first airplane. Sketch what you imagine to have been a typical logbook page from the project.
7. Look up the history of the famous 20th-century American inventor Lee DeForest. Sketch one page of a logbook that he might have kept during the time of his invention.
8. Sketch a logbook page that might have been recorded by Josephine Cochran, the inventor of the modern dishwasher.
9. Imagine what it was like to be Philo T. Farnsworth, the original inventor of the television. Sketch several logbook pages that summarize how the system works.
10. Develop some logbook pages related to the development of a personalized air transportation device.

6.3.7 The Importance of Logbooks: a Case Study

This case involves a small biomedical engineering company called Heartthrob[1] that was a leading contender for the development of a self-contained artificial human heart designed to be a permanent replacement for individuals who could otherwise stay alive only with a human heart transplant.

The major components of the heart system, depicted in Figure 6.16, include the central implanted pump and a system for transferring electrical power to run the pump through the skin from a battery pack worn outside the body. One key feature of the power transfer process, the lack of wires piercing the skin, is essential for the heart's long-term efficacy, because skin perforations are prime entry points for infection and require constant medical supervision in a skilled-care facility. Rather than requiring the patients to be tethered to a console, the fully self-contained device is designed to provide the patient some semblance of a normal, mobile, home life. The system for transferring electrical power, called the transcutaneous energy transfer device, or "TET," consists of two concentric AC magnetic coils. One is implanted under the skin, and the other is worn outside the skin. (Transcutaneous means "through the skin.") Heartthrob wished its engineers to fully focus on the daunting task of developing the heart pump itself, so it hired another biomedical engineering company, Tech-Heart, to design the energy transfer module.

After about four years of effort, Tech-Heart was still unable to produce a TET that met Hearthrob's stringent technical specifications. Although Tech-Heart claimed to be converging on a solution, Heartthrob was not convinced that a satisfactory TET device was imminent. Faced with an impending critical animal test that would determine future funding of its entire heart project from the National

[1]Names have been changed to protect privacy, but the story is entirely true.

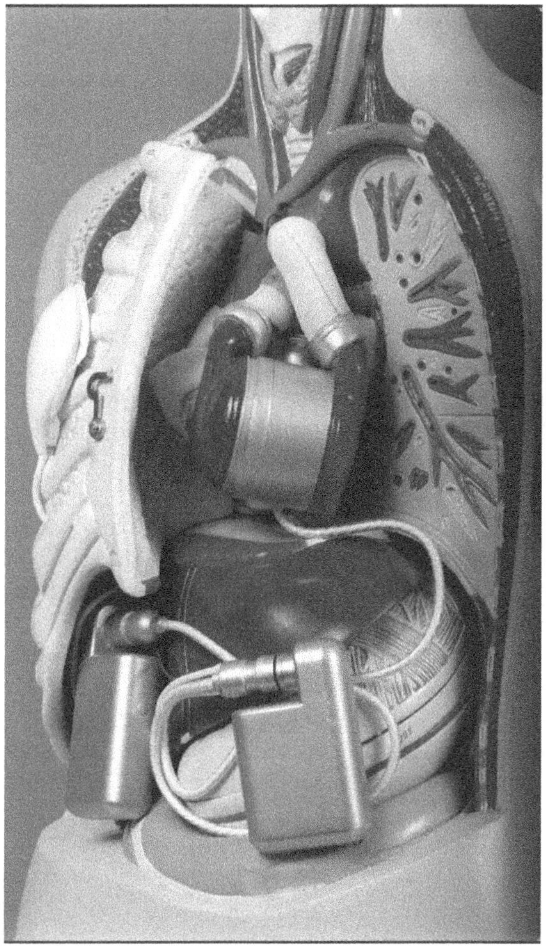

Figure 6.16
Heartthrob's total artificial heart system.

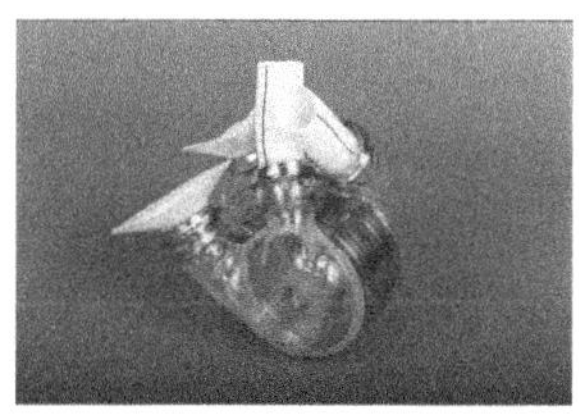

Institutes of Health, Heartthrob decided to sever its reliance on the Tech-Heart TET. The Heartthrob CEO instructed one of his engineers, Dr. Maven, to develop a home-grown TET device as quickly as possible. Dr. Maven was a very capable fellow, and after only four months of effort, he succeeded in designing and testing a fully working version of a TET device that met all the requirements of Heartthrob's impending heart-pump test.

This development led to a lawsuit by Tech-Heart who claimed that Heartthrob could not have developed its own TET device in the mere time span of four months without having stolen secrets and technology from Tech-Heart. After all, Tech-Heart claimed, its engineers had worked on the project for four years and were only just beginning to converge on a possible solution. How could the Heartthrob engineer have designed a superior TET in only four months? Heartthrob countered with a claim that its short path to success was due solely to the high competency of its engineer, and that, in fact, it had taken special precautions to ensure that nothing would be stolen from the previous unsuccessful Tech-Heart design effort. The suit went to court and, in the end, Heartthrob prevailed. The jury recognized that Heartthrob's TET design, while performing the same basic transcutaneous energy-transfer function as the Tech-Heart device, was completely different with regard to all details of implementation. The Heartthrob device, the jury concluded, used different circuits, materials, magnetic construction, and semiconductor components.

During the trial, a key component of Heartthrob's defense was the logbook kept by the engineer who designed the TET. His logbook was used to prove that Heartthrob had designed its own independent version of the TET. Following the company's customary logbook policy, as well as sound engineering practice, Dr. Maven had kept careful records of his TET design, having entered every design concept, schematic, circuit layout, and test result that emerged during the design process. He had even noted the various circuit configurations that had gone up in smoke on the test bench before his first working model emerged. Each page of his logbook had been dated, and each lab session involving successful tests had been signed and countersigned by another Heartthrob engineer. During trial, entire pages of Dr. Maven's logbook were reproduced and projected on a large screen for the jury to see. Dr. Maven's logbook played a crucial role in the success of Heartthrob's legal defense.

The presentation of Dr. Maven's logbook in court, while critical to the outcome of the trial, did not proceed flawlessly, however. Several pages of the logbook involving work done in January 1996 had been incorrectly dated with the year 1995. Following a common mistake that many individuals make when the year changes, Dr. Maven had, without thinking, hastily written the year from the previous month of December. The lawyers for Tech-Heart were quick to seize upon this error as prima facie evidence that something was "fishy." They claimed Dr. Maven had forged portions of his logbook in an attempt to present a false picture of his accomplishments. In the end, the jury was not convinced and realized that Dr. Maven's error was nothing more than a common calendar mistake. Nevertheless, this seemingly small lack of attention to detail in logbook procedure put the case against Heartthrob in jeopardy for a time.

Professional Success: Developing Good Logbook Habits

Most of us follow routines without thinking. When we wake up in the morning, we brush our teeth. When we eat, we instinctively reach for a clean plate. When we get into a car, we (hopefully) buckle our seatbelts automatically. An as engineer, the urge to write things down in a logbook should become as instinctive as these other common tasks. In contrast to these personal procedures that enhance our own well being, we engineers are not trained from childhood to record our experiences in a notebook. Developing this instinct requires practice, but it can become part of your routine over time. When personal computers and the Internet first came into being, most people did not think very much about the novelty of e-mail. The same was true for cell-phone text messaging. Now these activities have become daily routines for most. You developed these unnatural skills by practicing them over time. It should be the same with your engineer's logbook. Force yourself to get into the habit of using your logbook whenever you practice design. Over time, it will become as natural as brushing your teeth.

6.4 LEGAL ISSUES: INTELLECTUAL PROPERTY, PATENTS, AND TRADE SECRETS

It would be nice if engineers could focus solely on technical issues. After all, engineering students spend enormous amounts of time learning about fundamentals and applications, plus many technical skills. Of all the non-technical topics addressed in engineering curricula, arguably the most important of these is the role that legal issues play in the design process.

The juxtaposition between legal issues and design practice falls squarely within the domain of project management. Central to this juxtaposition is the issue of patent protection. The worldwide system of patents and licensing, part of the broader world known as *intellectual property,* has as its noble goal the protection of new ideas and inventions. The patent system originated as a means of encouraging innovation by preventing large companies with lots of money from gobbling up the ideas of smaller inventors and dominating the market. While well intentioned in concept, the patent system has evolved into a maze of regulations and practices that can be most perplexing to engineers uninitiated in the world of legal procedures. For this reason, most large companies employ attorneys whose job it is to navigate the company through the often bewildering world of intellectual property. If you come up with an idea in the course of your company work, a resident attorney will likely guide you through the patent process. The patent will be assigned to your company with you listed as the inventor.

Another important arena in which legal issues intersect engineering is in the area of product liability. The western world, particularly the United States, has become a litigious society in which companies can be sued for all sorts of real or perceived product defects leading to injury, loss of property, or loss of life. The more dramatic stories, particularly those related to automobile accidents, are revealed to the general public via the media. But countless other cases that have filled civil courtrooms everywhere are not as widely disseminated. Understanding how the management of a design project impacts the issue of product liability is a critical engineering skill.

This section introduces some of the basic legal concepts that underscore the relationship between engineering and the law. While the material presented here merely skims the surface of intellectual property law, it will introduce you to the language of the legal profession and build your awareness of legal issues as they relate to engineering design.

6.4.1 Patents

A patent can be issued for any invention provided that certain criteria are met. The invention must introduce a new concept or way of doing something. The product must be something that can actually be produced, and it must have been *reduced to practice* (i.e., produced in at least prototype form) prior to the initial patent application. Also, the subject of the patent must not simply be an obvious synthesis of preexisting ideas.

A patent is valid for twenty years (in the United States) during which time the inventor has sole rights to produce or license the invention. A patent *license* is a legal contract binding the inventor, who may not have the resources to actually produce the invention, with a company that has the resources and desires to manufacture and market the invention. In the United States, patents are awarded, or *issued* by the United States Patent and Trademark Office (USPTO, or simply "Patent Office"). This organization maintains a comprehensive public website (*www.uspto.gov*) from which anyone may download patents dating back as far as 1976. Earlier patents may also be ordered in hardcopy from the patent archive.

6.4.2 Patent Jargon

Lawyers have their own vocabulary for describing the patent process. The concept of *prior art* refers to the body of knowledge that existed at the time of the patented invention. A device cannot be patented if it is described anywhere in the prior art. The invention must be reproducible, and the language of the patent must be sufficiently clear that it can be understood by an individual *skilled in the art*. The latter phrase is used to describe someone who, at the time of the invention, would have been familiar with the field of endeavor to which the invention applies and would have had the requisite education typical of those working in the general area of technology. To be patentable, a new invention must pass the tests of *obviousness* and *anticipation*. In legal terms, an idea is "obvious" if elements of prior art can be combined to produce the invention, *and* if, at the time of the patent application, the general literature provided suggestions for doing so. An invention is "anticipated" if a single piece of prior art describes the invention in its entirety. These tests are applied as part of the application process, known simply as the patent's *prosecution history*. If an invention is deemed obvious or anticipated in the prior art by the USPTO, then it is not patentable.

KEY TERMS

Documentation	Patent	Teamwork
Gantt chart	PERT chart	Time line
Organizational chart	Project management	

PROBLEMS

6.1 Develop a simple checklist for building a fence.

6.2 Prepare a checklist of tasks required to tune a bicycle for optimum performance.

6.3 Prepare a checklist that will help guide you in the design of a child's safety seat.

6.4 Imagine that you work for a company that is designing a hybrid gas-electric car. Create an organizational chart for the company and a Gantt chart for designing the vehicle's drive train.

6.5 Choose an engineering company with which you are familiar. Develop an organizational chart for the company. Information about a company's personnel can often be found on the company's website.

6.6 Imagine that you wish to start your own company to write software tools for others developing online businesses. Create an organizational chart that outlines the positions you'll need to fill in order to get the company started.

6.7 Develop a timeline for the completion of the prototype of an automobile powered from fuel cells rather than an internal combustion engine.

6.8 Develop a timeline for completing your course requirements over the time span of 4 academic years.

6.9 Suppose that you've been given the assignment to develop a human-powered airplane. Develop a timeline for completing this ambitious assignment.

6.10 Create a Gantt chart for an entry into the national solar-powered vehicle design competition, First Solar (*www.firstsolar.com*).

6.11 Prepare a Gantt chart for the design of a retrofit fire escape for a 10-storey city building.

6.12 Prepare your own version of a Gantt chart that might have been used for the construction of the Golden Gate Bridge in San Francisco, California.

6.13 Develop a Gantt chart for housing an educational conference on engineering design. Consider all needed arrangements, including food, transportation, lodging, and meeting facilities.

6.14 Develop a Gantt chart for the design of a cell phone tower.

6.15 Consider the case of four engineers who are designing an entry into a material handling competition (see *www.mhiq.org*). Develop a Gantt chart for the design of a system based on a complete list of ideas that you develop. The vertical axis of your Gantt chart should reflect members of the design team, not specific project tasks.

6.16 Construct a "fuzzy" Gantt chart depicting the tasks required to design a software system for tracking inventory at a hardware store. Each time allocation should include an error estimate—an increment of fuzzy time—that allows for possible early or late completion of every task on the chart. Based on your fuzzy estimates, what are the longest and shortest possible time durations of the design effort? The vertical axis of your chart should indicate software developers, not individual design tasks.

6.17 Consider the fuzzy Gantt chart scenario of the previous problem. Develop one that might apply to the design of a pilotless model surveillance airplane.

6.18 Can you create a Gantt chart for the creation of the Earth? Your chart can be based on either a religious, evolutionary, or cosmological perspective, as you prefer.

6.19 Do some research into the steps required to build a two-storey commercial-grade building. Then construct a PERT chart applicable to the design and construction of a public library.

6.20 Construct a PERT chart that you imagine to be applicable to the design of the International Space Station.

6.21 Imagine that you lead a team of engineers designing a breakwater for a coastal sea port. Develop a PERT chart for the completion of this project. Consider all elements, from basic research and data analysis to final construction. Identify any critical pathways in your chart.

6.22 Consider the case of a car design competition in which the object is to be the first to drop a small beanbag (a "Hacky Sack®") into a 10-cm square hole cut into a 3-m square tabletop. Develop a PERT chart for the competition of a successful design effort. Identify any critical pathways in your chart and then adjust time allocations in the branches so that the project minimizes the former but results in a project that can be completed within one academic semester.

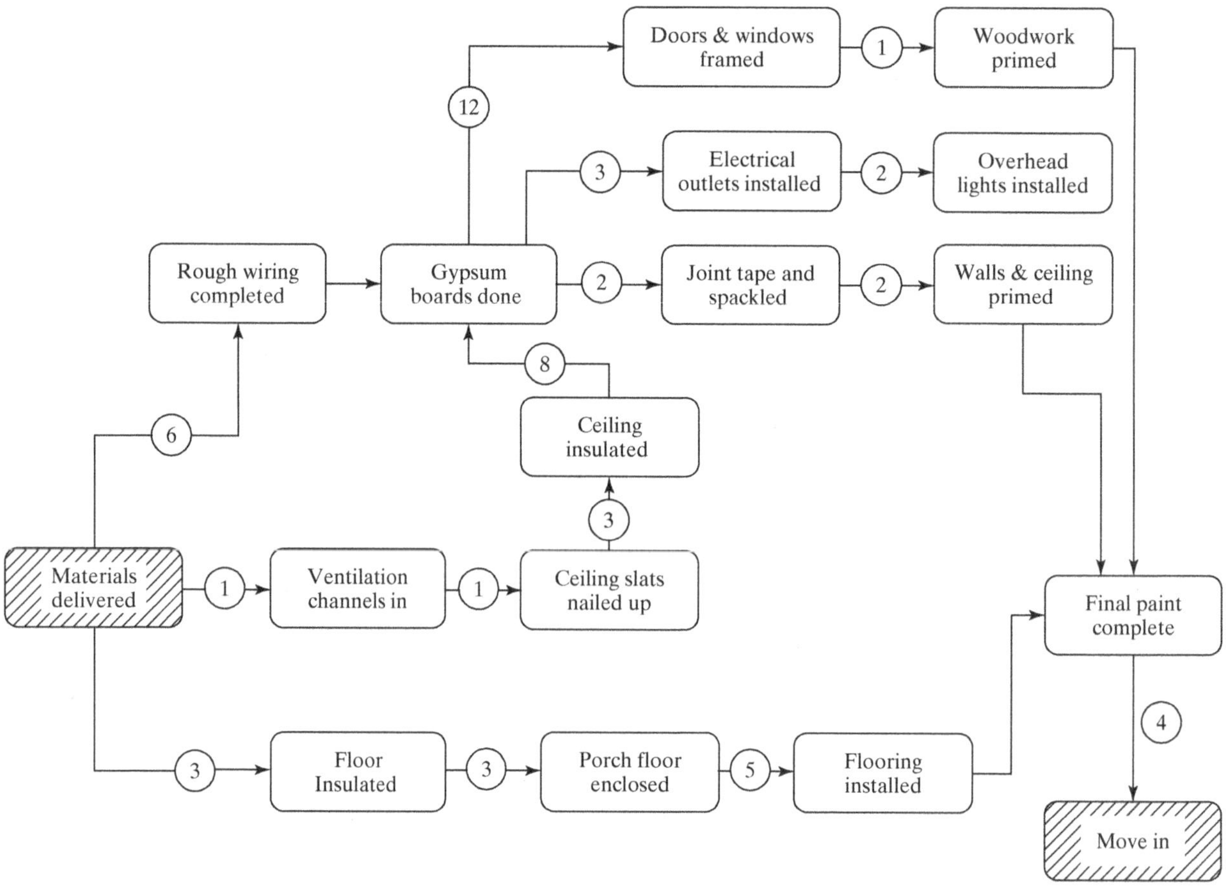

Figure 6.17
PERT chart for Problems 24–26.

6.23 Make a list of conditions under which you might choose a Gantt chart over a PERT chart for time management. Then make a similar list for the opposite case.

6.24 Examine the PERT chart of Figure 6.17. Make a list of all critical pathways as well as the location and duration of all slack time intervals.

6.25 For the PERT chart of Figure 6.17, what is the total time allocated to the completion of the entire design task?

6.26 For the PERT chart of Figure 6.17, identify all the different pathways and their total allocated path intervals.

6.27 Develop a PERT chart for the design of a tunnel linking Beijing to New York City.

6.28 Outline a PERT chart for the completion of the first self-sustaining lunar exploration base.

6.29 Develop a PERT chart for university commencement exercises.

6.30 Begin to keep a logbook of your class activities. Enter sketches and records of design assignments, inventions, and ideas.

6.31 Sketch a logbook page that describes a design concept for a recumbent bicycle.

6.32 Prepare a logbook page that describes the inner workings of a common DVD player.

6.33 Develop a logbook entry describing a machine that shells and stores peanuts.

6.34 Develop a logbook page for the design of the cooling system for an indoor ice-hockey rink.

6.35 Pretend that you are Alexander Graham Bell, the inventor of the telephone. Prepare several logbook pages that describe your invention.

6.36 Pretend that you are Marie Curie, the person who discovered the radioactive element radium. Prepare several logbook pages that describe the activities leading to your discovery.

6.37 Pretend that you are Dr. Zephram Cockrane, the inventor of plasma warp drive on the television and movie series *Star Trek*. Prepare several logbook pages that describe your invention.

6.38 Imagine that you are Elias Howe, the first inventor to perfect the sewing machine by putting the eye of the needle in its tip. This innovation made possible the bobbin system still in use in sewing machines today. Prepare several logbook pages that describe your invention and its initial tests.

6.39 Reconstruct logbook pages as they might have been written by Johan Vaaler, the Norwegian inventor of the common paperclip. (Write your entries in English.)

6.40 Imagine that you are Dr. Maven, the engineer involved in the Heartthrob case of Section 6.3.8. Sketch the logbook pages that outline the basic operating concept of your transcutaneous energy transfer device.

6.41 The invention of the incandescent light bulb is largely attributed to the famous American inventor, Thomas Edison. Reconstruct the logbook pages that Edison may have kept describing his classic design efforts.

6.42 The first computerized telephone system was invented by Erna Hoover in the 1950s. Her switching system used a computer to monitor incoming calls and adjust their acceptance rate by the central switching station. This process helped eliminate system overloads problems. Her U.S. Patent No. 3,623,007, entitled "Feedback Control Monitor for Stored Program Data Processing System," was one of the first software patents awarded in the United States. Look up this patent by its number at *www.uspto.gov*, then develop a logbook page that describes the kernel of Hoover's invention.

6.43 Patricia Bath, an ophthalmologist from New York, invented a method for removing cataracts. This invention transformed the art of eye surgery by using a laser to improve accuracy. Look up U.S. Patent No. 4,744,360 and reconstruct what you imagine to have been logbook pages pertinent to Bath's invention.

6.44 The cotton gin was developed by an American inventor, Eli Whitney, around 1800. This invention had a profound effect on the economic history of the early United States. Reconstruct logbook pages in which Whitney outlines the basic features and development of his invention. Note that patent law in America was in its infancy around the time that Whitney did his work on the cotton gin.

6.45 Famed actress Hedy Lamarr was also a pioneer in wireless communications. She helped develop a secret communication system to help the allies in World War II. Her system was based on manipulating radio frequencies at irregular intervals between transmission and reception. This method of transmitting formed an unbreakable code that prevented classified messages from being intercepted and decoded by hostile listeners. Develop a logbook page that describes the kernel of Lamarr's invention.

6.46 Samuel F.B. Morse, inventor of the telegraph in the 1830s and the pioneer who launched the world's first "information age," was actually an artist by profession when he developed his classic invention. During a long voyage home from study in France, he passed his time thinking about conversations he had heard concerning ongoing experiments in Europe on electricity and magnetism. He developed his ideas for the telegraph while returning to the United States. Sketch the logbook pages that Morse may have kept during his long sea voyage.

6.47 The first pocket calculator was designed by Jack Kilby, an engineer for Texas Instruments. Look up the history of this inventor, and see if you can reconstruct the probable appearance of one or more pages from his logbook.

6.48 Imagine that you are Bessie Blount, a physical therapist who worked with soldiers injured during the Second World War. In 1951, she patented a device that allowed amputees to feed themselves. The device allowed an individual sitting in a wheelchair to bite down on a tube to control the delivery of mouthfuls of food. She later invented a portable device of the same type worn around the patient's neck. Look up U.S. Patent No. 2,550,554 and develop several logbook pages pertinent to Blount's invention.

6.49 Sarah Boone patented an improved ironing board in 1892. The new board was designed to permit better ironing of sleeves and ladies' garments. Her narrow board was reversible, making it easy to iron both sides of a sleeve. Look up U.S. Patent No. 473,653 and reconstruct what may have been one of Boone's logbook pages.

6.50 Augustus Jackson was a candy confectioner from Philadelphia who created several ice cream recipes and invented an improved method of manufacturing ice cream around 1832. Look up the details of Jackson's invention and reconstruct what might have been one of his logbook pages.

6.51 Kevin Woolfolk is the listed inventor on U.S. Patent 5,649,503, entitled "Squirrel Cage Having a Cyclometer and Method for Monitoring the Activity of an Animal." In essence, this patent title describes an improved "hamster workout wheel" that records a pet's mileage or wheel revolutions. Look up U.S. Patent No. 5,649,503 and reconstruct what may have been one of Woolfolk's logbook pages.

6.52 Rachel Zimmerman is the inventor of a software program that has helped people who have difficulty communicating. While only twelve years old she created her software program using Blissymbols. The latter are symbols that enable non-speaking people, such as those having severe physical disabilities like cerebral palsy, to communicate with others. Develop a logbook page as Rachel may have kept it (even at age twelve) in which you outline the flowchart of the program.

6.53 The common automobile windshield wiper was invented by Mary Anderson around 1905. She received a patent for her window cleaning device in 1903.

Anderson conceived of her invention when she noticed that streetcar drivers had to open their car windows in order to see in the rain. To help solve this problem, she conceived of a swinging arm holding a rubber blade that could be operated by the driver from inside the streetcar via a lever. At first, the public was wary of her invention, thinking it would distract drivers, but by about fifteen years later, windshield wipers had become standard equipment on most vehicles. Imagine that you are Ms. Anderson, and sketch out logbook pages that describe your new invention.

CHAPTER 7

Engineering Design

Objectives

In this chapter, you will learn about:

- The engineering design process
- The difference between design, analysis, and replication
- The difference between good and bad design
- The elements of the design cycle
- How to generate ideas through brainstorming

Engineers have made tremendous contributions to the quality of life since the dawn of the industrial age. Automobiles, bicycles, airplanes, space exploration, digital communication, Internet, lifesaving medical devices, all forms of entertainment, global shipping networks, international transportation systems, fuel infrastructure, the national power grid, cellular-telephone networks, fax machines, and the global positioning system all began with one common feature: engineering *design*. One way to define design is that it's any activity whose objective is to meet a need. The object of design might be a physical device, such as a machine or building. Alternatively, the objective might be something less tangible, such as a software program, operating system, web infrastructure, manufacturing method, or process control. In the engineering profession, the word "design" simply answers the question, "What do engineers do?"

7.1 THE USE OF THE WORD "DESIGN"

The common dictionary definition of the word "design" may be found, for example, in Merriam-Webster's online dictionary[1]:

> *design:* to create, fashion, execute, or construct according to plan.

In this book, however, the word "design" will be assigned specific meanings appropriate to the engineering profession. It may be used as a verb (*design a widget that can slice a pizza into five even pieces*), or it might be used as a noun that defines the creation process itself (*learning design is an important part of engineering education*). Alternatively, the word "design" may be used as a noun that describes the end product of an engineer's efforts (*the design was a success and met the customer's specifications*). The word also may be used as an adjective (*this book will help you learn the design process*).

---

[1]Merriam-Webster Online (*www.m-w.com*)

Sometimes, we'll need another word to describe the end goal of a particular design process. For this purpose, we may use the word "product" in its generic sense, even if the item being designed is not a product for sale. Similarly, the word "device" may be used to describe the results of a design effort, even if the entity is not a physical apparatus. Thus, the words *product* and *device* may refer not only to tangible objects, but also to systems, procedures, processes, and software.

7.2 THE DIFFERENCE BETWEEN ANALYSIS, DESIGN, AND REPLICATION

Students of engineering are often confused by the distinction between *analysis*, *design*, and *replication*. In science classes, such as chemistry and physics, students are asked to take and evaluate data and then report the results. This process is known as *analysis*. The word "analysis" can also be applied whenever mathematics is used to predict or confirm the results of an experiment. In contrast, students of engineering are often asked to engage in *design*. Design is an open-ended process where more than one feasible solution may exist. The goal of design is to converge on the best possible solution. In this context, "best" encompasses many factors, including cost, accuracy, robustness, safety, and feasibility. The engineer seeks to meet a set of predetermined specifications, rather than to uncover the secrets behind some physical phenomenon. The difference between analysis and design might be defined in the following way: If the answer can be found by putting together the pieces of a puzzle, then the activity is probably analysis. For example, processing data and using it to test a theory is analysis. On the other hand, if more than one solution exists, and if deciding upon a suitable path demands creativity, choice, testing, iterating, and evaluating, then the activity is most certainly design. Design often includes analysis as one of its component steps, but it always involves creativity and choice as key ingredients.

Noun:
THE DESIGN ⇒

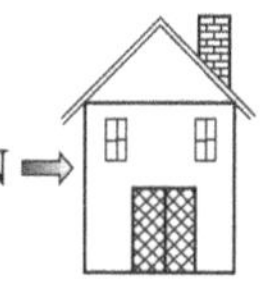

Adjective:
"The Design Product"

One example of the difference between analysis and design can be found in the weather monitoring buoys shown in Figure 7.1. These remote-controlled stations, maintained by the National Oceanic and Atmospheric Administration over all U.S. coastlines and waterways, provide vital data to mariners. Processing data from these buoys and using them to forecast the weather are examples of analysis. Deciding how to *build* the buoys so that they meet the specifications of NOAA is an example of design.

Figure 7.1
Building NOAA weather buoys requires engineering design. Interpreting the data from them involves analysis. *(Photo courtesy of the National Oceanic and Atmospheric Admisnistration.)*

Figure 7.2
The Lyle gun was used to launch rescue lines to vessels that ran aground along the rocky coasts of the Atlantic.

Another example that illustrates the difference between analysis and design can be found in the apparatus of Figure 7.2. The Lyle gun was invented at the behest of the U.S. government by an engineer named David A. Lyle in the late 1800s. This device was used to launch rescue lines to vessels that ran aground along the rocky coasts of the Atlantic. A crude cannon of sorts, the *Lyle gun* saved numerous lives at a time when ships sailed without the benefit of modern navigational aids. The Lyle gun worked by launching a tethered rescue buoy toward a stricken ship. The tether was then used to string a strong rescue cable from ship to shore. Sailors were pulled ashore in a basket hanging from the breeching cable. Of paramount importance was ensuring that the launched buoy projectile successfully reached its target.

Analysis

The development of the Lyle gun no doubt involved much engineering design. Determining the x–y trajectory of its launched, tethered projectile—a critical part of the effort—is an analysis problem.

The Lyle gun exerts an impulse force on the buoy only at the moment of launch, imparting to the buoy initial x and y velocity components v_x and v_y. Thereafter, the buoy follows a free trajectory, subject only to the forces of gravity, air resistance, and drag of the trailing tether. To help get a feeling for the problem, let's simplify it by ignoring air resistance and tether drag in the analysis to follow. (In a real engineering project, one would never ignore these secondary forces without firmly establishing their effects to be negligible.) With these secondary forces ignored, it's a fact of physics that after initial launch, the buoy will feel no x-directed forces, but only the y-directed force of gravity. Hence, the x-component of the buoy's velocity will remain unchanged. Conversely, the y-component of its velocity will be altered by the effect of gravity.

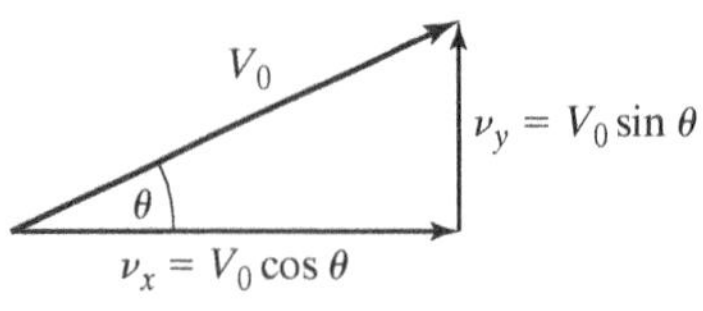

If the initial velocity components of the projectile upon exit from the Lyle gun are designated V_{x0} and V_{y0}, the situation for $t > 0$, after the launch at $t = 0$, can be described by the equations

$$\mathbf{F} = m\mathbf{a} \tag{7.1}$$

$$v_x = V_{x0} \tag{7.2}$$

and

$$v_y = V_{y0} - gt \tag{7.3}$$

If you've had some calculus already, you know that Eq. (7.3) may be found by integrating Newton's Law of Motion: $\mathbf{F} = m\mathbf{a}$. (If you've not yet had calculus, don't worry. Our analysis will proceed directly from Eqs. 7.2 and 7.3).

Here, g is the gravitational constant (9.8 m/s^2), and t is in seconds. In a practical situation, the capabilities of the Lyle gun are known; hence, it is possible to adjust the buoy's total initial velocity V_0 as well as its launch angle θ relative to the horizontal. These values can be used to find the x and y components of the initial launch velocity:

$$V_{x0} = V_0 \cos \theta \tag{7.4}$$

and

$$V_{y0} = V_0 \sin \theta, \tag{7.5}$$

where V_{x0} and V_{y0} at $t = 0$ are related to V_0 by the equation:

$$V_0^2 = V_{x0}^2 + V_{y0}^2 \tag{7.6}$$

With this information, it's possible to fully analyze the buoy's trajectory and predict its landing point. The launch speed V_0 and the launch angle θ of the buoy are set by the user of the Lyle gun by adjusting the amount of explosive charge and cannon inclination, respectively. As suggested by Figure 7.3, the user must first choose

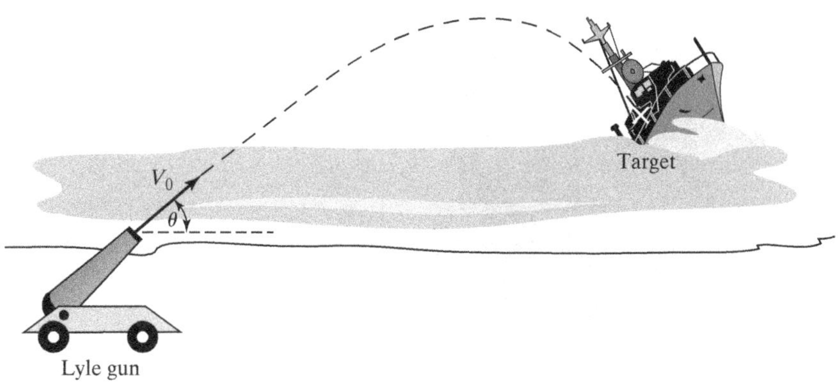

Figure 7.3
A Lyle gun sends a projectile along a trajectory toward a target. Computing the parameters needed to hit the target is an example of analysis. Determining how to best build the device is an example of design.

the target point (presumably a waiting recipient on the stricken ship), then adjust V_0 and θ so that the rescue buoy hits the desired target. Although this process involves making decisions (*for which target point shall I aim?*) and setting parameters (*which* V_0 *and* θ *shall I choose?*), and although the problem has more than one possible solution, it requires analysis only and involves no design.

Design

In contrast to the problem of analyzing the projectile's trajectory, determining *how to build* the buoy catapult system most certainly involves design. Such a system can be built in more than one way, and the designer must decide which method is best. Should the carriage of the cannon be made from wood or metal? The former will be lighter and easier for the rescue brigade to bring to the closest shore, but the latter will be stronger and less likely to fail. Wood can rot, but steel can rust. Should the support for the breeching cable be H-shaped or X-shaped? How large should they buoy be? Should the cannon rest on wheels or skids? Answering these questions requires experimentation, analysis, testing, evaluation, revision—and of course, creativity—all of which are elements of the design process.

Replication

The preceding example illustrates the difference between analysis and design. Specifically, analysis embodies the mathematics needed to determine the buoy's trajectory, while design involves the creation of the Lyle gun itself to meet the user's needs and specifications. In contrast, the word *replication* refers to the process of re-creating something that has already been designed. Replication may involve an exact reproduction, or it may involve minor revisions whose consequences have already been determined. For instance, assembling the laptop computer of Figure 7.4 from premade, constituent components—a circuit board, disk drive, memory module, and display screen—may involve some choices, but the bulk of the design work has been previously performed by other engineers. The task is principally a form of replication. Similarly, building a bird house from a set of purchased plans involves replication, but no design. Replication is an important part of engineering and lies at the core of manufacturing, but it does not require the same set of skills and tools as does true design.

Figure 7.4
Engineered devices, such as this laptop, are mass produced using the process of replication.

EXAMPLE 7.1

Trajectory Analysis[2]

Equations (7.2) and (7.3) describe the motion of the Lyle gun buoy after its initial launch. Given an initial velocity V_0 and launch angle θ, these equations can be solved for the buoy's position coordinates $x(t)$ and $y(t)$ as functions of time. The techniques for obtaining an analytical solution to the problem involve the use of calculus wherein the projectile coordinates $x(t)$ and $y(t)$ are related to the velocities by the derivatives

$$v_x(t) = \frac{dx}{dt} \tag{7.7}$$

and

$$v_y(t) = \frac{dy}{dt} \tag{7.8}$$

Equations of this sort are covered in the math courses taken by most students of engineering; however, many first-year students do not yet have the math skills

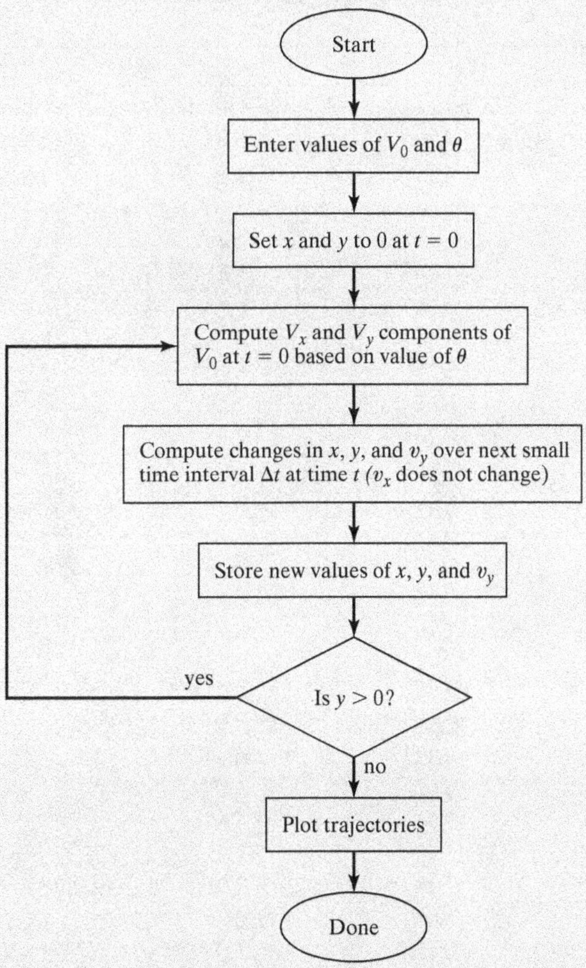

Figure 7.5 Flowchart for the iterative solutions of Equations (7.7) and (7.8). The solution requires that the initial velocity V_0 and launch angle θ be known ahead of time.

[2] If you've not yet had calculus, you can skip this example without loss of continuity.

needed to solve them. Hence, the integral solutions for $x(t)$ and $y(t)$ are simply stated here:

$$x(t) = V_{x0}\, t \tag{7.9}$$

and

$$y(t) = V_{y0}\, t - \frac{1}{2} g t^2 \tag{7.10}$$

Another method by which solutions for the projectile coordinates $x(t)$ and $y(t)$ can be found without differential calculus is to solve for them iteratively by computer. This second method is illustrated in this example. The flowchart of Figure 7.5 illustrates the basic roadmap for writing such a program.

	A	B	C	D	E	F
1	Buoy Trajectory Calculations					
2	V_0	15	launch velocity (m/s)			
3	theta	60	launch angle (degrees)			
4	dt	0.1	time increment in ms			
5	V_{x0}	7.51	m/s			
6	V_{y0}	12.99	m/s			
7	g	−9.8	gravitational constant			
8	pi	3.14	π			
9					**Cell Contents**	
10	time (ms)	x-position	y-position		x-position	y-position
11	0.00	0.00	0.00		V_{x0}*A11	V_{y0}*A11 + 0.5*g*A11^2
12	0.10	0.75	1.25		V_{x0}*A12	V_{y0}*A12 + 0.5*g*A12^2
13	0.20	1.50	2.40		V_{x0}*A13	V_{y0}*A13 + 0.5*g*A13^2
14	0.30	2.25	3.45			
15	0.40	3.00	4.41		:	:
16	0.50	3.75	5.27			
17	0.60	4.50	6.03			
18	0.70	5.25	6.69			
19	0.80	6.01	7.25			
20	0.90	6.76	7.72			
21	1.00	7.51	8.09			
22	1.10	8.26	8.36			
23	1.20	9.01	8.53			
24	1.30	9.76	8.60			
25	1.40	10.51	8.58			
26	1.50	11.26	8.45			
27	1.60	12.01	8.23			
28	1.70	12.76	7.92			
29	1.80	13.51	7.50			
30	1.90	14.26	6.99			
31	2.00	15.01	6.37			
32	2.10	15.76	5.66			
33	2.20	16.52	4.85			
34	2.30	17.27	3.95			
35	2.40	18.02	2.94		:	:
36	2.50	18.77	1.84		V_{x0}*A36	V_{y0}*A36 + 0.5*g*A36^2
37	2.60	19.52	0.64		V_{x0}*A37	V_{y0}*A37 + 0.5*g*A37^2
	2.70	20.27	−0.66			

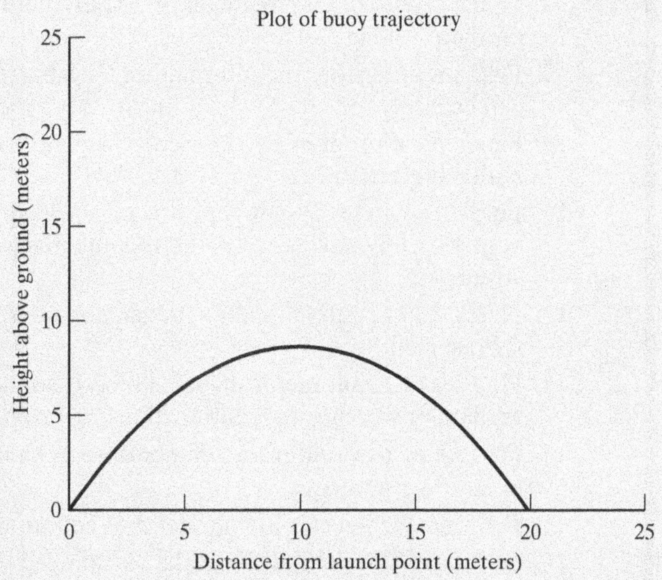

Figure 7.6
Result of the computation of Figure 7.5 for the case $V_0 = 15$ m/s and $\theta = 60°$.

A code listing that implements the flowchart of Figure 7.5 can be constructed using a spreadsheet such as Excel. The problem could also be solved using any of a number of programming languages such as MATLAB, Mathematica, or C^{++}. One possible approach using a spreadsheet is shown below. The value of the cell contents of columns B and C are shown in columns E and F. Each row of the spreadsheet represents a calculation based on the iterated value of time in column A. A plot of the trajectory, based on the x- and y-position values, is shown in Figure 7.6.

Practice!

For Exercises 1 to 25, determine whether the indicated task involves analysis, design, or replication.

1. Find the best land travel route between two cities.
2. Find the shape of soda bottle that will be able to withstand the highest pressure from carbonation.
3. Find a way to prevent people from cutting their hands while slicing a bagel.
4. Find the best dimensions of a 16-ounce soup can so that a packing box for 24 cans has the smallest volume.
5. Find a way to mount a cell phone on a bicycle to permit safe hands-free operation.
6. Find a way to produce individualized bar-coded badges that can be distributed to the attendees of a technical conference.
7. Find a way to seat passengers on an airplane so as to minimize loading time.

8. Find a way to distribute meals on a transcontinental airplane that does not rolling carts.
9. Find a way to store the information for a building directory so that it can be easily retrieved from a touch screen located in the lobby.
10. Find a hack-proof method for counting and tallying votes in a national election.
11. Find a way to use a global positioning system (GPS) receiver to automatically navigate street sweeping machine around city streets.
12. Find a way to automate the oil-changing operation at a drive-in car-service facility.
13. Find a way to automatically label blood samples taken from a hospital patient wearing an identification wrist band.
14. Find a way to monitor the temperature of an unoccupied vacation home via telephone.
15. Develop a children's toy that teaches counting.
16. Develop a system that can make customized shirts for individual customers based on a database of previously acquired body measurements.
17. Develop a system for optimizing a newspaper delivery route for minimum time.
18. Develop a system for loading folded newspapers into protective plastic bags for daily home delivery.
19. Find a way to spread grass seed evenly over an area of lawn.
20. Develop a system for downloading and updating the contact list in a cell phone from a wireless laptop.
21. Develop a system for a telemarketing system that automatically deletes entries from the National Do Not Call Registry (*www.donotcall.gov*).
22. Find a way to assess the condition of the shingles or tiles on a pitched roof without requiring that a person climb on the roof.
23. Develop a system for fulfilling drive-up orders at a fast food establishment.
24. Find a way to determine the height of trees in need of trimming without requiring that the trees be scaled.
25. Find a way to produce custom-printed books that allow a customer to select various available chapters for an individualized text.
26. *Challenge Exercise.* Show by direct solution that Equations (7.2) and (7.3) yield a parabolic trajectory.
27. *Challenge Exercise.* Why does the mass of the projectile not affect the parabolic trajectory of Figure 7.6?
28. *Challenge Exercise.* For a given initial velocity V_0, find the launch angle θ that will yield the furthest target distance x.
29. *Challenge Exercise.* A projectile-buoy target lies 100 m away. Determine at least one possible set of values for V_0 and θ in Figure 7.3 such that a 5-kg projectile hits the target. How much energy must the Lyle

gun impart to the projectile in order to meet this objective? (Hint: When the buoy first leaves the muzzle, its kinetic energy is $mV_0^2/2$.)

30. *Challenge Exercise.* Modify the spreadsheet of Example 7.1 so that a program based in its pathways will plot a) the height of the projectile as a function of time, and b) the angle of the total projectile velocity relative to the horizontal as a function of time.

7.3 GOOD DESIGN VERSUS BAD DESIGN

Anyone who has taken a car in for repair recognizes the difference between a good mechanic and a bad mechanic. A good mechanic diagnoses the problem in a timely manner and makes repairs that last. A bad mechanic fails to find the real problem and masks the symptoms with expensive, unnecessary repairs. Engineers are a bit like auto mechanics in this respect. The world is full of both good engineers and bad engineers. Just because an engineer has produced something does not mean it has been designed well. Just because the product works initially doesn't mean that it will last over time. Although the criteria by which a product is judged varies with the nature of the product, the success of most design efforts can be judged by the characteristics summarized in Table 7.1.

The contrast between good and bad design is readily illustrated in the context of a product made to help rescue individuals from ships run aground. Built in the days before helicopters, the device worked by shooting a tether-carrying projectile out of a small cannon toward the stricken ship, and then using the tether to run a "breeches buoy" line and basket from ship to shore. Imagine that you work for the Apex Rescue Catapult Corporation (ARCC) which has been asked to produce this device for a shoreline rescue brigade. The buyers will judge the worthiness of the apparatus based on the considerations of good and bad design, as illustrated by the following discussion.

1. Does the product meet technical requirements?

It might seem simple to decide whether a buoy launcher meets its technical requirements. Either the tethered buoy hits its target, or it does not. But success can be judged in many ways. A launcher that is well designed will accommodate a wide range of buoy weights, sizes, and shapes. It will require the efforts of only one or two people to operate and will repeatedly hit its target, even in strong wind or rain. A poorly designed launching system may work under ideal conditions but may accommodate buoys of only a single weight or shape. It may fail in strong winds, or it may not produce repeatable trajectories. The launcher might work fine in clear weather, only to fail in a storm.

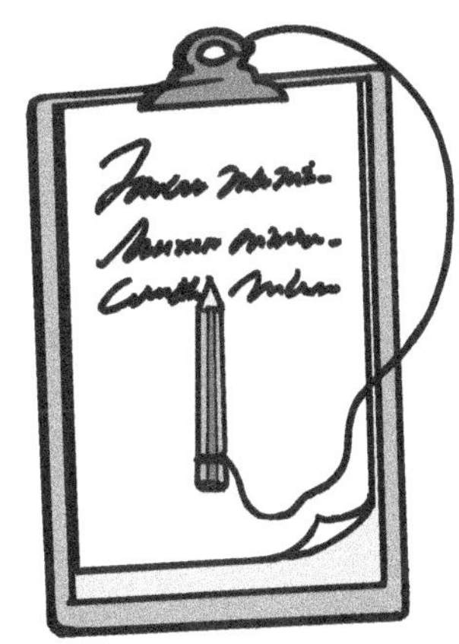

Table 7.1 Characteristics of Good Design versus Bad Design

Good design	Bad design
1. Meets all technical requirements	1. Meets only some technical requirements
2. Works all the time	2. Works initially, but stops working after a short time
3. Meets cost requirements	3. Costs more than it should
4. Requires little or no maintenance	4. Requires frequent maintenance
5. Is safe	5. Poses a hazard to users
6. Creates no ethical dilemma	6. Raises ethical questions

2. Does the product work?

During the development stage, the product cannot be expected to work flawlessly the very first time it is tested. However, it *must* work perfectly before it can be delivered to the customer. It must be durable, and it must not fail in the field. The buoy launcher example provides an excellent illustration of this second principle. Even a bad designer could produce a launcher capable of meeting its specifications upon initial delivery to the customer. The Apex Company could make the a cheap cannon from inexpensive aluminum and construct a simple carriage from wood and nails. A bad designer would also build the launcher in an ad hoc fashion, adding new features on top of old ones without examining how each feature interacts with those before it. Such a launcher might pass inspection upon delivery and be able to launch projectiles during practice, only to fray a line, crack a cannon barrel, or break its trigger mechanism during an actual rescue. After a short period of use, the ill-designed frame might weaken, causing the device to fail during a particularly difficult rescue. Such shortcomings could prove disastrous to a ship's crew awaiting rescue.

A good designer would develop a robust launcher capable of many long hours of service even under the most adverse weather conditions. This conscientious engineer would test different building materials, carriage constructions, trigger mechanisms, and launch devices before choosing materials and design strategies. The good engineer would design the launcher as a whole, carefully considering how its various parts interact. The resulting product might require stronger and more expensive materials and perhaps more assembly parts, but it would prove more reliable and allow the user to hit the target repeatedly under the stress of the most severe rescue operations.

3. Does the product meet cost requirements?

Some design problems can be approached without regard to cost, but in most cases, cost is a major factor that affects design decisions. Often a tradeoff exists between adding features and adding cost. A cannon carriage made from iron will be much less expensive than one made from stronger stainless steel. Will the consumer be willing to pay the higher price? Durable titanium will last even longer and be much lighter in weight, but the monetary cost might be exhorbitant. Will the consumer absorb the cost of the more durable material? Painting the launcher might make it visually more attractive and longer lasting, but will not enhance performance. Will the customer want an attractive piece of machinery at a higher price? Will an attractive appearance make the rescue brigade feel more professional? An engineer must face questions such as these as part of the design process.

4. Will the product need extensive maintenance?

A durable product will provide many years of flawless service. Durability is something that must be considered as part of the design process, even when the cost of the final product is of concern. At each step, in the design process, the designer must decide whether cutting corners to save money will lead to component failure in the future. A good designer will eliminate as many latent weaknesses as possible. A bad designer will ignore them as long as the product can pass its initial inspection tests. If the ARCC wishes to make a long-lasting rescue device worthy of its company's name, it must design durability into its product from the very start.

5. Is the product safe?

Safety is a quality measured in relative terms. No product can be made completely hazard free, so a "safe" product simply has a significantly smaller probability of causing injury than does an "unsafe" product. Assigning a safety value to a product is one of the harder aspects of engineering design because adding safety features usually requires adding cost. Also, accidents are subject to chance, and it can be difficult to identify a potential hazard until an accident occurs. An unsafe product might never cause harm to a particular user, but statistically, it would be more likely to cause injury to some fraction of a large group of users. The cannon-based buoy launching system provides an example of the tradeoff between safety versus cost. Can a launcher be designed to facilitate rescue without endangering people? When a buoy is launched at a stricken vessel, there is a small probability that it will hit a person instead. Perhaps designing a buoy that disperses its mass during flight will reduce the potential for human injury, but such an approach may add to the cost of designing and producing a more complex projectile. Features also could be added to the cannon system to protect its primary users. Guards, safety shields, and interlocks could prevent accidental misfirings, but they would increase the cost and inconvenience of the finished product.

6. Does the product create an ethical dilemma?

Designing a device that can save lives seems like an unquestionably altruistic goal. Yet even such a task raises ethical questions. Although one of its primary goals is to help people, ARCC must ultimately make money if it is to stay in business. Imagine yourself as an employee of this company. If your boss asked you to use cheaper materials but not tell the customer, would you comply with these instructions or defy your employer? If you discovered a serious safety flaw in the system that might lead to human injury, would you insist on costly revisions that would reduce the profitability of the product? Or would you say nothing and hope for the best? Even the very use of the product raises questions of societal ethics. Some communities might be better able to afford a rescue system than would others. Will you advocate more expensive pricing to service affluent communities, or keep the price low so that everyone can afford it? These questions are never simple, but engineers face similar questions all the time. As part of your training as an engineer, you must learn to apply your own ethical standards to problems that you encounter on the job. This aspect of design will be one of the hardest to learn, but is it one that you must master if you wish to be an engineer.

Professional Success: Choose a Good Engineer to be Your Mentor

Will you become a good engineer or a bad engineer? Practicing engineers of both types can be found in the profession, and you must learn to distinguish between the two. As you make the transition from student to professional engineer, you are likely to seek a mentor at some point in your career. Be certain that the individual you choose practices good engineering. Seek an engineer who has an intrinsic feeling for why and how things work. Find someone who adheres to ethical standards that are consistent with your own. Avoid "formula pluggers" who memorize equations and blindly plug in numbers to arrive at design decisions but have little feeling for what the formulas actually mean. Avoid engineers who lack vision and creativity. Likewise, shun engineers who take irresponsible shortcuts, ignore safety concerns, or choose design solutions without thorough testing. In contrast, emulate engineers who are well respected, experienced, and practiced in the art of design. A mentor who excels in explaining things to you is a big plus.

7.4 THE DESIGN CYCLE

Design is an iterative process. Seldom does a finished product emerge from the design process without undergoing changes along the way. Sometimes, an entire design approach must be abandoned, and the product must be redesigned from the ground up. The sequence of events leading from idea to finished product is called the *design cycle*. Although the specific steps of the design cycle may vary with the product and field of engineering, as well as with the instructor of any given undergraduate design course, most cycles resemble the sequence depicted in Figure 7.7. The following section explores this diagram in more detail.

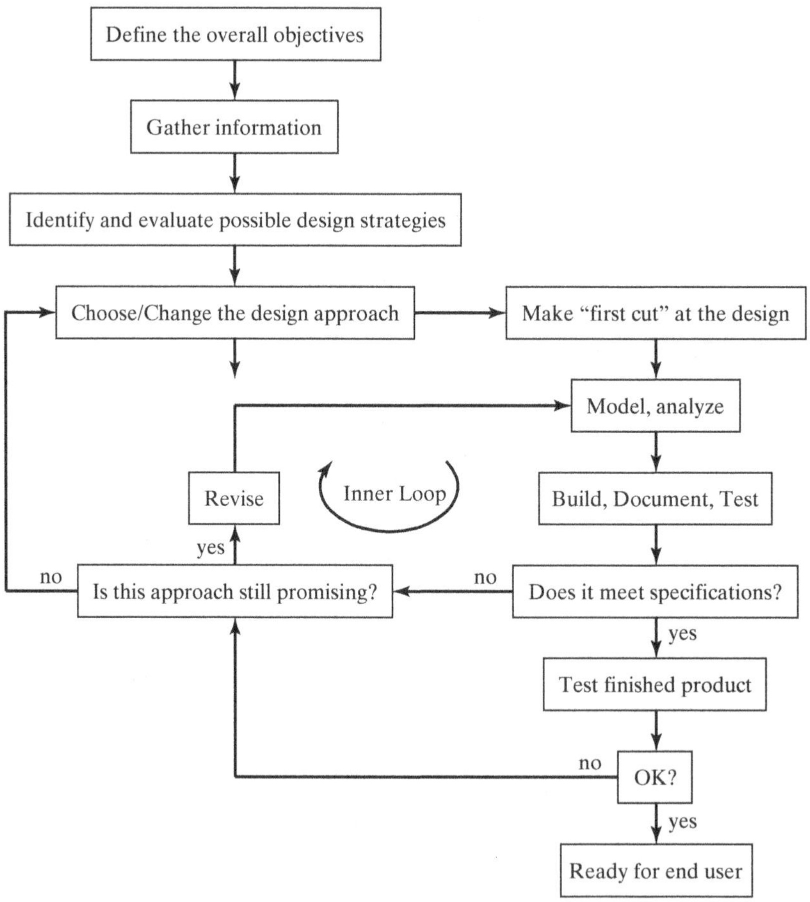

Figure 7.7
One version of the design cycle. Successful design often requires many cycles around the inner loop. Sometimes, the cycle must be reentered when the finished product does not meet design goals.

7.4.1 Define the Overall Objectives

A design team should begin any new project by defining the overall objectives. This step may seem like a nuisance to the student eager to build and test, but it is one of the most important steps in the process. Only by viewing the requirements from a broad perspective can an engineer determine all factors relevant to the design effort. Good design involves more than just making technical choices. Consider, for example, a marine engineer who must design a new sailboat. In addition to technical constraints, the designer must consider the aesthetic, safety, and cost factors that go into a successful design. The engineer must thus ask the following questions: Who will use the product? Will it be a seasoned mariner or a first-time boater? What are the needs of the end user? Will it be used, for example, for business or recreation? What are the cost factors? Which features are critical, and which are only desirable? For example, should the boat have power winches, GPS navigation, and a stereo sound system, or are these needless, expensive frivolities? What are the manufacturing constraints? Must the boat be made of fiberglass, or are other materials acceptable? What are the safety factors? How much risk is acceptable? Answering these questions at the outset will help at each subsequent stage of the design process.

7.4.2 Gather Information

In the early stages of a new project, a large amount of time should be devoted to gathering information. Learn as much as possible about the relevant technology. Do similar solutions already exist? What can you learn from them? Can you identify off-the-shelf modules or components that can be incorporated directly into your design, so that you don't have to "reinvent the wheel?" Look for product descriptions, data sheets, and application notes on the Web. Keep this information in a file folder (either as a hard copy or on a computer) where you'll be able find it easily. Also look for reports or project descriptions in the same general area as your own project. Detailed specifications about most component parts and devices are available on company websites. Such electronic databases have largely supplanted printed catalogs as the primary source of information for design engineers.

Perusing advertisements in trade magazines and journals can be a good way to learn about commercial products that might be relevant to your project. Each field of engineering has many such publications, but the list in Table 7.2 provides a

Table 7.2 Some Engineering Trade Publications

Bioscience Technology	Hydraulics and Pneumatics	Miscellaneous Publications
Compliance Engineering	*Machine Design*	*Advanced Manufacturing*
Computer Technology Review	*Medical Design Technology*	*Aviation Week and Space Technology*
Design News	*Plastics Technology*	*Chemical Engineering/Progress*
EDN (Electronic Design News)		*Environmental Science and Technology*
Electronic Design		*Industrial Engineer Magazine*
		Medical Electronics Magazine

representative set. A more comprehensive directory of technical magazines can be found at *www.techexpo.com.*

7.4.3 Identify and Evaluate Possible Design Strategies

When working with the design cycle, the next step is to identify possible strategies for meeting your design objectives. At this stage, the design team will typically run a brainstorming session to help identify them. Other decisions, such as whether preexisting components will be incorporated into the product, or whether the product will be designed from the ground up, are often made at this time. The team then evaluates the merits of each possible solution, choosing one or more for detailed investigation. The chosen approach is the one deemed most likely to succeed. This choice may change later on, depending on the outcome of each step of the design process.

7.4.4 Make a First Cut at the Design

Once the team has identified a likely design strategy, it's time to make a "first cut" (initial attempt) at the design. This step typically involves rough approximations and estimates. If the product is to be a physical entity, tentative choices are made for dimensions, weight, user controls, construction materials, part numbers, and component values—whatever parameters might be relevant. If the objective is a software product, its outer shell, modular pathways, and user interface are laid out as part of the first-cut attempt. If the design involves a system—for example, a manufacturing procedure—then the overall flowchart might be specified at this time.

If the system is complex, it should be divided into smaller pieces that can be designed by subdivisions of the team and later combined to form the complete product. These subsections, or modules, should be designed so that they can be tested individually before the entire system is assembled. Dividing a large job into several more manageable tasks simplifies synthesis, testing, and evaluation. In a team design effort, the modular approach is essential. For example, the various components of the automobile in Figure 7.8—the engine, cooling system, electrical

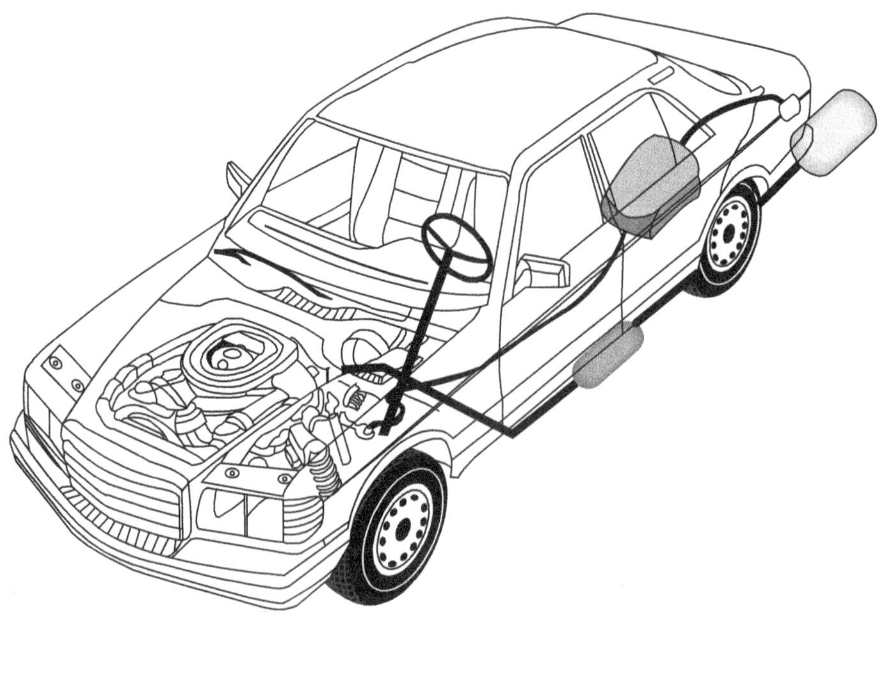

Figure 7.8
Component systems of a typical automobile are designed as interacting modules.

system, suspension, braking system, chassis, and drive train—are each designed and tested individually before the entire automobile is assembled. Constant communication between individuals is vital to ensuring that the automobile's constituent components work together in harmony. The final integrated product will be more than the sum of its individual parts. This same modular strategy applies to large software systems, where each section should be designed as an independently testable module.

The design strategy should also consider similar efforts that may have been attempted in the past. Does a new technology exist that will improve upon an existing design? Perhaps a partial solution is already available in commercial form. For example, suppose that your design for a talking clock includes the amplification of voice or music. You could design your own sound amplifier. Alternatively, many inexpensive off-the-shelf kits exist for constructing board-level amplifiers (see, for example, *www.rainbowkits.com*), so designing your own amplifier would be an unnecessary duplication of effort. The wise engineer uses existing products and components to simplify the design task. There is no shame in using off-the-shelf ingredients if they can help you achieve your design objectives more quickly and inexpensively. Typically, labor is the most expensive part of any development effort, so it's often cheaper (and sometimes more reliable) to buy something ready made and previously debugged than it is to design it from scratch. Imagine how needlessly complex it would be to design a house without using prefabricated windows, doors, electrical panels, insulation, radiators, and air conditioners that are available from other vendors. Be certain, however, that using another company's product does not create patent infringement problems if your product is destined for commercial sale.

7.4.5 Model and Analyze

This activity goes hand in hand with making the first design cut. In some design situations, it could even precede the latter. In the modeling phase of the cycle, a combination of mathematics and/or software modeling tools is used to fine-tune design parameters to shorten the path to the finished product. Modeling and analysis is usually much less expensive (in terms of person-hours of labor) than building an actual prototype for testing.

During the modeling phase of design, commercially available computer simulation tools (such as AutoCAD, ProENGINEER, Solidworks, MATLAB, Orcad, and Simulink) can save time and expense by allowing the designer to predict performance before the actual construction of the prototype. These software packages can help identify hidden flaws before the product is built and provide some indication as to the success of the design approach. Whenever practical, however, computer simulations should never be used as a substitute for actual physical testing unless the product is very similar to one that you have previously designed. "Glitches," "bugs," and other anomalies caused by physical effects not modeled by the simulator have a nasty habit of appearing when a new product undergoes actual testing. Despite the usefulness of computer-aided design tools and simulators, there is simply no substitute for testing a real physical prototype. The use of scale models—for example, an aircraft model made one-tenth of actual size inside a wind tunnel—can be useful for testing large products.

7.4.6 Build, Document, and Test

After the design team has reached a consensus on a first cut, its next goal is to build a working prototype. In some situations, the cost of fabrication is so high that an initial

Figure 7.9
A prototype of the finished product made from wood and other easy-to-fabricate materials.

prototype may not be practical. An oil platform or space station, for example, would fall into this category. In such cases, modeling and analysis can help close the loop between the design concept and the final product.

The typical first prototype is destined to be revised many times before the design cycle has been completed. This first prototype should be functional, but it does not need to be visually attractive. Its primary purpose is to provide a starting point for evaluation and testing. If the product is mechanical, for example, it can be built as a mock-up from easy-to-modify materials, such as wood or pre-punched metal, as shown in Figure 7.9. For instance, a prototype for a new self-standing ATM machine might be built inside an open wooden box made from plywood. Such a structure would permit easy access to the inner machinery during testing, but obviously would never be used in a commercial installation. If the product is electrical, its prototype might be wired on a temporary circuit breadboard, such as the one shown in Figure 7.10. If the product is software, the prototype might consist of the main calculation sections without the fancy graphical interfaces that the user will expect in the final version.

Documentation

Note that documentation is part of the inner loop of the design cycle of Figure 7.7. The typical engineer faces considerable temptation to leave documentation to the very end of the design process. Pressed with deadlines and project milestones, many inexperienced engineers think of documentation as an annoying intrusion rather than an integral part of the design process. After working diligently on a design project, the unseasoned engineer may panic at the reality of documentation. ("*Now I have to write all this up*?") Documentation added as an afterthought is often incomplete or

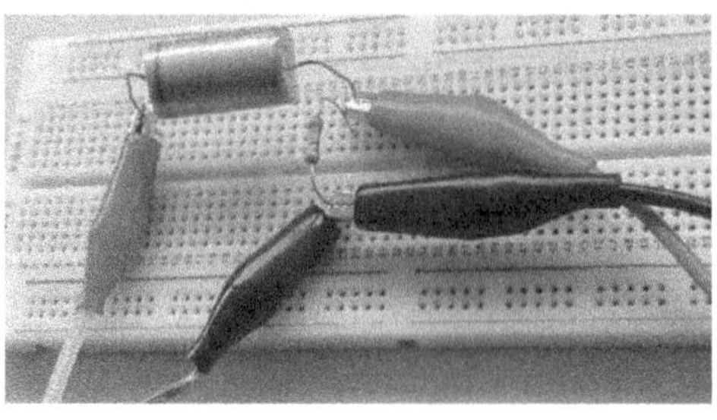

Figure 7.10
Temporary circuit breadboard for electronic prototyping.

substandard, because most of the relevant facts and steps have been forgotten by the time the writing takes place. Haphazard, after-the-fact documentation is the province of the bad design engineer. Many a product, developed at great cost but delivered with grossly inadequate documentation, has found its way to the trash heap of engineering failures because no one could figure out how to use or repair the product. Poor documentation also leads to a duplication of effort and reinvention of the wheel, because no one can remember the results of previous work.

Project Documentation is an important part of engineering design.

A good engineer recognizes that documentation is absolutely critical to every step of the design process. The good engineer will plan for documentation from the very beginning, keeping careful records of everything from initial feasibility studies to final manufacturing specifications. As the design progresses, it's a good idea to write everything down, even if it seems unimportant at the time. Information should be written in such a way that another engineer with the same technical background could continue your work at any time by simply reading your documentation. Careful documentation will also aid in writing product literature and technical manuals should the product be destined for commercial sale. Good documentation provides the engineer with a running record of the design history and the answers to key questions that were addressed as part of the design process. It provides vital background information for patent applications, product revisions, and redesign efforts, and it serves as insurance in cases of product liability. Above all, documentation is part of an engineer's professional responsibility. Its importance to engineering design cannot be overemphasized.

7.4.7 Revise and Revise Again

One of the characteristics that distinguishes design from replication is that the finished product may be totally different from what was envisioned at the beginning of the design cycle. Elements of the system may fail during testing, forcing the engineer to rethink the design strategy. The design process may lead the engineer down an unexpected path or into new territory. A good engineer will review the status of a product many times, proceeding through numerous revisions until the product meets its specifications. In truth, this revision process constitutes the principal work of the engineer. An experienced engineer recognizes it as a normal part of the design process and does not become discouraged when some aspect of the product fails on the first or second try. The revision cycle may require many iterations before success is achieved.

7.4.8 Test the Product Thoroughly

As the design process converges on a probable solution, the product should be thoroughly tested and retested. Performance should be assessed from many points of view; the design should be modified, then tested again, whenever problems are identified. If, at any stage, the product does not meet design requirements, the designer should reenter the iterative loop once again.

Sometimes, though the end may seem near, it is unreachable due to an unforeseen late-breaking but fatal flaw in the design. The product may thus be deemed a failure, even if it seems ready for use. In such cases, the design loop must be reentered with an alternative solution as its basis.

If the product is a physical entity, it should be tested—if relevant—for the effects of temperature, humidity, loading, and other environmental factors, as well as the effects of repeated and prolonged use. Physical products, such as each one of the instruments in Figure 7.11, should be subjected to a "burn-in" (extended use) period

Figure 7.11
Mass-produced electronic instruments for sale undergo a "burn-in" period prior to final shipment.

to help identify latent defects that might cause the product to fail in the field. The human response to the product should also be assessed, because no two people are exactly alike. Exposing the product to many different individuals will help identify problems that may not have been apparent during the development phase, when the product was examined only by the design team. Only after a comprehensive test period is the product ready to be put into actual service. Nothing will discourage consumers faster than a new product that malfunctions in the field.

Like their physical counterparts, software products should also be tested by a variety of different users who can discover hidden bugs. Different individuals will exercise a software product in very different ways; hence, extensive testing by a multitude of users is essential if all software bugs are to be discovered. One way to discover hidden bugs is to release the software program to a control group of customers before widespread distribution. This control group understands that bugs may exist in the preliminary version, and the group is usually given incentives (e.g., a reduced cost or a jump on competitors) to serve as real-world testers. This type of trial is sometimes called a *beta test.*

Professional Success: Being Realistic About the Design Cycle

The design loop shown in Figure 7.7 is typical of what an engineering design team might follow, but it is not inclusive of all situations. Sometimes, steps may be omitted, while others might be added. For example, if the problem is so well defined and its goals immediate, one might pass over the information gathering steps without detriment. Similarly, the analysis step might be so trivial as to be unnecessary. A good structural engineer, for example, would know by rote the rules of thumb for the strengths of commonly used beams. An electrical designer would be familiar with the ampacity of commonly available wire gauges.

In some cases, the prototype step inherit to the "first cut" might be impractical, or not feasible at all. This latter scenario typically applies to large public works projects, for example. The key thing to remember in studying the design cycle is its test and retest loop – an essential element of good design, whether it be by perfecting a physical prototype, or constantly reiterating over a simulation or paper design.

Practice!

1. Consider the questions posed under "Define the Overall Objectives" in Section 7.4.1. Answer these questions in the context of designing a recreational sailboat.
2. Consider the questions posed under "Define the Overall Objectives" in Section 7.4.1. Answer these questions in the context of designing a high-performance bicycle.
3. Suppose that you are asked to build a recumbent bicycle (a low-profile bicycle in which the pedals are in front of the seat.) Make a list of the various ways in which you might gather information as part of the design cycle.
4. Draw a modified design cycle, similar to Figure 7.7, that includes feedback from a test group of individual users (i.e., a "beta" test group).
5. Define the design strategy that might have been used to invent the first personal computer. Imagine yourself before the days of hard disk drives, graphical user interfaces, color monitors, compact disks, USB ports, and inexpensive memory chips. (In the early days of computers, random-access memory chips were one of the most expensive components of the PC.)
6. Describe the various stages of the design cycle for the development of the ballpoint pen.
7. Write a chronology of the design cycle for a paper clip.
8. Specify the steps in Figure 7.7 as they might apply to the design of a microwave oven.
9. Discuss the elements of the design cycle that might apply to a heart rate monitor to be used as part of a medical diagnostic system.
10. Imagine that you were the American engineer who first invented the pinball machine. Write a short essay on your experience, trials, and errors in developing this widely used entertainment device.
11. In-flight refueling is a technique that makes most military air operations possible. Prepare a chronological table that shows how the design cycle may have proceeded when the technique was developed. Label each event in your table with the corresponding step in Figure 7.7.
12. The automobile airbag has become an indispensable safety device, but its history is rather short compared with the evolution of the automobile. Describe what you imagine to have been the first airbag prototype, and then document each of the changes that led to today's airbag design.
13. Imagine that your design team has developed a personal human rocket transportation device. List the design changes that helped you converge on a successful product.
14. Research the history of invention; identify one product that did not require design revisions before becoming successful. (Hint: some inventions that *did* undergo extensive, iterative revisions during the design process include the airplane, the automobile, the sewing machine, and the ballpoint pen.)

15. Write a short essay that describes the evolution of the sewing machine. This product underwent numerous design changes from its inception to the first commercially viable machine.
16. Research the history of the modern dishwasher; draw a detailed design cycle that chronicles the development of this device by its inventor, Josephine Cochran.
17. How do the design steps for a production process differ from those of a physical product? List several examples.
18. Compare the design cycle of a commercial aircraft with the design cycle of a radio-controlled model airplane. How might the various elements differ?
19. The design of a large-scale solvent ore extraction plant will differ markedly from the design of a similar system that classifies ores on a laboratory scale. List several ways in which the design process will differ for these two products.
20. What differences might exist between the design of a reading lamp and the design of a light tower for a football stadium?
21. Consider the issues involved in interfacing a large solar-cell plant that produces hydrogen for fuel cells with the national power grid. What design issues would be of paramount importance
22. Consider the design of a large-scale wind turbine farm. What project-specific elements do you imagine filling the boxes of the design cycle of Figure. 7.7?
23. Consider the design of a tethered buoy for harnessing ocean energy. What project-specific elements do you imagine filling the boxes of the design cycle of Figure. 7.7?

7.5 GENERATING IDEAS

The element of creativity is one of the more salient characteristics that distinguishes design from analysis or replication. Creativity is quintessential to the human experience, particularly for engineers. Without it, the requisite iterations versions of the design cycle would not be possible.

When engineers gather to solve design problems, they can arrive at ideas through a variety of methods. One of these techniques is *brainstorming*. Brainstorming helps engineers channel their creative energies by requiring a spontaneous mode of thinking that frees the mind from traditional boundaries. All too often, we limit our problem-solving approach to obvious solutions that have worked in the past, or perhaps to the first thing we think of. Responsible engineering often requires the consideration of other design alternatives. A good engineer will never settle on a solution just because it's the first one to come to mind. Brainstorming provides one good method for generating a plethora of ideas.

In the early stages of the design process, creativity should proceed spontaneously, unfettered by concerns that any proposed ideas are "way out," "ridiculous," or otherwise impractical. When the constraints of traditional paradigms are removed, new solutions often emerge. Hearing the ideas of others can tap ideas in the subconscious mind. Promising, but different, ideas can eventually be discarded as unfeasible, but only after study, analysis, and comparison with competing ideas.

Brainstorming allows engineers to consider as many options as possible before choosing a design path.

Brainstorming follows many forms. It can be done informally, or it can follow one of several time-tested formal methods. The latter are appropriate for large group settings where organization is needed to avoid chaos. Less formal brainstorming methods are reserved for groups of, say, one to four people who wish to generate ideas. Although they differ in execution, formal and informal brainstorming methods share the same set of core principles. The primary goal is to foster the uninhibited free exchange of ideas by creating a friendly, nonjudgmental environment. Brainstorming is an art. It requires practice, but any team member who has an open mind and imagination can learn this important skill.

7.5.1 Ground Rules for Brainstorming

When a team decides to brainstorm, member should agree ahead of time on a set of rules of behavior. These ground rules should be chosen to create a friendly, nonthreatening environment that encourages the free flow of ideas. Although the specific rules depend on the version of brainstorming, the following list can serve as a guideline:

1. No holding back. Any idea may be brought to the floor at any time.
2. No boundaries. An idea is never too outrageous or "way out" to mention.
3. No criticizing. An idea may not be criticized until the final discussion phase.
4. No dismissing. An idea may not be discounted until after group discussion.
5. No limit. Another idea is never one too many.
6. No restrictions. Participants may generate ideas from any field of expertise.
7. No shame. A team participant should never be made to feel embarrassed for contributing an idea.

7.5.2 Formal Brainstorming Method

When a large group gets together to brainstorm, formal method can be helpful. Without such structure, a flood of competing ideas, all brought to the floor simultaneously, can create chaos. Instead of thinking creatively, participants can become confrontational as they strive to gain a voice in the conversation. With so many randomly competing opinions, each person's creative process is inhibited, and the brainstorming session becomes unproductive. This effect is sometimes called *idea chaos*. Adding formal structure to a brainstorming session restricts the flow of ideas to a manageable rate without restricting the number of ideas generated. In fact, adding formal structure to large group settings can enhance the brain's creative process by preventing aggressive individuals from dominating the conversation and by providing time for people to think.

When a brainstorming session is in progress, one person should act as the facilitator, and another should record everyone's ideas. It's also possible to use video or audio tape in lieu of a human secretary.

Many formal brainstorming techniques exist, but the *idea trigger* method has been well tested and acknowledged to be an effective way to generate ideas in large groups. The idea trigger method enhances the brain's creativity via a process of alternating tension and relaxation to tap the brain's inner resources. By listening to other people's ideas and being forced to respond with counter ideas, a participant's behavioral patterns, personality constraints, and narrow modes of thinking can be broken momentarily. This allows ideas hidden in the recesses of the brain to come to the foreground. A shy participant who is reluctant to offer seemingly silly ideas, for example, may be more willing to speak under the alternating tension and relaxation of the purge-trigger sequence.

The idea trigger method requires a leader, at least four participants, and a piece of paper divided into four or five columns.

A group of students holding a brainstorming session.

Phase 1: Idea Generation Phase

The problem or design issue is summarized by the leader. Without talking, participants each rapidly write down as many ideas or solutions as possible under the first column on their papers. Key words suffice; whole sentences are not necessary. During the idea-generation phase, participants open their minds, consider many alternatives, and do not worry if ideas seem too trivial or ridiculous. "Pie in the sky," radical, or impossible ideas should definitely be included. In short, participants write down any relevant ideas that come to mind. Since the ideas are written down silently, the element of intimidation is removed from the idea generation process.

After the first 2 minutes of the session, the group takes a break and then attempts to write additional ideas under Column 1 for another 1 minute. This *tension and relaxation* sequence has been shown to enhance creativity. It helps to extract all ideas from the brain's subconscious memory, much like squeezing and releasing a sponge several times to extract all the water.

Phase 2: Idea Trigger Phase

After the idea generation phase, participants take turns reading their entries from Column 1. As people recite their Column 1 entries, others silently cross out the duplicates on their own lists. Hearing other ideas will trigger new ideas, which should be immediately written under Column 2. This process is called *idea triggering*. Listening to others causes hidden thoughts stored in the subconscious to surface. The purpose of the idea trigger phase is not to discount the ideas from Column 1, but rather to amplify them, modify them, and add to them.

After all members have read their Column 1 entries and have completed their Column 2 entries, the idea trigger process is repeated. This time, entries from Column 2 are read, and any new ideas are entered under Column 3. The process is repeated, with entries added to Columns 4, 5, and so on, until all ideas are exhausted. Complex problems may require as many as five rounds of the idea trigger phase.

The entries that appear under the second and third columns (and the fourth and fifth columns if the problem is complex) are usually the most creative. Such richness results from several factors. Often participants are secretly angry because their ideas were stolen by others. This simple competitive pressure can propel a person toward new unexplored territory. Conversely, when ideas have not been duplicated by others, participants can receive positive reinforcement, which helps them create even better ideas. Some individuals may respond to their own unduplicated entries with a desire to produce more as a way of hoarding the good ideas. Yet others may subconsciously think that augmenting previously discussed ideas fosters group cooperation.

Phase 3: Compilation Phase

When the idea trigger phase has been completed, the leader compiles everyone's sheets and makes one master list of all the ideas that have been generated. The group then proceeds to discuss all ideas, discarding the ones that probably will not work, and deciding which of the remaining ideas are appropriate for further consideration and development.

EXAMPLE 7.2

A Formal Brainstorming Session

Let's illustrate the formal idea trigger method with an example. Four engineers, Morris, Lawrence, Cindy, and Shana are designing an entry for a design competition. The overall objective is to design a self-propelled vehicle that can climb one side of a trapezoidal, carpet-covered ramp, stop at the top, and prevail over an opposing vehicle that is climbing up the ramp from the other side. The basic elements of the challenge are as follows:

1. The vehicle must be self-contained. No wires or tethers are permitted.
2. The vehicle must fit inside a 25 × 25 × 25-cm cube.
3. The vehicle can be powered in only one of the following ways: a single 9-V battery, any number of rubber bands, or a mousetrap that measures 3 cm × 6 cm or less.
4. The vehicle's weight may not exceed 2 kg.

The specifications of the competition's ramp are shown in Figure 7.12.

The team recently held a brainstorming session using the idea trigger method. They addressed all elements of the design, including the issues of propulsion, offensive and defensive strategies, and a stopping mechanism. The following discussion chronicles their brainstorming session. Moe acted as the leader and timed the first 2 minutes, the break, and the subsequent 60-second idea generation phase. At the end of the phase, Larry's page showed the following ideas:

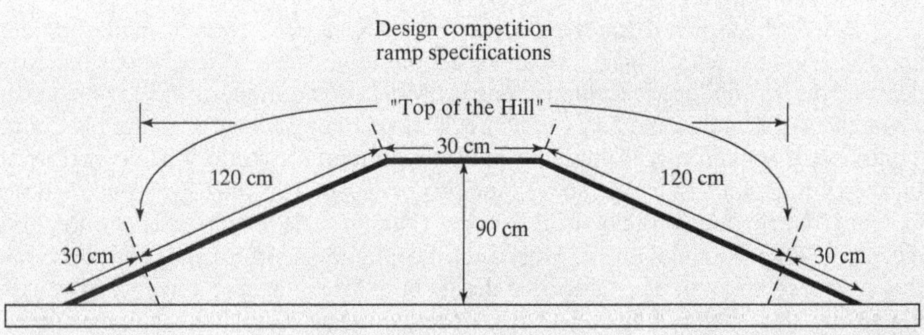

Figure 7.12
Ramp specifications for the vehicle design competition.

Larry	**Idea Generation Phase Column 1 (2 Minutes)**
	Support structure = wood (easy to make)
	Use angle irons from Meccano™
	Plastic body for lighter weight
	Zinc air batteries (lightweight)
	Wheels taken from my radio-controlled car
	Rubber band for chain drive
	Small car will be harder for opponent to deflect
	1 Minute
	Ramming device
	Wedge-shaped body

Larry read his entries. As Cindy listened, she crossed out her own duplicate entries. When Larry was finished, Cindy's first column, including cross outs, showed the following ideas:

Cindy	**Idea Generation Phase Column 1 (2 Minutes)**
	~~No heavy batteries (use zinc air~~)
	Larger wheels for slower turning speed
	Gear box
	Higher torque (harder for opponent to push backwards)
	~~Use plastic for body~~
	Electronic timer for stopping mechanism
	Rechargeable batteries
	~~Wedge shaped design~~
	1 Minute
	~~Buy wheels from hobby shop for radio controlled car~~
	Sense speed, determine distance traveled
	Aluminum frame

Next Shana read her entries that had not been duplicated by Larry. As Cindy listened, an idea flashed into her head: *we can make the drive shaft from a threaded rod.* Cindy reasoned that the threaded rod could screw a sliding nut toward a cut-off switch. The method would not be foolproof, because slipping wheels could ruin the system's ability to track distance. It seemed worth discussing, though, so she wrote "threaded rod" under her Column 2 entries.

When Shana heard Larry read his "ramming device" entry, it made her think about using an ejected object as part of an offensive strategy. She wrote the words "ejected device" under her Column 2 entries. Moe reacted similarly to Larry's idea and wrote the words, "lob something on the track ahead of opposing car" under his Column 2 entries. The spoken trigger phase made its way around the group. When everyone had finished, Moe, acting as leader, started the process again. This time, everyone read their Column 2 entries and wrote new ideas under Column 3. As Shana read her entry about ejected devices from Column 2, Moe had a fleeting image of a Lyle gun breech buoy that he'd read about in an engineering design book. He imagined a flying arrow with a barbed tip shot ahead of the vehicle over the top of the hill. *After hitting the carpet in front of the opposing vehicle,* he thought, *the barbed tip will dig into the carpet, blocking the other car. This spear will be very difficult to dislodge.* Moe wrote "harpoon" under his Column 3 entries.

The second idea trigger round progressed, and Moe started a third. After about 45 minutes, the entire Phase 2 session was finished. Moe suggested a break so that he could compile everyone's lists of ideas. His combined list of entries from everyone's three columns showed the following ideas:

Shape

Small car = harder for opponent to deflect
Wedge-shaped vehicle having same width as track
Rolling can design
Snowplow-shaped wedge

Structure

Support structure = wood (easy to make)
Aluminum frame
Plastic body for light weight
Use angle irons
Hot-melt glue balsa wood

Power

Zinc air batteries (lightweight)
Rechargeable batteries
Change batteries after every run
Electronic timer for stopping mechanism
Microprocessor-controlled car with onboard sensors
Microprocessor speed determines distance traveled

Propulsion

Wheels from radio-controlled car purchased at hobby shop
Large wheels
Rubber band for chain drive
Plastic-linked chain from junked radio-controlled car chassis
Single large mousetrap with mechanical links
Wind up large rubber band

Strategies

Ramming device
Flying barbed harpoon
Pickup arm
Throw jacks in front of oncoming opponent
Roll over opponent with large roller

After the break, Moe reconvened the team to discuss the list of ideas. They weeded out the ones that did not seem feasible and compared ideas that looked promising. Finally, they combined multiple ideas and converged on the concept of a slow-moving, wedge-shaped vehicle for the prototype stage. They also decided to try Moe's flying harpoon strategy, where the harpoon is designed to dig into the carpet and block the path of the opposing vehicle. This strategy constituted their "first cut" entry into the inner loop of the design cycle.

7.5.3 Informal Brainstorming

As discussed in the previous section, the formal brainstorming method requires organization and planning. In contrast, informal brainstorming can be done anywhere. As a technique for engineering design, informal brainstorming in a round table format is appropriate for small groups of people. Ideas can be contributed in random order by any participant. The lack of tension inherent in large groups is absent, and the randomness of interjection can be tolerated if the group is small. The flow of ideas need not be logical, and new proposals can be offered whenever they come to mind. The ground rules introduced in Section 7.5.1 should still be enforced during an informal brainstorming session.

EXAMPLE 7.3

Informal Brainstorming

The following example of an informal brainstorming session describes a hypothetical conversation between two structural engineers who were asked to design a crossbeam for a renovated building. The renovation involves gutting the building interior and tearing out existing internal walls. However, after the initial demolition, the designers discover a support column hidden inside a wall in the middle of a large room. The architectural plans require that the vertical column be removed to create open space, as illustrated in Figure. 7.13. This column holds up the mid-span of a wooden crossbeam that had previously been hidden inside the ceiling. The engineers must find some new way to support the crossbeam and the upper floor it sustains, so that the column can be removed permanently. In its present form, the wooden crossbeam is too weak to be supported only at its two ends. If left unsupported in the middle of its span, it will surely break and cause the floor above it to fall. The engineers,

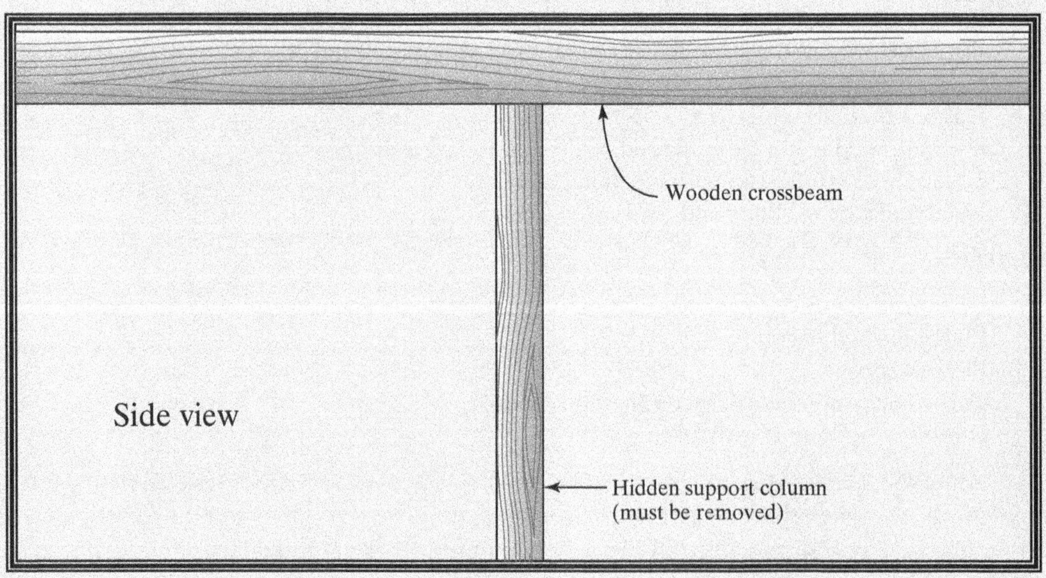

Figure 7.13
Demolition plan showing a previously hidden vertical support column and crossbeam (side view).

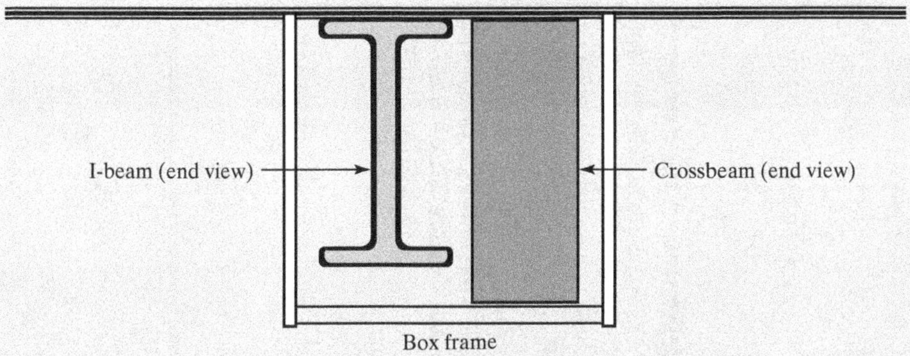

Figure 7.14
I-beam and wooden crossbeam side by side. The box encasing the two beams is very wide and ugly.

Robert and Ernest, discuss the problem using the informal brainstorming method. Note the ebb and flow of ideas between the two engineers. They do not immediately fixate on the first idea that comes to mind; instead, they allow the flow of ideas to lead them to a feasible solution.

Bert: "Let's just install a new steel I-beam under the upper floor, running it side-by-side next to the existing wooden one. A steel beam is much stronger than a wooden beam, so we could remove the supporting column without worry."

Ernie: "OK, we could." *He thought for a while.* "But we'll have to enclose the existing beam in a box casing when we finish the ceiling. Adding another beam side-by-side will make this box *very* wide." *He drew the sketch shown in Figure 7.14.* "This wider-than-expected box will compromise the artistic interior design of the building and will also interfere with the planned layout of the hanging light fixtures.

"What if we *take out* the existing wooden beam and replace it with a steel I-beam? Thus, we can minimize the impact of the box casing."

Bert: "That might work, but we would need to make a temporary beam to support the upper floor while we remove the old beam; that would require a temporary column as well. We'd need to build a false wall in the basement to support the temporary column. It would take lots of time and lots of money. It may not be worth it." *Bert showed Ernie the sketch in Figure 7.15.*

Ernie: "Yeah, it would cost a lot. But the client will understand that we have no choice."

Bert: "But she would not be very happy about it. We need a better solution." *Bert thought for a while.* "I've got it. We can drill holes through the existing beam, and then create thin steel plates with holes in the same places. Then we can bolt the plates on either side of the wooden beam with these huge bolts." *He made a sizing gesture with his hands.* "The composite beam would be strong enough to support the upper floor without the column." *Bert drew the sketch of Figure 7.16 to illustrate his idea.*

Ernie: "I have a better idea. Let's have the plates made up first with the holes. We'll clamp the plates in place, and *then* drill the holes in the wooden beam, using the steel-plate sandwich as a template."

Bert: "Yeah! Good idea. The beam sandwich will be nice and strong and only a little bit wider than the wooden beam alone." *He thought a moment more.* "But we should do some analysis to be sure that the reinforced beam will be

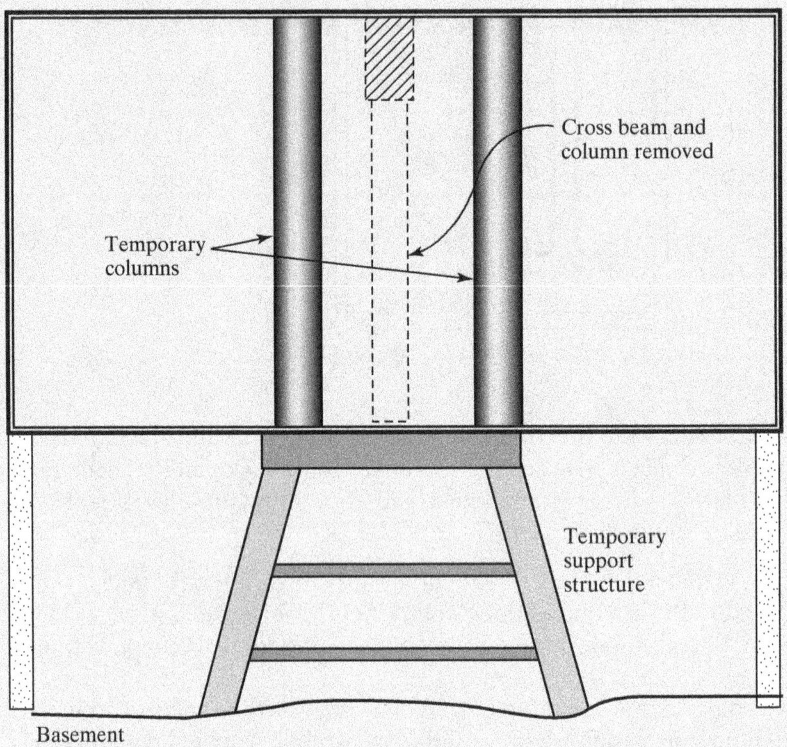

Figure 7.15
Replacing the existing wooden beam requires building a temporary structure in the basement so that the upper floor can be supported while the column and beam are removed.

strong enough. I'll do some rough calculations to figure out how thin the plates can be, and then do some more accurate computer modeling."

Ernie: "Good. In the long run, it will be cheaper to do it this way. I like this idea much better than the first two. It's a good compromise. "

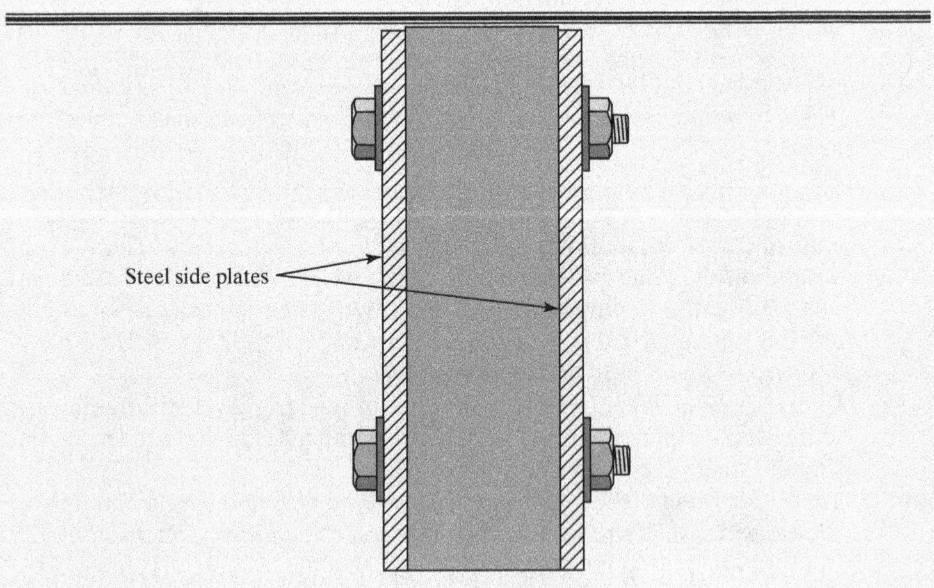

Figure 7.16
Bert's idea for two side plates to strengthen the existing wooden beam.

Professional Success: What to do When One Individual Dominates a Brainstorming Session

Suppose that you are the leader of a brainstorming session, and one member of your team dominates the conversation. That person may criticize participants, dismiss unconventional ideas, cut off speakers, or otherwise break the rules. When this situation occurs, it's your responsibility to keep the offender in line. You might say to the group, "Hey, we need to stick to the formal rules of brainstorming. Let's institute a don't-speak-until-called-upon rule." This approach will tactfully short-circuit the behavior of the dominant person and maintain harmony among team members.

Practice!

1. Examine the final list of ideas compiled by Morris and his group during their idea trigger session. Conduct a one-person "mini-brainstorming session" and add as many ideas as you can. Allow yourself 4 minutes to compile your ideas.
2. A non-engineering friend complains about a pair of ear-bud headphones that keep falling out. Give yourself 5 minutes of brainstorming time, and compile a large list of ideas to solve your friend's problem. After the 5 minutes are over, take a short break, then sort your list, categorizing each idea with a "feasibility" rating of 1 to 5, with 5 being the most feasible.
3. In 2 minutes, write down as many ways as you can for safely confining a dog to a yard. Then rate your ideas in rank order, from least costly to most costly.
4. In 2 minutes, write down as many ideas as you can for designing a hands-free water faucet.
5. To save energy, you'd like to devise a method for reminding people to turn off lights. Devise as many methods as you can to achieve this objective. Allow yourself 2 minutes of brainstorming time.
6. Form a team of three to five colleagues. Using brainstorming method of your choice, come up with as many methods as you can for alerting city officials that parking meters are full and need to be emptied of coins (e.g., emptied "on demand", rather than on a schedule.
7. Form a team of three to five colleagues. Using brainstorming method of your choice, come up with as many concepts as you can for a system that will automatically feed pellets to a pet hamster. Allow yourself 3 minutes of brainstorming time. If possible, work with a group of up to four people.
8. Get together with a teammate to conduct an informal brainstorming session. Take turns contributing a single new idea, one after the other, until one of you has run out of ideas. Then the other should declare all remaining ideas. Each of you should write down ideas as soon as they come to mind. Use this method to develop a concept for designing an automatic bagel-slicing machine.

9. Imagine that you are part of design team that is developing a method to synchronize an MP3 personal music player with a companion program on a desktop computer. Compose the idea trigger session sheets that four teammates might contribute to such a session. These sheets should have the form shown in Figure 7.12. Also, compile a composite sheet that summarizes the session.
10. In two time spans of 2 minutes each, with a 1 minute break in between, think of as many ways as possible to determine the depth and breadth of a newly discovered underground oil deposit.
11. The following exercise involves a futuristic problem. Meet with one or more classmates for a brainstorming session (either formal or informal). Write multiple ways to transfer information from a human into a computer without a keyboard or mouse. Provide details of each possible method.
12. Imagine that you have a toy rocket, a baseball, and a helium balloon. Get together with three to five fellow students, and run a brainstorming session to determine different ways of using these items, and these items only, to determining the height of Mount Washington in New Hampshire, USA.
13. Run a brainstorming session to determine how one might use a barometer, stopwatch, and tape measure to determine the height of the Washington Monument in Washington, DC?
14. Reconstruct a brainstorming sheet of the type discussed in Example 7.2 as it might evolve during a session to develop a sensing mechanism for measuring the speed of a bicycle.
15. Suppose that your design team has been given an egg, some tape, and several plastic drinking straws. Run a brainstorming session to devise a system to prevent the egg from breaking when dropped from a height of 6 ft (2 m). You are to use only the aforementioned materials.
16. Get together with one classmate. Run an informal brainstorming session to devise as many different methods as you can for using your desktop computer as a stop watch to time a road race.
17. Think about a large aquarium (the kind the public visits to see large sea creatures). Suppose that your team has been given the task of designing a system for washing the inside surfaces of the windows from the outside. Reconstruct a brainstorming sheet of the type discussed in Example 7.2 as it might evolve during a brainstorming session.
18. Reconstruct a brainstorming sheet of the type discussed in Example 7.2 related to the design a system to help a quadriplegic turn the pages of a book.
19. Reconstruct a brainstorming sheet of the type discussed in Example 7.2 related to the design of a system for automatically raising and lowering the flag at dawn and dusk each day.
20. Reconstruct a brainstorming sheet of the type discussed in Example 7.2 related to the design of a system that will automatically turn on a car's windshield wipers when needed.

21. Reconstruct a brainstorming sheet of the type discussed in Example 7.2 related to the design of a device that can alert a blind person when a pot of water has boiled.
22. Reconstruct a brainstorming sheet of the type discussed in Example 7.2 related to the design of a system to line up screws on an assembly-line conveyor belt. All screws must point in the same direction.
23. Reconstruct a brainstorming sheet of the type discussed in Example 7.2 related to the design of a method for detecting pinhole leaks in latex surgical gloves during the manufacturing process.
24. Reconstruct a brainstorming sheet of the type discussed in Example 7.2 related to the design of a method for deriving an electrical signal from a magnetic compass. The compass must interface with a computer running navigational software.
25. Imagine that you have been given a coil of rope and eight poles. Run an informal brainstorming session to devise a method for building a temporary emergency shelter in the wilderness.
26. Run an informal brainstorming session to devise an alarm system to prevent a thief from stealing memory chips from inside a personal computer.
27. Imagine custodial workers who are in the habit of yanking on the electrical cords of vacuum cleaners to unplug them from the wall. Run an informal brainstorming session to devise a system or device to prevent damage to the plugs on the ends of the cords.
28. Reconstruct a brainstorming sheet of the type discussed in Example 7.2 related to the design of a system for automatically dispensing medication to an elderly person who has difficulty keeping track of schedules.
29. Reconstruct a brainstorming sheet of the type discussed in Example 7.2 related to the design of a system that will agitate and circulate the water in an outdoor swimming pool so that a chlorine additive will be evenly distributed. Assume that a ground-fault protected (GFCI) electrical outlet is available at the pool site.
30. Reconstruct a brainstorming sheet of the type discussed in Example 7.2 related to the design of a system that will allow truck drivers to check their tires' air pressure without getting out of the vehicle.
31. Can brainstorming be used to solve math problems? Why or why not?

7.6 DESIGN EXAMPLES

In this section, the principles of engineering design are illustrated by three specific examples. The approach in each case emulates key elements of the design process.

7.6.1 Model Vehicle Design Competition

Imagine that you've entered a vehicle design competition in which the sponsor, your engineering college alumni association, offers textbook gift certificates to the college bookstore to the winners. Given the high cost of textbooks, you're eager to win.

The goal of the competition, as previously shown in Figure 7.12, is to design and construct a vehicle that can climb a ramp under its own power, stop at the top of the ramp, and sustain its position against an opposing vehicle coming up the other side of the ramp. The vehicle deemed "on top of the hill" is the one closest to the centerline after a 15-second time interval. Multiple runs against different pairs of competitors will determine the final winner.

The rules state that the vehicle can be powered by just one of the following energy sources:

- A battery of up to 9 volts
- A rubber band (4 mm × 10 cm maximum size in its unstretched state)
- A mousetrap (3 cm × 6 cm maximum spring size).

Imagine that you have entered this design competition and wish to design a competitive vehicle. Let's examine the problem using the design cycle of Figure 7.7. Remember that the problem can be addressed in many different ways. The solution presented here is but one of many.

1. Gather Information

The rules for the competition comprise a self-evident source of information about the design objectives. At this stage, it would be a good idea to gather information about battery life and power output, rubber band statistics, commercially available mousetraps, and, of course, easily obtainable electric motors. Only after compiling a comprehensive set of such information can you proceed with confidence to the next step of choosing a design strategy.

Choose a Design Strategy. Many different design strategies will lead to a vehicle capable of competing. Building a *winning* design, however, requires careful consideration of several key issues. How can you know ahead of time what the right choices will be? In truth, you can't, especially if you've never built such a vehicle before. You can only start by making educated guesses based on your experience and intuition, plus the information you've gathered in the previous step. You then can rely on the iterative nature of the design process to help you converge on a workable solution. The following list defines some possible design strategies that you might adopt:

1. *Design for speed.* The vehicle with the fastest speed will not necessarily win, but one strategy could be to race to the top as fast as possible, then defend against a slower opponent by blocking access to the top of the ramp using a proper defensive strategy.
2. *Design for strength.* Alternatively, you could design a strong, steady, slower-moving vehicle capable of pushing away the opposing car as it "bulldozes" its way to the center of the ramp.
3. *Design for easy changes.* The rules state that modifications to the vehicle are permitted between runs. Adopting an easy-to-change construction strategy will facilitate "on-the-fly" changes to your vehicle.
4. *Design for durability.* During the competition, your vehicle must endure many trips up the contest ramp. Opposing vehicles and accidents can damage a fragile design. You must weigh the issue of durability against your desire to produce a vehicle that's flexible and easy to modify.

Note that strategies (3) and (4) are not independent of one another. For example, designing for easy changes may conflict with building a durable vehicle. Engineers typically face such tradeoffs when making design decisions. Deciding which pathway to take requires experience and practice, but making any decision at all means that you've begun the design process.

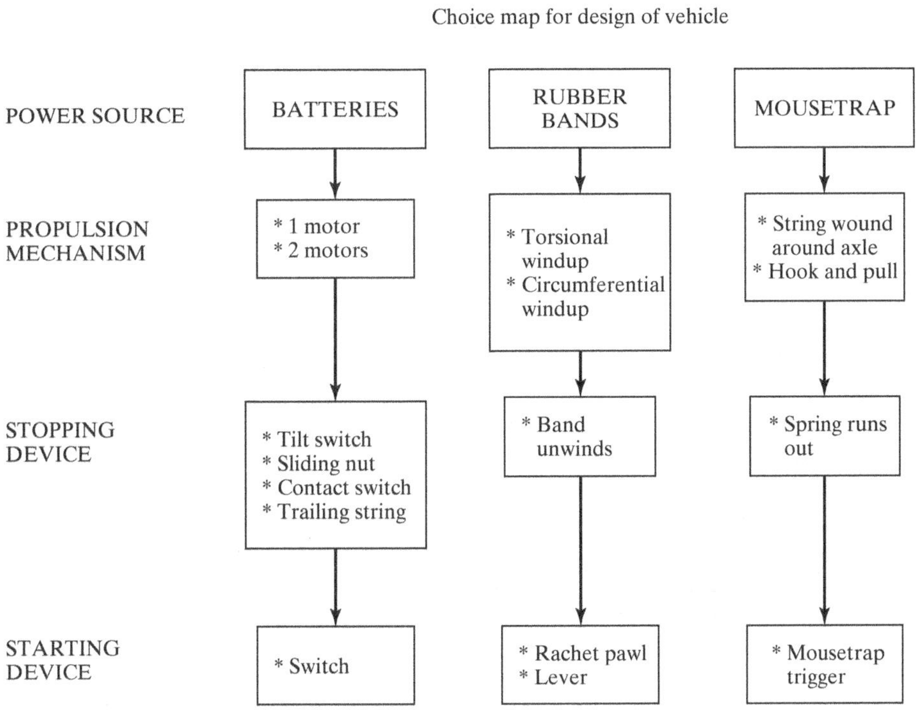

Figure 7.17
Choice map that outlines the decision tree for the first phase of the design process in the vehicle design competition.

The rules of this particular competition provide for many alternatives in vehicle design. Regardless of the details, however, all vehicles will require the same basic components: an *energy source*, a *propulsion mechanism*, a *stopping method*, and a *starting device*. After some discussion with your teammate, you develop the *choice map* shown in Figure 7.17.

Although the choice map does not provide an exhaustive list, it serves as an excellent starting point for your design effort. The following paragraphs outline some of the thought processes that might accompany your design choices.

Energy Source. Batteries are attractive as an energy source because they require no winding or preparation. They will need frequent replacement, however, and will thus be more expensive than the mechanical alternatives. Rubber bands will require much less frequent replacement, but will store the smallest amount of energy among the three choices. Like a rubber band, a mousetrap does not need frequent replacing. It stores more energy than a rubber band, but because of its physical form, it offers the fewest options for harnessing its stored energy.

Propulsion Mechanism. Your choice of the propulsion device, or *energy converter*, will depend on your choice of an energy source. If you decide to use batteries, an electric motor will be the obvious choice for turning the vehicle's wheels. Rubber bands can be stretched to provide linear motion or twisted for torsional energy storage to turn an axle or power shaft. Alternatively, a rubber band can be stretched around a shaft or spool like a fishing reel and used to propel the vehicle's wheels. A mousetrap can provide only one kind of motion. When released, its bale will retract in an arc, as depicted in Figure 7.18. In principle, this motion can be harnessed and used to propel the vehicle.

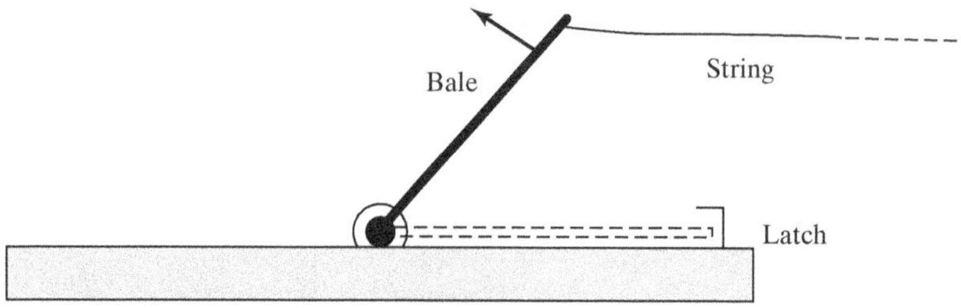

Figure 7.18
Harnessing the stored mechanical energy of a mousetrap. The bale retracts in an arc when the mousetrap is released.

Stopping Method. According to the competition rules, your vehicle must stop when it arrives at the top of the ramp. This requirement can be met by interrupting propulsion power precisely at the right moment and relying on a combination of gravity and friction to stop the vehicle. A braking device to augment these forces might also be considered. If the vehicle is powered by batteries, there are many ways to interrupt power flow to the vehicle. A simple tilt switch can disconnect the battery when the vehicle is level, but connect the battery when the vehicle is on a slope. A metal ball that rolls inside a small cage and makes contact with two electrodes, as illustrated in Figure 7.19, might serve as a suitable tilt switch. Other choices include a spring-loaded contact switch, such as the one shown in Figure 7.20, or a system that cuts off power to the wheels after the car has traveled a preset distance as measured by wheel rotations. This latter scheme will work well only if the wheels do not lose traction and slip on the track.

One interesting alternative to a mechanical switch would be to use an electronic timer circuit that shuts off power from the battery after a precise time interval. Through trial and error, you could set the elapsed time so that the vehicle stops precisely at the top of the ramp. One problem inherent to this open-loop timing system is that the vehicle does not actually sense its own arrival at the top of the

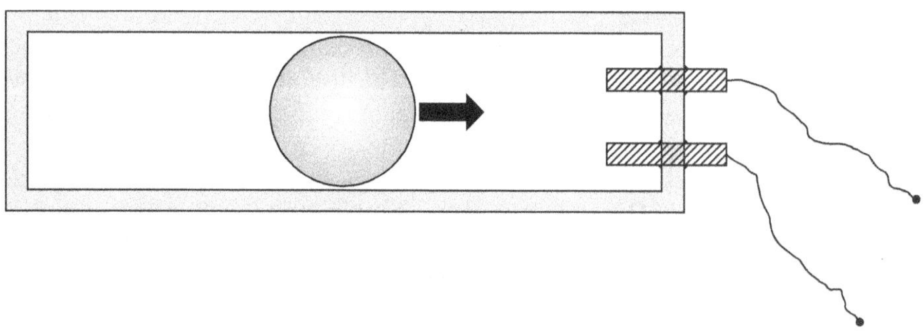

Figure 7.19
Tilt switch made from a small enclosure, a metal ball bearing, and two contact points.

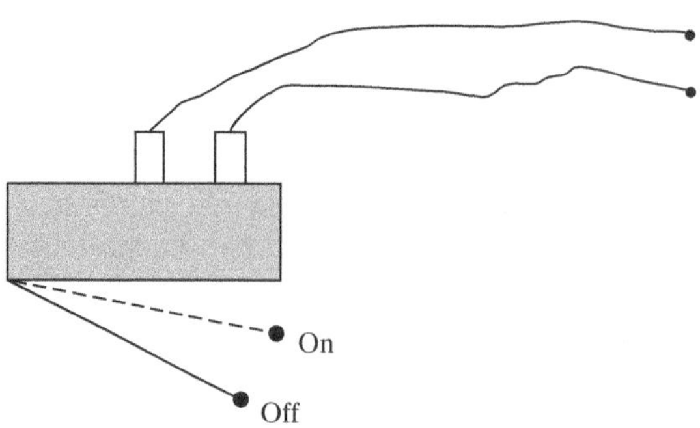

Figure 7.20
Spring-loaded contact switch.

ramp, but rather infers it by interval timing. Because the speed of the vehicle may decrease with each successive run as battery energy is depleted, this timing scheme might cause problems. The method will also fail if the wheels slip on the track. On the other hand, it is likely to be more reliable than solutions that involve mechanical parts.

If a rubber band or mousetrap is chosen as the power source, then stopping the vehicle will require something other than an electrical switch. One way of stopping a vehicle propelled by mechanical energy storage is simply to allow the primary energy source to run out (e.g., allow the rubber band to completely unwind). While this method is crude, it is reliable, because power input to the vehicle will *always* cease when the source of stored energy has been depleted.

Starting Device. If the vehicle is powered by a battery, then an electrical switch becomes the most feasible starting device. You could, however, design the vehicle so that the motor is brought up to speed prior to the start, then mechanically engaged at the starting time. This latter approach might enable a strategy involving rapid ascent up the ramp. A rubber-band power source will require a mechanical device such as a trip lever to initiate power flow to the wheels. A mousetrap can use its built-in trigger mechanism or any other starting mechanism that you might devise.

2. Make a First Cut at the Design

The first design iteration begins with rough estimations of the dimensions, parameters, and components of the vehicle to make sure that the design is technically feasible. After discussing the long list of design choices, you and your teammate decide to use a battery-powered vehicle. This decision allows many choices for a stopping device. You feel that the flexibility inherent to this design choice far outweighs the advantages of mechanical propulsion schemes. You decide upon a defensive strategy and agree to build a slower-moving, wedge-shaped vehicle powered by a small electric motor. The advantage of this design approach is that the motor can be connected to the wheels using a large gear ratio, thereby providing higher torque at the wheels and a mechanical advantage that would be unavailable to a very fast vehicle. Because your vehicle will be slower than the others, it may not reach the top of the ramp first, but its wedge-shaped design will help to dislodge any opposing vehicle that does arrive first at the top of the ramp. On the other hand, if your vehicle arrives first, its defensive wedge shape will cause your opponent's car to ride over your car's body, allowing you to maintain your position at the top.

A rough preliminary sketch of your car is shown in Figure 7.21. You've entered this sketch into a notebook that contains all information relevant to the project, including design calculations, parts lists, and sketches of various pieces of the car. Shown in Figure 7.21 are the car's wedge-shaped design; a single drive shaft driven by a motor, belt, and pulleys; and a single switch to turn off the motor when the vehicle arrives at the top of the ramp.

3. Build, Document, Test, and Revise

The sketch of Figure 7.21 represents a beginning, but it is not the finished product. You still have many hurdles to overcome and tests to run before your vehicle will be ready to compete. The next step in the design process should probably be the building and testing of a "first-cut" prototype. To help you in this phase of the design process, you've built a test ramp that mimics one side of the official test ramp. You begin by constructing a chassis shell in the form of a wedge, but without a motor drive or stopping mechanism.

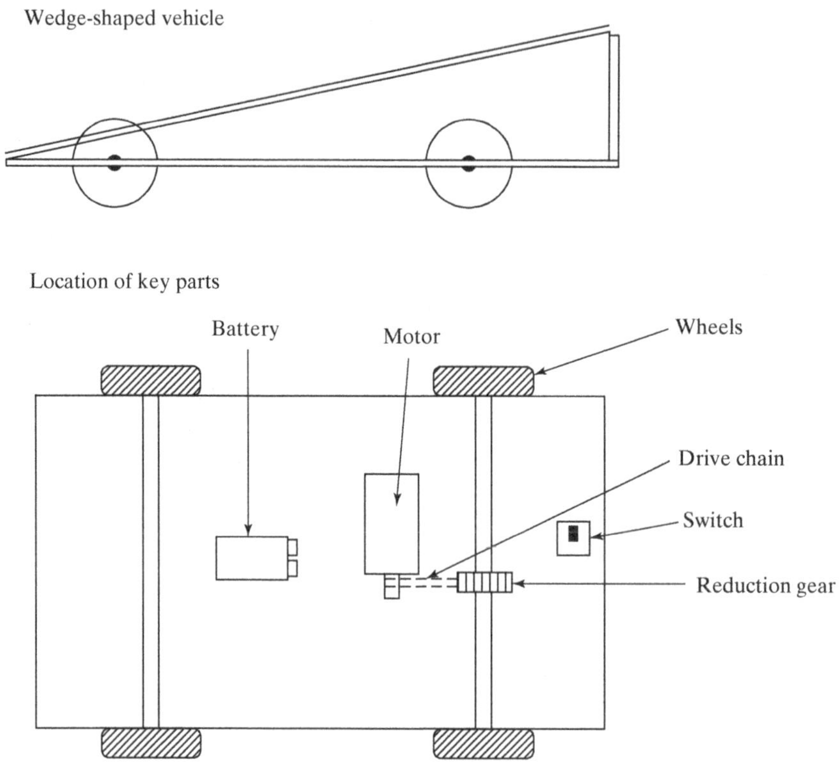

Figure 7.21
Rough, preliminary sketch of a car for the vehicle design competition.

You run your wedge-shaped vehicle up the ramp slope by hand. You soon discover that the bottom of the vehicle hits the ramp at the top of the hill, as depicted in Figure 7.22(a). The change in the angle of the ramp is large, and all four wheels do not always maintain contact with the track surface. You discuss several solutions to this problem with your teammate. One solution would be to increase the diameter of the wheels, as shown in Figure 7.22(b). This change would decrease the mechanical advantage between the motor and the wheels, requiring you to recalculate the torque required from the motor. Another solution would be to make the vehicle shorter, as in Figure 7.22(c), but you realize that this solution would lead to a more acute angle of your wedge shape and reduce its effectiveness as a defensive strategy. (The thinner the wedge, the more capable the car of wedging itself under opposing vehicles. However, the rear of the wedge must be the same thickness to leave space for the motor and gear box.)

4. Revise Again

Your teammate suggests keeping the wheels and shape of the wedge the same and moving the rear wheels forward, as depicted in Figure 7.23. You rebuild the vehicle by moving the rear shaft mount forward, and you test your vehicle again. The redesigned vehicle no longer bottoms out on the track, and you claim success. Your professor sees your design changes and suggests that you test your vehicle under more realistic conditions. For example, what will happen when another vehicle rides over the top of your wedge-shaped body? You proceed to simulate such an event by placing a weight at various positions on the top of the car. The results of these additional tests suggest that moving the wheel locations may not be the best solution to your problem. When you move the rear wheels forward, you change the base of

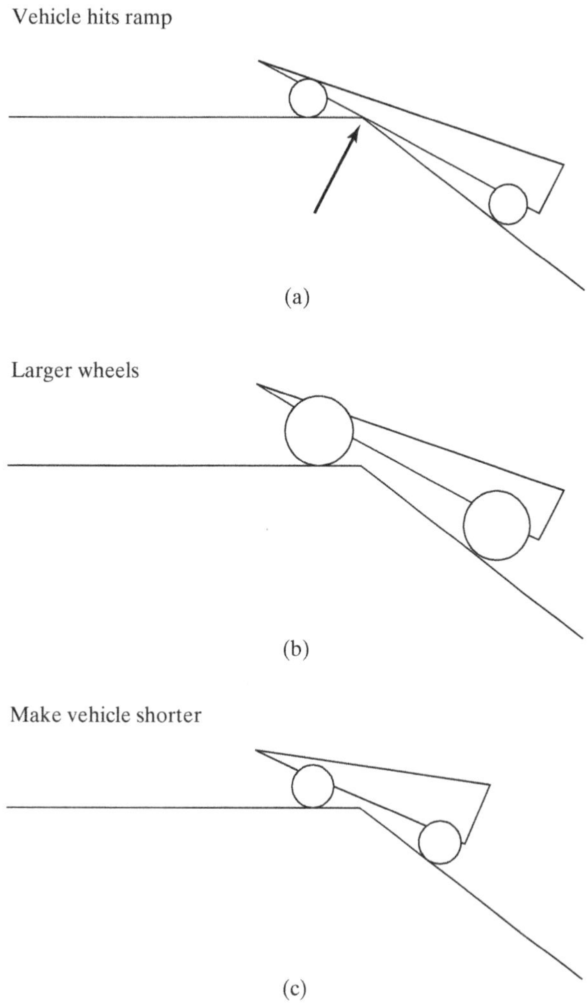

Figure 7.22
Vehicle at the top of the ramp. (a) Bottom of vehicle hits the ramp; (b) vehicle with larger wheels; (c) a shorter vehicle.

support for the car's center of gravity. You discover that if an opposing vehicle rides over the top of your car, the net center of gravity moves toward the rear, eventually causing your car to topple backwards, as depicted in Figure 7.24.

5. Reality Check

This latest discovery may seem like a setback, but it's a normal part of the iterative design process. Some things work the first time, while others do not. By observing

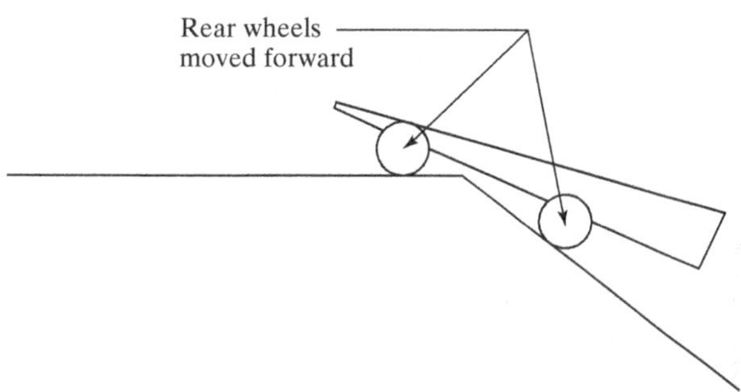

Figure 7.23
Moving the rear wheels forward.

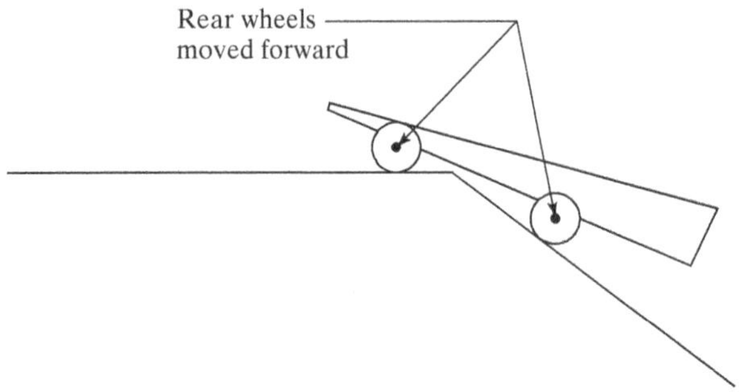

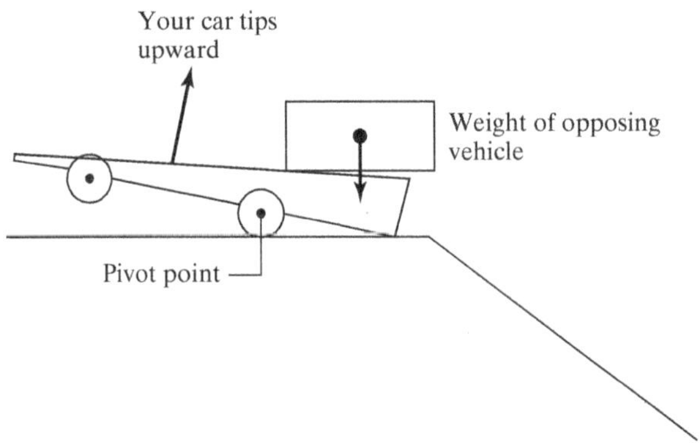

Figure 7.24
Weight of opposing vehicle on top of rear end causes car to topple backwards.

and learning from failure and by building, testing, revising, and retesting, you will be able to converge on the best solution to the problem.

6. More Revisions

After some thought, you decide that increasing the diameter of the wheels may be the best option after all. Your teammate points out that you can simply change the ratio of the gear box to preserve the net mechanical advantage between the motor and the wheels. This change will allow you to accommodate larger wheels. You buy some new wheels and try them with success. With the rear axle moved to its original location and the larger wheels in place, your car no longer bottoms out on the ramp.

Practice!

1. Make a two-column list that outlines the advantages of the various power sources for traveling up the ramp of Figure 7.12 in the vehicle design competition.
2. Make a list of additional propulsion mechanisms that could be used to drive a design competition vehicle.
3. Make a two-column list that outlines the advantages of using gravity and friction versus an applied brake as a stopping mechanism for a design competition vehicle.
4. Determine the minimum energy needed to drive a 2-kg vehicle from the bottom of a ramp to the top, a net vertical distance of 1 meter.

5. How much electrical energy (in joules) is needed to exert one newton of force over a distance of 1 meter (m)?
6. How much electrical power (in watts) is needed to exert one newton of force on a body over a distance of 1 m for 10 seconds?
7. Determine the number of turns per centimeter (cm) of wheel diameter that will be required to move a vehicle from the bottom to the top of a 1-m ramp in a vehicle design competition.
8. For a 1-m design competition ramp, determine the wheel diameter needed to move a vehicle from the bottom of the ramp to the top with 50 turns of the drive axle.

Professional Success: Parallel Paths to Different Design Solutions

One of the interesting features of the design process is that more than one perfectly acceptable solution may exist to the same engineering problem. The endpoint of your design efforts may be affected more by arbitrary design decisions than by the intrinsic worth of one solution over another. This scenario is not unlike that found in any of several science fiction stories involving time travel, in which parallel time travel paths, each originating at the same point event, lead to drastically different outcomes. In the classic movie *Back to the Future*, for example, Marty McFly (played by Michael J. Fox) inadvertently travels back in time to his parents' 1955 high school. With the help of his scientist companion, Dr. Emmett Brown (played by Christopher Lloyd), he repeatedly travels back to the present, each time returning to a different reality based on small changes he made in his parents' past. The parallel time pathways depicted in the story all lead to valid, believable, but very different outcomes.

This same scenario exists in the world of engineering design. Small changes in a design approach can lead to very different, but equally valid, design solutions. If you find yourself in such a situation, know that the decision may be left to your whim and fancy—provided that each outcome will truly meet design objectives. In any given situation, take heed that the latter is really true. Sometimes, hidden factors (such as the availability of raw materials, public reaction to a particular style, or the ability to market one design over another) may, in fact, make the choice far from arbitrary.

7.6.2 DVD Production Facility

In this example, we'll examine the design cycle for the case where the product is a manufacturing process, rather than a tangible physical entity. Although the means of execution of the various stages of the cycle may differ, the basic principles will be identical to those used in the previous example.

The case involves a software interface that must be written for a company that makes digital video disks (DVDs). The disks are sold online by several vendors, including Amonia.com, BestVids.com, CircuitTown.com, and WebFlix.com. Each of these vendors wishes to minimize its physical inventory while maximizing its ability to fulfill orders quickly. A machine that can mass produce plastic DVDs from etched metal master disks is very expensive—on the order of a few million dollars—and none of the DVD vendors alone can afford that sort of capital investment. They each would like a subcontractor who can manufacture DVDs on demand as they are ordered from each vendor's individual website. In recognition of this need, your company, Disk Stamper Inc., has raised money from venture capitalists, purchased a stamping machine, and has contracted with each of the four vendors separately to fulfill their orders on demand. The job of your design team is to determine a manufacturing system than can accomplish this task. Let's discuss this design problem in the context of the design cycle of Figure 7.7.

1. Define the Overall Objectives

The overall objectives of this design problem are clear. First, you desire to keep your clients happy. You must stamp and ship their DVD orders as quickly as possible and strive for 100% accuracy. You must provide for contingencies should your machine (or machines) be out of service for repairs, and you must regularly back up data. In addition, you need to fulfill your prime directive as a small company: you must make money. This last objective is neither whimsical nor callous. If your company does not make money, you will not be able to sustain your operations, and you and your employees will be out of jobs.

2. Gather Information

Your first step should be to interview your clients to determine their requirements. If you're lucky, their needs will coincide, and your design task will be simplified. In practice, however, their needs are apt to differ somewhat. One element of your information-gathering goal will be to seek common threads among your customers' aggregate needs. Perhaps it will be possible to persuade them to alter their stated requirements so that you can converge on a workable solution.

You should also investigate the operations of other companies involved in the order fulfillment business. You'll find that some, like you, actually manufacture the products they ship, while others simply act as wholesale distributors for other manufacturers—in essence, they act as the interface between the manufacturing company and online sales vendor. Some online bookstores, for example, work in this latter way, sending their orders to central distributors, called *fulfillment* companies, who in turn order the books from publishers and ship them.

3. Choose a Design Strategy

At this stage of the design cycle, you outline one or more approaches to the overall problem. In one plan, you would allot a daily manufacturing time slot to each DVD seller. During each vendor's allocated time slot (6 hours total, for around-the-clock operations), its accumulated online orders would be fulfilled, and DVDs of the same title would be processed together. This approach, illustrated graphically in Figure 7.25, would require that you change the master stamping disk several times during the vendor's time window, but it would enable you to group together each vendor's orders for shipment. An alternative strategy, shown in Figure 7.26, would collect DVD orders by title regardless of vendor and then stamp all the DVDs of the same title together, as a single batch. This approach would reduce the need to change stamp masters frequently, but it would require that you sort orders by seller at periodic intervals. After some discussion, you decide to investigate the latter approach: You will collect orders for all the vendors, sort them by DVD title, and then stamp single titles as a batch, changing masters only when it's time to change the DVD title. There is some motivation for this approach, because it's faster and

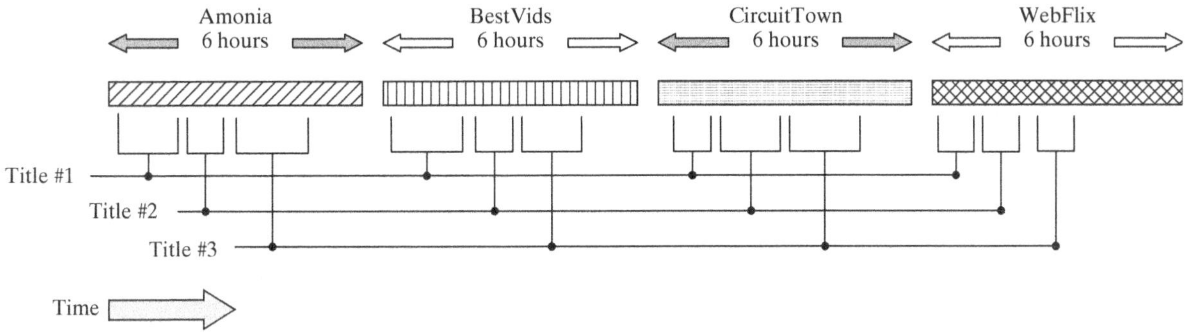

Figure 7.25
One possible implementation of the DVD order fulfillment process. Each of the four vendors is allocated one 6-hour time slot, and individual orders for the same DVD title are processed as a batch within each vendor's time slot.

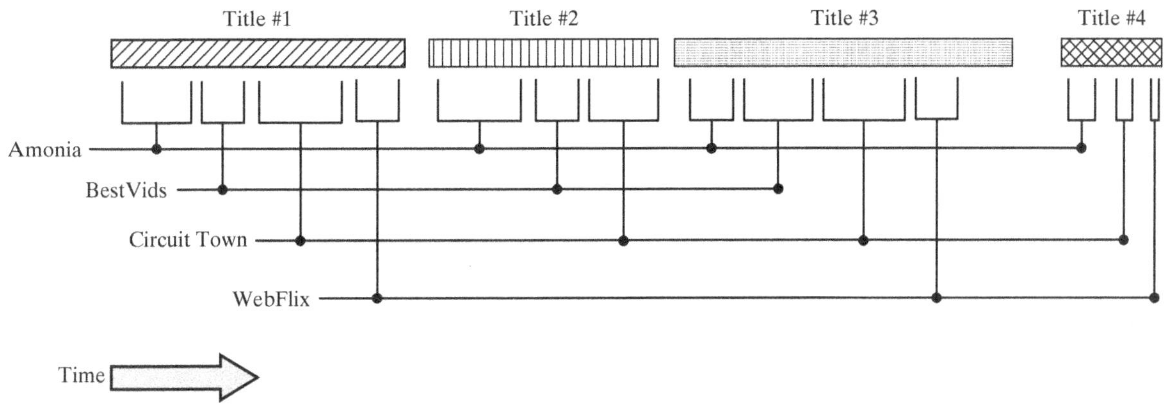

Figure 7.26
Another possible implementation of the DVD order fulfillment process. In this scheme, DVD titles are processed in batches independent of vendor.

will reduce labor costs. Once your business becomes operational, you'll need to pay machine operators by the hour. Given the low profit margin for DVDs (most of the profit in the movie industry goes to the movie production company), reducing labor costs seems like a good idea.

4. Make a First Cut at the Design

A block diagram of your proposed system is shown in Figure 7.27. At the present time, the only definite piece of hardware is the actual stamping machine; all other components will ultimately have to be designed and integrated into the system. Incoming data from the vendors is stored in an "orders" database, and the master disks are stored in a digitally controlled carousel. When the stamping machine becomes free for stamping, the controller retrieves a packet of orders from the input queue, then sends a signal to the carousel to load the correct master into the stamping machine. As each fabricated DVD exits the machine, it is labeled, then, its carrier vessel is tagged with a bar code that links that specific disk to the shipping information stored in the orders database. Packaging occurs down the production line, where the shipping information is printed on the outside of the package.

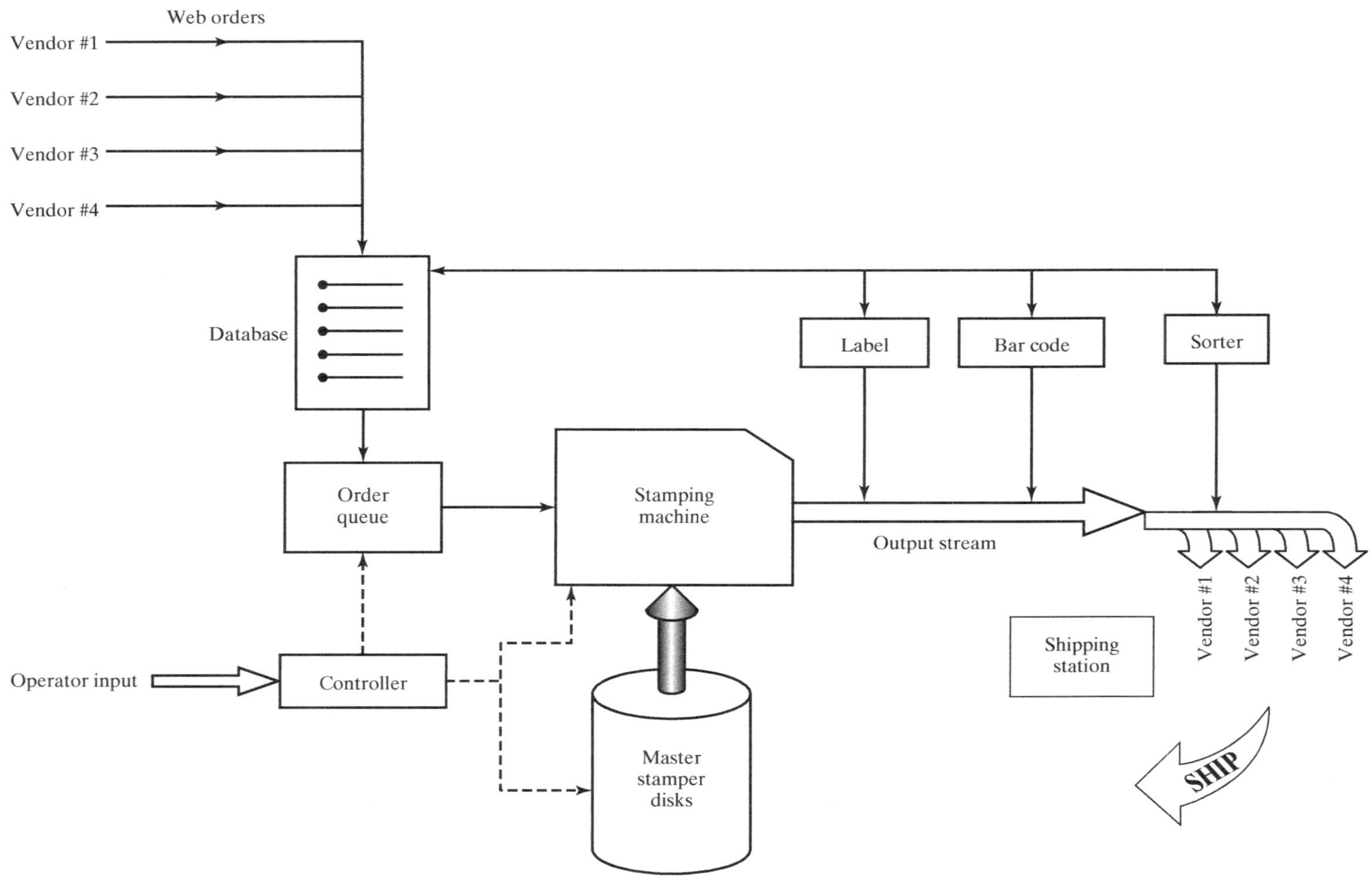

Figure 7.27
Block diagram of a system that can implement the production scheme of Figure 7.26.

5. Model and Analyze

In order to ensure that the system will work as expected, some engineering analysis is appropriate. Each physical transition of a disk from one station to another will require some time delay. If the system is to operate efficiently and provide the fastest overall throughput, proper coordination of events is necessary. Thus, several questions arise. For example, is it better to wait until the disk has been tagged with a bar code before loading the next blank disk in the stamping machine? Or should the blank disk be loaded immediately after its predecessor has left the machine, without regard for the tagging operation? Similarly, what steps must be taken to ensure that incoming orders do not arrive simultaneously, thereby confusing the system? Alternatively, if they do, what techniques can be incorporated into the system to deal with the conflict? A systems-level analysis, in which the dynamics of the queuing process are examined in detail, will help to optimize the system. Answering questions of this sort falls in the domain of *systems engineering,* usually included under the umbrella of *manufacturing* or *industrial engineering* in most college curricula.

6. Build, Document, and Test

Full-scale construction of your proposed system will be costly; hence, you decide (wisely) to attempt the build-and-document phase of the design cycle using simulation. Specifically, you choose to model the entire system using a commercial systems-analysis software package. Several excellent sources of such programs exist; you decide upon Simulink,™ a toolbox in the MATLAB software suite.

Your simulation and testing reveal no debilitating problems, but a major bottleneck develops at the packaging stage. Your vendors have specified different shipping methods (e.g., United Parcel Service, FedEx Ground, and the U.S. Postal Service). These different shipping methods require that you change shipping labels from order to order. Given that you have decided to process orders by DVD title, rather than by vendor, changing labels must occur often, possibly causing unneeded delays.

7. Revise and Revise Again

You solve the labeling problem by adding a package sorting station to the system downstream of the stamping operation. This feature will identify each packaged DVD by its bar code, then divert it to one of three shipping stations (one for each method of shipping). This addition to the system will increase capital costs—you will need to purchase and install the components of the additional sorting station—but the overall improvement in efficiency will more than pay for the added startup cost. Your systems-level analysis estimates the payback period for the additional feature to be about 4 months at a 100% production level. Your team decides to adopt this design revision.

8. Thoroughly Test the Finished Product

After extensive testing of your simulated system, you will then build the actual devices and connect them together on the factory floor. You will be able to approach this high-cost portion of the project with some confidence, having analyzed and revised the system prior to construction. Full-scale testing using a simulated order queue will be the next step before you open your company for business. Limited trials of actual orders will be next, followed by ramp-up to full-scale production. Even in this final phase of implementation, you should expect to uncover previously hidden problems and be prepared to return to the design cycle for revisions and further tests.

7.6.3 Automatic Pipette Machine

In this example, we examine a real-world design problem from the combined fields of biomedical, mechanical, and electrical engineering. It provides a good example of a multidisciplinary problem. Our discussion will highlight the various design choices that the engineers faced during the "Choose a Design Strategy" step of the design cycle of Figure 7.7.

In the typical biomedical engineering laboratory, one may find one or more versions of the calibrated pipette dispenser. This essential tool delivers a precisely calibrated volume of liquid ranging from microliters to hundreds of milliliters. Often these calibrated liquid doses are dispensed into the cells of a standardized 96-well microplate such as the one shown in Figure 7.28. The user sets the dials of the pipette to the desired volume, then picks up a disposable tip. Examples of single-tip, manually operated pipettes are shown in Figure 7.29. Samples of disposable tips that fit in the ends of the pipettes are shown in Figure 7.30.

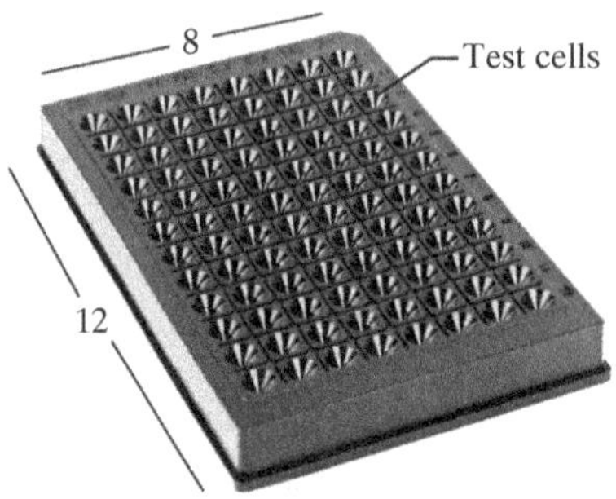

Figure 7.28
Standard 96-well microplate. The volume of each cell is about 100 microliters.

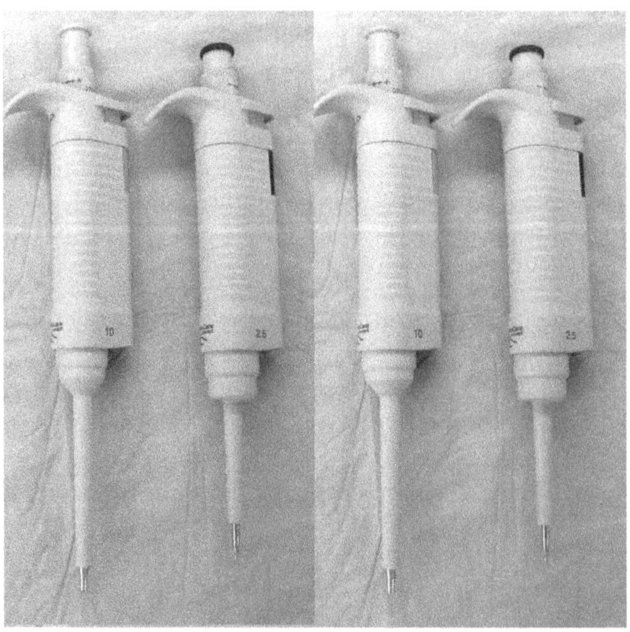

Figure 7.29
Variety of hand-held single-tip micropipettes (disposable tips not attached)

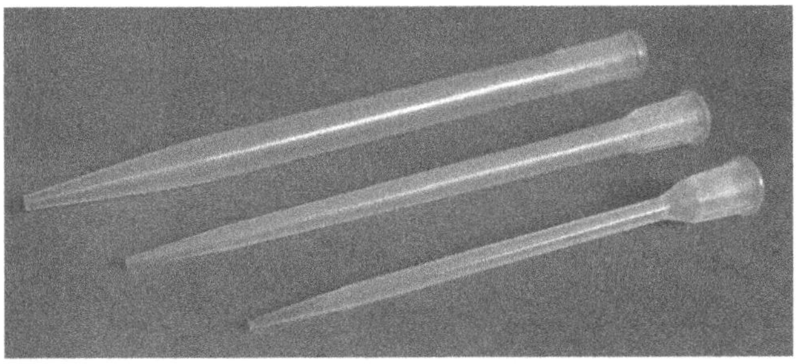

Figure 7.30
Examples of disposable pipette tips.

To use a pipette, the user picks up a disposable tip which is replaced each time the pipette contacts a new liquid or cell culture. The tip is dipped into the liquid to be dispensed. The user pushes, then withdraws a thumb activated plunger to draw up the designated liquid volume into the pipette's tip. The liquid can then be dispensed into any cell of the target microplate by pressing the plunger again. The latter operation causes the liquid to be expelled from the tip. The disposable tip can be changed at any time (for example, to avoid cell-to-cell contamination) by pushing a tip ejection lever.

The application in question involves a Public Health professor who needs to assess the effect of multiple chemical compounds on living cell cultures. Both the compounds and target cultures reside in the cells of 96-well microplates. Because thousands of different chemicals must be tested, the professor has asked a colleague in mechanical engineering to build a robotic machine to perform the tests automatically.

The following table lists the requirements of the system:

- Accommodate standard 96-well microplates (12 rows of 8 cells each spaced 9 mm apart).
- Pick up 10 μL of solvent from reservoir.
- Dispense solvent to each of 96 wells in a target microplate carrying the different compounds, then mix.
- Pick up 3 μL of chemical-laden solvent, then dispense 1 μL into each of the 96 wells in three adjacent microplates. Each well carries 100 μL of cell culture.
- Eject used tips between dispense operations into a disposal bin.

Designing this machine presents the engineer with a myriad of choices that must be resolved as part of the design process. The depth of design choices is illustrated in the following discussions.

Decision: Manual or Electronic Pipette?

Issue. Pipettes are available from several manufacturers in either manual or electronic form. In the former, all operations are fully mechanical and are operated by hand

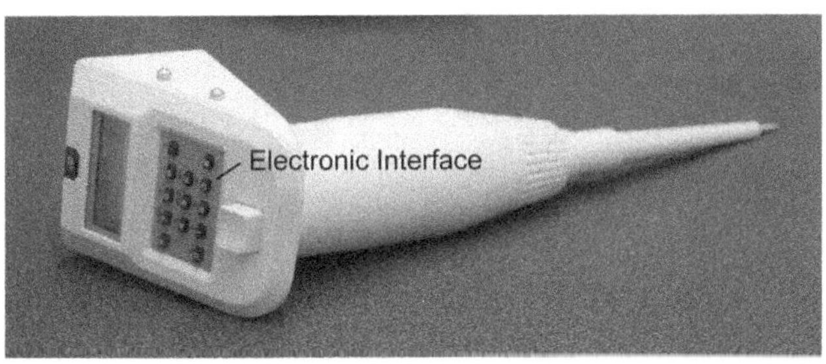

Figure 7.31
Electronic pipette

manipulation. In the latter, the pipette's dispensing syringe is operated by an internal motor and servo control system; the user interface consists of buttons and a display on the pipette's small operating panel. One such pipette is shown in Figure 7.31. Some electronic pipettes can be interfaced to a computer via a cable.

Pipette Comparison		
	Manual Pipette	**Electronic Pipette**
Cost	Inexpensive	Very expensive
Reliability	Simple; no circuitry or motors to malfunction	Complicated electronics and internal motor must function consistently
Design of Interface	Must design complicated mechanical holders, pistons, and actuators for operation in an automated robot system.	Can be interfaced with computer via a single data wire using a standard connector.
Accuracy	Adequate	Marginally better than manual

Decision: Single or Multi-Channel Pipette?

Issue. All pipettes, whether manual or electronic, are available with a single tip or in multi-channel versions having four, eight or twelve parallel tips each tied to the same dispensing mechanism. All volumes must be set to the same value, and all syringes are operated simultaneously, whether drawing up liquid or dispensing it. An example of a multi-channel pipette is shown in Figure 7.32.

Figure 7.32
Multi-channel pipette

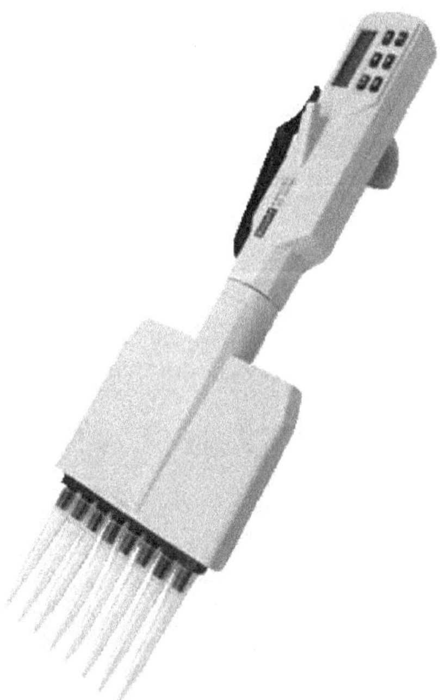

Number of Channels Comparison		
	Single Channel	**Multi-Channel**
Mechanical Complexity	Requires two-axis (X-Y) motion to reach all cells of the microplate	Only one direction of motion (x-axis) is needed. Can dose an entire microplate row simultaneously if a 8- or 12-channel pipette is chosen.
Overall Speed	Slower	Faster; Can dispense all the cells in microplate row at the same time.
Cost	Least Expensive	Cost rises with number of channels.
Tip Pickup	Easy; requires pressure at only a single point.	Difficult; requires that pressure be applied evenly across an entire row of tips.
Accuracy	Same as multi-channel	Same as single channel
Small droplet pickup	Feasible; requires contact of only one tip with the droplet	Difficult; requires precise alignment of 4 to 12 tips with their respective droplets.

Decision: Aluminum, Plain Steel, or Stainless-Steel Frame?

Issue. The material for constructing the basic support frame must be chosen in consideration of strength, cost, ease of fabrication, and ease of "on-the-fly" design changes. The material chosen for the prototype may differ from that used in the finished product.

Materials Comparison			
	Aluminum	**Plain Steel**	**Stainless Steel**
Ease of fabrication and revision	Easy to machine	Harder to machine	Hardest to machine
Biocompatibility	Can interact with some chemicals; oxides readily	Rusts; can interact with some chemicals	More resistant to corrosion.
Weight	Lightest of feasible metals	Heaviest of feasible metals	Relatively heavy
Cost	Least expensive	Fairly inexpensive	Most expensive of feasible metals
Strength	Weakest	Strong	Strong

Decision: Types of Motion Required to Move Pipette Horizontally and Vertically, and also to Activate its Eject Button

Issue #1. The pipette must be clamped in a secure holder, and the latter must be mounted on a "carriage" capable of motion along one or two axes. The number of axes depends on whether a multi-channel or single channel pipette is chosen. The carriage must be positioned so that the tip (or tips) of the pipette align with the centers of the wells of the microplate. Moving from cell to cell, and also from microplate to microplate, requires long-distance linear motion.

Issue #2. Prior to accessing any microplate well, the pipette syringes must be fitted with disposable tips from a carrier. The spacing between tips in the carrier is identical to that of the inter-well spacing on the microplate.

Issue #3. In order to pick up tips that will reliably remain on the pipette, extra force must be applied during tip pickup. The amount of force required may be large, because a good seal must be achieved between tip and syringe.

Issue #4. Once the pipette has been positioned above a given microplate cell, it must be lowered so that its tip(s) can make contact with the contents of the cell(s). This action requires short distance linear motion.

Issue #5. When a given dosing sequence is completed, the apparatus must push the tip-eject lever. This action requires very short (1 cm) linear motion, but the force required can be large, depending on how forcefully the tips have been picked up previously.

General considerations regarding motion

Type of Actuation. Electric power is readily available almost everywhere via standard wall plugs. Electrically derived forces, however, are generally weaker per unit device volume than those derived from compressed air. Sources of the latter are less ubiquitous than electricity, but compressed air outlets can generally be found in most well equipped scientific laboratories. A stand-alone compressor also can be used to supply compressed air, but this is a generally costly and noisy alternative. The strongest of all actuators are those operated by hydraulic fluids. These types of devices are generally found in the realm of heavy machinery and are probably not well matched to the design requirements of the pipette robot.

Rotary motion versus Linear Motion. The natural role of an electric motor is to provide rotary motion. Many different types and sizes of commercial off-the-shelf (COTS) motors are available from numerous vendors. Common types include stepper, servo, and synchronous motors. Linear electric motors are available but tend to be specialty devices.

Deriving linear motion from an electric motor requires the use of a screw-gear apparatus. A small number of electric motors capable of direct linear motion may be found, but these tend to be large specialty devices. Linear motion may be easily derived from compressed air or hydraulic fluid using simple, inexpensive piston devices.

Deriving rotary motion from compressed air requires an expensive and noisy air turbine. Simple air pistons, on the other hand, are compact, cheap, and easy to locate. Figure 7.33 illustrates some of these common types of motors and actuators. The various functions in the pipette robot may involve one, some, or all of these types of devices.

Design Choice: Relative or Absolute Position Sensing

Issue. Regardless of whether one-axis or two-axis motion is chosen, the location of the pipette must be controlled to a tolerance of about 0.5 mm (the approximate diameter of the tip orifice). Many robotic systems utilize a relative registration system in which all positions are determined by the distance traveled from a known home location (e.g., the *x-y* coordinate 0,0). Position increments are measured by devices such as the rotary encoders shown in Figure 7.34. Each of these devices produces an electronic pulse for each 2.5 degrees of shaft rotation. The designer can set the relationship between this angular increment and a distance increment by appropriate gear ratios.

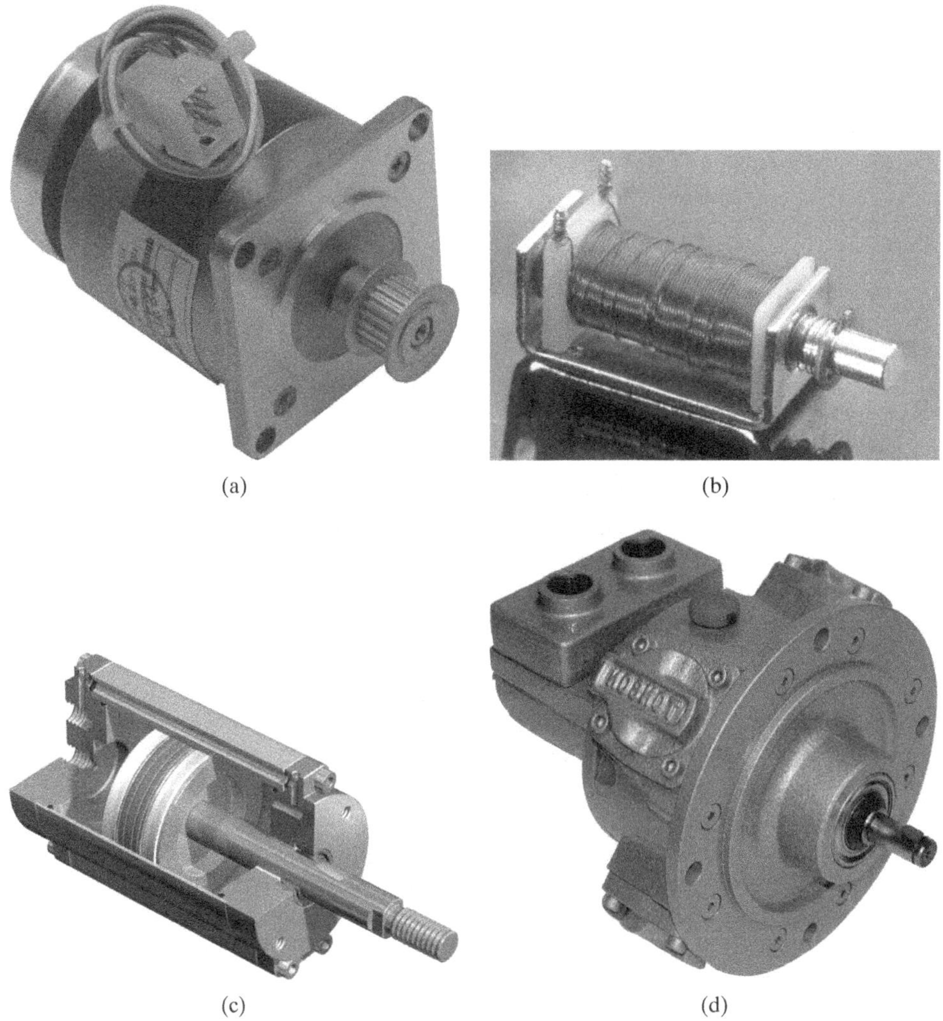

Figure 7.33
Various types of actuators: a) stepper motor with belt pulley; b) magnetic solenoid; c) pneumatic cylinder; d) air turbine motor.

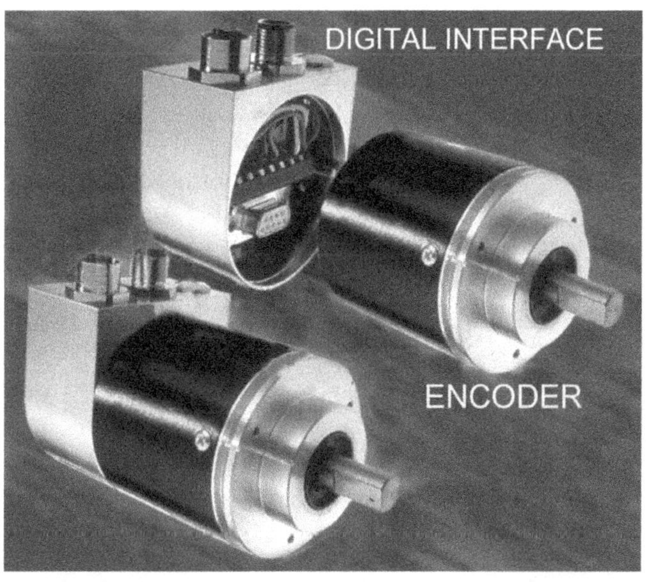

Figure 7.34
Examples of rotary shaft encoders.

Other robotic systems use absolute positioning systems in which a location is determined by the arrival of the moving member at a fixed sensor. With this method, a sensor is required at each location at which stopping is desired.

Position Sensing Comparison		
	Relative	**Absolute**
Cost	Only one shaft encoder required	Many position sensors required
Design Flexibility	Critical stopping positions are easily changed in software.	Changing stopping positions requires movable sensors, hence more mechanical complexity
Decoding	Entirely a software solution	Requires data acquisition hardware with an input for each sensor.
Reliability	One gear slip throws off position tracking.	Any slippage in system irrelevant. Position sensors are fixed.

Epilogue. The above problem actually emerged at a major research university. The design effort for the automatic pipette system took place over the span of approximately two years. It involved numerous iterations around the design loop. The first design, based on a single-channel pipette and all electrical actuation, had to be abandoned at a stage of near completion for two reasons. First, the *x*-*y* motion designed to move the pipette between cell and microplates resulted in a system that required four hours for each experimental run. This duration exceeded the viability time of the living cells in the microplate wells. Second, electric actuation of the pipette in the downward direction proved insufficient to reliably pick up tips between dispense operations. Ultimately, a multi-channel, one-axis design was employed. The final iteration of the design is shown in Figure 7.35. It uses a stepper motor with low pitch drive screw to move the pipette between well plates, a two-way pneumatic piston to move the pipette vertically, and a short-stroke air piston to push the tip-eject button. The electronic pipette is tethered to a laptop computer which controls all functions via a MATLAB program and digital interface board.

EXAMPLE 7.4

Sailboat Autopilot

Homer S. has decided to embark on a solo sailing voyage across the Atlantic. He plans to leave from his home town of Springfed, arriving in Brest, France sometime thereafter. For this purpose, he has built his own wooden sailboat complete with watertight sleeping quarters. As he prepares for the trip, he realizes that it will definitely take longer than one day, hence he needs some arrangement by which he can sleep while the boat is underway. It would not be proper to simply let the boat drift while he sleeps.

Not being very mechanically inclined, he has asked you, his engineering pal, to convene a team to design an autopilot for the boat. You graciously accept, hoping that the story will someday be featured in a television episode or major animated movie chronicling Homer's journey. Let's once again analyze the problem in the context of the design cycle of Figure 7.7.

Several commercial solutions to this problem exist, so the analysis to follow simply highlights some of the choices that one might make in addressing Homer's navigational needs.

Figure 7.35
Final iteration of the pipette robot design.

Gather Information

You quickly confirm the obvious: a sailboat is propelled by the force of the wind on its sails. When a sail is properly set, its cross section resembles that of an airplane wing pointing up toward the sky. The "lift" in this case is applied horizontally. A small component of the lift extends toward the front of the boat and moves the boat forward. The remaining lift force is balanced by the hydrodynamic and gravitational forces of the boat's keel. The complete balance of forces acting on a moving sailboat is complicated, but it works quite well. A vector diagram illustrating the various forces on a sailboat is shown in Figure 7.36.

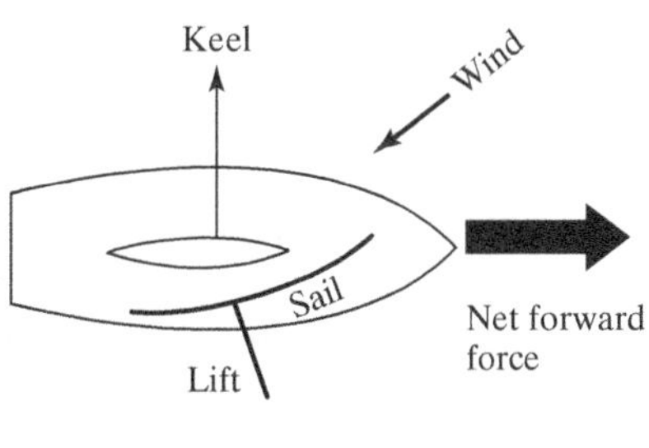

Figure 7.36
Vector forces acting on a sailboat. The net force is in the forward direction.

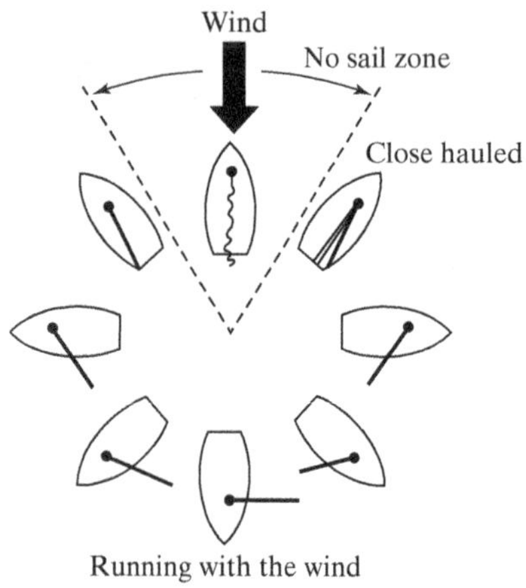

Figure 7.37
The points of sail. Inside the "no sail" zone, the sail cannot fill with wind.

You also learn that a sailboat cannot travel directly into the wind. Rather, its forward motion is confined to the region outside an arc of approximately 90° centered about the wind direction, as shown in Figure 7.37. If one tries to steer a sailboat inside this arc, the sail cannot fill with wind, the delicate balance of forces is disturbed, and the boat ceases to progress. The 90° arc is sometimes called the "no-sail" zone. When a sailboat is directed as close as possible to the wind direction, its sailing position is called "close hauled." A boat traveling in the same direction as the wind is called "running free" or "running with the wind." It seems counterintuitive, but a sailboat on a close-hauled course moves faster than one running free.

The direction of a sailboat is determined by its rudder which extends into the water at the rear of the boat. Turning the rudder to the right or left causes the rear of the boat to swing in the same direction, in turn altering the direction of travel. The rudder is controlled by either a long steering pole called the "tiller," or, if the boat is large, by a mechanically linked steering wheel similar to that found in a car. Homer has chosen the much simpler tiller method of steering for his boat.

Choose a Design Strategy

There are three possibilities for keeping the sailboat on course. One would be to measure the prevailing wind direction, then maintain the angle between it and the desired course direction. A simple wind vane is ideal for this purpose. As long as the chosen direction does not extend into the no-sail zone, the boat will move forward. This method has the advantage that the steering system will always choose a direction in which the boat can actually move. Its principal disadvantage is that the wind direction may change as the boat travels along. If the boat's travel direction is oriented to the wind direction, it will veer off course when the wind changes.

Another possibility is to orient the boat relative to the Earth's magnetic north. A compass must be employed for this purpose. The main advantage of this method is that the boat will always be pointed in the desired course direction. It is possible, however, for the autopilot to point the boat into the no-sail zone, in which case it will not move at all.

Most sailboat compasses are simple in design, consisting of a magnetized disk floating on a pin joint. The disk is marked with the points of the compass (e.g., North, South, East, West). If the compass orientation method for the autopilot is chosen, however, then a more expensive electronic compass will be the obvious choice. Interfacing with the delicate floating magnet of a conventional compass is a very difficult task and not one easily accomplished. Electric power on a sailboat is usually derived from solar cells, or sometimes from a small windmill mounted on the rear of the boat. In either case, the availability of electric power can be unpredictable and assured only if the helmsman is diligent about harvesting the energy from sunlight or wind.

A third possibility is to use the global positioning system (GPS) to determine the boats course. This approach has the same advantages and disadvantages as the compass method, except that it would be possible to keep a record of the boat's actual course while on autopilot.

Gather More Information

At this point, you realize that it's impossible to choose between the wind and compass methods of orientation without knowing more about the nature of the wind along Homer's projected path. It is likely to depend on his ports of embarkation and arrival, and possibly the season in which the trip is planned. (Given that Springfed lies somewhere in Indiana, you highly recommend to Homer that he choose a more convenient embarkation point; you suggest Montauk Point, an isolated peninsula at the tip of Long Island, NY. Leaving from this state park location will allow access to whatever curious spectators may wish to endure the long car ride to see him off.)

Your research leads you to the National Ocean Data Center maintained by the US National Oceanic and Atmospheric Administration (NOAA). Your data compilation, summarized by the global map of Figure 7.38, reveals the that in the months of October through December, the prevailing winds from Montauk to Brest

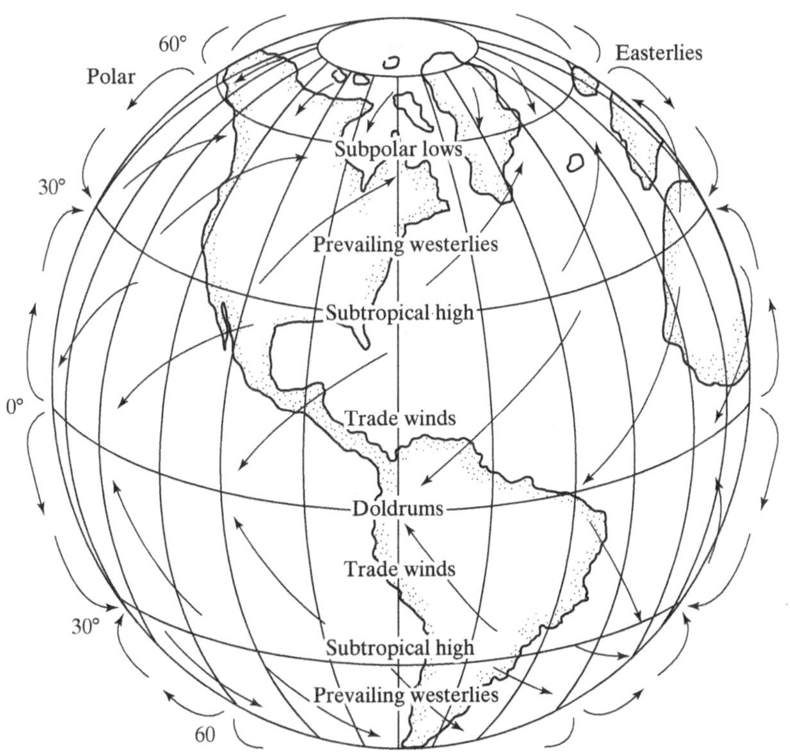

Figure 7.38
Global wind vector map.

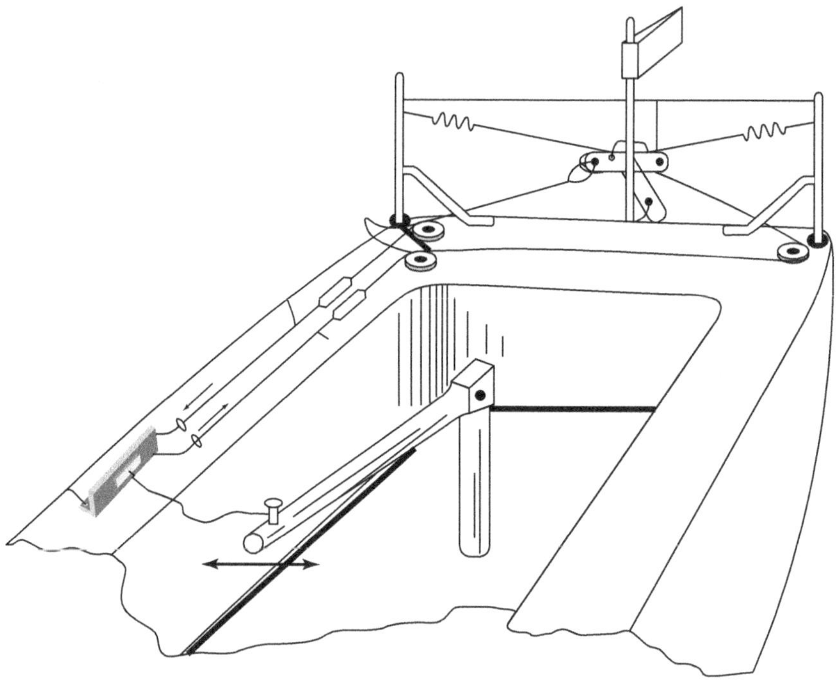

Figure 7.39
Possible solution to Homer's autopilot requirement.[3]

should be westerly and steady. Thus Homer's boat will not likely have to deal with the no-sail zone inherent to sailing close-hauled.

Make a First Cut at the Design

From the additional information you have gathered, you decide that an orientation system tied to wind direction is probably the best choice. This decision is based, in part, on human factors. Homer is insistent on making the trip in the colder fall months, when sunlight will be less prevalent. You also realize that he may forget to charge his boat batteries when sunlight is available. Additionally, if the boat is going to be traveling in the same direction as the prevalent west-to-east wind, then the apparent wind it will feel (the difference between the actual wind and the boat's velocity) may be small. You realize that if a system based on wind direction is chosen, then a fully mechanical steering system could be built, thus eliminating entirely the need for electricity. This feature would not be true of a compass oriented system, which would require constant electric power.

Based on the above considerations, you sketch the system shown in Figure. 7.39[3]. A wind vane sitting atop a pole mounted on the back of the boat moves cables through pulleys to make slight course correction adjustments to the tiller, thus keeping the boat moving at a selected angle relative to the wind.

Build, Document, Test, Revise

Testing must take place once Homer has towed his boat from Springfed to Montauk. He must first figure out how to get the boat through the much smaller garage door.

[3]Figure derived from U.S. Patent No. 4,366,767.

SUMMARY

In this chapter, the essential elements of the design process have been outlined at a very basic level. As an engineering endeavor, design differs from analysis and replication because it involves multiple paths to solution, plus decision making, evaluation, revision, testing, and retesting. The *design cycle* is an important part of engineering problem solving, as are knowledge, experience, and intuition. *Documentation* is critical to the success of a product and should be an integral part of the design cycle.

KEY TERMS

Analysis	Evaluation	Simulation
Decision making	Iteration	Testing
Design	Replication	Brainstorming
Design cycle	Revision	Idea Generation

Professional Success: How to Tell a Good Engineer from a Bad Engineer

If you pursue engineering as a career, you will encounter many colleagues along the way. Some will be good engineers, and others will be bad engineers. In your quest to identify and emulate only good engineers, you should learn the differences between the two. The following list highlights the traits of both types of engineers:

A Good Engineer	A Bad Engineer
• Listens to new ideas with an open mind. • Considers a variety of solution methodologies before choosing a design approach. • Does not consider a project complete at the first sign of success, but insists on testing and retesting. • Is never content to arrive at a set of design parameters solely by trial and error. • Uses phrases such as "I need to understand why" and "Let's consider several possibilities."	• Seldom listens to the ideas of others. • Has tunnel vision and only pursues the first design approach that comes to mind. • Ships the product without thorough testing. • Uses phrases such as "Good enough" and "I don't understand why it won't work. So-and-so did it this way." • Equates pure trial and error with engineering design.

PROBLEMS

The following problem statements can be used to practice problem-solving skills and idea generation. Some of them involve paper designs, while others are suitable for actual fabrication and testing.

7.1 Develop a design concept for a mechanical device that will allow hands-free operation of a standard wired telephone. Outline its basic form, key features, proposed method of construction, and prototyping plan. Consider size, weight, shape, safety factors, and ease of use.

7.2 Develop a concept for a device that will allow hands-free use of a cellular telephone without the addition of accessories such as an earphone, or microphone, or Blue-Tooth® headset. Your design should be based solely on a mechanical solution.

7.3 Design a device for recording design ideas that enter your mind while you are riding your bicycle. Your contraption should enable the rider to record notes while holding onto the handlebars with both hands.

7.4 Design a device for securing a cell phone near the driver's seat of an automobile. It should be universally adaptable to a wide variety of vehicles. Address safety and liability issues as part of your design.

7.5 Develop at least three design concepts for a non-lethal mousetrap. Your device should be cost-competitive with an ordinary, spring-bale mousetrap.

7.6 Devise a concept for a powered device that can dig holes in the ground for the installation of fence posts. Sketch a prototype and outline a test plan for your design concept.

7.7 Design a device that will allow the inside and outside surfaces of windows to be cleaned from the inside only. Compare the projected cost with that of a simple, handheld, squeegee-type window cleaner.

7.8 Design a system for feeding a pet lizard automatically when the owner is out of town.

7.9 Develop a design concept for a spill-proof table candle. Such an item is required by some fire departments for use at banquets, etc. If knocked over, it should right itself without spilling any wax or extinguishing its flame.

7.10 Develop a design concept for a warning system that monitors the inside temperature of a food storage freezer.

7.11 Design a device for conveying bricks to the top of a house for chimney repair. (The alternative is to carry them up a ladder by hand.)

7.12 Design at least three different methods for measuring the height of a tall building.

7.13 Design a system that will enable a self-propelled, unattended lawn mower to cut the grass in a yard.

7.14 Design a system for minimizing the number of red lights encountered by cars traveling east and west through a major city. The system should not unduly impede north–south traffic flow.

7.15 Devise a method for managing the flow of two-way railroad traffic on a one-track system. The track may have parallel spur tracks at regular intervals. These parallel sections of track allow one train to wait as another passes in the opposite direction. Note, however, that each parallel section adds cost to the system.

7.16 Develop a concept for a public transportation system in which every traveler can ride a private vehicle on demand from any one station to any other.

7.17 Design a transportation system based on situating free-use bicycles at strategic points around a large city.

7.18 Design a transportation system based on keeping fee-for-use, electric automobiles at strategic points around a large city. Your system must include a means for recharging the vehicles

7.19 Most large airports provide carts for travelers to transport luggage from baggage claim areas to taxi stands, transit stations, and parking lots. Develop a system for locating and reclaiming carts from various places around the airport for return to the baggage claim area.

7.20 Develop a design concept for a system to measure the number of people passing by a storefront. Separate your data into those who enter and those who do not.

7.21 Design a system that will assist flight attendants in feeding passengers on a wide-body aircraft without blocking the aisles with food and beverage carts.

7.22 Design a system for automatically turning off a small electric baking oven when a cake is done. The system must include a means for assessing the status of the cake.

7.23 Design a system that will help a person locate misplaced eyeglasses.

7.24 Develop a concept that will assist individuals in finding keys that have been misplaced around the house.

7.25 Design an electric light switch that will turn off lights if the room is vacated, but not sooner than some user-specified time interval.

7.26 Design a device that will enable a quadriplegic to change the channels on a television set.

7.27 Devise a system for turning on security lights at dusk and turning them off at dawn. These lights are to be installed throughout a large factory. Your system should have a single master override for all lights.

7.28 Design a system for turning off a stove-top burner when a pot of water has boiled. Such a system would be valuable for cooking rice or preparing hard-boiled eggs.

7.29 Design a method or concept for dispensing transparent adhesive tape in small precut lengths.

7.30 Design a system that will automatically water houseplants when they are in need of moisture.

7.31 Design a system that will selectively water sections of a garden based on the moisture content of the soil at the location of each plant.

7.32 Design an irrigation system that will bring water from a nearby pond to your vegetable garden.

7.33 Develop a design concept for a system that will automatically pick apples from the trees of a commercial apple orchard.

7.34 Design a system that can aerate the pond in a city park so that algae growth will not overtake other forms of wildlife.

7.35 Design a device that will allow a one-armed individual to properly use dental floss.

7.36 Design a system that will prevent the user of a battery charger from inserting the batteries the wrong way. The unit charges four AAA batteries simultaneously, but each battery must be inserted with its positive (+) end in the correct direction.

7.37 Design a method for counting the number of people who attend a football game. The stadium has a maximum capacity of 40,000 fans and eight entry gates.

7.38 A big problem in nursing homes for the elderly is the misplacement of false teeth. Some dentures are inevitably discarded with dining room debris. Devise a system that will enable kitchen staff to detect dentures that appear in the waste stream.

7.39 Design a system for automatically steering a hot-air balloon along a desired compass heading.

7.40 Design a weatherproof mail slot for a house that will keep cold air out but permit the insertion of mail from outside.

7.41 Most small sailboats have a tiller (steering stick) in lieu of a steering wheel. Design a system for automatically steering a tiller-controlled sailboat on a course that lies along a user-specified compass direction. Design for a simple boat that has no onboard electricity.

7.42 Specify the components of a system that will allow a group of six to listen to music from a single MP3 player or iPod. All listeners will use headphones.

7.43 *Challenge Problem.* One of the problems with recycling post-consumer waste is the sorting of materials, for example, the various plastics and metal containers placed into curbside recycling bins. Consumers and homeowners do not always sort correctly, yet only a small amount of erroneously sorted material can ruin a large batch of recycled material when it is melted to raw material. At the present time, most municipalities resort to manual labor to sort recyclable materials. Devise a concept that will sort metal cans, plastic bottles, and plastic containers at a recycling plant. Develop a plan for modeling and testing your system.

7.44 Design a kitchen device that will crush aluminum and steel cans in preparation for recycling. Such a device would be helpful for households that practice recycling but have limited storage space.

7.45 Devise a system for painting car bodies automatically by robot. You must include a method for training the robot for each painting task.

7.46 Devise a plan for a campus-wide information system that allows any professor to access the grades of any student, while maintaining the privacy of the system to other users. Students also should be able to obtain their own grades, but not those of others.

7.47 Design an automatic pencil sharpener that will turn itself off when the pencil has been sharpened to a desired tip sharpness.

7.48 Develop a design concept for a device that will automatically close a skylight window when rain falls. The window should be closed partway when rainfall is light but should be closed completely when rainfall is moderate to heavy.

7.49 Design an apparatus that will keep a telescope pointed at a distant star despite the rotation of the earth.

7.50 Design a system for keeping solar panels pointed directly toward the sun.

7.51 Devise a system for transferring personnel in flight from one airplane to another.

7.52 Devise a system for automatically collecting tolls from cars traversing a major interstate highway. Note that many such systems exist throughout the United States. (See, for example, *www.mtafastlane.com* or *www.ezpass.com.*) This problem asks you to imagine (or find out) the details of how these systems work.

7.53 *Challenge Problem.* Laser communication, or "laser-com," is a system by which digital data is sent from one location to another via a modulated laser beam. Laser-com systems are used whenever connections via wires or fiber-optic cables are too expensive, not possible, or not desirable. One of the principal drawbacks of laser-com systems is the difficulty in maintaining beam alignment when the sender and receiver are moving. Design a system that will automatically direct the communication beam sent by one vehicle toward another receiving vehicle.

7.54 *Challenge Problem.* Design a concept for a two-way communication system that works via a modulated laser beam. In this system, only a fixed ground station will emit a beam. The other participant in the link will be a moving person who will receive the beam and reflect a modified version back to the base station.

7.55 *Challenge Problem.* Design a system consisting of several emergency buttons that will be installed at each of several workshop fabrication stations on a factory floor. Pressing any one of these buttons will activate a signal at a central control console and identify the location of the activated button. Voice communication over the system would also be a desirable feature. One matter to consider is whether a wired or a wireless system is preferable.

7.56 *Challenge Problem.* An elementary school teacher needs a calendar-teaching system to help young students learn about dates, appointments, and scheduling events. The basic system should be a large pad over which a monthly calendar can be placed. The underlying pad should have touch-sensitive sensors that can detect a finger placed on each day block in the calendar. The entire unit should interface with a computer which will run a question-and-answer game or program. Typical questions might include "You have a dentist appointment two weeks from today. Point to the day on the calendar on which you should go to the dentist," or "Sara's birthday is on February 11. Point to that day on the calendar." An appropriate acknowledgement, either visual, auditory, or both, should be issued by the computer for correct answers. A non-intimidating signal should be issued for incorrect answers. Outline the key features of your system and devise a development plan.

7.57 *Challenge Problem.* The rules of a particular design competition require an autonomous vehicle to be placed behind a starting line located 30 cm up the side of a 1.5-m inclined ramp. After the starting signal has been given, contestants may release their vehicles. Any vehicle that travels over the starting line prior to the "go" signal loses the race. Currently, the starting sequence is initiated orally by a judge and timed by stopwatch. This system leads to great variability among judges, as many use different starting signals (e.g., "on your mark, get set, go!" or "one, two, three, go!"), and any one judge may be lax in timing or checking for starting-line violations.

Design a system consisting of starting-line sensors, a start signal, and starting-line violation signals for each side of a double ramp. The judge should have a button that initiates the start sequence. A series of periodic beeps that mimic the rhythm of the words, "Ready, set, go!" should sound, with the final "go" being a loud and clearly distinguishable tone or buzzer. In addition, a green light or LED should illuminate when the "go" signal is sounded. The system should run for 15 seconds, then sound another tone or buzzer to indicate the end of the 15-second time interval signifying the end of the contest. If a vehicle crosses the starting line prior to the "go" signal, a red light should appear on the violating vehicle's side of the ramp, and a special "violation" signal should be sounded to alert the judge.

7.58 *Challenge Problem.* Teams in a model airplane design competition are called to the central runway when it is their turn to compete. After the initial call, each team has 3 minutes to arrive at its starting line. A team that does not arrive at the starting line after 3 minutes loses that run. Warnings are given 2 minutes and 1 minute before the deadline. Traditionally, the announcer has issued these warnings orally over a public address amplification system. There are several problems with this method, however. Acoustics are poor, three races run simultaneously, with different starting times, and only some teams require the full 3 minutes to arrive. Thus, the proper issuing of these cues has been lax. The judges need you to design an automated system that will inform a given team how much of its 3-minute sequence has elapsed. The system must send an appropriate signal (oral, auditory, or visual) only to the relevant team, and the timing sequence must be initiated from the judges' bench. As many as 80 teams may compete on a given day, and each is assigned one work table from a large array of 3 × 8-foot tables on the competition field. All too often, the team being called is delayed, because it is repairing or modifying its vehicle. One of the key design issues is whether a wireless or wired system is better, given the logistical constraints of the competition environment. Because the event operates on a strict budget, final cost also is an important factor.

7.59 The vehicles entering a student design competition must utilize batteries with a voltage no higher than a specified limit. Each vehicle is checked once with a voltmeter at the start of the day by the head judge. Having a standardized voltage-checking device would shorten the time for voltage checking. Design a unit that has a rotary (or other type of) switch that can select a predetermined battery voltage. If the measured battery falls within the acceptable range, a green light should appear. If the voltage falls below or above the range, yellow or red lights, respectively, should appear.

7.60 Outline the design of a general purpose software system for a track meet. Your system should enable judges to automatically pair up runners for matches, randomly at first, but by demonstrated ability thereafter (best runners against best runners, worst runners against worst runners). The program should display match sets before each round of the competition, and allow the recording judge to enter the result of each match after its winner has been determined. Assume a competition of up to 140 runners and six sets of matches between contenders.

7.61 Design a system for projecting matches on a display board so that audiences watching a tennis tournament can keep track of who is matched against whom, and monitor the results of each match. Assume a competition of 20 players with six sets of matches between contenders.

7.62 Design a system for detecting start-line violations for a horse race.

7.63 Design a system for determining the winner of each match in a swimming race.

7.64 Design a software system that will assist in registration for a fund-raising walk. Registration information includes the name, address, and age of the participants. Registrants must pay a small fee, sign photograph permission and liability waiver forms, and receive an event T-shirt and assigned walker number. Several volunteers will work simultaneously at the event to register participants as quickly as possible so that the event can begin on time, but a common stored database is needed.

7.65 Design a can opener that performs two tasks: a) it must not allow metal slivers to fall into the can, and b) it must catch and hold the cut lid for subsequent hands-free disposal.

7.66 Design a system for measuring and reporting accumulated snowfall at a remote monitoring station.

7.67 Design an ice tray that allows the user to easily extract a single cube at a time.

7.68 Design a system that can help a one-armed person tie a necktie.

7.69 Design a snow shovel for a one-armed individual.

7.70 Develop a design concept for trimming tall trees using a remote-controlled robot.

7.71 Design a method for automatically walking a dog in a safe and reliable manner. The system should require no human intervention other than setup.

7.72 Develop a concept for a scuba diving mask that is self-clearing. "Clearing" refers to the act of expelling water from inside the mask without surfacing.

7.73 Develop a system that can be worn by a downhill ski racer and can send the skier's traveling speed to an observing coach.

7.74 One of the problems with may small, recreational boats is that they fill up with rainwater which must be pumped out by the owner after a storm. Design a system that will automatically pump out such a watercraft. Your system may use an onboard storage battery, but it must then include a means for charging the battery during period of sunlight.

7.75 Design a set of wireless, waterproof headphones that can be used by a recreational swimmer. The source of music can be attached to the swimmer, or it can placed in a location out of the water.

7.76 If a common household toilet fails to fill its tank after flushing and is left "running," it can waste a considerable amount of water. Design an inexpensive device that can alert a homeowner to such a problem.

7.77 Design a system that will allow a blind man to properly trim sideburns with an electric shaver.

7.78 Design a method for alerting a driver of the fraying and imminent breakage of an automobile's timing belt.

7.79 Design an automated potato harvest system.

7.80 Devise a method for applying pesticides on crops that does not involve aerial spraying. Your goal should be to minimize the amount of chemicals dispersed into the environment, putting only what is needed on the plants. In the ideal case, the pesticide should be applied to the underside of the plant leaves, where most parasites reside.

7.81 Many supermarkets now offer online ordering and delivery. Often the orders are gathered by hand and put into delivery bins. Design a system, either semi- or fully automated, for filling online food shopping orders. The output should be a set of bins ready for delivery.

7.82 Devise an automatic dog walking system based on GPS navigational devices.

7.83 Devise a method for closing ornamental shade awnings when the wind becomes so high that it might damage the awnings.

7.84 Devise a method for lowing ornamental shade awnings when the sun comes out, and raising them with the sun is not present (e.g., behind clouds).

7.85 Devise an advanced, water-saving drip irrigation system that senses the moisture content of the soil near each patch of plants.

7.86 Devise a system for tracking and reporting at all times the location of spent nuclear fuel rods being transported for long-term storage.

7.87 The earliest "self-winding" watches used the motion of the wearer to wind the mechanical spring of the timepiece. Develop a similar concept to be used as a backup source of power for portable electronic devices such as cell phones and MP3 players.

7.88 Design a way to charge your iPod from the pedal power of your bicycle.

7.89 Design a system for alerting park maintenance personnel when remotely located trash cans are full and in need of emptying.

7.90 Design a system for alerting custodial workers when a given restroom needs soap and/or paper towels.

7.91 Design a system for automatically turning on an automobile's windshield washer when the windshield is dirty.

7.92 Design a system for automatically turning on an automobile's windshield washer when it begins to rain.

7.93 Design a "just in time" system for distributing home heating oil via delivery truck. Current systems used by oil delivery companies rely on statistics of "heating degree days" and histories of customer consumption. It would be preferable to have a sensor in each home's oil tank that can communicate with the central office.

7.94 Design a system to improve car mileage by measuring the vehicle's tire pressure, then automatically inflating or deflating the tires as needed. Note that recommended tire pressure depends on both tire size and the loaded weight of the vehicle.

7.95 *Challenge Problem.* A teacher wants a clock system that can help young students learn the relationship between time displayed by digital clocks and time displayed by analog clocks. The system should have a console that contains a large analog clock face, as well as a digital clock display with large digits. In operation, the teacher will set either clock, then ask the student to set the other clock to the same time. If the student sets the time correctly, the unit should signal the student appropriately. If the student fails to set the time correctly, the unit should also issue an appropriate response. Outline the salient mechanical and electrical features of such a system.

7.96 *Challenge Problem.* Develop a design concept for a computer-interfaced electronic display board that can be placed in the lobby of an office building to display messages of the day, announce upcoming seminars, or indicate the location of special events. The objective of the problem is to use a matrix of addressable light-emitting diodes (LEDs) rather than a video display. The system should accept messages by wire from a remote site. One approach might be to design your display board system so that it is capable of independently connecting to a local-area computer network. Alternatively, you could build a separate remote device that could be connected to a desktop computer and then brought down to the display board to load in the data. A wireless solution might also be possible. These examples are suggestions only. In general, any means for getting data to the board is acceptable, but a separate computer (PC) cannot become a dedicated part of the finished display.

7.97 *Challenge Problem.* Design a handheld medicine dispenser for dispensing pills at specific times of the day. The unit is to be carried by an individual and must have sufficient capacity to hold medication for at least 1 day. The unit should open a compartment and should emit an audible or visual signal when it dispenses medication. The unit must be easy to load and should be easy to program.

7.98 *Challenge Problem.* One perennial problem with radio-controlled model airplanes concerns the lack of knowledge about the flight direction and orientation of the airplane when it is far from the ground-based operator. When the airplane is too far away to be seen clearly, the operator loses the ability to correctly control its motion. Develop a design concept for a roll, pitch-, and compass-heading indicator system that can be mounted on the model airplane and used to radio the information to the operator's control console. Your system should sense the pitch and roll of the airplane over the range +90 degrees to −90 degrees and be able to withstand a full 360-degree roll or "loop-de-loop."

7.99 *Challenge Problem.* Design a remote readout system for a vacation home to be interrogated by a remote computer over a modem and telephone line or cellular telephone. The unit in the vacation home should answer the phone after 10 rings, provide means for an entry password, and then provide the following information: inside and outside temperatures, presence of any running water in the house, presence of any loud noises or unusual motion, and status of alarm switches installed on doors and windows. Discuss the design specifications for such a unit and develop a block diagram for its design and implementation.

7.100 *Challenge Problem.* Design a system for identifying which circuit breaker is associated with a given electrical outlet. The best system will be one that does not require the operator to move back and forth between the basement, where circuit breakers are typically found, and the locations of the outlets on the upper floors of the building.

7.101 *Challenge Problem.* Design a system for personalized airport baggage handling. Luggage is to be directed to each waiting individual, or at least to small clusters of people waiting at specified stations.

7.102 *Challenge Problem.* The future may bring a host of hydrogen-powered vehicles that run on either direct combustion or fuel cells. If such vehicles are to come into widespread use, a practical system must be put into place to distribute the fuel. Given that hydrogen, the lightest of the elements, is also extremely volatile and is the gas most prone to accidental ignition, this task is not an easy one. Design a system for nationwide distribution of hydrogen that will gradually replace the country's current network of gasoline stations. Consider cost, convenience, and safety factors in your deliberations. Design from the top down, beginning with the production of the fuel itself, and ending with proposed equipage for the refueling stations.

7.103 *Challenge Problem.* Modern technology has vastly extended the sensory capability of humans to the point where machines can talk, listen, touch, and see. Devise a system which might heretofore have been unthinkable: a means for a blind person to drive an automobile.

7.104 *Challenge Problem.* Design a system for harvesting the energy contained in ocean waves. Note that many such solutions have been proposed (see, for example, ocsenergy.anl.gov). Most of these systems, however, are oriented toward large-scale production tied to the national power grid. Your system is to be used by the owner of a small vacation home on an island near Puget sound in Washington state. The island is not connected to any external source of electric power.

7.105 *Challenge Problem.* Space debris has become a problem of note as we probe the orbital space of our planet. (See, for example, *www.orbitaldebris.jsc.nasa.gov*). Devise a system for an orbital spacecraft that can alert astronauts to an impending collision with a rogue object, and also devise a means to prevent the collision without altering the course of the spacecraft.

8 MANUFACTURING ENGINEERING

8.1 INTRODUCTION

Manufacturing engineering may be defined as *designing the production process for a product.* Although there is a large difference of opinion on exactly what is included in the design of a production process, almost all people would generally agree with the above definition. Production or manufacturing engineering includes all considerations pertaining to the *process* of production. This includes such functions as the following:

1. Evaluating the manufacturability of the product.
2. Selecting processes and setting process parameters such as cutting tool material, size, and shape; cutting speed; depth of cut; and so on.
3. Designing work-holding devices (jigs and fixtures) to secure and control the position of the workpiece during manufacture.
4. Estimating the cost of manufacturing the part.
5. Assuring the quality of the part produced.

The rate of technological change in manufacturing engineering is phenomenal and likely to increase. "Keeping up" is already a very difficult task. The future is likely to see a progressive manufacturing engineer spending 15 to 25% of his or her time simply studying new technology. Computerization, integration of controls, robotization, and coordination of manufacturing activities are areas where change is likely to be most rapid. It is important to note, however, that in spite of the new technology, the basics still apply. For example, the most sophisticated computer-controlled machine center still performs its functions according to basic metal cutting theory. It is vital for the manufacturing engineer to know and understand the basics of manufacturing engineering.

This chapter discusses some of the primary areas of manufacturing engineering. Since much industrial engineering has been applied to the metal working industry, we concentrate on that particular industry. The reader should be aware that industrial engineering techniques can be applied in any operating system, such as manufacturing, service, and governmental activities. Fully 60% of the American work force functions in service and governmental areas; but the limited length of this chapter prohibits any detailed discussion of those areas.

8.2 PRODUCT-PRODUCTION DESIGN INTERACTION

Product design requires that a person develop and evaluate the ability of the part to perform its intended function. Part characteristics such as size, shape, strength, reliability, safe operating range, and so on, are evaluated using knowledge of physics, strength of materials, tribology, and so on, often using computerized analysis. Manufacturing engineering develops and evaluates the cost of producing the part and uses knowledge of the relative cost, capabilities, and limitations of the various alternative processing methods available to produce the particular part shape, as well as detailed knowledge of cutting tools, machine tools, skill levels of workers, similarity to other parts being produced, and so on. Unfortunately, few industrial engineers are trained in product design and few product designers are trained in manufacturing engineering. This makes their interaction exceptionally important.

In every manufacturing operation, some variation in the size of the individual parts produced by the process will inevitably occur due to a variety of causes, such as tool wear, operator error, and material variations. The range of part sizes that can be used without compromising part function or reliability, that is, the variation in each dimension of the product that can be tolerated, is referred to as the tolerance for the product. The part designer, who is most concerned with product function, would want the tolerances set as small as possible to assure that the part will function without any problems. The manufacturing engineer, who is most concerned with product cost, wants the largest possible tolerances to be specified, as this often gives him or her a wider choice of processes to be used in manufacturing the part. This wider choice will often result in a reduction in product cost. Sometimes, a product designer may specify very tight tolerances on the product because he or she does not realize the cost of machining to exceptionally tight tolerances or the inability of machines to produce unusual configurations. Often, the manufacturing engineer will assume that these tight tolerances or unusual configurations are necessary (when in fact they are not) and design the process to produce them. This adds unnecessary cost to the product.

Ideally, a manufacturing engineer should work with the product designer from the very beginning to ensure producibility. If it is not possible to have this early interaction, the manufacturing engineer should inform the design engineer of unusually costly operations. By having this information, the design engineer can frequently avoid certain costly operations. This interaction must happen and should occur as early as possible in the design of products.

Employee participation is also vital in this interaction phase. No one knows the details of a job better than the person doing it, so this employee feedback input must be encouraged for industry to remain competitive. In fact (as will be seen later), the use of employee skills in some type of participative management may be the most exciting trend today.

> It is difficult to overemphasize the value of input from the hourly shop floor workers in planning and implementing automation systems. No one knows the nitty-gritty of a job better than the person who does it.[1]

---

[1]Frank Curtin, "Automating Existing Facilities: GE Modernizes Dishwasher, Transportation Equipment Plants," *Industrial Engineering*, Sept. 1983, pp. 32–38.

8.3 PROCESS ENGINEERING

Process engineering is concerned with the design of the actual process to be used in the manufacture of the product. In designing the processes to be used, a six-step sequence should be undertaken: defining the product structure and specifications, assessing each component's manufacturability, listing the different processes capable to manufacturing the component, evaluating the cost of each of the alternative processes, determining the sequence in which the operations are to be performed, and documenting the process.

8.3.1 Defining Product Structure and Specifications

Product structures are often shown in a hierarchical chart that shows all of the subassemblies, sub-subassemblies, components, and raw materials that comprise the product. Figure 8.1 shows the product structure for a product assembled from two subassemblies (S_i), each of which is composed of sub-subassemblies (SS_j), components (C_k), and raw materials (R_l).

This type of chart clearly defines "what goes into what" and on which of the five levels the item belongs. More complex products would have more levels in their product structures. In Figure 8.1 each horizontal line indicates that an assembly operation must be designed to join its constituent parts. For example, the horizontal line linking subassemblies 1 and 2 indicates that a process to connect these two subassemblies must be designed. Note that sub-subassemblies, components,

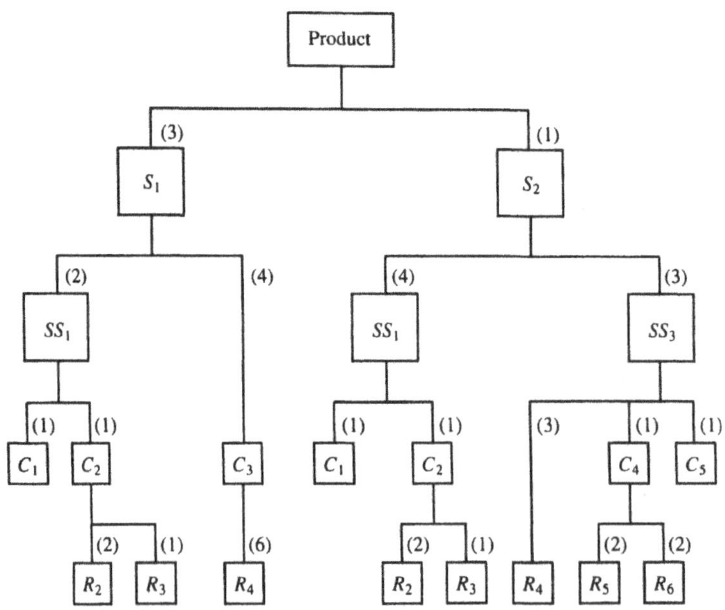

Fig. 8.1. Example product structure.

and/or raw materials can "go into" the subassembly; the hierarchy of the product structure dictates only that the horizontal line be above the highest-level item (subassembly, component, raw material) in the assembly. Also shown in Figure 8.1 are components that are purchased already fabricated and/or assembled and thus have no raw materials that must be processed by this facility. Vertical lines in the product structure indicate that the lower-level item is modified by some operation to transform or process it into the higher-level item. In industrial applications, each item in the product structure is identified by a unique number and incorporated in this number is the level of the item. This numbering is helpful when scheduling the production of the item.

The number of each item required to make one of the next-higher-level item is also shown in the product structure. Figure 8.1 shows that three of subassembly 1 are required to make one unit of the product and that two sub-subassembly Is are needed for each subassembly 1. This means that six units of sub-subassembly 1 are needed to make our product. These data are used when determining the number of each item that will be required to meet the sales forecast for our product.

This information can also be shown in bills of material. A *bill of material* contains the number and name of all parts, the source of each part (purchase or manufacture), and the number and name of the completed product. Normally, only those parts that go directly into that product are shown. For example, if a subassembly is used, the number and name of that subassembly are shown and a separate bill of material is developed for that subassembly. Figure 8.2 is a sample bill of material. In this figure, if the shelf (stock number 3) were a subassembly consisting of perhaps the basic shelf, a laminated walnut top, and molding, then the bill of material for the shelf would show these three items.

Nowhere have the powers of computers been more evident than in the generation of these forms (process sheets, route sheets, bills of material) and their subsequent use in production control. Many on-line computerized control systems generate the route sheets and trace the product as it flows through the shop. Recent research has even demonstrated the possibility of the computer going from product design through process engineering, through recognition of shapes and forms and "artificial intelligence." In artificial intelligence, the computer learns from past experiences until it actually is capable of designing the production process. The manufacturing engineer should watch this development carefully.

Once all of the items that "go into" the product have been enumerated using the product structure or bill of material, the specifications associated with each item must be developed. These specifications would detail the desired material, length, width, height, diameter, material, surface finish, and tolerances for the item, which will assure that it will perform its intended function. This will enable the evaluation of the item's manufacturability and development of a list of the processes that are capable of producing this item.

BILL OF MATERIAL			
Product Description: Bookcase, metal, 3 shelves. *Stock Number* : 1			
Component		Quantity (Amount) Required	Source
Stock No.	Description		
3	Shelf	3	Manufacturing
4	Leg	4	Manufacturing
5	Inserts	8	Purchasing
6	Screws	12	Purchasing

BILL OF MATERIAL			
Product Description : Bookcase, metal, 6 shelves. *Stock Number* : 2			
Component		Quantity (Amount) Required	Source
Stock No.	Description		
3	Shelf	6	Manufacturing
4	Leg	8	Manufacturing
5	Inserts	8	Purchasing
6	Screws	24	Purchasing
7	Connectors	4	Purchasing

BILL OF MATERIAL			
Product Description : Shelf *Stock Number* : 3			
Component		Quantity (Amount) Required	Source
Stock No.	Description		
8	Sheet metal	3 square feet	Purchasing

BILL OF MATERIAL			
Product Description : Leg *Stock Number* : 4			
Component		Quantity (Amount) Required	Source
Stock No.	Description		
8	Sheet metal	2 square feet	Purchasing

Fig. 8.2. Bill of material. [Taken with permission, from Lawrence E. Doyle et al., *Manufacturing Processes and Materials for Engineers*, 2nd ed. (Englewood Cliffs, N.J.: Prentice Hall, 1969).]

8.3.2 Assessing Manufacturability

The manufacturability of an item is a measure of the relative ease or difficulty of producing the item within tolerance. The tighter the tolerances, in general, the more expensive the item will be to produce (see Figure 8.3). Also contributing to the manufacturability of the part is the incorporation of standard-sized features in the part. For example, if a hole in a part is specified to be a size that can be produced by a commonly available drill size, its cost to drill would be relatively low compared to the cost to drill a nonstandard hole size. Similarly, if at all possible, the part material should be specified to be a commonly available material. Only when particular part characteristics are not available from a commonly used material should an exotic material be specified. Each feature of each item in the product must be examined to determine if its function can be achieved in a manner that will allow consideration of a greater number of processing alternatives, some of which may be less expensive.

8.3.3 Determining Processes Capable of Producing the Part

By examining the specifications for each item and the capabilities of the various processes, a list of candidate processes can be developed. In developing this list, the following must be considered:

1. The compability of the material and the process.
2. The ability of the process to produce the desired tolerances.
3. Part (re)design required to make it easier to produce by this process.
4. The availability of the process.
5. The tooling, and so on, necessary to make the part using the process.

Use of tables such as those shown in Figures 8.4 through 8.6 enables the manufacturing engineer to quickly generate a complete list of available processes.

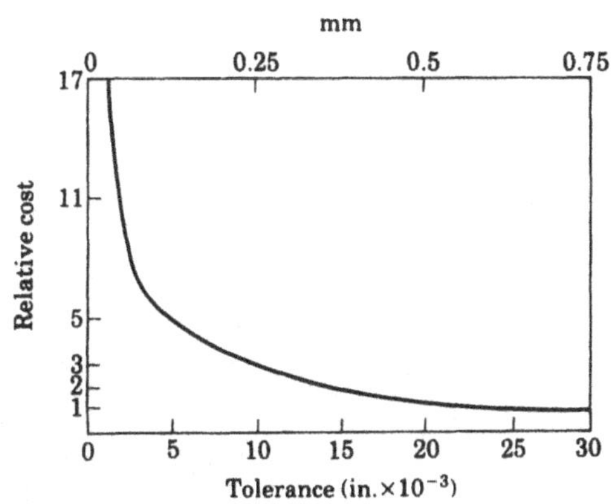

Fig. 8.3. Relationship of cost and tolerance. [From Serope Kalpakjian, *Manufacturing Engineering and Technology*, © 1989, by Addison-Wesley Publishing Company, Inc. Reprinted with permission of the publisher.]

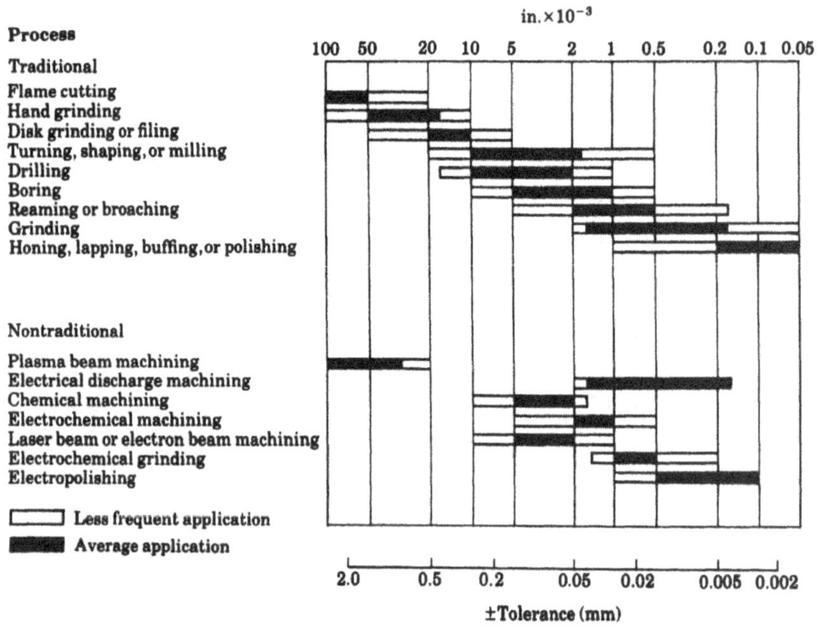

Fig. 8.4. Tolerances produced by various processes. [From Serope Kalpakjian, *Manufacturing Engineering and Technology*, © 1989, by Addison-Wesley Publishing Company, Inc. Reprinted with permission of the publisher.]

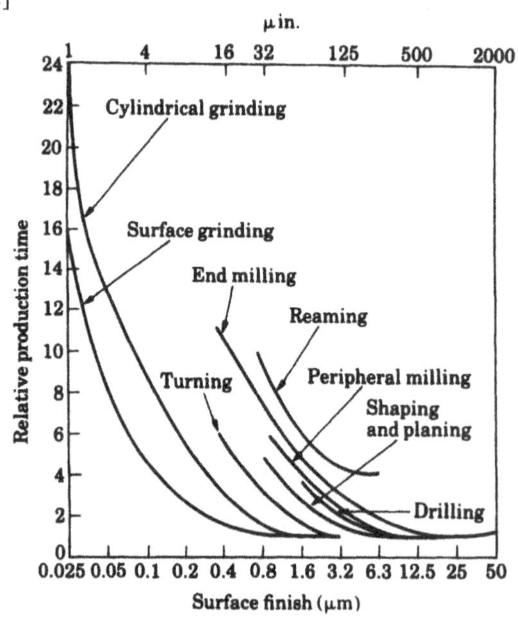

Fig. 8.5. Relative production time as a function of surface finish produced by various manufacturing methods. [From Serope Kalpakjian, *Manufacturing Engineering and Technology*, © 1989, by Addison-Wesley Publishing Company, Inc. Reprinted with permission of the publisher.]

Process	*Scrap (%)*
Machining	10–60
Hot closed-die forging	20–25
Sheet-metal forming	10–25
Cold or hot extrusion, forging	15
Permanent-mold casting	10
Powder metallurgy	5

Fig. 8.6. Amount of scrap produced by various processes. [From Serope Kalpakjian, *Manufacturing Engineering and Technology*, © 1989, by Addison-Wesley Publishing Company, Inc. Reprinted with permission of the publisher.]

8.3.4 Evaluating the Cost of Each Process

To evaluate the cost of using each process, two types of costs must be determined: fixed costs and variable costs. Fixed costs are those costs that are constant regardless of the number of parts made using the process. Examples of fixed costs are the purchase cost of the machine(s) used in the process, the cost of installing the machine(s), the cost of designing and fabricating the work-holding devices, and the cost of the space that the machine(s) occupy. Variable costs are those costs that vary directly (i.e., proportionally) with the number of parts produced. Examples of variable costs include the cost of the machine operator's time while making the part, the cost of running the machine(s) (e.g., electricity, natural gas, other fuels), the cost of all cutting tools that are dulled or otherwise consumed during the manufacture of the part, and the cost of the material used in manufacturing the part.

Fixed costs can be estimated by contacting vendors and/or users of the various equipment that is being considered as well as cost data developed by the firm's accounting department. Estimates of the variable cost rate (i.e., dollars/hour) can be obtained from the same sources. Expressing this in an equation, we obtain

$$\text{total cost} = \text{fixed cost} + \text{variable cost/unit} \times \text{number of units}$$

This relationship is also shown in Fig. 8.7.

If a number of alternative processes are being considered, their costs can be plotted on the same graph and the range of production quantities for which each process is less expensive can be determined. For example, suppose that three processes are under consideration and the following costs apply:

	Process A	*Process B*	*Process C*
Fixed cost	$ 0	$2,000	$8,000
Cost/unit	10	8	5

If we graph the costs, as shown in Figure 8.8, we can see that if we need to produce fewer than 1,000 units, process A should be used. If more than 2,000 units are needed, process C should be used. For production requirements between 1,000 and

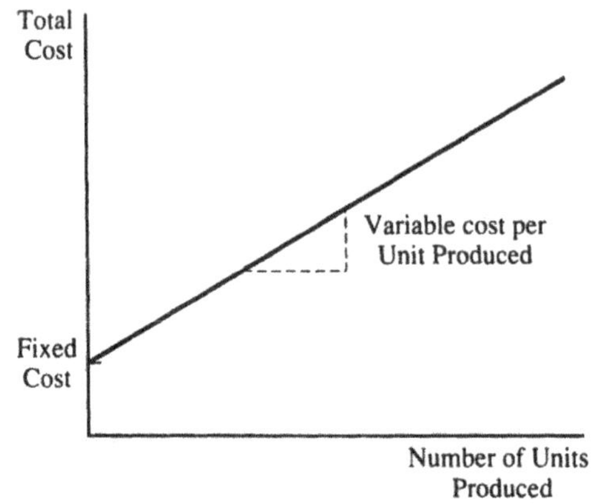

Fig. 8.7. Total cost as a function of the number produced.

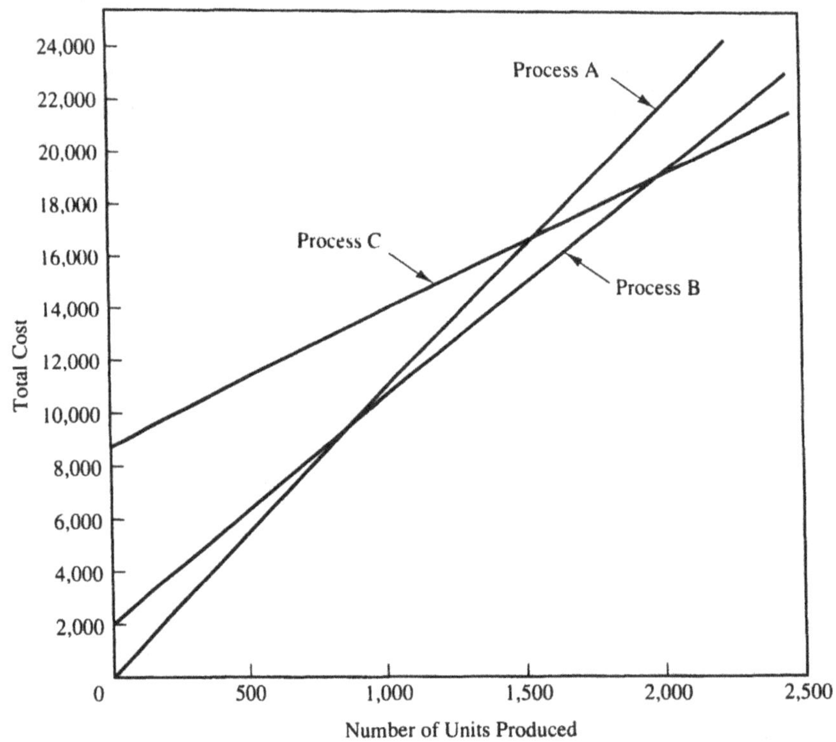

Fig. 8.8. Total cost versus quantity for three processes.

2,000, process B should be used. The production requirement of 1,000 is said to be the break-even point between process A and process B and is the production volume at which the total manufacturing costs are equal for the two processes. Production quantities below this break-even point can be produced at a lower cost using process A, and quantities above this break-even point can be produced at a lower total cost using process B. Similarly, 2,000 is the break-even point for processes B and C.

We could also have used the two total cost equations to determine the breakeven points:

$$\text{total cost}_A = 0 + 10X$$
$$\text{total cost}_B = 2{,}000 + 8X$$

where X is the quantity to be produced. Realizing that total costs for both processes must be equal at the break-even point, we set the equations equal to each other and solve for X, the break-even point.

$$0 + 10X = 2{,}000 + 8X$$
$$2X = 2{,}000$$
$$X = 1{,}000$$

Similar calculations can be used to verify that the break-even point between process B and process C is 2,000.

The variable cost per part reflects the time that it takes to manufacture one part using the process and the costs that are incurred during this period of time.[2] This variable cost for a metal cutting operation has four components: the cost of labor (operator) and the machine while the machine is cutting, the cost of the tool, the cost of labor and the machine while sharp tools are being mounted, and the cost of labor and the machine while the workpiece is being loaded and unloaded. These costs can be expressed as

$$\text{variable cost} = T_c \times r_{l,m} + r_t \times \frac{T_c}{T} + T_{dt} \times r_{l,m} \times \frac{T_c}{T} + r_{l,m} \times T_{dw} + r_{l,m} \times T_a$$

where T_c = cutting time/part (min)
T = tool life (i.e., amount of time a tool can cut before dulling) (min)
$r_{l,m}$ = cost (dollars/min) of running the machine, including operator cost
T_{dt} = time (min) it takes to put a sharp tool on the machine (downtime for tool change)
r_t = cost of one sharp tool
T_{dw} = time (min) to load/unload one workpiece (downtime for workpiece change)
T_a = time (min) to adjust or position tool on each workpiece

This cost is dependent on the process parameters that are selected for the operation. The need to define the relationship of the cutting speed and tool life was addressed by Federick W. Taylor, the father of industrial engineering, early in the twentieth century.[3]

---

[2]Vidosic, Joseph P., *Metal Machining and Formating Technology* (New York Ronald Press, 1964), p.329.

[3]Taylor, F. W., "On the Art of Cutting Metals," *Transactions of ASME*, 28 (1906), p. 31.

He knew that the time it takes to machine a part can be determined by calculating the length of the cut(s) that are required to machine the part to the desired size. For example, if a shallow groove is to be cut into a 36-inch-long workpiece using a single-point cutting tool operating at 20 ft/min, it would take 9 seconds to complete the cut. If 20 grooves were required on the part, 3 minutes of cutting would be required to manufacture the part. If one wanted to reduce the processing time, one could increase the cutting speed. The problem with this approach is that when the cutting speed is increased, the tool wears out or dulls much more quickly. To determine the optimum cutting speed, the relationship between the speed at which a cutting tool was being operated and how long the tool would cut until it was too dull to use was needed. This relationship is described by the following formula, known as *Taylor's tool life equation*:

$$C = VT^n$$

where C = cutting speed (ft/min) at which the tool would dull after one minute of use

V = cutting speed (ft/min) at which the process will be operated

T = tool life (min) that can be expected when the tool is operated at cutting speed V

n = a constant that is dependent on the material being cut and the cutting tool material

If we define the length of the cuts on a particular workpiece to be L, $T_C = L/V$, and if we rearrange the terms in the Taylor tool life equation,

$$T = \left(\frac{C}{V}\right)^{1/n}$$

we can substitute these two variables into our variable cost equation:

$$\text{cost/part} = \frac{L}{V} \times r_{l,m} + r_t \times \frac{L}{V} \times \left(\frac{C}{V}\right)^{-1/n} + T_{dt} \times r_{l,m} \times \frac{L}{V} \times \left(\frac{C}{V}\right)^{-1/n} + r_{l,m} \times T_{dt} + r_{l,m} \times T_a$$

Since this is now an equation involving only the single variable V, we can differentiate it with respect to V, set the first derivative equal to 0, and solve for V. When we do,

$$V = C\left(\frac{n}{1-n}\right)^n \left(\frac{r_{l,m}}{r_{l,m}T_{dt} + r_t}\right)^n$$

which is the optimum cutting speed, that is, the cutting speed that minimizes the cost of cutting each workpiece and hence minimizes the cost of producing the product on the machine in question.

Suppose, for example, that a 2-foot section of a 3-in.-diameter 1020 steel part is to be machined using a high-speed steel tool. Assume that:

$$C = 225 \text{ ft/min} \qquad r_{l,m} = 1 \text{ \$/min} \qquad r_t = \$4$$
$$n = 0.105 \qquad T_{dt} = 3 \text{ min}$$

$$V = 225\left(\frac{0.125}{1 - 0.125}\right)^{0.105}\left[\frac{1}{(1)(3) + 4}\right]^{0.105}$$
$$= (225)(0.7976)(0.8152)$$
$$= 146.2 \text{ ft/min}$$

Further, the time to cut the part using a feed rate (the translation of the tool per revolution of the workpiece) of 0.05 in./rev:

$$\text{No. revolutions to cut 24 in.} = \frac{24}{0.05} = 480$$

$$\text{rev/min at cutting speed of 146 ft/min} = \frac{146 \text{ ft/min}}{\pi\left(\frac{3}{12}\right) \text{ ft/rev}} = 186 \text{ rev/min}$$

$$\text{time to cut } (T_c) = \frac{480}{186} = 2.58 \text{ min}$$

$$\text{tool life } (T) = \sqrt[n]{\frac{C}{V}} = \sqrt[0.105]{\frac{225}{146}} = 61.5 \text{ min}$$

$$\text{length of the cut} = \pi(3)\left(\frac{24}{0.05}\right) = 4{,}523 \text{ in.} = 377 \text{ ft.}$$

and if $T_{dt} = 3$ min, $T_{dw} = 1$ min, $r_{l,m} = \$1$/min, $T_a = 2$ min, and $r_t = \$1$/edge:

$$\text{total cost} = \frac{L}{V} r_{l,m} + r_t \frac{L}{V}\left(\frac{C}{V}\right)^{-1/n}$$
$$+ T_{dt} r_{l,m} \frac{L}{V}\left(\frac{C}{V}\right)^{-1/n}$$
$$+ r_{l,m} T_{dt} + r_{l,m} T_a$$
$$= \$7.87$$

8.3.5 Determining the Sequence of Operations

The determination of the sequence in which the operations that will be used to transform the material into its desired final shape depends on a number of considerations, and a full discussion of them is beyond the scope of this book. The student should be aware, however, that the sequence is developed based on (1) minimization of part handling (the part is routed, if possible, along the shortest path from machine to machine without backtracking); (2) assuring that no succeeding operation adversely affects previous operations by leaving burrs, debris, scratches, and so on; and (3) performing as many operations on each machine as possible so that close tolerances can be maintained and quality assured.

8.3.6 Documenting the Process

After the manufacturing engineer has reviewed the various processes and has decided on the most economical and effective procedure or set of processes, he or she must then find a way of recording this information for later use in the shop. One way is to construct an *operations process chart*.

An operations process chart depicts the flow of the material through the various processes. It shows only operations and inspections; it does not show transportations, storages, or delays. These are shown in flow process charts that will be discussed later. Using the following symbols, a manufacturing engineer can construct an operations process chart (see Fig 8.29 for an example.)

In the operations process chart, a complete unit is shown on a page (whenever possible). The chief component starts at the top and subsidiary components are added as the flow progresses to the right and down.

The next step is to develop a route sheet that shows the sequence of operations, a description of each, and the various machining information necessary. A *route sheet* is a set of instructions to the plant showing how to manufacture the product. Figure 8.9 is an example of a route sheet. One such sheet would be required for each part in the operations process chart. The exact information shown varies between companies, but a good rule of thumb is to provide enough information to be able to produce the part including machine numbers, tool numbers, directions, and perhaps times for setup of the equipment and actual run time required. Figure 8.9 is not only a routing sheet, but also a production order for use in dispatching an order to the shop.

8.4 INDUSTRIAL PROCESSES

There is a wide variety of processes available to the manufacturing engineer to transform material into the desired size and shape. While a large number of materials are used in products, we will limit our discussion to those used for production of metal parts. These processes include refining and alloying of the metal, casting, forming, metal cutting, welding, assembly, and finishing.

8.4.1 Refining and Alloying

The general procedure for almost all metals is to begin with ore in its natural state and send it through some process to improve its usefulness. Steel, for example, is extracted as iron ore, is sent through a blast furnace to produce pig iron, and then it

TR

TOTAL COPIES

W-WORK ORDER
R-ROUGH STORES
F-FINISHED STORES

SHEET NO.

PRODUCTION WORK ORDER

MIN. LOT 72 PRACT. LOT 4400 PART NO. 8B-3381

OPER. NO.	MACHINE LOC. CLASS	MACHINE M. NO.	BASE TIME HRS. PER SET-UP	BASE TIME HRS. PER PIECE	BURDEN HRS.	PERFORM. HRS.	EFFECTIVE DATE IF NOT STARTED OPERATION
1	X18B A2C	1939	1.00	.0009 1111			CHAMFER & CUT OFF
2	X23D G2R	1383	.10	.0025 400			REMOVE BURRS
3	X19C G20C	1997	.50	.0061 164			GRIND O.D. (2) PASSES
4	X19C M1B	7080	.70	.0161 62			STRADDLE MILL TONGUE
5	X19C M1B	7080	.80	.0141 71	.80 .0141	.80 .0141	MILL KEYWAY
6	X23D G2R	1383	.10	.013 77			REMOVE BURRS & CHAMFER TONGUE
7	X19C D6S	6076	.30	.0167 60			DRILL HOLE
8	X22D E1B	B131	.20	.0050 200			REMOVE BURRS

PARTS GROUP NO. 49

FILE	HT	I	C	W	R	F

COMP. PARTS: 4B-588, 4B-608, 4B-3429, 7B-4920, 3B-6794, 4B-8735

MATERIAL: C.T. #1E-40 C.F. STEEL 5/8" RD. M.L. (N.C.)

PART NAME: OIL PUMP GEAR SHAFT

QTY. PER PC.: .505 FT.

DELIVER TO

ORDER DATE	SCHED. LOT SIZE	SCHEDULE	CSTGS. REQ'D

8B-3381
PART NO.

GG:MKG:CK

Fig. 8.9. Sample route sheet. [Taken with permission, from Lawrence E. Doyle et al., *Manufacturing Processes and Materials for Engineers*, 2nd ed. (Engle-wood Cliffs, N.J.: Prentice Hall, 1969).]

is sent to the steel mill where it is made into steel by the open hearth, basic oxygen, or electric furnace process. The type of steel produced is determined by the temperature of the furnace and its chemical composition.

Since all metals do not have the same properties, the proper choice of the metal to be used in a product depends on the functions the part must perform. This choice occurs in the production-product design interaction. To further compound the problem, metals emerging from the primary operations usually do not have the properties necessary. This means that some sort of metallurgical transformation is required to improve the hardness, strength, workability, or other properties of the metal. Much of this is done by heat treating the metal to alter its properties. In some cases, it may be necessary to combine two or more metals to obtain better properties. These combinations are called *alloys.* The production engineer must be aware of the different types of metals available and the particular alloys that can be made and their properties.

Because the properties of the various metals are very important, testing for these properties is of primary importance. Some of the properties that are often tested are:

1. *Tensile Strength*—The ability of the metal to withstand elongation forces.
2. *Hardness*—The ability of the metal to withstand penetration forces.
3. *Impact Resistance*—The ability of the metal to absorb energy.
4. *Malleability*—The ability of the metal to be shaped or extended.
5. *Fatigue Resistance*—The ability of the metal to withstand repeated applications of a force.
6. *Corrosion Resistance*—The ability of the metal to withstand corrosive forces.

The production engineer must understand these and other tests in order to select the correct material.

8.4.2 Casting

The first step in production is to obtain an approximate shape that, if necessary, can later be refined or finished more smoothly. One way of doing this is through casting. Casting is the process of forming objects by pouring liquid material into a mold and allowing the material to solidify. Casting is an old process, but it is still the most economical and effective method for obtaining many desired shapes.

Mold design is one of the most important parts of casting and the correct design of the mold is both a science and an art. Although many types of molds exist, by far the most predominant is a sand mold, especially for iron and steel.

Figure 8.10 illustrates a typical sand mold with its principal parts. Each part is described below.

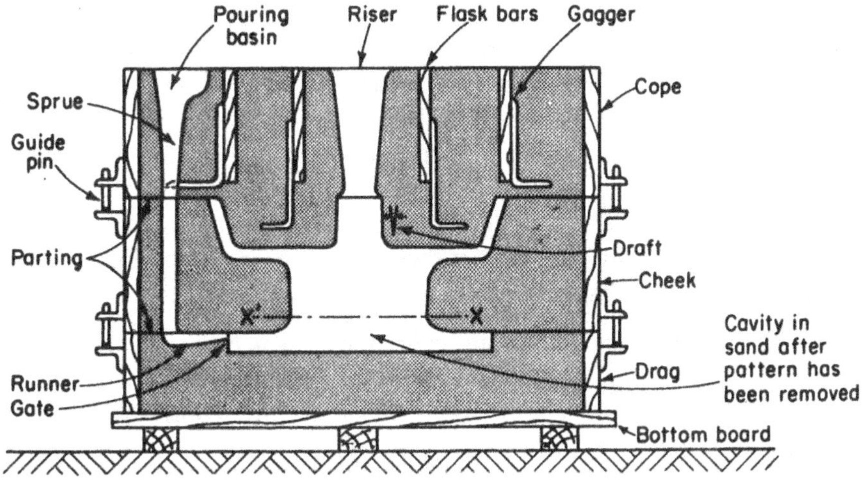

Fig. 8.10. Cross-sectional view of a three-part mold. [Taken with permission, from Lawrence E. Doyle et al., *Manufacturing Processes and Materials for Engineers*, 2nd ed. (Englewood Cliffs, N.J.: Prentice Hall, 1969).]

Flask—Frame in which the mold is made.
Cope—Top section of the flask.
Drag—Bottom section of the flask.
Cheek—Intermediate sections of the flask (when required).
Riser—Reservoir of liquid metal to supply additional liquid metal as the metal solidifies and contracts.
Pouring Basin—Area into which molten metal is poured.
Sprue—Channel for vertical delivery of molten metal.
Runner—Channel for horizontal delivery of molten metal.
Gate—Opening through which metal enters the mold.

The correct relative and absolute shapes and dimensions of each of the above parts are highly critical in the design of a good mold. If the manufacturing engineer is to be working with castings, he or she should be cognizant of these and should be able to design good molds.

Obviously, this type of sand casting is expensive when large quantities of products are to be made, as the mold is destroyed each time a casting is made. Consequently, many specialized types of casting with permanent molds have been developed. The subject is too long and complex for this book, but some of the specialized types are centrifugal casting, die casting, gravity or permanent mold casting, pressed casting, ceramic shell casting, plaster mold casting, and shell molding casting.

8.4.3 Metal Forming

Another type of primary metal working process is metal forming. Here, the metal is worked under pressure to form a desired shape, to improve the physical properties, or both. This working can be either *hot* or *cold.* In hot working, the metal is worked at a temperature above its recrystallization temperature so that certain properties are enhanced. For example, in hot working, the metal working requires less pressure, and unusual shapes can be made more easily. Cold working takes place at a temperature below the metal's recrystallization temperature. This is done to hold close tolerances, to produce good surface finishes, and to enhance other physical properties.

Some of the processes that are classified as metal forming operations are listed below with short descriptions. Many of these processes can be performed in either hot or cold states, depending on the needs.

Rolling—Rolling is a squeezing operation in which the metal is elongated and/or widened as it passes through two or more rollers, as depicted in Figure 8.11. Shaped rolls can be used to manufacture a variety of shapes, for example, I-beams.

Wire Drawing—Wire drawing is an operation in which a wire or rod is reduced in cross-sectional diameter by being drawn (pulled) through a die. Figure 8.12 depicts the process.

Forging—Forging is metal forming by a single or a series of intermittent applications of pressure instead of by continuous pressure as in rolling. A blacksmith's hammering of a horseshoe is an example of forging.

Extrusion—Extruding a metal is accomplished by compressing it beyond its elastic limit and forcing it to flow through and to take on the shape of an opening. Figure 8.13 shows examples of extrusions.

Bending—Bending is applying force to permanently distort the metal to a preconceived shape. It may be done by a tube bending machine, a press with V dies, or several other means. Figure 8.14 depicts some typical sheet metal bends.

Drawing and Stretching—These operations produce seamless vessels by applying pressure and forcing the sheet metal to take on the shape of a preconceived form. As in wire drawing, some stretching of the material occurs. Figure 8.15 is an example of a simple drawing operation.

8.4.4 Metal Cutting

After the primary shape forming, the next step is often metal cutting. Metal cutting is the removal of metal by using a cutting tool. The objective is to obtain desired shapes, close tolerances, and specified surface finishes. There are many types of metal cutting operations. Some of the better known are listed below with short descriptions:

Shearing—Shearing is the cutting of sheet metal by applying pressure and forcing the metal between two sharp edges. Blanking, parting, punching, nibbling, notching, and shearing along a straight line are all shearing operations. Examples are shown in Figure 8.16.

Turning—Turning is the process of rotating the workpiece and cutting material by feeding a cutting tool into the workpiece and/or the workpiece into the tool. These operations are generally performed on a lathe and may include facing, contour turning, threading, parting, drilling, knurling, and others.

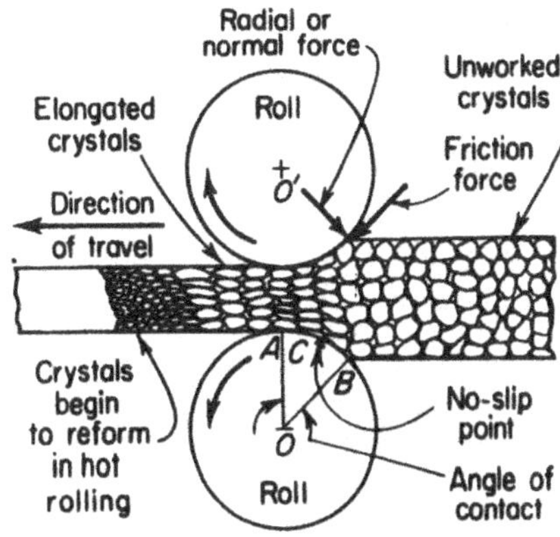

Fig. 8.11. Rolling operations. [Taken with permission, from Lawrence E. Doyle et al., *Manufacturing Processes and Materials for Engineers*, 2nd ed. (Englewood Cliffs, N.J.: Prentice Hall, 1969).]

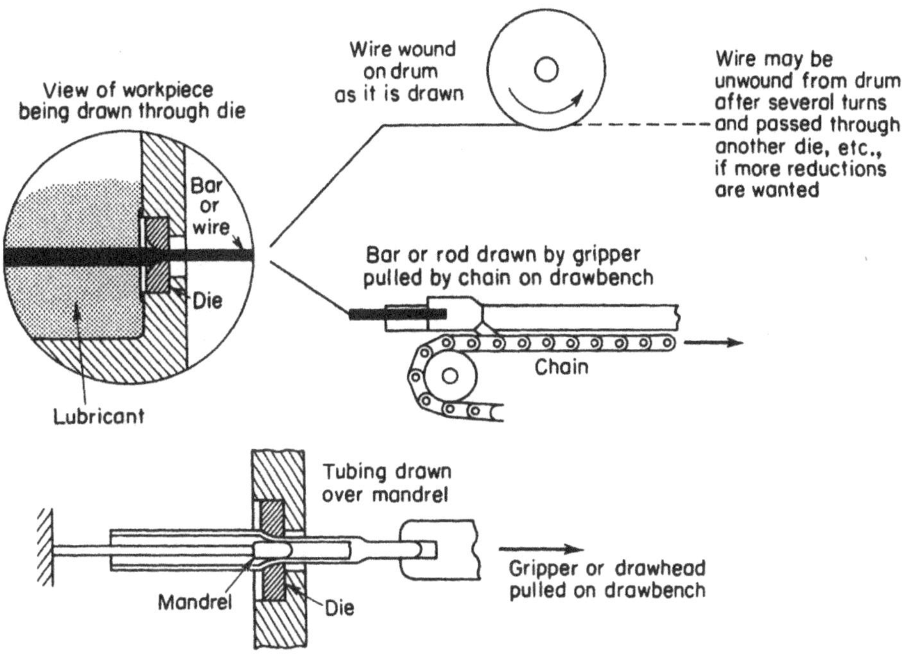

Fig. 8.12. Drawing operations. [Taken with permission, from Lawrence E. Doyle et al., *Manufacturing Processes and Materials for Engineers*, 2nd ed. (Englewood Cliffs, N.J.: Prentice Hall, 1969).]

Drilling—Drilling is the opening, enlarging, or finishing of a hole. It is generally associated with the process in which the tool turns and the workpiece is stationary. When the workpiece turns, it is known as a *turning operation* (see preceding page). The operation is generally performed on a drill press or boring machine.

Shaping and Planing—These are operations in which surfaces are cut by a reciprocating action. Generally, the surfaces are flat, but curved surfaces can

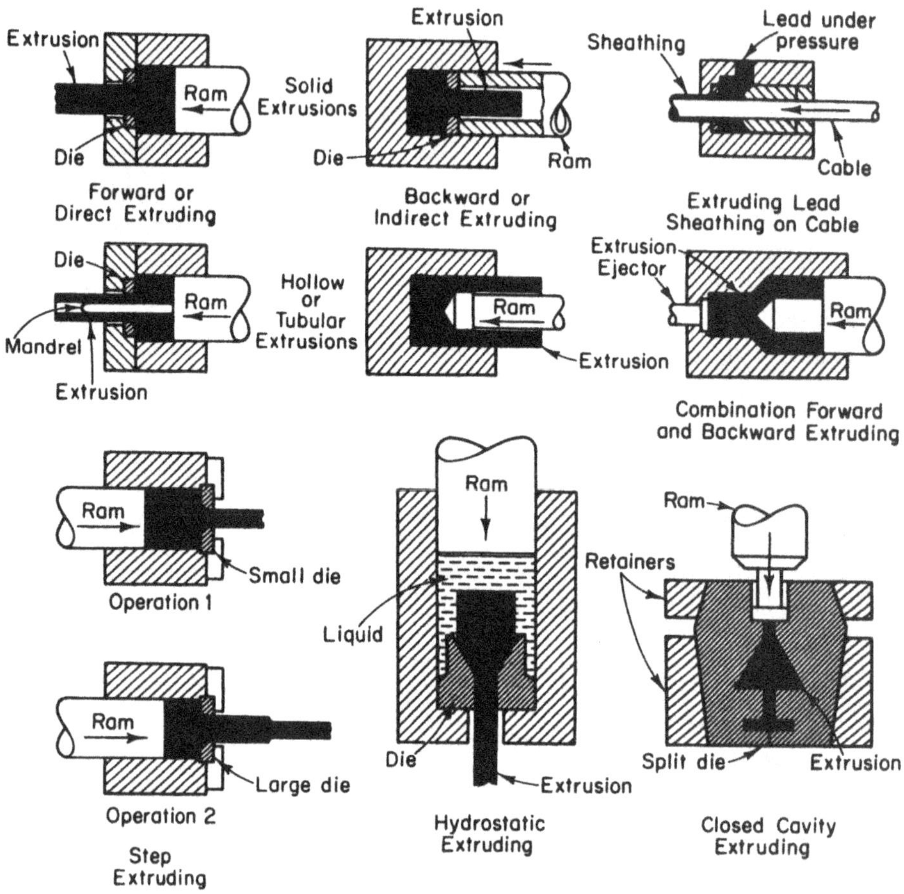

Fig. 8.13. Extruding. [Taken with permission, from Lawrence E. Doyle et al., *Manufacturing Processes and Materials for Engineers*, 2nd ed. (Englewood Cliffs, N.J.: Prentice Hall, 1969).]

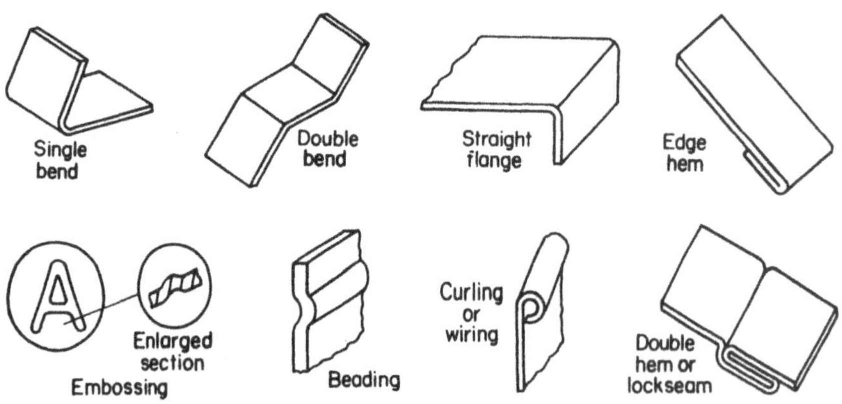

Fig. 8.14. Examples of sheet metal bends. [Taken with permission, from Lawrence E. Doyle et al., *Manufacturing Processes and Materials for Engineers*, 2nd ed. (Englewood Cliffs, N.J.: Prentice Hall, 1969).]

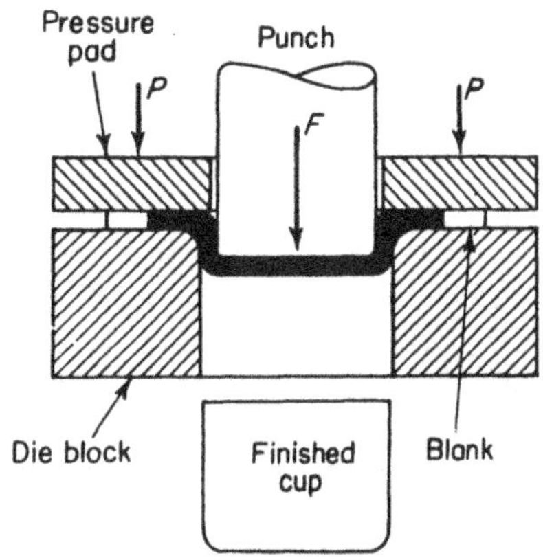

Fig. 8.15. Drawing of a cup. [Taken with permission, from Lawrence E. Doyle et al., *Manufacturing Processes and Materials for Engineers*, 2nd ed. (Englewood Cliffs, N.J.: Prentice Hall, 1969).]

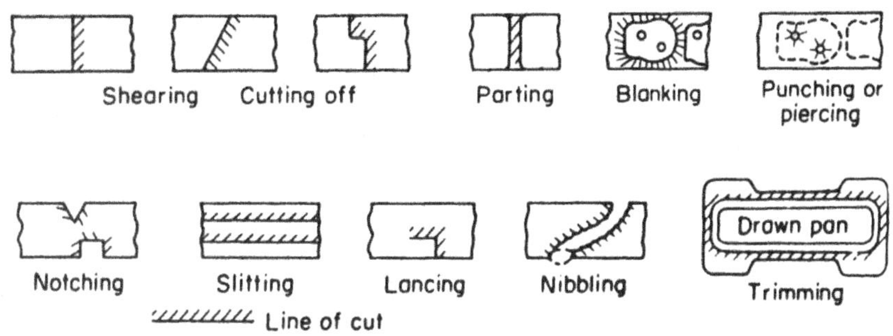

Fig. 8.16. Typical shearing operations. [Taken with permission, from Lawrence E. Doyle et al., *Manufacturing Processes and Materials for Engineers*, 2nd ed. (Englewood Cliffs, N.J.: Prentice Hall, 1969).]

also be machined by the processes. In shaping, the tool reciprocates while the workpiece is stationary. In planing, the workpiece reciprocates while the tool is stationary. Figure 8.17 is an illustration of a shaper in operation.

Milling—Milling is the process by which a revolving cutting tool with a number of teeth takes intermittent, successive cuts. Usually the workpiece is fed into the tool, but this is sometimes reversed. Figure 8.18 depicts some common milling operations.

Broaching—Broaching is similar to milling in that a series of teeth are used to take successive cuts on the workpiece. The broaching tool does not revolve, however; it is pushed or pulled. Figure 8.19 shows the basics of a broaching operation.

Sawing-Filing—Sawing and filing also remove metal by taking cuts with successive teeth on a tool that is pushed or pulled. In sawing, however, large amounts of metal can be removed without reducing it to chips. Figure 8.20 is a picture of a power hacksaw in operation.

Grinding—Grinding is the removal of metal in small pieces by the continuous action of an abrasive material on the workpiece. Grinding may be necessary on materials too hard for other methods, to improve surface finishes, or to hold very close tolerances. The small grinder in a garage used to sharpen hand tools is a familiar example.

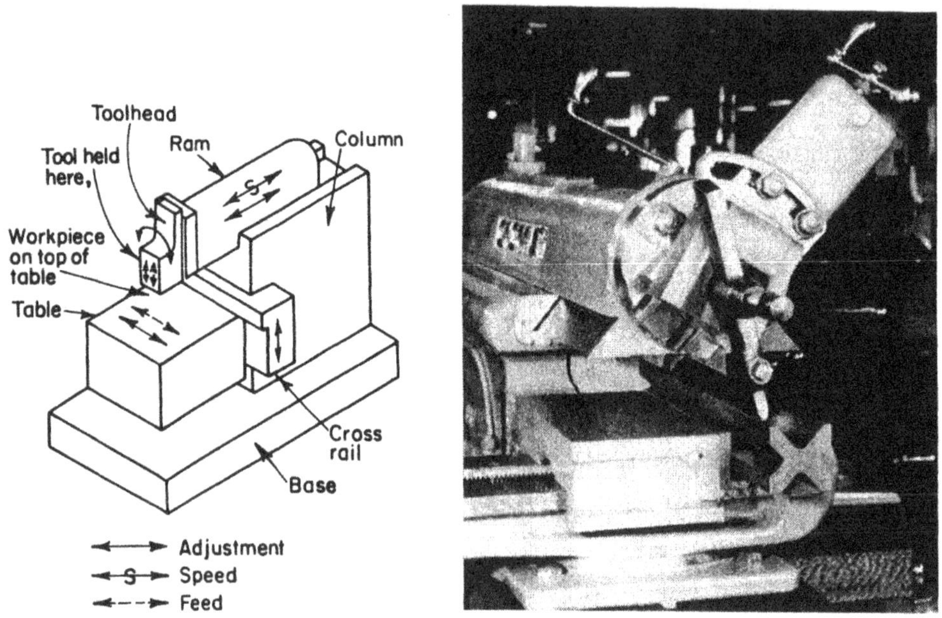

Fig. 8.17. Shaper in operation. [Taken with permission, from Lawrence E. Doyle et al., *Manufacturing Processes and Materials for Engineers*, 2nd ed. (Englewood Cliffs, N.J.: Prentice Hall, 1969).]

Other Operations—There has been a great deal of work directed toward the development of more sophisticated methods of metal removal. Examples are electrodischarge machining, electrochemical machining, electrolytic grinding, photoetching, and chemical milling. These will not be discussed in this chapter.

8.4.5 Welding

Welding is a process by which two pieces of the same metal are made to bond through the application of heat or pressure or both. The process actually fuses metals together by forcing the two parent parts to melt and coalesce. Some of the broad classifications and more popular types of welding processes are listed on the following pages.

Electric Arc—Electric arc welding is a process by which an electric arc is maintained between an electrode and the workpiece or between two electrodes. The arc generates immense quantities of heat causing the metal to melt and, upon cooling, cause the weld to occur. In some cases, the electrode provides the filler metal and sometimes a slag or protective covering over the weld.

Resistance Welding—In resistance welding, an electric current is passed through two pieces of metal that are held together under pressure. The electric resistance across the surface discontinuity generates heat and the pressure causes the two metals to bond. Spot and seam welding are two popular examples of resistance welding. Figure 8.21 is a diagram of a spot welder.

Beam Welding—This process supplies the heat by bombarding the workpiece with a concentrated beam of electrons (electron beam welding) or by

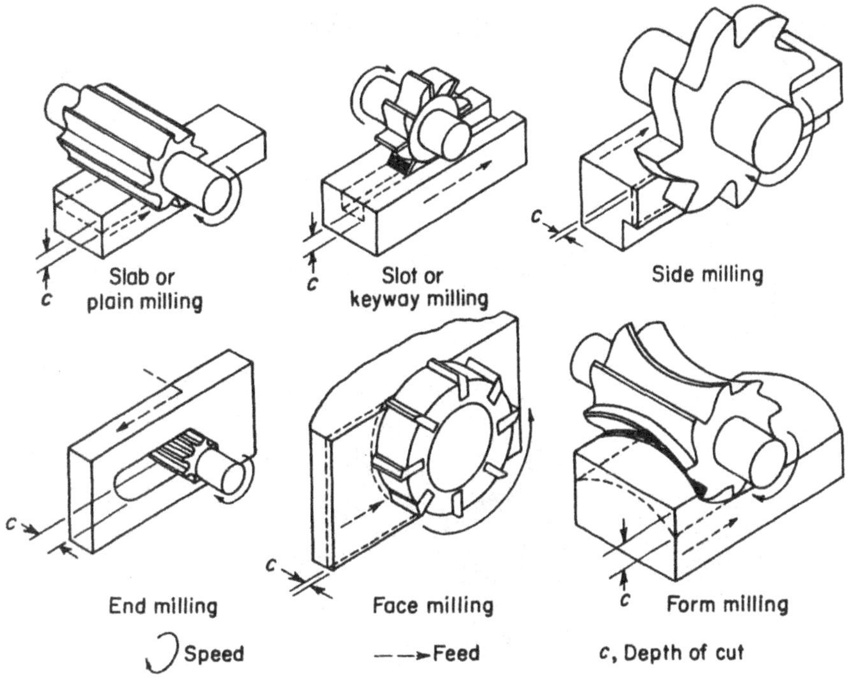

Fig. 8.18. Common milling operations. [Taken with permission, from Lawrence E. Doyle et al., *Manufacturing Processes and Materials for Engineers*, 2nd ed. (Englewood Cliffs, N.J.: Prentice Hall, 1969).]

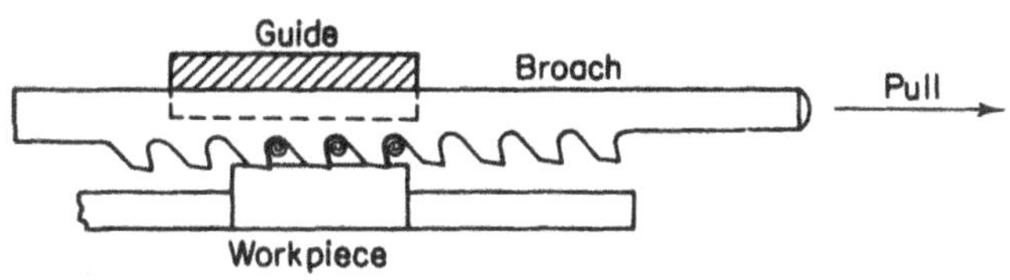

Fig. 8.19. Broaching operation. [Taken with permission, from Lawrence E. Doyle et al., *Manufacturing Processes and Materials for Engineers*, 2nd ed. (Englewood Cliffs, N.J.: Prentice Hall, 1969).]

concentrating a high-energy light beam on the workpiece (laser welding). Both are available commercially but have somewhat limited applications at this point in time.

Thermit Welding—Thermit welding is a process by which a cavity is formed by the pieces to be welded and molten metal poured into the cavity to make the weld. The molten metal is usually iron, steel, or copper and the welding process is particularly useful for large parts.

Pressure Welding—In pressure welding, the two parts are made to join by applying sufficient pressure with or without an external heat source. The pressure must be high enough to squeeze out all imperfections. Friction welding is a good example of pressure welding.

Fig. 8.20. Power hacksaw. [Taken with permission, from Lawrence E. Doyle et al., *Manufacturing Processes and Materials for Engineers*, 2nd ed. (Englewood Cliffs, N.J.: Prentice Hall, 1969).]

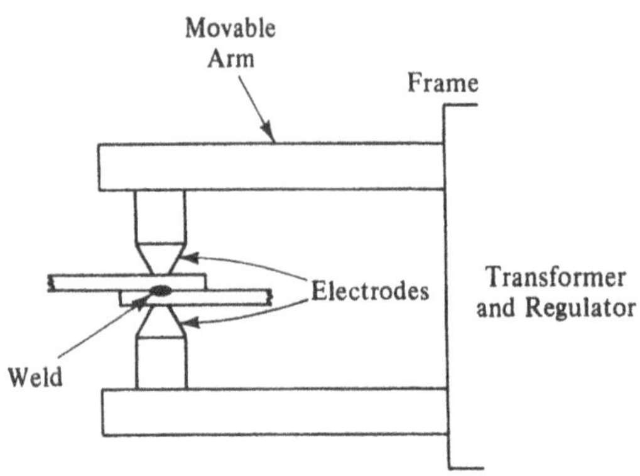

Fig. 8.21. Diagram of a spot welder. [Taken with permission, from Lawrence E. Doyle et al., *Manufacturing Processes and Materials for Engineers*, 2nd ed. (Englewood Cliffs, N.J.: Prentice Hall, 1969).]

Gas Welding—In gas welding, a combustible gas is burned with air or oxygen in a concentrated flame. This flame is used to heat the parent pieces and filler for the actual weld.

Brazing and Soldering—In brazing and soldering, the filler metal is melted and added to the weld area. The filler is chosen so that its melting point is below that of the parent metal. Consequently, the parent pieces never melt, and when the filler material cools and solidifies, it acts as a (metallic) glue to hold the parts together.

8.4.6 Assembly

Assembly is the process by which the various parts and subassemblies are brought together to form a completed assembly or product. Since there are numerous methods of assembly operations, they will not be covered in detail.

For example, welding is sometimes called an assembly process although assembly usually implies less permanent bonds. Gluing, riveting, screwing, and force fitting are familiar examples of assembly processes. The manufacturing engineer must be familiar with these various processes and understand the relative advantages and disadvantages of each process.

Organizationally, the assembly may be done in a *batch* process whereby assemblies are made intermittently or in a *continuous* process such as assembly lines. The manufacturing engineer must be familiar with these concepts; he or she must know how to set up and balance assembly lines, and so on.

8.4.7 Finishing

Finishing is a set of processes in which the material, either as a completed product, subassembly, or component part, is made more effective or presentable through the external application of energy and/or other materials.

Honing and lapping are two processes in which an abrasive material is rubbed over the metal to produce a fine surface. Honing and lapping are similar to grinding except that the speed is slower and the part is never overheated. The purpose is to improve the surface finish. Other more familiar finishing operations are polishing, buffing, brushing, and tumbling.

Another form of finishing is surface cleaning and coating. The purpose of cleaning is to remove dirt and scale that may be contaminating the workpiece, usually prior to coating. Familiar cleaning operations are spraying, vapor degreasing, and dip baths (such as pickling with an acid solution).

The purpose of coating is to provide decoration, protection, texture, and other desirable properties. Painting and plating are two of the most common coating techniques. Metallizing (spraying of the surface with an atomized molten metal) and phosphatizing and chromating (formulation of a noncorrosive surface over the metal that is also a good paint bond) are examples of less familiar coating operations.

The manufacturing engineer must be familiar with these cleaning and coating techniques and the advantages and disadvantages of each. He or she must know how to set up the processes and estimate the cost of each.

8.5 ANCILLARY FUNCTIONS

The production or manufacturing engineer performs many other functions other than the process selection just discussed. Some of these are tool, fixture, and jig design, cost estimating, maintenance systems design, and packaging systems design. Each of these is discussed below.

8.5.1 Tool, Jig, and Fixture Design

Tool design is the designing of tools for the most effective and economical operation. Although it is one of the most important areas of production-manufacturing engineering, tool design is overlooked (except in passing) by almost all industrial engineering curricula because it is so large and complex (and often an art as opposed to a science) that it would be very difficult to cover this subject adequately. Correct die design for a stamping operation could be a course in itself. Obviously, it would be practically impossible to thoroughly cover tool design for all metalworking operations, not to mention the basic operations in other industries.

Figure 8.22 gives an indication of the many parameters involved in designing a very simple cutting tool. Each of the angles shown must be selected based on the material being cut, the material of the cutting tool, and the cutting conditions. Changes of one degree in these angles can significantly affect the cutting operation. Imagine, then, the complexities and problems in designing a progressive (several stages or operations sequentially performed by one die) blanking, piercing, and forming die such as shown in Figure 8.23. Many industrial engineers become tool designers who learn almost all their skills on the job.

Similarly, jig and fixture design is an important phase of production-manufacturing engineering passed by in many industrial engineering curricula for the same reasons discussed above. A fixture is a device used to hold and locate a workpiece. The vise found in home workshops is a simple example of a fixture. In highly repetitive and exacting work, fixture design can be very important. A jig is similar to a fixture except that it also guides the tool into the workpiece. Figure 8.24 is an example of a jig. Jigs and fixtures must be designed for each unique application. Consequently, the designer must be highly creative and knowledgeable in the area.

8.5.2 Cost Estimating

Of primary importance to the success of any manufacturing endeavor is cost estimating. *Cost estimating* is the determination of the total manufacturing cost and consists of the following three items:

- *Material—Direct material* is any material whose cost is directly chargeable to a product. Examples are the wood that goes into a wooden pencil and into the paper that makes up the pages of this book. *Indirect material* is all other material cost, for example, cleaning compounds for sweeping floors and paint thinner for cleaning brushes.
- *Labor—Direct labor* is any labor whose cost is directly chargeable to a product. Examples are a lathe operator who performs an operation on the product to be sold and the painter who actually paints the product. *Indirect labor* is all other factory labor, for example, maintenance personnel, foremen, and inspectors.
- *Overhead*—Overhead is all cost of production other than direct labor or direct material. Indirect material and indirect labor are overhead items.

The purpose of cost estimating is to estimate the product's cost prior to actual production. This can be done by several means or methods, but usually some type of planning sheet is used. A *planning sheet* is a form designed for detailed processing of the product through the various stages of manufacture, including all equipment and tool requirements. Costs are added to the planning sheet.

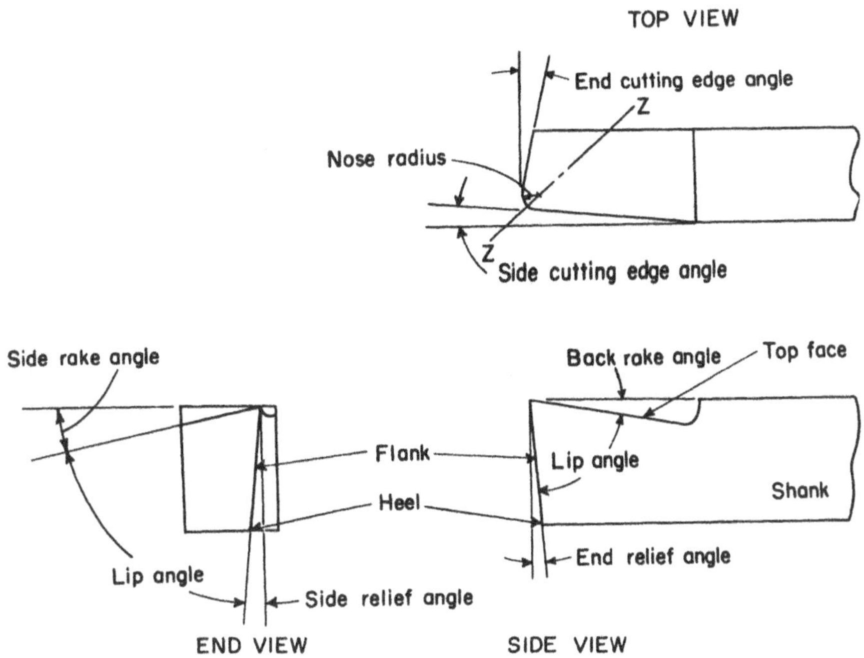

Fig. 8.22. Cutting tool design considerations

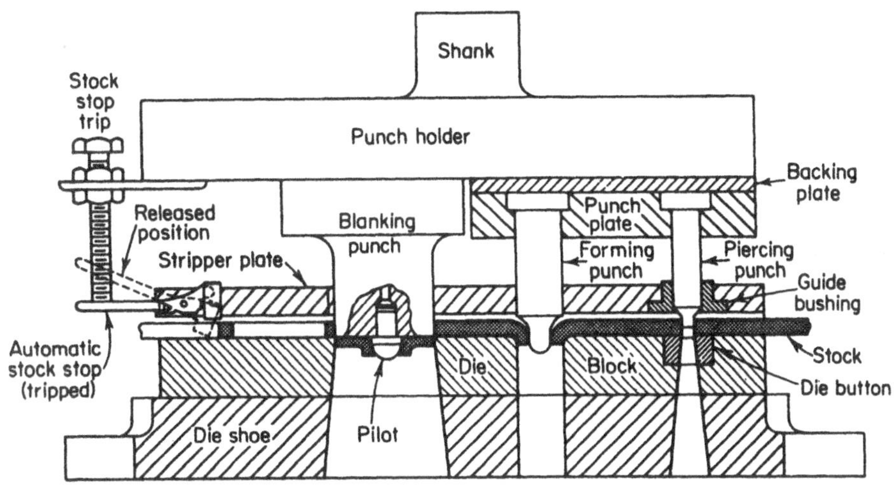

Fig. 8.23. Progressive die example. [Taken with permission, from Lawrence E. Doyle et al., *Manufacturing Processes and Materials for Engineers*, 2nd ed. (Englewood Cliffs, N.J.: Prentice Hall, 1969).]

The actual form used varies widely between companies. Usually, enough detail is provided to be able to generate the routing sheets. Figure 8.25 is one example of such a form. Notice that no provisions are made for material handling or other indirect costs because these can be added on at the end by use of an overhead factor (e.g., 150% of direct labor cost) or space can be provided in the form and actual overhead cost calculated. The latter is more accurate, but it is also more difficult.

Fig. 8.24. Jig to line-bore six holes in a master rod for an aircraft engine.

Date ______________ Part Name ______________ Part Number ______________

For use in final product ______________ Number per unit of final product ________

Operation Number	Operation Description	Equipment	Tooling	Direct Material	Direct Labor	Equipment Availability Cost
			Totals			

Fig. 8.25. Planning sheet form.

8.5.3 Maintenance Systems Design

Almost all the machines necessary for the processes discussed in Section 8.3 are very expensive and many are sensitive. Periodic care and maintenance of these machines and other equipment are exceptionally important. Periodic maintenance is called *preventive maintenance.* Some of the equipment is certain to break down occasionally and repair is necessary. This repair is called *emergency maintenance.*

It should be obvious that emergency maintenance could be the more costly of the two because of production interruptions, rescheduling of jobs and labor, and so on. The primary purposes of preventive maintenance, then, are to prolong the life of the equipment and to minimize the amount of emergency maintenance. Proper preventive maintenance such as lubricating and bearing inspection can minimize the number of emergency shutdowns.

Figure 8.26 is an example checklist used for periodic preventive maintenance of automotive vehicles. The design of these checklists and the determination of frequency of their use are often functions of the production-manufacturing engineer.

Automotive Vehicles
3,000 Miles or 3-Months
Preventive Maintenance Service

Vehicle No. ______________

Date	Make and Model	Miles

Service:	Fill Battery	Radiator Liquid	Fill Windshield Washer.	Check Engine Oil Level	Tire Pressure	Anti Freeze

Item No.	Description	Status	Corrective Action Including Material	Initials	Hours
A.	Interior				
1.	Foot and Hand Brake				
2.	Ignition Switch				
3.	Starter				
4.	Clutch Pedal				
5.	All Dash Instruments				
6.	Horn				
7.	Windshield Wipers and Washers				
8.	Gear Shift Loose or Broken				
9.	Steering Loose Motion				
10.	Fire Extinguisher Check Seal and Bracket				
B.	Exterior				
1.	Mirror and Brackets				
2.	Doors and Latches				
3.	Chock Blocks				
4.	Tires–RR LR RF LF				
	Underneath Check Exhaust System				
	Under Hood Check Leaks, Hoses, Water, Fuel, Oil and Brake Fluid				
	Lights Check all Lights and Reflectors				

Foreman Signed ______________________ Date __________

Fig. 8.26. Automotive vehicles 3,000 miles or 3-month preventive maintenance service.

In both types of maintenance, the determination of the maintenance staff and the equipment required is exceptionally important. Here, the production-manufacturing engineer may use forecasting methods, operations research (to be discussed), and other tools of industrial engineering to design the system.

8.5.4 Packaging Systems

Packaging is the last phase of what we have been calling production-manufacturing engineering. Here, the packaging of the finished product must be designed to protect it during transit and to ensure that the product will arrive undamaged. The package designer must be familiar with the various materials for packaging, the techniques of package design, and the cost of various types of packages. He or she must also know how to set up workstations in order to obtain maximum effectiveness and efficiency.

Packaging is becoming a large area in itself. This may be seen by observing the interest in packaging programs throughout the country and the high demand for good package engineers. "The best package is no package" is an often heard quote. This implies that when economically feasible the product should be designed to withstand the perils of handling during shipment. When this becomes infeasible, a protective package is required. Other purposes of packaging include ease and aesthetics of product display.

8.6 EXAMPLE

The Wheatley Company, a progressive pump and valve manufacturing company located in Tulsa, Oklahoma, has decided to start production on a new check valve, pictured in Figure 8.27.

The steps in the process engineering of this product are depicted below.[4] Only the bill of material, operation process chart, and a sample routing sheet are shown. The bill of material (highly simplified) is shown in Figure 8.28. The operations

Fig. 8.27. Check valve.

[4]Routing information was supplied by the Wheatley Company. The forms used are those of the authors, not the company. The authors are deeply indebted to the Wheatley Company for their aid and generosity in providing this example.

Bill of Material

Model:	Check Valve	Type X Design	Date__________	Sheet 1 of 1
Item	Quantity	Description	Source	Part No.
1	1	Body	Manufacturing	227
2	1	Cover	Manufacturing	275
3	8	Stud	Purchasing	271
4	8	Nut	Purchasing	243
5	1	Gasket, RTJ	Purchasing	353
6	1	Seat	Purchasing	236
7	1	O-Ring	Purchasing	201
8	1	O-Ring (Seat Backing)	Purchasing	301
9	1	Clapper	Manufacturing	237
10	1	Washer	Purchasing	220
11	1	Pin, "Roll Pin"	Purchasing	406
12	1	Arm	Manufacturing	221
13	1	Arm Pin	Purchasing	406
14	1	Plug, Arm Pin	Purchasing	420
15	1	Name Plate	Purchasing	282
16	2	Screw, Drive	Purchasing	320

Fig. 8.28. Bill of material for check valve.

process chart is developed next. It is shown in Figure 8.29 using fictitious machine numbers.

After the operations process chart has been prepared, the planning sheet is usually completed. Here, all details on material handling equipment, cost and time estimates, and so on are given. Next, a routing sheet is prepared for each part requiring processing. The routing sheet for the body only is shown in Figure 8.30. The form is such that it can be reproduced later and used as a production order, which instructs the shop to initiate production. Separate routing sheets would also be prepared for the cover, clapper, and arm, but they are not shown here.

The next step is to be sure that all necessary tools, jigs, and fixtures are available, design them when necessary, and issue production orders. Of course, packaging and maintenance systems must also be designed.

8.7 COMPUTER APPLICATIONS

Computers have changed the nature of production processes in many ways. They are often used to control the production process. For example, a minicomputer can be programmed to automatically adjust the process through a feedback mechanism. This is common in some process industries, such as paper mills. Although numerical control (the use of punched tape to control the operation of a machine) is not a computerized process, computer programs do exist that will translate the part's requirements into a numerical control language and automatically punch the tape (a very tedious process when done by hand).

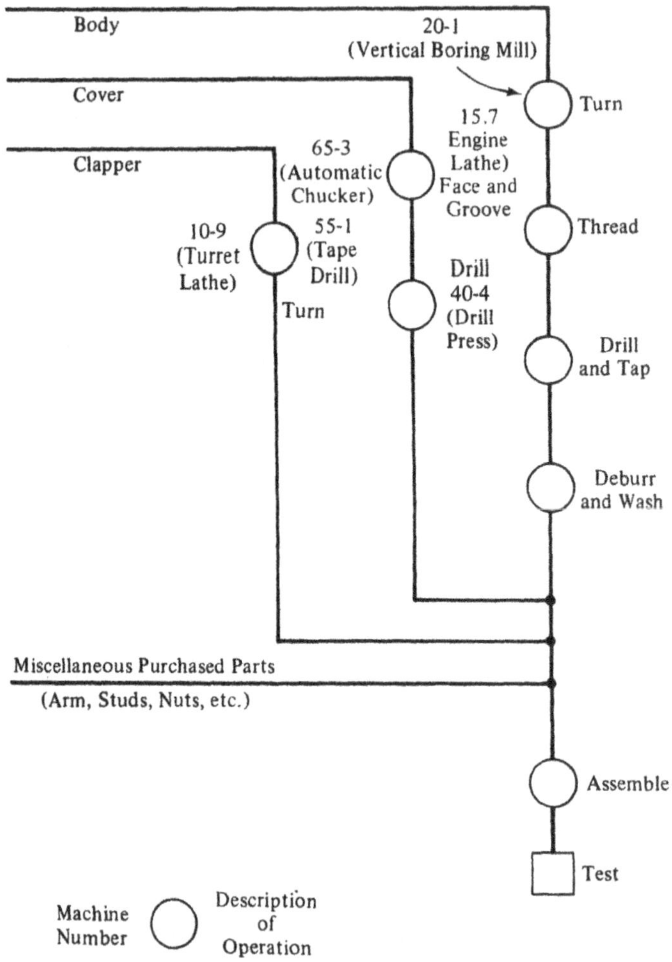

Fig. 8.29. Operations process chart for check valve.

It is often helpful to use a computer to set the parameters of the production process. At the extreme, the computer can be told what the part is, its dimensions, requirements, and so on, and it can completely design the process, including the order of operations, all feeds and speeds, and so forth. This last application is currently being researched and has, in fact, been done for some parts by using a mathematical programming formulation.

The trend today is to use computers and production control techniques to integrate islands of automation in search of a totally automated factory. This means that computers will schedule, sequence, and control machine centers, robots, and automatic machining centers. The future is bright and unlimited for computer usage in manufacturing engineering.

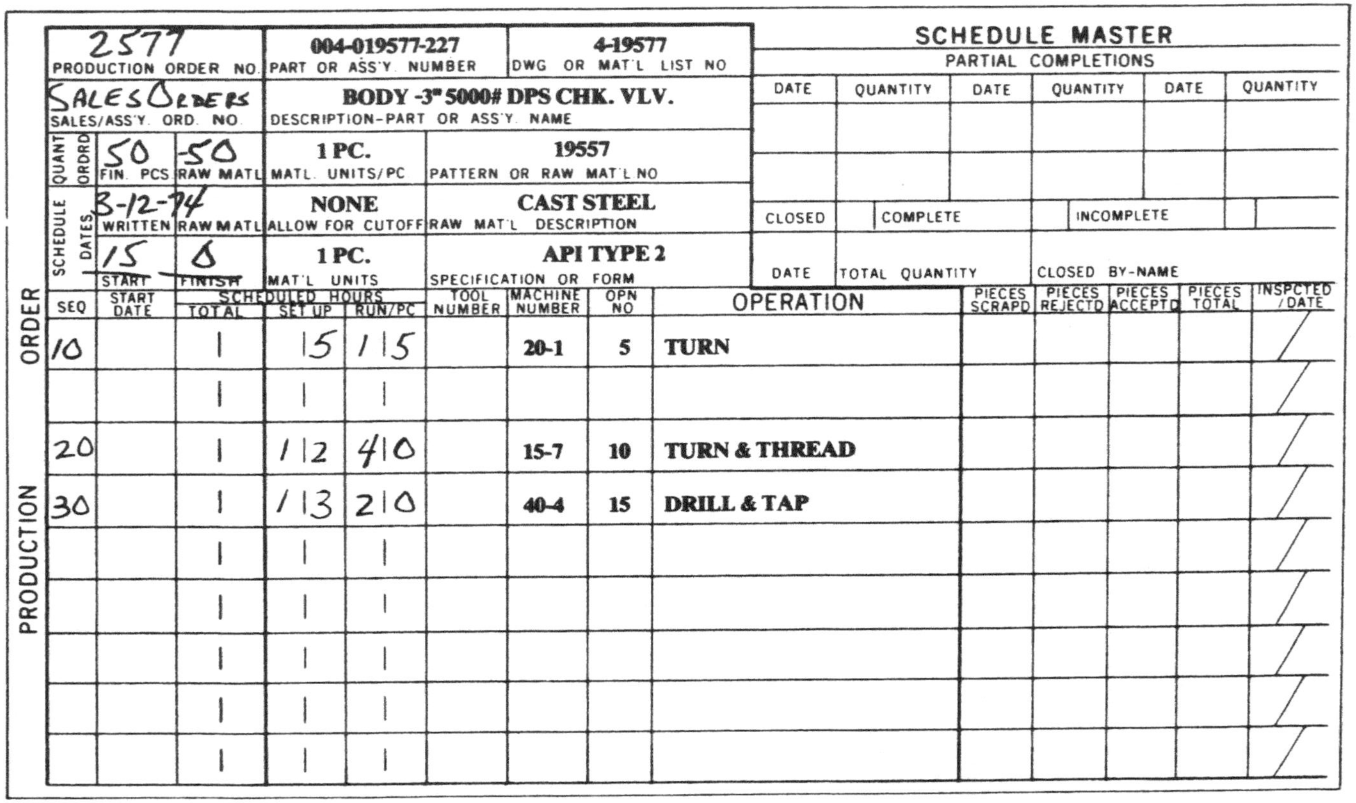

SCHEDULE MASTER

PRODUCTION ORDER NO.	PART OR ASS'Y. NUMBER	DWG OR MAT'L LIST NO
2577	004-019577-227	4-19577

SALES/ASS'Y. ORD. NO.: SALES ORDERS
DESCRIPTION-PART OR ASS'Y. NAME: BODY -3" 5000# DPS CHK. VLV.

ORDER QUANT ORDRD — FIN. PCS: 50; RAW MATL: 50
MATL. UNITS/PC: 1 PC.
PATTERN OR RAW MAT'L NO: 19557
SCHEDULE DATES — WRITTEN: 3-12-74; RAW MATL
ALLOW FOR CUTOFF: NONE
RAW MAT'L DESCRIPTION: CAST STEEL
START: 15; FINISH: 8
MAT'L UNITS: 1 PC.
SPECIFICATION OR FORM: API TYPE 2

PARTIAL COMPLETIONS

DATE	QUANTITY	DATE	QUANTITY	DATE	QUANTITY

CLOSED — COMPLETE — INCOMPLETE
DATE — TOTAL QUANTITY — CLOSED BY-NAME

SEQ	START DATE	SCHEDULED HOURS TOTAL	SET UP	RUN/PC	TOOL NUMBER	MACHINE NUMBER	OPN NO	OPERATION	PIECES SCRAPD	PIECES REJECTD	PIECES ACCEPTD	PIECES TOTAL	INSPCTED /DATE
10		1	1 5	1 15		20-1	5	TURN					
20		1	1 12	4 10		15-7	10	TURN & THREAD					
30		1	1 13	2 10		40-4	15	DRILL & TAP					

PRODUCTION ORDER

Fig. 8.30. Routing sheet—body.

DISCUSSION QUESTIONS

1. In the discussion of metal working processes we mentioned that there were many other processes in industry and government. List at least 10 other processing areas (e.g., plastics, refuse collection, landfill operations).
2. Discuss what product-production design interaction is. Why is it necessary and how should it occur?
3. List the basic types of metal working discussed. Give an example of each.
4. We implied that production-manufacturing engineering is a part of industrial engineering. Defend or disagree with this viewpoint.
5. Distinguish between wire drawing and drawing of a cup-shaped object.
6. Give an example of all types of metal cutting operations (shearing, turning, drilling, shaping and planing, milling, broaching, sawing-filing, and grinding).
7. Why are brazing and soldering popular in the electronics field?
8. Find an example of a jig or a fixture and describe it.
9. Discuss the trade-offs between emergency maintenance and preventive maintenance.
10. What is artificial intelligence? How is it being utilized (researched) in production engineering?
11. Discuss the concept of integrated islands of automation.

PROBLEMS

1. Draw an operations process chart for the preparation of a strawberry refrigerator pie. The recipe is given below (courtesy of Mrs. David F. Byrd).

    Ingredients:

    1 cup sugar

    1 package frozen strawberries (1½ Vi cups)

    2 well-beaten eggs

    1 package strawberry Jello

    1 can evaporated milk

    Thaw and drain strawberries to get about 3A cup liquid. Blend liquid of strawberries, with sugar and eggs. Boil at least 1 minute. Add Jello, mix, and let cool in refrigerator. Add strawberries to mixture. Whip cold milk and fold into mixture. Pour into two 9-inch pie shells, place in refrigerator to cool, and serve.

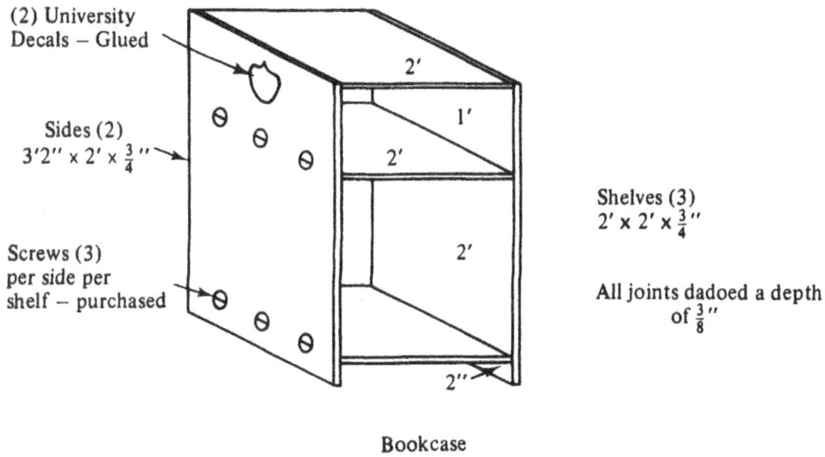

Fig. P8.2.

2. Prepare a bill of material for the product shown in Figure P8.2.

3. Prepare planning sheets for the bookcase in Problem 2. Assume that a skillsaw is used to rip the sides and shelves out of a 4-ft × 8-ft plywood sheet. A good-quality glue and 2-in. roundhead woodscrews are used in the assembly. A radial arm saw is available for dadoing, and all screws and decals are purchased. Do not attempt to derive time or cost figures.

4. Prepare routing sheets for each of the parts in Problem 2.

5. Write a simple FORTRAN program that calculates the total cost of a product given the direct material and direct labor in each of its component parts or assemblies (allow for multiple parts per product unit). Assume that the overhead cost is a fixed percentage of the total direct labor cost.

6. Either use the program in Problem 5 or manually determine the total cost of the completed product, the parts of which are listed below. Assume that the final assembly cost is \$5.00 per unit (all direct labor) and that the overhead is 135% of direct labor cost.

Part	*Parts/Unit*	*Direct Material (Cost per Part)*	*Direct Labor per Part*
A	1	\$1.00	\$5.00
B	2	2.00	1.00
C	1	1.50	0.50
D	3	5.00	0.75
E	1	3.00	0.60
F	1	1.50	10.00
G	2	0.75	7.00
H	1	0.50	6.00

7. Find a simple metal part or product, and using the information presented in this chapter, develop the following:
   (a) Bill of material
   (b) Operations process charts.
   (c) Route sheets (use fictitious machine numbers).
8. Prepare an operations process chart for the task of washing your car. Start with obtaining a sponge, a bucket, a hose, and soap. Use your present procedure. By examining the procedure, can you find a better way?
9. By asking graduate students or faculty, find one way that your department is applying a computer to production engineering.

CHAPTER

9 Control Systems

PREVIEW

In this chapter, we discuss open- and closed-loop feedback control systems. A control system consists of interconnected components to achieve a desired purpose. We examine examples of control systems through the course of history. These early systems incorporated many of the same ideas of feedback that are employed in modern manufacturing processes, alternative energy, complex hybrid automobiles, and sophisticated robots. A design process is presented that encompasses the establishment of goals and variables to be controlled, definition of specifications, system definition, modeling, and analysis. The iterative nature of design allows us to handle the design gap effectively while accomplishing necessary trade-offs in complexity, performance, and cost. Finally, we introduce the Sequential Design Example: Disk Drive Read System. It represents a very important and practical control system design problem while simultaneously serving as a useful learning tool.

DESIRED OUTCOMES

Upon completion of Chapter 9, students should:

- ❑ Possess a basic understanding of control system engineering and be able to offer some illustrative examples and their relationship to key contemporary issues.
- ❑ Be able to recount a brief history of control systems and their role in society.
- ❑ Be capable of discussing the future of controls in the context of their evolutionary pathways.
- ❑ Recognize the elements of control system design and possess an appreciation of controls in the context of engineering design.

9.1 INTRODUCTION

Engineering is concerned with understanding and controlling the materials and forces of nature for the benefit of humankind. Control system engineers are concerned with understanding and controlling segments of their environment, often called **systems**, to provide useful economic products for society. The twin goals of understanding and controlling are complementary because effective systems control requires that the systems be understood and modeled. Furthermore, control engineering must often consider the control of poorly understood systems such as chemical process systems. The present challenge to control engineers is the modeling and control of modern, complex, interrelated systems such as traffic control systems, chemical processes, and robotic systems. Simultaneously, the fortunate engineer has the opportunity to control many useful and interesting industrial automation systems. Perhaps the most characteristic quality of control engineering is the opportunity to control machines and industrial and economic processes for the benefit of society.

Control engineering is based on the foundations of feedback theory and linear system analysis, and it integrates the concepts of network theory and communication theory. Therefore control engineering is not limited to any engineering discipline but is equally applicable to aeronautical, chemical, mechanical, environmental, civil, and electrical engineering. For example, a control system often includes electrical, mechanical, and chemical components. Furthermore, as the understanding of the dynamics of business, social, and political systems increases, the ability to control these systems will also increase.

A **control system** is an interconnection of components forming a system configuration that will provide a desired system response. The basis for analysis of a system is the foundation provided by linear system theory, which assumes a cause–effect relationship for the components of a system. Therefore a component or **process** to be controlled can be represented by a block, as shown in Figure 9.1. The input–output relationship represents the cause-and-effect relationship of the process, which in turn represents a processing of the input signal to provide an output signal variable, often with a power amplification. An **open-loop control system** uses a controller and an actuator to obtain the desired response, as shown in Figure 9.2. An open-loop system is a system without feedback.

An open-loop control system utilizes an actuating device to control the process directly without using feedback.

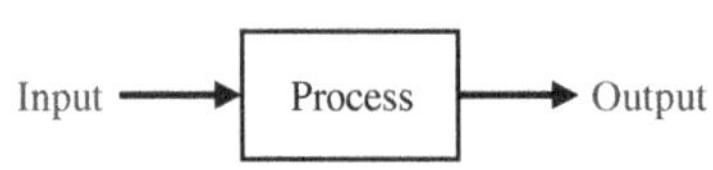

FIGURE 9.1 Process to be controlled.

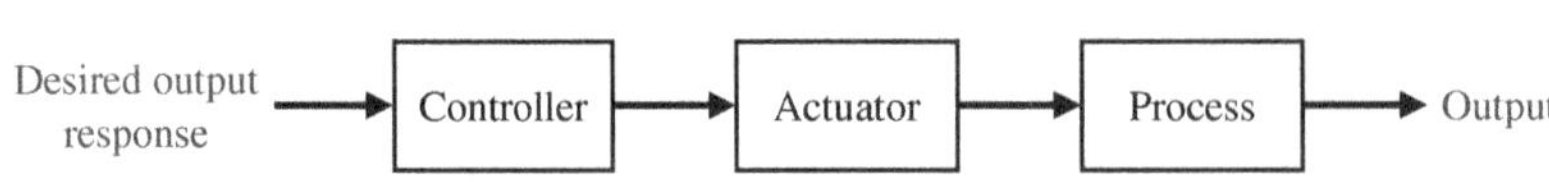

FIGURE 9.2 Open-loop control system (without feedback).

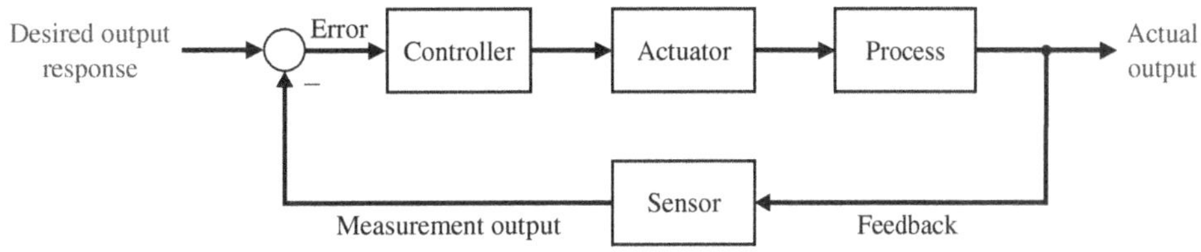

FIGURE 9.3 Closed-loop feedback control system (with feedback).

In contrast to an open-loop control system, a closed-loop control system utilizes an additional measure of the actual output to compare the actual output with the desired output response. The measure of the output is called the **feedback signal**. A simple **closed-loop feedback control system** is shown in Figure 9.3. A feedback control system is a control system that tends to maintain a prescribed relationship of one system variable to another by comparing functions of these variables and using the difference as a means of control. With an accurate sensor, the measured output is a good approximation of the actual output of the system.

A feedback control system often uses a function of a prescribed relationship between the output and reference input to control the process. Often the difference between the output of the process under control and the reference input is amplified and used to control the process so that the difference is continually reduced. In general, the difference between the desired output and the actual output is equal to the error, which is then adjusted by the controller. The output of the controller causes the actuator to modulate the process in order to reduce the error. The sequence is such, for instance, that if a ship is heading incorrectly to the right, the rudder is actuated to direct the ship to the left. The system shown in Figure 9.3 is a **negative feedback** control system, because the output is subtracted from the input and the difference is used as the input signal to the controller. The feedback concept has been the foundation for control system analysis and design.

A closed-loop control system uses a measurement of the output and feedback of this signal to compare it with the desired output (reference or command).

Closed-loop control has many advantages over open-loop control including the ability to reject external **disturbances** and improve **measurement noise** attenuation. We incorporate the disturbances and measurement noise in the block diagram as external inputs, as illustrated in Figure 9.4. External disturbances and measurement noise are inevitable in real-world applications and must be addressed in practical control system designs.

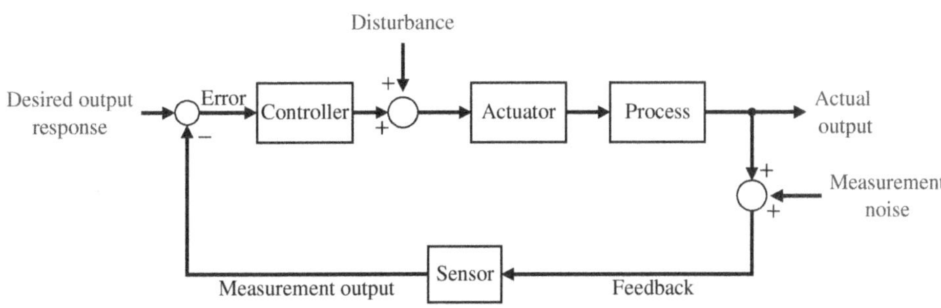

FIGURE 9.4 Closed-loop feedback system with external disturbances and measurement noise.

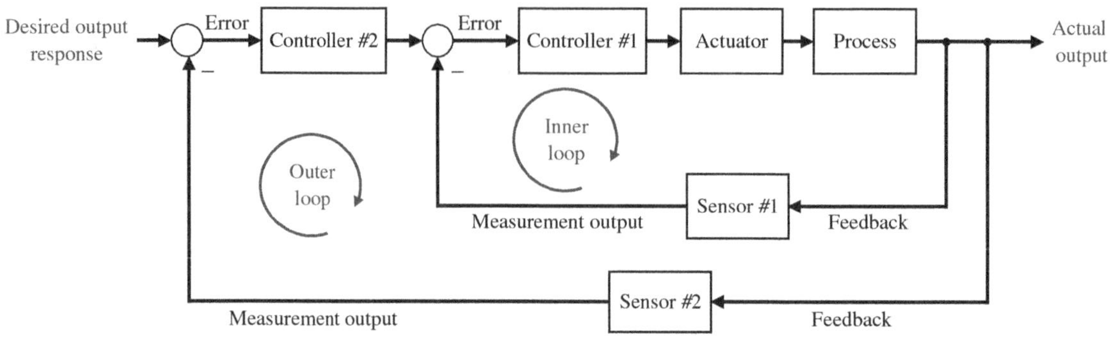

FIGURE 9.5 Multiloop feedback system with an inner loop and an outer loop.

The feedback systems in Figures 9.3 and 9.4 are single-loop feedback systems. Many feedback control systems contain more than one feedback loop. A common **multi-loop feedback control system** is illustrated in Figure 9.5 with an inner loop and an outer loop. In this scenario, the inner loop has a controller and a sensor and the outer loop has a controller and sensor. Other varieties of multiloop feedback systems are considered throughout the book as they represent more practical situations found in real-world applications. However, we use the single-loop feedback system for learning about the benefits of feedback control systems since the outcomes readily extend to multiloop systems.

Due to the increasing complexity of the system under control and the interest in achieving optimum performance, the importance of control system engineering has grown in the past decade. Furthermore, as the systems become more complex, the interrelationship of many controlled variables must be considered in the control scheme. A block diagram depicting a **multivariable control system** is shown in Figure 9.6.

A common example of an open-loop control system is a microwave oven set to operate for a fixed time. An example of a closed-loop control system is a person steering an automobile (assuming his or her eyes are open) by looking at the auto's location on the road and making the appropriate adjustments.

The introduction of feedback enables us to control a desired output and can improve accuracy, but it requires attention to the issue of stability of response.

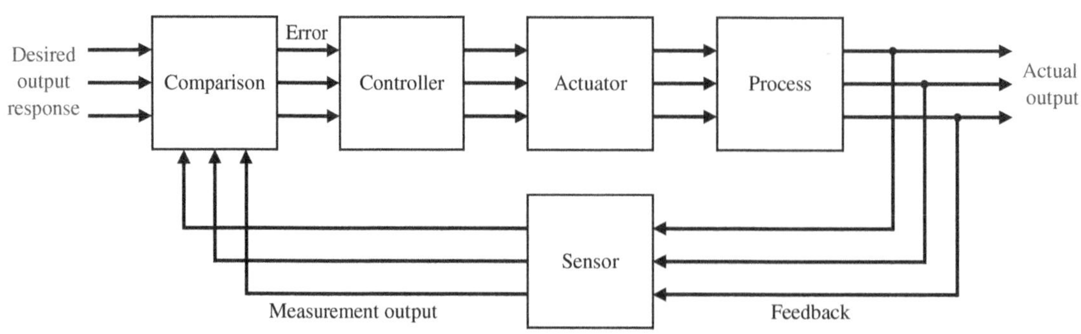

FIGURE 9.6 Multivariable control system.

9.2 BRIEF HISTORY OF AUTOMATIC CONTROL

The use of feedback to control a system has a fascinating history. The first applications of feedback control appeared in the development of float regulator mechanisms in Greece in the period 300 to 1 B.C. [1, 2, 3]. The water clock of Ktesibios used a float regulator (refer to Problem P9.11). An oil lamp devised by Philon in approximately 250 B.C. used a float regulator in an oil lamp for maintaining a constant level of fuel oil. Heron of Alexandria, who lived in the first century A.D., published a book entitled *Pneumatica*, which outlined several forms of water-level mechanisms using float regulators [1].

The first feedback system to be invented in modern Europe was the temperature regulator of Cornelis Drebbel (1572–1633) of Holland [1]. Dennis Papin (1647–1712) invented the first pressure regulator for steam boilers in 1681. Papin's pressure regulator was a form of safety regulator similar to a pressure-cooker valve.

The first automatic feedback controller used in an industrial process is generally agreed to be James Watt's **flyball governor**, developed in 1769 for controlling the speed of a steam engine [1, 2]. The all-mechanical device, shown in Figure 9.7, measured the speed of the output shaft and utilized the movement of the flyball to control the steam valve and therefore the amount of steam entering the engine. As depicted in Figure 9.7, the governor shaft axis is connected via mechanical linkages and beveled gears to the output shaft of the steam engine. As the steam engine output shaft speed increases, the ball weights rise and move away from the shaft axis and through mechanical linkages the steam valve closes and the engine slows down.

The first historical feedback system, claimed by Russia, is the water-level float regulator said to have been invented by I. Polzunov in 1765 [4]. The level regulator system is shown in Figure 9.8. The float detects the water level and controls the valve that covers the water inlet in the boiler.

The next century was characterized by the development of automatic control systems through intuition and invention. Efforts to increase the accuracy of the

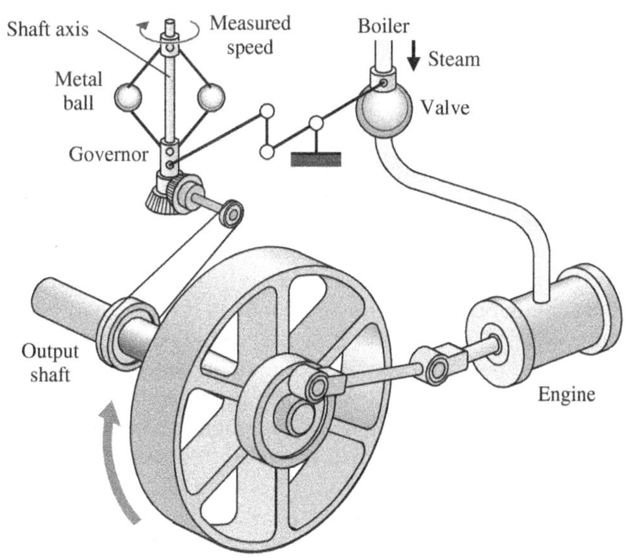

FIGURE 9.7 Watt's flyball governor.

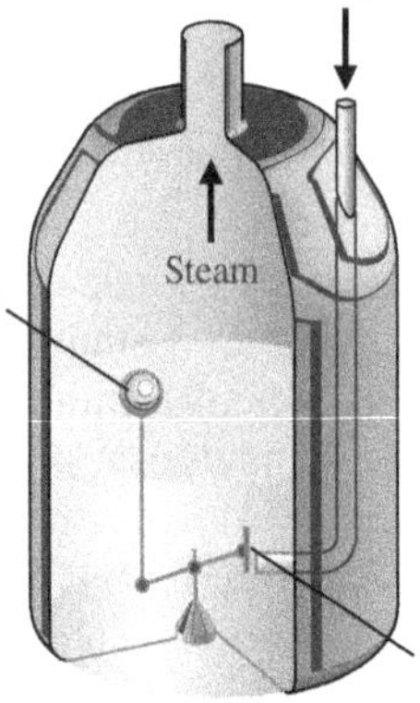

FIGURE 9.8 Water-level float regulator.

control system led to slower attenuation of the transient oscillations and even to unstable systems. It then became imperative to develop a theory of automatic control. In 1868, J. C. Maxwell formulated a mathematical theory related to control theory using a differential equation model of a governor [5]. Maxwell's study was concerned with the effect various system parameters had on the system performance. During the same period, I. A. Vyshnegradskii formulated a mathematical theory of regulators [6].

Prior to World War II, control theory and practice developed differently in the United States and western Europe than in Russia and eastern Europe. The main impetus for the use of feedback in the United States was the development of the telephone system and electronic feedback amplifiers by Bode, Nyquist, and Black at Bell Telephone Laboratories [7–10, 12].

Harold S. Black graduated from Worcester Polytechnic Institute in 1921 and joined Bell Laboratories of American Telegraph and Telephone (AT&T). In 1921, the major task confronting Bell Laboratories was the improvement of the telephone system and the design of improved signal amplifiers. Black was assigned the task of linearizing, stabilizing, and improving the amplifiers that were used in tandem to carry conversations over distances of several thousand miles.

Black reports [8]:

> Then came the morning of Tuesday, August 2, 1927, when the concept of the negative feedback amplifier came to me in a flash while I was crossing the Hudson River on the Lackawanna Ferry, on my way to work. For more than 50 years I have pondered how and why the idea came, and I can't say any more today than I could that morning. All I know is that after several years of hard work on the problem, I suddenly realized that if I fed the amplifier output back to the input, in reverse phase, and kept the device from oscillating (singing, as we called it then), I would have exactly what I wanted: a means of canceling out the distortion in the output. I opened my morning newspaper and on a page of *The New York Times* I sketched a simple canonical diagram of a negative feedback amplifier plus the equations for the amplification with feedback. I signed the sketch, and 20 minutes later, when I reached the laboratory at 463 West Street, it was witnessed, understood, and signed by the late Earl C. Blessing.
>
> I envisioned this circuit as leading to extremely linear amplifiers (40 to 50 dB of negative feedback), but an important question is: How did I know I could avoid

> self-oscillations over very wide frequency bands when many people doubted such circuits would be stable? My confidence stemmed from work that I had done two years earlier on certain novel oscillator circuits and three years earlier in designing the terminal circuits, including the filters, and developing the mathematics for a carrier telephone system for short toll circuits.

The frequency domain was used primarily to describe the operation of the feedback amplifiers in terms of bandwidth and other frequency variables. In contrast, the eminent mathematicians and applied mechanicians in the former Soviet Union inspired and dominated the field of control theory. Therefore, the Russian theory tended to utilize a time-domain formulation using differential equations.

The control of an industrial process (manufacturing, production, and so on) by automatic rather than manual means is often called **automation**. Automation is prevalent in the chemical, electric power, paper, automobile, and steel industries, among others. The concept of automation is central to our industrial society. Automatic machines are used to increase the production of a plant per worker in order to offset rising wages and inflationary costs. Thus industries are concerned with the productivity per worker of their plants. **Productivity** is defined as the ratio of physical output to physical input [26]. In this case, we are referring to labor productivity, which is real output per hour of work.

The transformation of the U.S. labor force in the country's brief history follows the progressive mechanization of work that attended the evolution of the agrarian republic into an industrial world power. In 1820, more than 70 percent of the labor force worked on the farm. By 1900, less than 40 percent were engaged in agriculture. Today, less than 5 percent works in agriculture [15].

In 1925, some 588,000 people—about 1.3 percent of the nation's labor force—were needed to mine 520 million tons of bituminous coal and lignite, almost all of it from underground. By 1980, production was up to 774 million tons, but the work force had been reduced to 208,000. Furthermore, only 136,000 of that number were employed in underground mining operations. The highly mechanized and highly productive surface mines, with just 72,000 workers, produced 482 million tons, or 62 percent of the total [27].

A large impetus to the theory and practice of automatic control occurred during World War II when it became necessary to design and construct automatic airplane piloting, gun-positioning systems, radar antenna control systems, and other military systems based on the feedback control approach. The complexity and expected performance of these military systems necessitated an extension of the available control techniques and fostered interest in control systems and the development of new insights and methods. Prior to 1940, for most cases, the design of control systems was an art involving a trial-and-error approach. During the 1940s, mathematical and analytical methods increased in number and utility, and control engineering became an engineering discipline in its own right [10–12].

Another example of the discovery of an engineering solution to a control system problem was the creation of a gun director by David B. Parkinson of Bell Telephone Laboratories. In the spring of 1940, Parkinson was a 29-year-old engineer intent on improving the automatic level recorder, an instrument that used strip-chart paper to plot the record of a voltage. A critical component was a small potentiometer used to control the pen of the recorder through an actuator.

Parkinson had a dream about an antiaircraft gun that was successfully felling airplanes. Parkinson described the situation [13]:

> After three or four shots one of the men in the crew smiled at me and beckoned me to come closer to the gun. When I drew near he pointed to the exposed end of the left trunnion. Mounted there was the control potentiometer of my level recorder!

The next morning Parkinson realized the significance of his dream:

> If my potentiometer could control the pen on the recorder, something similar could, with suitable engineering, control an antiaircraft gun.

After considerable effort, an engineering model was delivered for testing to the U.S. Army on December 1, 1941. Production models were available by early 1943, and eventually 3000 gun controllers were delivered. Input to the controller was provided by radar, and the gun was aimed by taking the data of the airplane's present position and calculating the target's future position.

Frequency-domain techniques continued to dominate the field of control following World War II with the increased use of the Laplace transform and the complex frequency plane. During the 1950s, the emphasis in control engineering theory was on the development and use of the s-plane methods and, particularly, the root locus approach. Furthermore, during the 1980s, the use of digital computers for control components became routine. The technology of these new control elements to perform accurate and rapid calculations was formerly unavailable to control engineers. There are now over 400,000 digital process control computers installed in the United States [14, 27]. These computers are employed especially for process control systems in which many variables are measured and controlled simultaneously by the computer.

With the advent of Sputnik and the space age, another new impetus was imparted to control engineering. It became necessary to design complex, highly accurate control systems for missiles and space probes. Furthermore, the necessity to minimize the weight of satellites and to control them very accurately has spawned the important field of optimal control. Due to these requirements, the time-domain methods developed by Liapunov, Minorsky, and others have been met with great interest in the last two decades. Recent theories of optimal control developed by L. S. Pontryagin in the former Soviet Union and R. Bellman in the United States, as well as recent studies of robust systems, have contributed to the interest in time-domain methods. It now is clear that control engineering must consider both the time-domain and the frequency-domain approaches simultaneously in the analysis and design of control systems.

A notable recent advance with worldwide impact is the U.S. space-based radionavigation system known as the Global Positioning System or GPS [82–85]. In the distant past, various strategies and sensors were developed to keep explorers on the oceans from getting lost, including following coastlines, using compasses to point north, and sextants to measure the angles of stars, the moon, and the sun above the horizon. The early explorers were able to estimate latitude accurately, but not longitude. It was not until the 1700s with the development of the chronometer that, when used with the sextant, the longitude could be estimated. Radio-based navigation systems began to appear in the early twentieth century and were used in World War II. With the advent of Sputnik and the space age, it became known that radio signals from satellites could be used to navigate on the ground by observing the Doppler shift of the received radio signals. Research and development culminated in the

1990s with 24 navigation satellites (known as the GPS) that solved the fundamental problem that explorers faced for centuries by providing a dependable mechanism to pinpoint the current location. Freely available on a continuous worldwide basis, GPS provides very reliable location and time information anytime, day or night, anywhere in the world. Using GPS as a sensor to provide position (and velocity) information is a mainstay of active control systems for transportation systems in the air, on the ground, and on the oceans. The GPS assists relief and emergency workers to save lives, and helps us with our everyday activities including the control of power grids, banking, farming, surveying, and many other tasks.

A selected history of control system development is summarized in Table 9.1.

Table 9.1 Selected Historical Developments of Control Systems

1769	James Watt's steam engine and governor developed. The Watt steam engine is often used to mark the beginning of the Industrial Revolution in Great Britain. During the Industrial Revolution, great strides were made in the development of mechanization, a technology preceding automation.
1800	Eli Whitney's concept of interchangeable parts manufacturing demonstrated in the production of muskets. Whitney's development is often considered to be the beginning of mass production.
1868	J. C. Maxwell formulates a mathematical model for a governor control of a steam engine.
1913	Henry Ford's mechanized assembly machine introduced for automobile production.
1927	H. S. Black conceives of the negative feedback amplifier and H. W. Bode analyzes feedback amplifiers.
1932	H. Nyquist develops a method for analyzing the stability of systems.
1941	Creation of first antiaircraft gun with active control.
1952	Numerical control (NC) developed at Massachusetts Institute of Technology for control of machine-tool axes.
1954	George Devol develops "programmed article transfer," considered to be the first industrial robot design.
1957	Sputnik launches the space age leading, in time, to miniaturization of computers and advances in automatic control theory.
1960	First Unimate robot introduced, based on Devol's designs. Unimate installed in 1961 for tending die-casting machines.
1970	State-variable models and optimal control developed.
1980	Robust control system design widely studied.
1983	Introduction of the personal computer (and control design software soon thereafter) brought the tools of design to the engineer's desktop.
1990	Export-oriented manufacturing companies emphasize automation.
1994	Feedback control widely used in automobiles. Reliable, robust systems demanded in manufacturing.
1995	The Global Positioning System (GPS) was operational providing positioning, navigation, and timing services worldwide.
1997	First ever autonomous rover vehicle, known as Sojourner, explores the Martian surface.
1998–2003	Advances in micro- and nanotechnology. First intelligent micromachines are developed and functioning nanomachines are created.
2007	The Orbital Express mission performed the first autonomous space rendezvous and docking.

9.3 EXAMPLES OF CONTROL SYSTEMS

Control engineering is concerned with the analysis and design of goal-oriented systems. Therefore the mechanization of goal-oriented policies has grown into a hierarchy of goal-oriented control systems. Modern control theory is concerned with systems that have self-organizing, adaptive, robust, learning, and optimum qualities.

Feedback control is a fundamental fact of modern industry and society. Driving an automobile is a pleasant task when the auto responds rapidly to the driver's commands. Many cars have power steering and brakes, which utilize hydraulic amplifiers for amplification of the force to the brakes or the steering wheel. A simple block diagram of an automobile steering control system is shown in Figure 9.9(a).

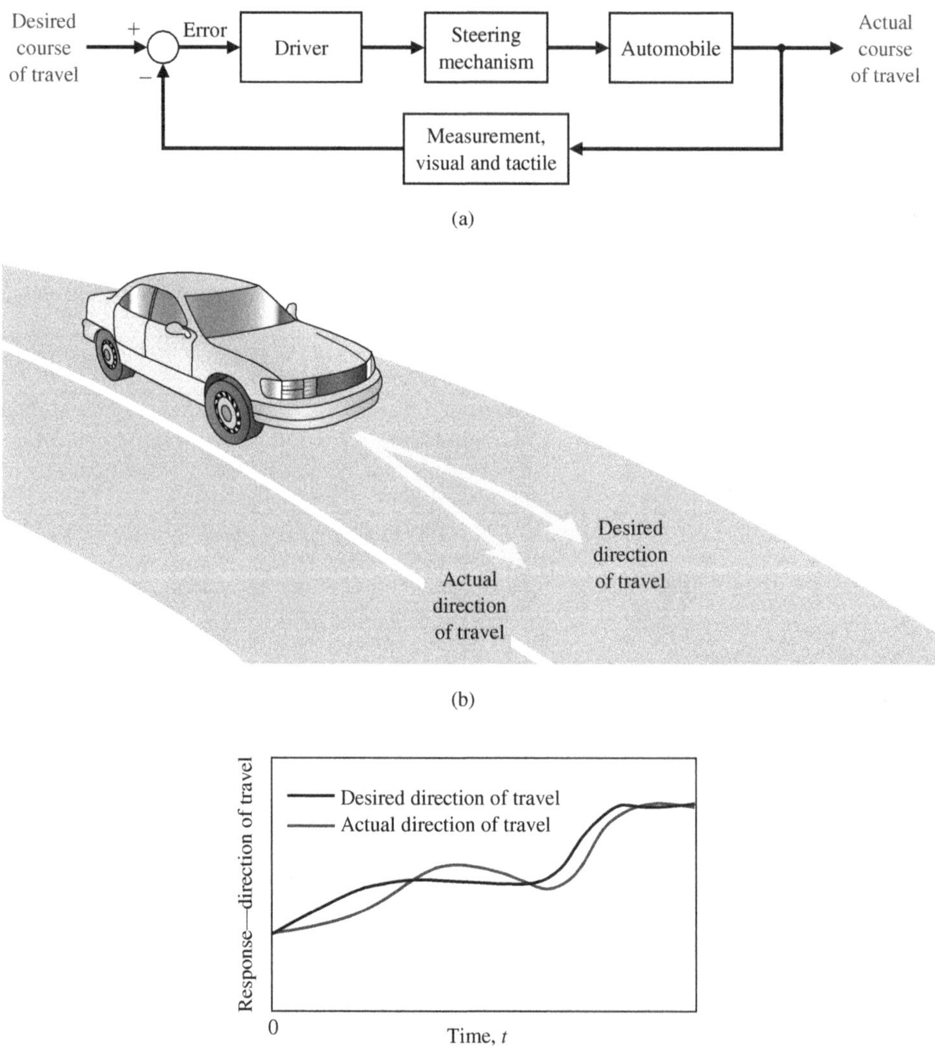

FIGURE 9.9
(a) Automobile steering control system. (b) The driver uses the difference between the actual and the desired direction of travel to generate a controlled adjustment of the steering wheel.
(c) Typical direction-of-travel response.

The desired course is compared with a measurement of the actual course in order to generate a measure of the error, as shown in Figure 9.9(b). This measurement is obtained by visual and tactile (body movement) feedback, as provided by the feel of the steering wheel by the hand (sensor). This feedback system is a familiar version of the steering control system in an ocean liner or the flight controls in a large airplane. A typical direction-of-travel response is shown in Figure 9.9(c).

A basic, manually controlled closed-loop system for regulating the level of fluid in a tank is shown in Figure 9.10. The input is a reference level of fluid that the operator is instructed to maintain. (This reference is memorized by the operator.) The power amplifier is the operator, and the sensor is visual. The operator compares the actual level with the desired level and opens or closes the valve (actuator), adjusting the fluid flow out, to maintain the desired level.

Other familiar control systems have the same basic elements as the system shown in Figure 9.3. A refrigerator has a temperature setting or desired temperature, a thermostat to measure the actual temperature and the error, and a compressor motor for power amplification. Other examples in the home are the oven, furnace, and water heater. In industry, there are many examples, including speed controls; process temperature and pressure controls; and position, thickness, composition, and quality controls [14, 17, 18].

In its modern usage, automation can be defined as a technology that uses programmed commands to operate a given process, combined with feedback of information to determine that the commands have been properly executed. Automation is often used for processes that were previously operated by humans. When automated, the process can operate without human assistance or interference. In fact, most automated systems are capable of performing their functions with greater accuracy and precision, and in less time, than humans are able to do. A semiautomated process is one that incorporates both humans and robots. For instance, many automobile assembly line operations require cooperation between a human operator and an intelligent robot.

Feedback control systems are used extensively in industrial applications. Thousands of industrial and laboratory robots are currently in use. Manipulators can pick up objects weighing hundreds of pounds and position them with an accuracy of one-tenth of an inch or better [28]. Automatic handling equipment for home, school, and industry is particularly useful for hazardous, repetitious, dull, or simple tasks.

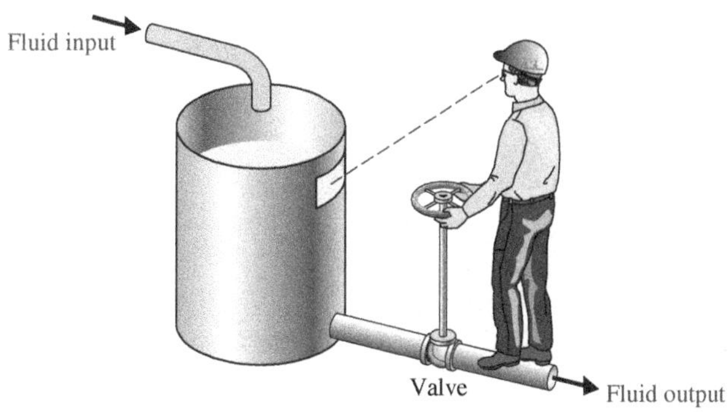

FIGURE 9.10 A manual control system for regulating the level of fluid in a tank by adjusting the output valve. The operator views the level of fluid through a port in the side of the tank.

Machines that automatically load and unload, cut, weld, or cast are used by industry to obtain accuracy, safety, economy, and productivity [14, 27, 28, 38]. The use of computers integrated with machines that perform tasks like a human worker has been foreseen by several authors. In his famous 1923 play, entitled *R.U.R.* [48], Karel Capek called artificial workers *robots*, deriving the word from the Czech noun *robota*, meaning "work."

A **robot** is a computer-controlled machine and involves technology closely associated with automation. Industrial robotics can be defined as a particular field of automation in which the automated machine (that is, the robot) is designed to substitute for human labor [18, 27, 33]. Thus robots possess certain humanlike characteristics. Today, the most common humanlike characteristic is a mechanical manipulator that is patterned somewhat after the human arm and wrist. Some devices even have anthropomorphic mechanisms, including what we might recognize as mechanical arms, wrists, and hands [14, 27, 28]. An example of an anthropomorphic robot is shown in Figure 9.11. We recognize that the automatic machine is well suited to some tasks, as noted in Table 9.2, and that other tasks are best carried out by humans.

Another very important application of control technology is in the control of the modern automobile [19, 20]. Control systems for suspension, steering, and engine control have been introduced. Many new autos have a four-wheel-steering system, as well as an antiskid control system.

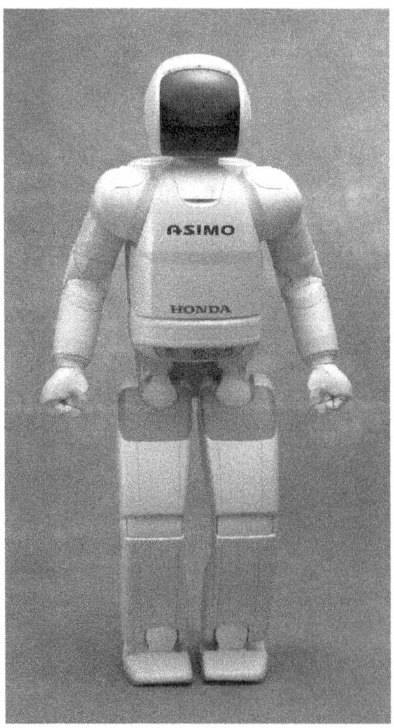

FIGURE 9.11 The Honda ASIMO humanoid robot. ASIMO walks, climbs stairs, and turns corners. Photo courtesy of American Honda Motor Co., Inc.

Table 9.2 Task Difficulty: Human Versus Automatic Machine

Tasks Difficult for a Machine	Tasks Difficult for a Human
Inspect seedlings in a nursery.	Inspect a system in a hot, toxic environment.
Drive a vehicle through rugged terrain.	Repetitively assemble a clock.
Identify the most expensive jewels on a tray of jewels.	Land an airliner at night, in bad weather.

A three-axis control system for inspecting individual semiconductor wafers is shown in Figure 9.12. This system uses a specific motor to drive each axis to the desired position in the x-y-z-axis, respectively. The goal is to achieve smooth, accurate movement in each axis. This control system is an important one for the semiconductor manufacturing industry.

There has been considerable discussion recently concerning the gap between practice and theory in control engineering. However, it is natural that theory precedes the applications in many fields of control engineering. Nonetheless, it is interesting to note that in the electric power industry, the largest industry in the United States, the gap is relatively insignificant. The electric power industry is primarily

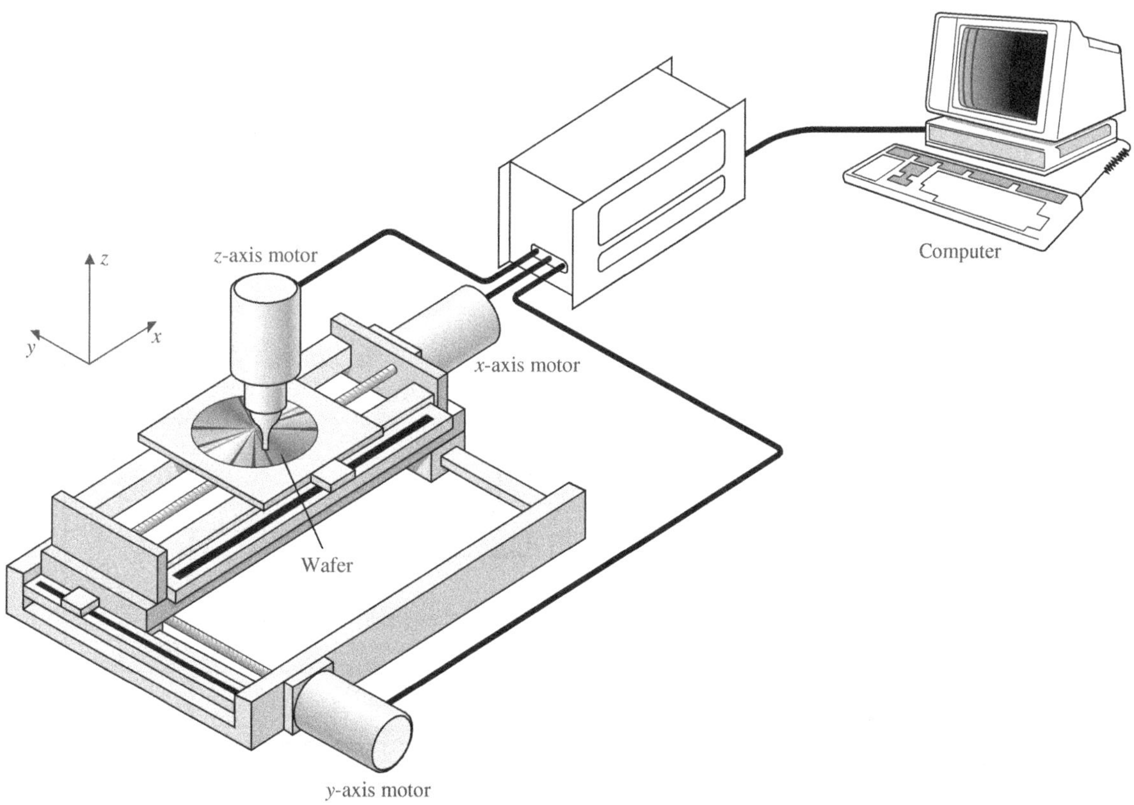

FIGURE 9.12 A three-axis control system for inspecting individual semiconductor wafers with a highly sensitive camera.

interested in energy conversion, control, and distribution. It is critical that computer control be increasingly applied to the power industry in order to improve the efficient use of energy resources. Also, the control of power **plants** for minimum waste emission has become increasingly important. The modern, large-capacity plants, which exceed several hundred megawatts, require automatic control systems that account for the interrelationship of the process variables and optimum power production. It is common to have 90 or more manipulated variables under coordinated control. A simplified model showing several of the important control variables of a large boiler–generator system is shown in Figure 9.13. This is an example of the importance of measuring many variables, such as pressure and oxygen, to provide information to the computer for control calculations.

The electric power industry has used the modern aspects of control engineering for significant and interesting applications. It appears that in the process industry, the factor that maintains the applications gap is the lack of instrumentation to measure all the important process variables, including the quality and composition of the product. As these instruments become available, the applications of modern control theory to industrial systems should increase measurably.

Another important industry, the metallurgical industry, has had considerable success in automatically controlling its processes. In fact, in many cases, the control theory is being fully implemented. For example, a hot-strip steel mill, which involves a $100-million investment, is controlled for temperature, strip width, thickness, and quality.

Rapidly rising energy costs coupled with threats of energy curtailment are resulting in new efforts for efficient automatic energy management. Computer controls are used to control energy use in industry and to stabilize and connect loads evenly to gain fuel economy.

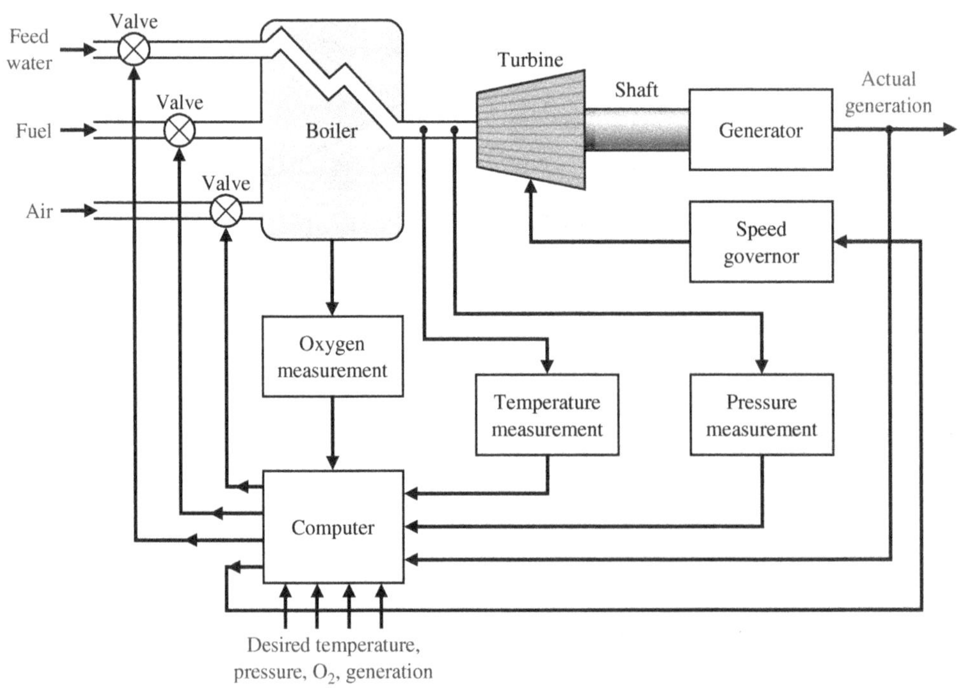

FIGURE 9.13 Coordinated control system for a boiler–generator.

There has been considerable interest recently in applying the feedback control concepts to automatic warehousing and inventory control. Furthermore, automatic control of agricultural systems (farms) is receiving increased interest. Automatically controlled silos and tractors have been developed and tested. Automatic control of wind turbine generators, solar heating and cooling, and automobile engine performance are important modern examples [20, 21].

Also, there have been many applications of control system theory to biomedical experimentation, diagnosis, prosthetics, and biological control systems [22, 23, 48]. The control systems under consideration range from the cellular level to the central nervous system and include temperature regulation and neurological, respiratory, and cardiovascular control. Most physiological control systems are closed-loop systems. However, we find not one controller but rather control loop within control loop, forming a hierarchy of systems. The modeling of the structure of biological processes confronts the analyst with a high-order model and a complex structure. Prosthetic devices that aid the 46 million handicapped individuals in the United States are designed to provide automatically controlled aids to the disabled [22, 27, 39]. The robotic hand shown in Figure 9.14 belongs to Obrero, a humanoid robot developed at MIT that is capable of sensitive manipulation. The Obrero robot is responsive to the properties of the object it holds and does not rely on vision as the main sensor. The hand has position and force control of the fingers employing very sensitive tactile sensors and series elastic actuators in its joints.

Finally, it has become interesting and valuable to attempt to model the feedback processes prevalent in the social, economic, and political spheres. This approach is undeveloped at present but appears to have a reasonable future. Society, of course, is composed of many feedback systems and regulatory bodies, such as the Federal Reserve Board, which are controllers exerting the forces on society necessary to maintain a desired output. A simple lumped model of the national income feedback control system is shown in Figure 9.15. This type of model helps the analyst to understand the effects of government control—granted its existence—and the dynamic effects of

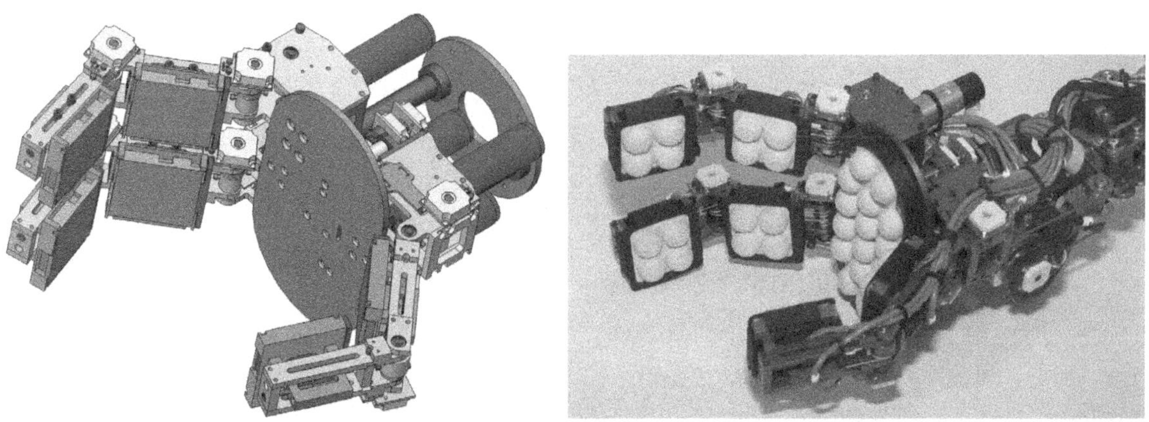

(a) Computer-aided drawing (Courtesy of Eduardo Torres-Jara). (b) The Obrero robotic hand (Photo by Iuliu Vasilescu).

FIGURE 9.14 The Obrero robot is responsive to the properties of the object it holds and does not rely on vision as the main sensor but as a complement. Obrero is part of the Humanoid Robotics Group at the MIT Computer Science and Artificial Intelligence Laboratory.

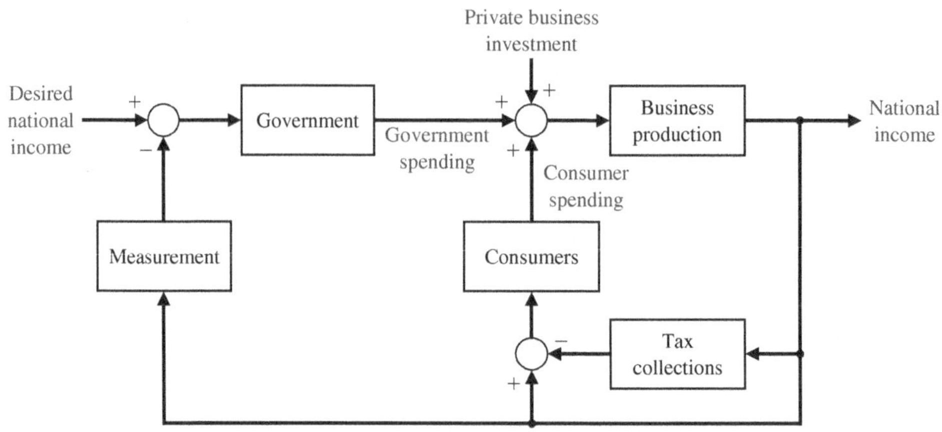

FIGURE 9.15
A feedback control system model of the national income.

government spending. Of course, many other loops not shown also exist, since, theoretically, government spending cannot exceed the tax collected without generating a deficit, which is itself a control loop containing the Internal Revenue Service and the Congress. In a socialist country, the loop due to consumers is de-emphasized and government control is emphasized. In that case, the measurement block must be accurate and must respond rapidly; both are very difficult characteristics to realize from a bureaucratic system. This type of political or social feedback model, while usually nonrigorous, does impart information and understanding.

The ongoing area of research and development of unmanned aerial vehicles (UAVs) is full of potential for the application of control systems. An example of a UAV is shown in Figure 9.16. UAVs are unmanned but are usually controlled by ground operators. Typically they do not operate autonomously and their inability to provide the level of safety of a manned plane keeps them from flying freely in the commercial airspace. One significant challenge is to develop control systems that will avoid in-air collisions. Ultimately, the goal is to employ the UAV autonomously in such applications as aerial photography to assist in disaster mitigation, survey

FIGURE 9.16
An unmanned aerial vehicle. (Used with permission. Credit: DARPA.)

work to assist in construction projects, crop monitoring, and continuous weather monitoring. In a military setting, UAVs can perform intelligence, surveillance, and reconnaissance missions [74]. Smart unmanned aircraft will require significant deployment of advanced control systems throughout the airframe.

9.4 ENGINEERING DESIGN

Engineering design is the central task of the engineer. It is a complex process in which both creativity and analysis play major roles.

Design is the process of conceiving or inventing the forms, parts, and details of a system to achieve a specified purpose.

Design activity can be thought of as planning for the emergence of a particular product or system. Design is an innovative act whereby the engineer creatively uses knowledge and materials to specify the shape, function, and material content of a system. The design steps are (1) to determine a need arising from the values of various groups, covering the spectrum from public policy makers to the consumer; (2) to specify in detail what the solution to that need must be and to embody these values; (3) to develop and evaluate various alternative solutions to meet these specifications; and (4) to decide which one is to be designed in detail and fabricated.

An important factor in realistic design is the limitation of time. Design takes place under imposed schedules, and we eventually settle for a design that may be less than ideal but considered "good enough." In many cases, time is the *only* competitive advantage.

A major challenge for the designer is writing the specifications for the technical product. **Specifications** are statements that explicitly state what the device or product is to be and do. The design of technical systems aims to provide appropriate design specifications and rests on four characteristics: complexity, trade-offs, design gaps, and risk.

Complexity of design results from the wide range of tools, issues, and knowledge to be used in the process. The large number of factors to be considered illustrates the complexity of the design specification activity, not only in assigning these factors their relative importance in a particular design, but also in giving them substance either in numerical or written form, or both.

The concept of **trade-off** involves the need to resolve conflicting design goals, all of which are desirable. The design process requires an efficient compromise between desirable but conflicting criteria.

In making a technical device, we generally find that the final product does not appear as originally visualized. For example, our image of the problem we are solving does not appear in written description and ultimately in the specifications. Such **design gaps** are intrinsic in the progression from an abstract idea to its realization.

This inability to be absolutely sure about predictions of the performance of a technological object leads to major uncertainties about the actual effects of the designed devices and products. These uncertainties are embodied in the idea of unintended consequences or **risk**. The result is that designing a system is a risk-taking activity.

Complexity, trade-off, gaps, and risk are inherent in designing new systems and devices. Although they can be minimized by considering all the effects of a given design, they are always present in the design process.

Within engineering design, there is a fundamental difference between the two major types of thinking that must take place: engineering analysis and synthesis. Attention is focused on models of the physical systems that are analyzed to provide insight and that indicate directions for improvement. On the other hand, **synthesis** is the process by which these new physical configurations are created.

Design is a process that may proceed in many directions before the desired one is found. It is a deliberate process by which a designer creates something new in response to a recognized need while recognizing realistic constraints. The design process is inherently iterative—we must start somewhere! Successful engineers learn to simplify complex systems appropriately for design and analysis purposes. A gap between the complex physical system and the design model is inevitable. Design gaps are intrinsic in the progression from the initial concept to the final product. We know intuitively that it is easier to improve an initial concept incrementally than to try to create a final design at the start. In other words, engineering design is not a linear process. It is an iterative, nonlinear, creative process.

The main approach to the most effective engineering design is parameter analysis and optimization. Parameter analysis is based on (1) identification of the key parameters, (2) generation of the system configuration, and (3) evaluation of how well the configuration meets the needs. These three steps form an iterative loop. Once the key parameters are identified and the configuration synthesized, the designer can **optimize** the parameters. Typically, the designer strives to identify a limited set of parameters to be adjusted.

9.5 CONTROL SYSTEM DESIGN

The design of control systems is a specific example of engineering design. The goal of control engineering design is to obtain the configuration, specifications, and identification of the key parameters of a proposed system to meet an actual need.

The control system design process is illustrated in Figure 9.17. The design process consists of seven main building blocks, which we arrange into three groups:

1. Establishment of goals and variables to be controlled, and definition of specifications (metrics) against which to measure performance
2. System definition and modeling
3. Control system design and integrated system simulation and analysis

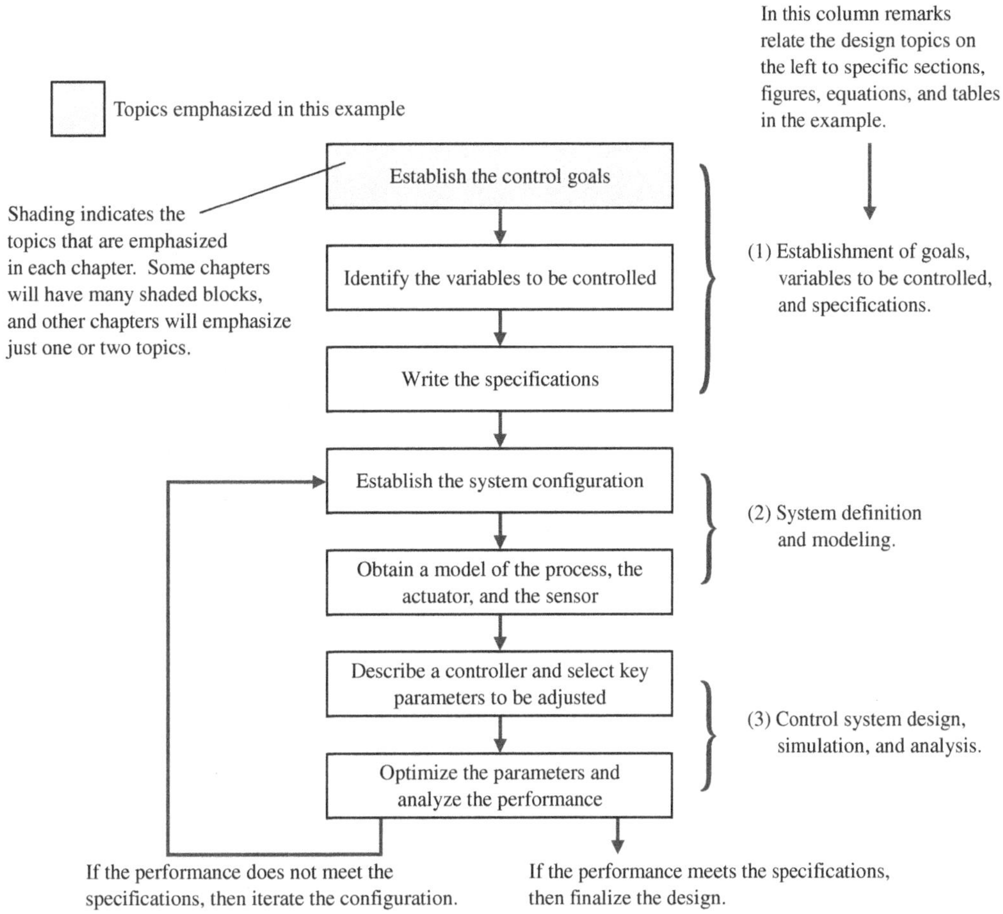

FIGURE 9.17 The control system design process.

The objective is to demonstrate different aspects of the design process through illustrative examples. We have established the design process block diagram:

1. Establishment of goals, control variables, and specifications.
2. System definition and modeling.
3. Control system design, simulation, and analysis.

The first step in the design process consists of establishing the system goals. For example, we may state that our goal is to control the velocity of a motor accurately. The second step is to identify the variables that we desire to control (for example, the velocity of the motor). The third step is to write the specifications in terms of the accuracy we must attain. This required accuracy of control will then lead to the identification of a sensor to measure the controlled variable. The performance specifications will describe how the closed-loop system should perform and will include (1) good regulation against disturbances, (2) desirable responses to commands, (3) realistic actuator signals, (4) low sensitivities, and (5) robustness.

As designers, we proceed to the first attempt to configure a system that will result in the desired control performance. This system configuration will normally consist of a sensor, the process under control, an actuator, and a controller, as shown in Figure 9.3. The next step consists of identifying a candidate for the actuator. This will, of course, depend on the process, but the actuation chosen must be capable of effectively adjusting the performance of the process. For example, if we wish to control the speed of a rotating flywheel, we will select a motor as the actuator. The sensor, in this case, must be capable of accurately measuring the speed. We then obtain a model for each of these elements.

Students studying controls are often given the models, frequently represented in transfer function or state variable form, with the understanding that they represent the underlying physical systems, but without further explanation. An obvious question is, where did the transfer function or state variable model come from? Within the context of a course in control systems, there is a need to address key questions surrounding modeling. To that end, we will provide insight into key modeling concerns and answer fundamental questions: How is the transfer function obtained? What basic assumptions are implied in the model development? How general are the transfer functions? However, mathematical modeling of physical systems is a subject in and of itself. We cannot hope to cover the mathematical modeling in its entirety, but interested students are encouraged to seek outside references (see for example [76–80]).

The next step is the selection of a controller, which often consists of a summing amplifier that will compare the desired response and the actual response and then forward this error-measurement signal to an amplifier.

The final step in the design process is the adjustment of the parameters of the system to achieve the desired performance. If we can achieve the desired performance by adjusting the parameters, we will finalize the design and proceed to document the results. If not, we will need to establish an improved system configuration and perhaps select an enhanced actuator and sensor. Then we will repeat the design steps until we are able to meet the specifications, or until we decide the specifications are too demanding and should be relaxed.

The design process has been dramatically affected by the advent of powerful and inexpensive computers and effective control design and analysis software. For example, the Boeing 777, which incorporates the most advanced flight avionics of any U.S. commercial aircraft, was almost entirely computer-designed [56, 57]. Verification of final designs in high-fidelity computer simulations is essential. In many applications, the certification of the control system in realistic simulations represents a significant cost in terms of money and time. The Boeing 777 test pilots flew about 2400 flights in high-fidelity simulations before the first aircraft was even built.

Another notable example of computer-aided design and analysis is the McDonnell Douglas Delta Clipper experimental vehicle DC-X, which was designed, built, and flown in 24 months. Computer-aided design tools and automated code-generation contributed to an estimated 80 percent cost savings and 30 percent time savings [58].

In summary, the controller design problem is as follows: Given a model of the system to be controlled (including its sensors and actuators) and a set of design goals, find a suitable controller, or determine that none exists. As with most of engineering design, the design of a feedback control system is an iterative and nonlinear process. A successful designer must consider the underlying physics of the plant under control, the control design strategy, the controller design architecture (that is, what type of controller will be employed), and effective controller tuning strategies. In addition, once the design is completed, the controller is often implemented in hardware, and hence issues of interfacing with hardware can appear. When taken together, these different phases of control system design make the task of designing and implementing a control system quite challenging [73].

9.6 MECHATRONIC SYSTEMS

A natural stage in the evolutionary process of modern engineering design is encompassed in the area known as **mechatronics** [64]. The term mechatronics was coined in Japan in the 1970s [65–67]. Mechatronics is the synergistic integration of mechanical, electrical, and computer systems and has evolved over the past 30 years, leading to a new breed of intelligent products. Feedback control is an integral aspect of modern mechatronic systems. One can understand the extent that mechatronics reaches into various disciplines by considering the components that make up mechatronics [68–71]. The key elements of mechatronics are (1) physical systems modeling, (2) sensors and actuators, (3) signals and systems, (4) computers and logic systems, and (5) software and data acquisition. Feedback control encompasses aspects of all five key elements of mechatronics, but is associated primarily with the element of signals and systems, as illustrated in Figure 9.18.

Advances in computer hardware and software technology coupled with the desire to increase the performance-to-cost ratio has revolutionized engineering design. New products are being developed at the intersection of traditional disciplines of

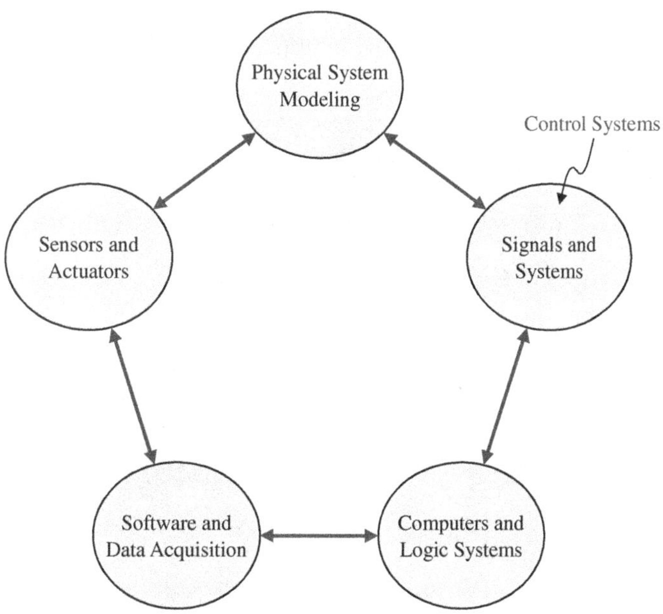

FIGURE 9.18 The key elements of mechatronics [64].

engineering, computer science, and the natural sciences. Advancements in traditional disciplines are fueling the growth of mechatronics systems by providing "enabling technologies." A critical enabling technology was the microprocessor which has had a profound effect on the design of consumer products. We should expect continued advancements in cost-effective microprocessors and microcontrollers, novel sensors and actuators enabled by advancements in applications of microelectromechanical systems (MEMS), advanced control methodologies and real-time programming methods, networking and wireless technologies, and mature computer-aided engineering (CAE) technologies for advanced system modeling, virtual prototyping, and testing. The continued rapid development in these areas will only accelerate the pace of smart (that is, actively controlled) products.

An exciting area of future mechatronic system development in which control systems will play a significant role is the area of alternative energy production and consumption. Hybrid fuel automobiles and efficient wind power generation are two examples of systems that can benefit from mechatronic design methods. In fact, the mechatronic design philosophy can be effectively illustrated by the example of the evolution of the modern automobile [64]. Before the 1960s, the radio was the only significant electronic device in an automobile. Today, many automobiles have 30–60 microcontrollers, up to 100 electric motors, about 200 pounds of wiring, a multitude of sensors, and thousands of lines of software code. A modern automobile can no longer be classified as a strictly mechanical machine—it has been transformed into a comprehensive mechatronic system.

EXAMPLE 9.1 Hybrid fuel vehicles

Recent research and development has led to the next-generation **hybrid fuel automobile**, depicted in Figure 9.19. The hybrid fuel vehicle utilizes a conventional internal combustion engine in combination with a battery (or other energy storage device such as a fuel cell or flywheel) and an electric motor to provide a propulsion system capable of doubling the fuel economy over conventional automobiles. Although

FIGURE 9.19 The hybrid fuel automobile can be viewed as a mechatronic system. (Used with permission of DOE/NREL. Credit: Warren Gretz.)

these hybrid vehicles will never be zero-emission vehicles (since they have internal combustion engines), they can reduce the level of harmful emissions by one-third to one-half, and with future improvements, these emissions may reduce even further. As stated earlier, the modern automobile requires many advanced control systems to operate. The control systems must regulate the performance of the engine, including fuel–air mixtures, valve timing, transmissions, wheel traction control, antilock brakes, and electronically controlled suspensions, among many other functions. On the hybrid fuel vehicle, there are additional control functions that must be satisfied. Especially necessary is the control of power between the internal combustion engine and the electric motor, determining power storage needs and implementing the battery charging, and preparing the vehicle for low-emission start-ups. The overall effectiveness of the hybrid fuel vehicle depends on the combination of power units that are selected (e.g., battery versus fuel cell for power storage). Ultimately, however, the control strategy that integrates the various electrical and mechanical components into a viable transportation system strongly influences the acceptability of the hybrid fuel vehicle concept in the marketplace. ■

The second example of a mechatronic system is the advanced wind power generation system.

EXAMPLE 9.2 **Wind power**

Many nations in the world today are faced with unstable energy supplies, often leading to rising fuel prices and energy shortages. Additionally, the negative effects of fossil fuel utilization on the quality of our air are well documented. Many nations have an imbalance in the supply and demand of energy, consuming more than they produce. To address this imbalance, many engineers are considering developing advanced systems to access other sources of energy, such as wind energy. In fact, wind energy is one of the fastest-growing forms of energy generation in the United States and in other locations around the world. A wind farm now in use in western Texas is illustrated in Figure 9.20.

In 2006, the installed global wind energy capacity was over 59,000 MW. In the United States, there was enough energy derived from wind to power over 2.5 million homes, according to the American Wind Energy Association. For the past 35 years,

FIGURE 9.20 Efficient wind power generation in west Texas. (Used with permission of DOE/NREL. Credit: Lower Colorado River Authority.)

researchers have concentrated on developing technologies that work well in high wind areas (defined to be areas with a wind speed of at least 6.7 m/s at a height of 10 m). Most of the easily accessible high wind sites in the United States are now utilized, and improved technology must be developed to make lower wind areas more cost effective. New developments are required in materials and aerodynamics so that longer turbine rotors can operate efficiently in the lower winds, and in a related problem, the towers that support the turbine must be made taller without increasing the overall costs. In addition, advanced controls will be required to achieve the level of efficiency required in the wind generation drive train. ■

EXAMPLE 9.3 **Embedded computers**

Many contemporary control systems are **embedded control** systems [81]. Embedded control systems employ on-board special-purpose digital computers as integral components of the feedback loop. Figure 9.21 illustrates a student-built rover constructed around the Compact RIO by National Instruments, Inc. that serves as the on-board embedded computer. In the rover design, the sensors include an optical encoder for measuring engine speed, a rate gyro and accelerometer to measure turns, and a Global Positioning System (GPS) unit to obtain position and velocity estimates of the vehicle. The actuators include two linear actuators to turn the front wheels and to brake and accelerate. The communications device permits the rover to stay in contact with the ground station.

Advances in sensors, actuators, and communication devices are leading to a new class of embedded control systems that are networked using wireless technology, thereby enabling spatially-distributed control. Embedded control system designers

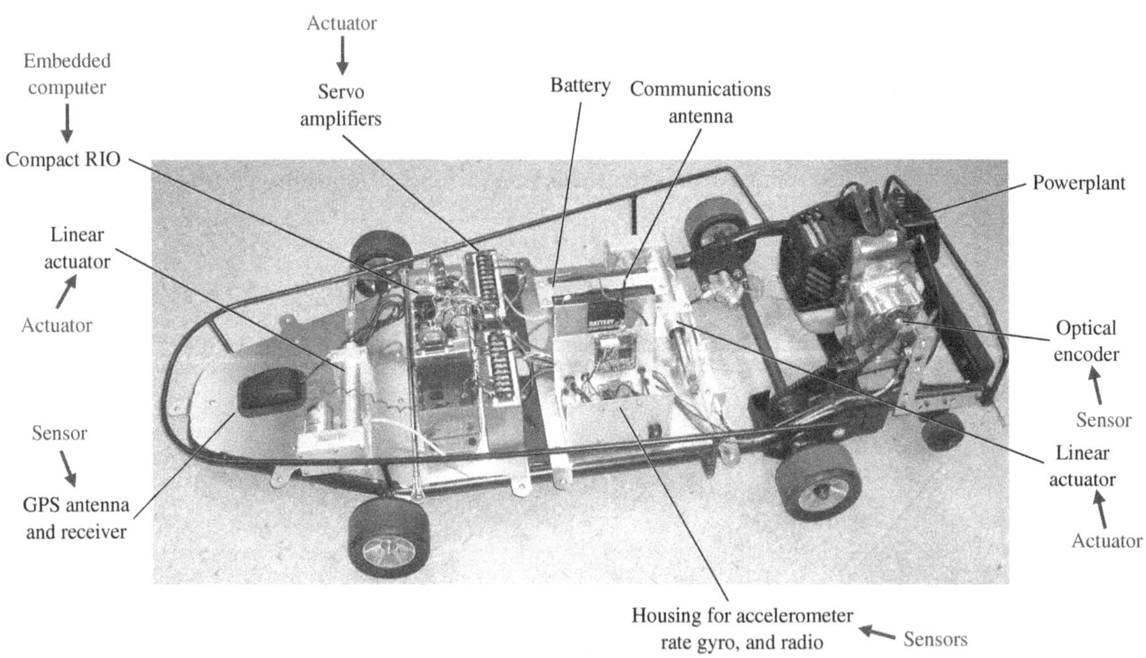

FIGURE 9.21 A rover using an embedded computer in the feedback loop. *(Photo by R.H. Bishop.)*

must be able to understand and work with various network protocols, diverse operating systems and programming languages. While the theory of systems and controls serves as the foundation for the modern control system design, the design process is rapidly expanding into a multi-disciplinary enterprise encompassing multiple engineering areas, as well as information technology and computer science. ■

Advances in alternate energy products, such as the hybrid automobile and the generation of efficient wind power generators, provide vivid examples of mechatronics development. There are numerous other examples of intelligent systems poised to enter our everyday life, including autonomous rovers, smart home appliances (e.g., dishwashers, vacuum cleaners, and microwave ovens), wireless network-enabled devices, "human-friendly machines" [72] that perform robot-assisted surgery, and implantable sensors and actuators.

9.7 GREEN ENGINEERING

Global issues such as climate change, clean water, sustainability, waste management, emissions reduction, and minimizing raw material and energy use have caused many engineers to re-think existing approaches to engineering design in critical areas. One outcome of the evolving design strategy is to consider an approach that has come to be known as "green engineering." The goal of green engineering is to design products that will minimize pollution, reduce the risk to human health, and improve the environment. The basic principles of green engineering are [86]:

1. Engineer processes and products holistically, use systems analysis, and integrate environmental impact assessment tools.
2. Conserve and improve natural ecosystems while protecting human health and well-being.
3. Use life-cycle thinking in all engineering activities.
4. Ensure that all material and energy inputs and outputs are as inherently safe and benign as possible.
5. Minimize depletion of natural resources.
6. Strive to prevent waste.
7. Develop and apply engineering solutions, while being cognizant of local geography, aspirations, and cultures.
8. Create engineering solutions beyond current or dominant technologies; improve, innovate, and invent technologies to achieve sustainability.
9. Actively engage communities and stakeholders in development of engineering solutions.

Putting the principles of green engineering into practice leads us to a deeper understanding of the power of feedback control systems as an enabling technology. For example, in Section 9.9, we present a discussion on smart grids. Smart grids aim to deliver electrical power more reliably and efficiently in an environmentally friendly fashion. This in turn will potentially enable the large-scale use of renewable energy sources, such as wind and solar, that are naturally intermittent. Sensing and feedback are key technology areas that enable the smart grids [87]. Green engineering applications can be classified into one of five categories [88]:

1. Environmental Monitoring
2. Energy Storage Systems
3. Power Quality Monitoring
4. Solar Energy
5. Wind Energy

As the field of green engineering matures, it is almost certain that more applications will evolve, especially as we apply the eighth principle (listed above) of green engineering to create engineering solutions beyond current or dominant technologies and improve, innovate, and invent technologies. In the subsequent chapters, we present examples from each of these areas.

There is a global effort underway to reduce greenhouse gases from all sources. To accomplish this, it is necessary to improve both thc quality and quantity of our environmental monitoring systems. An example is using wireless measurements on a cabled robotic controlled mobile sensing platform moving along the forest understory to measure key environmental parameters in a rain forest.

Energy storage systems are critical technologies for green engineering. There are many types of energy storage systems. The energy storage system we are most familiar with is the battery. Batteries are used to power most of the electronic devices in use today; some batteries are rechargeable and some are single-use throwaways. To adhere to green engineering principles, we would favor energy storage systems that are renewable. A very important energy storage device for green engineering systems is the fuel cell.

The problems associated with power quality monitoring are varied and can include leading and lagging power, voltage variations, and waveform harmonics. Many of the green engineering systems and components require careful monitoring of current and voltages. An interesting example would be the modeling of current transformers that are used in various capacities for measuring and monitoring within the power grid network of interconnected systems used to deliver electricity.

Efficiently converting solar energy into electricity is an engineering challenge. Two technologies for generation of electricity using sunshine are solar photovoltaic and solar thermal. With photovoltaic systems the sunlight is converted directly to electricity, and with solar thermal the sun heats water to create steam that is used to power steam engines. Designing and deploying solar photovoltaic systems for solar power generation is one approach employing green engineering principles to utilize the sun's energy to power our homes, offices, and businesses.

Power derived from wind is an important source of renewable energy around the world. Wind energy conversion to electric power is achieved by wind energy turbines connected to electric generators. The intermittency characteristic of wind energy makes the smart grid development (see Example 9.4) essential to bring the energy to the power grid when it is available and to provide energy from other sources when the wind dies down or is disrupted. The irregular character of wind direction and power also results in the need for reliable, steady electric energy by using control systems on the wind turbines themselves. The goal of these control devices is to reduce the effects of wind intermittency and the effect of wind direction change.

The role of control systems in green engineering will continue to expand as the global issues facing us require ever increasing levels of automation and precision.

9.8 THE FUTURE EVOLUTION OF CONTROL SYSTEMS

The continuing goal of control systems is to provide extensive flexibility and a high level of autonomy. Two system concepts are approaching this goal by different evolutionary pathways, as illustrated in Figure 9.22. Today's industrial robot is perceived as quite autonomous—once it is programmed, further intervention is not normally required. Because of sensory limitations, these robotic systems have limited flexibility in adapting to work environment changes; improving perception is the motivation of computer vision research. The control system is very adaptable, but it relies on human supervision. Advanced robotic systems are striving for task adaptability through enhanced sensory feedback. Research areas concentrating on artificial intelligence, sensor integration, computer vision, and off-line CAD/CAM programming will make systems more universal and economical. Control systems are moving toward autonomous operation as an enhancement to human control. Research in supervisory control, human–machine interface methods, and computer database management are intended to reduce operator burden and improve operator efficiency. Many research activities are common to robotics and control systems and are aimed at reducing implementation cost and expanding the realm of application. These include improved communication methods and advanced programming languages.

The easing of human labor by technology, a process that began in prehistory, is entering a new stage. The acceleration in the pace of technological innovation inaugurated by the Industrial Revolution has until recently resulted mainly in the displacement of human muscle power from the tasks of production. The current

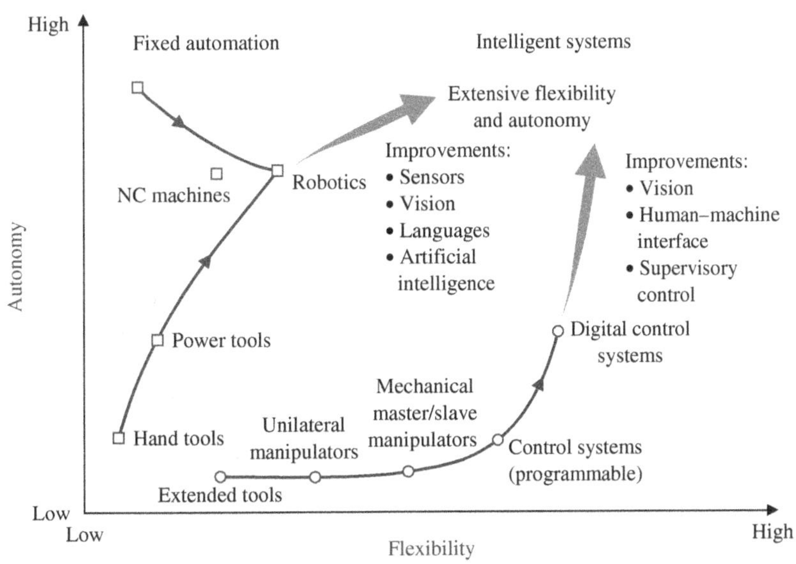

FIGURE 9.22 Future evolution of control systems and robotics.

revolution in computer technology is causing an equally momentous social change, the expansion of information gathering and information processing as computers extend the reach of the human brain [16].

Control systems are used to achieve (1) increased productivity and (2) improved performance of a device or system. Automation is used to improve productivity and obtain high-quality products. Automation is the automatic operation or control of a process, device, or system. We use automatic control of machines and processes to produce a product reliably and with high precision [28]. With the demand for flexible, custom production, a need for flexible automation and robotics is growing [17, 25].

The theory, practice, and application of automatic control is a large, exciting, and extremely useful engineering discipline. One can readily understand the motivation for a study of modern control systems.

9.9 DESIGN EXAMPLES

In this section we present illustrative design examples. This is a pattern that we will follow in all subsequent chapters. Each chapter will contain a number of interesting examples in a special section entitled Design Examples meant to highlight the main topics of the chapter. At least one example among those presented in the Design Example section will be a more detailed problem and solution that demonstrates one or more of the steps in the design process shown in Figure 9.17. In the first example, we discuss the development of the smart grid as a concept to deliver electrical power more reliably and efficiently as part of a strategy to provide a more environmentally friendly energy delivery system. The smart grid will enable the large-scale use of renewable energy sources that depend on the natural phenomenon to generate power and which are intermittent, such as wind and solar. Providing clean energy is an engineering challenge that must necessarily include active feedback control systems, sensors, and actuators. In the second example presented here, a rotating disk speed control illustrates the concept of open-loop and closed-loop feedback control. The third example is an insulin delivery control system in which we determine the design goals, the variables to control, and a preliminary closed-loop system configuration.

EXAMPLE 9.4 **Smart grid control systems**

A smart grid is as much a concept as it is a physical system. In essence, the concept is to deliver power more reliably and efficiently while remaining environmentally friendly, economical, and safe [89, 90]. A smart grid can be viewed as a system comprised of hardware and software that routes power more reliably and efficiently to homes, businesses, schools, and other users of power. One view of the smart grid is illustrated schematically in Figure 9.23. Smart grids can be national or local in scope. One can even consider home smart grids (or microgrids). In fact, smart grids encompass a wide and rich field of investigation. As we will find, control systems play a key role in smart grids at all levels.

One interesting aspect of the smart grid is real-time demand side management requiring a two-way flow of information between the user and the power generation system [91]. For example, smart meters are used to measure electricity use in the home and office. These sensors transmit data to utilities and allow the utility to transmit control signals back to a home or building. These smart meters can control and turn on

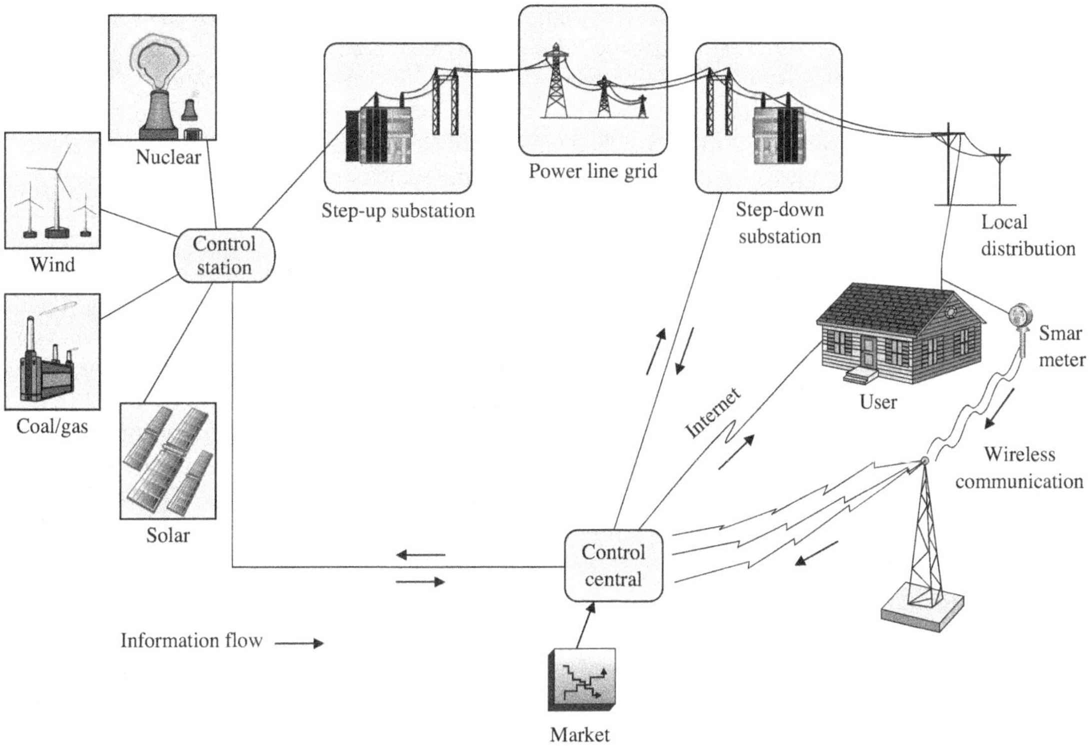

FIGURE 9.23 Smart grids are distribution networks that measure and control usage.

or off home and office appliances and devices. Smart home-energy devices enable the homeowners to control their usage and respond to price changes at peak-use times.

The five key technologies required to implement a successful modern smart grid include (i) integrated communications, (ii) sensing and measurements, (iii) advanced components, (iv) advanced control methods, and (v) improved interfaces and decision support [87]. Two of the five key technologies fall under the general category of control systems, namely (ii) sensing and measurements and (iii) advanced control methods. It is evident that control systems will play a key role in realizing the modern smart grid. The potential impact of the smart grid on delivery of power is very high. Currently, the total U.S. grid includes 16,000 power plants, about 3,300 utility companies, and 300,000 miles of power lines. A smart grid will use sensors, controllers, the Internet, and communication systems to improve the reliability and efficiency of the grid. It is estimated that deployment of smart grids could reduce emissions globally of CO_2 due to power systems by 14 percent by 2020 [91].

One of the elements of the smart grid are the distribution networks that measure and control usage. In a smart grid, the power generation depends on the market situation (supply/demand and cost) and the power source available (wind, coal, nuclear, geothermal, biomass, etc.). In fact, smart grid customers with solar panels or wind turbines could sell their excess energy to the grid and get paid as microgenerators [92]. In the subsequent chapters, we discuss various control problems associated with pointing solar panels to the sun and with prescribing the pitch of the wind turbine blades to manage the rotor speed thereby controlling the power output.

Transmission of power is called power flow and the improved control of power will increase its security and efficiency. Transmission lines have inductive, capacitive, and resistive effects that result in dynamic impacts or disturbances. The smart grid must anticipate and respond to system disturbances rapidly. This is referred to as self-healing. In other words, a smart grid should be capable of managing significant disturbances occurring on very short time scales. To accomplish this, the self-healing process is constructed around the idea of a feedback control system where self-assessments are used to detect and analyze disturbances so that corrective action can be applied to restore the grid. This requires sensing and measurements to provide information to the control systems. One of the benefits of using smart grids is that renewable energy sources that depend on intermittent natural phenomena (such as wind and sunshine) can potentially be utilized more efficiently by allowing for load shedding when the wind dies out or clouds block the sunshine.

Feedback control systems will play an increasingly important role in the development of smart grids as we move to the target date. It may be interesting to recall the various topics discussed in this section in the context of control systems as each chapter in this textbook unfolds new methods of control system design and analysis.

EXAMPLE 9.5 **Rotating disk speed control**

Many modern devices employ a rotating disk held at a constant speed. For example, a CD player requires a constant speed of rotation in spite of motor wear and variation and other component changes. Our goal is to design a system for rotating disk speed control that will ensure that the actual speed of rotation is within a specified percentage of the desired speed [40, 43]. We will consider a system without feedback and a system with feedback.

To obtain disk rotation, we will select a DC motor as the actuator because it provides a speed proportional to the applied motor voltage. For the input voltage to the motor, we will select an amplifier that can provide the required power.

The open-loop system (without feedback) is shown in Figure 9.24(a). This system uses a battery source to provide a voltage that is proportional to the desired speed. This

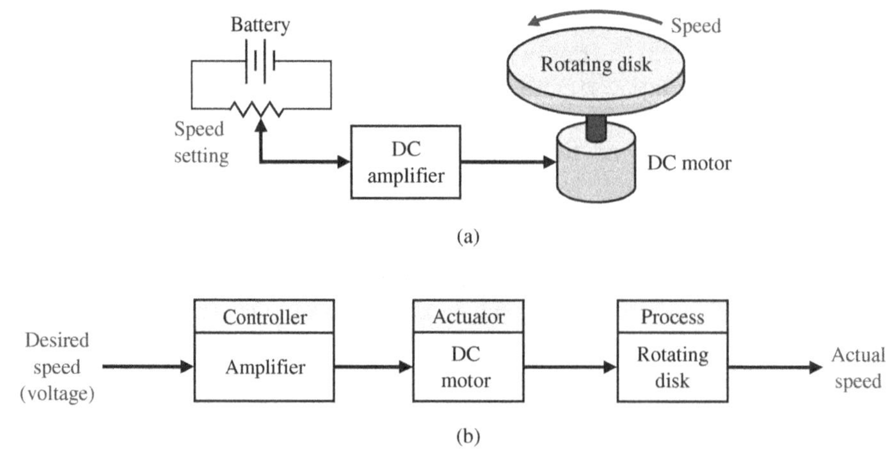

FIGURE 9.24 (a) Open-loop (without feedback) control of the speed of a rotating disk. (b) Block diagram model.

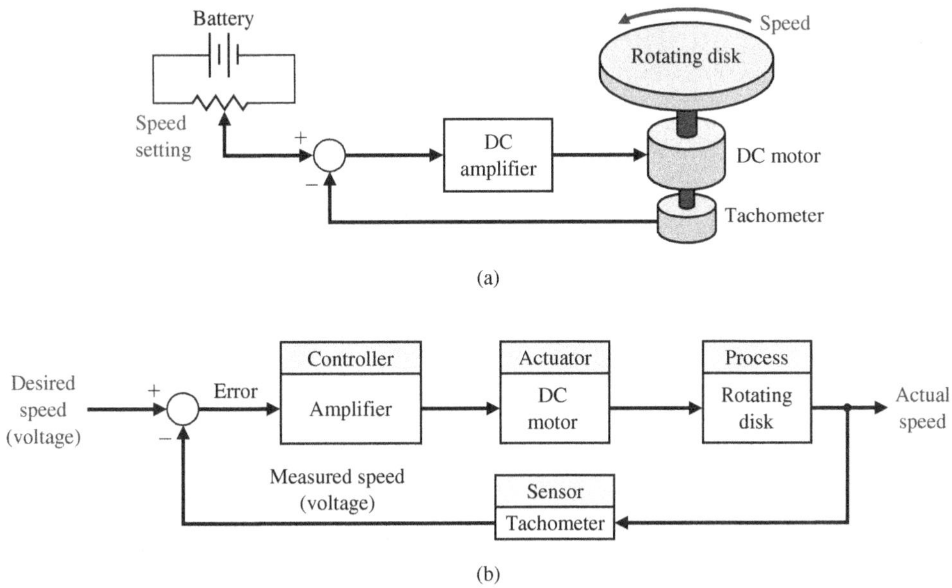

FIGURE 9.25 (a) Closed-loop control of the speed of a rotating disk. (b) Block diagram model.

voltage is amplified and applied to the motor. The block diagram of the open-loop system identifying the controller, actuator, and process is shown in Figure 9.24(b).

To obtain a feedback system, we need to select a sensor. One useful sensor is a tachometer that provides an output voltage proportional to the speed of its shaft. Thus the closed-loop feedback system takes the form shown in Fig. 9.25(a). The block diagram model of the feedback system is shown in Fig. 9.25(b). The error voltage is generated by the difference between the input voltage and the tachometer voltage.

We expect the feedback system of Figure 9.25 to be superior to the open-loop system of Figure 9.24 because the feedback system will respond to errors and act to reduce them. With precision components, we could expect to reduce the error of the feedback system to one-hundredth of the error of the open-loop system. ■

EXAMPLE 9.6 Insulin delivery control system

Control systems have been utilized in the biomedical field to create implanted automatic drug-delivery systems to patients [29–31]. Automatic systems can be used to regulate blood pressure, blood sugar level, and heart rate. A common application of control engineering is in the field of open-loop system drug delivery, in which mathematical models of the dose–effect relationship of the drugs are used. A drug-delivery system implanted in the body uses an open-loop system, since miniaturized glucose sensors are not yet available. The best solutions rely on individually programmable, pocket-sized insulin pumps that can deliver insulin according to a preset time history. More complicated systems will use closed-loop control for the measured blood glucose levels.

The blood glucose and insulin concentrations for a healthy person are shown in Figure 9.26. The system must provide the insulin from a reservoir implanted within the diabetic person. Therefore, the control goal is:

Control Goal

Design a system to regulate the blood sugar concentration of a diabetic by controlled dispensing of insulin.

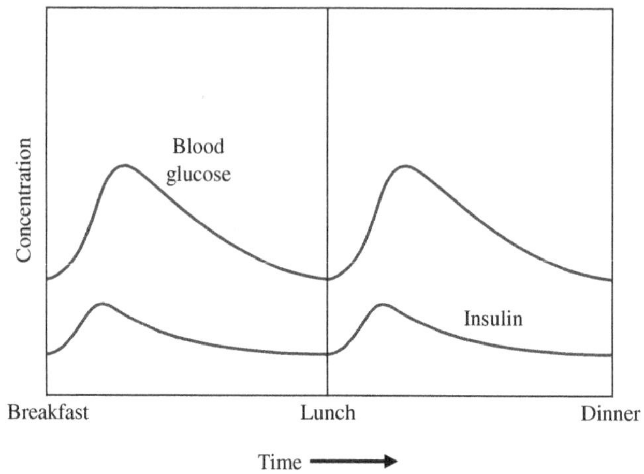

FIGURE 9.26 The blood glucose and insulin levels for a healthy person.

Referring to Figure 9.26, the next step in the design process is to define the variable to be controlled. Associated with the control goal we can define the variable to be controlled to be:

Variable to Be Controlled

Blood glucose concentration

In subsequent chapters, we will have the tools to quantitatively describe the control design specifications using a variety of steady-state performance specifications and transient response specifications, both in the time-domain and in the frequency domain. At this point, the control design specifications will be qualitative and imprecise. In that regard, for the problem at hand, we can state the design specification as:

Control Design Specifications

Provide a blood glucose level for the diabetic that closely approximates (tracks) the glucose level of a healthy person.

Given the design goals, variables to be controlled, and control design specifications, we can now propose a preliminary system configuration. An open-loop system would use a preprogrammed signal generator and miniature motor pump to regulate the insulin delivery rate as shown in Figure 9.27(a). The feedback control system would use a sensor to measure the actual glucose level and compare that level with the desired level, thus turning the motor pump on when it is required, as shown in Figure 9.27(b). ■

9.10 SEQUENTIAL DESIGN EXAMPLE: DISK DRIVE READ SYSTEM

This design example, identified by the arrow icon, will be considered sequentially in each chapter. We will use the design process of Figure 9.17 in each chapter to identify the steps that we are accomplishing. For example, in Chapter 9 we (1) identify the control goal, (2) identify the variables to control, (3) write the initial specifications for the variables, and (4) establish the preliminary system configuration.

Information can be readily and efficiently stored on magnetic disks. Disk drives are used in notebook computers and larger computers of all sizes and are essentially all standardized as defined by ANSI standards [50, 63]. The worldwide total available market for disk drives is greater than 650 million units [51]. In the past, disk drive designers

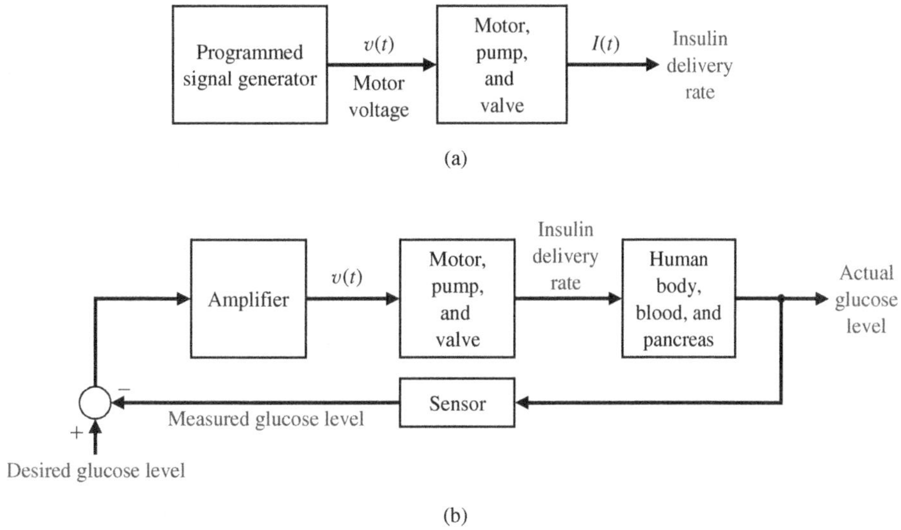

FIGURE 9.27 (a) Open-loop (without feedback) control and (b) closed-loop control of blood glucose.

have concentrated on increasing data density and data access times. Recent trends show that hard disk drive densities are increasing at about 40 percent per year [62]. Figure 9.28 shows the disk drive density trends. Designers are now considering employing disk drives to perform tasks historically delegated to central processing units (CPUs), thereby leading to improvements in the computing environment [63]. Three

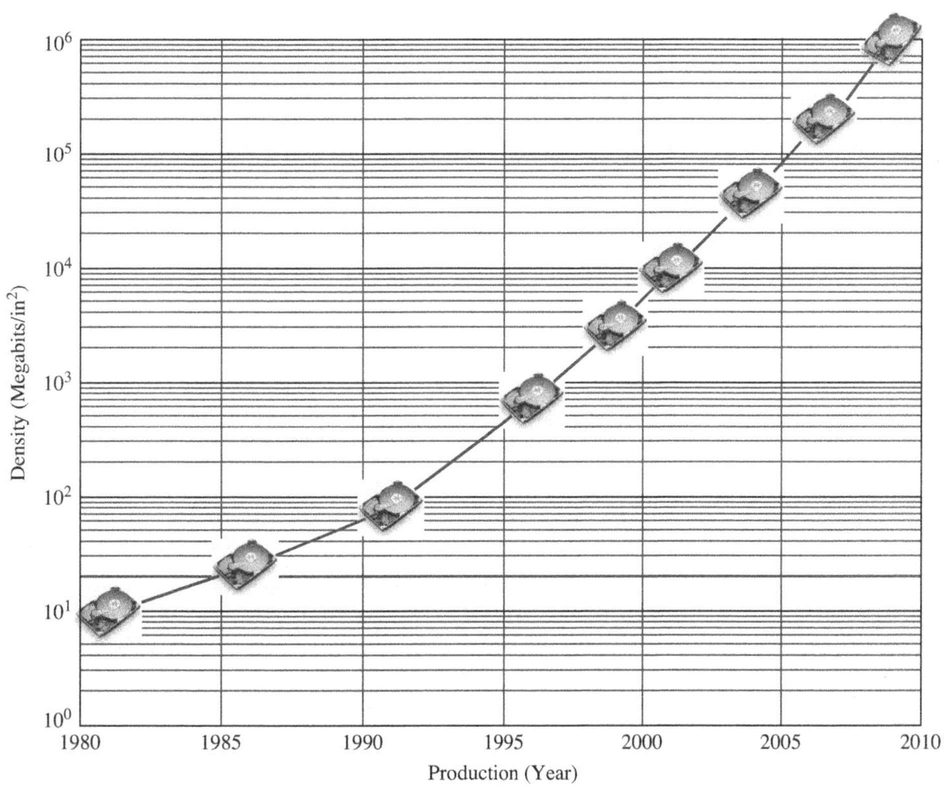

FIGURE 9.28 Disk drive data density trends (Source: IBM).

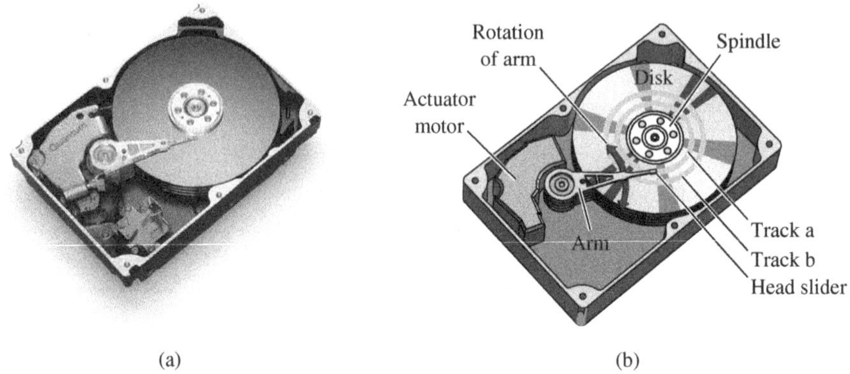

FIGURE 9.29 (a) A disk drive © 1999 Quantum Corporation. All rights reserved. (b) Diagram of a disk drive.

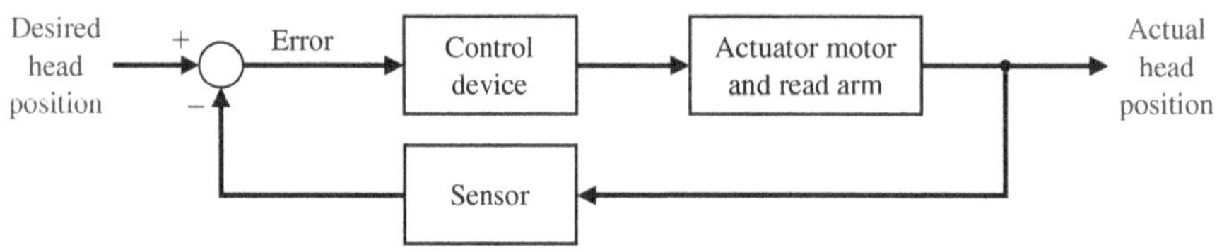

FIGURE 9.30 Closed-loop control system for disk drive.

areas of "intelligence" under investigation include off-line error recovery, disk drive failure warnings, and storing data across multiple disk drives. Consider the basic diagram of a disk drive shown in Fig. 9.29. The goal of the disk drive reader device is to position the reader head to read the data stored on a track on the disk. The variable to accurately control is the position of the reader head (mounted on a slider device). The disk rotates at a speed between 1800 and 7200 rpm, and the head "flies" above the disk at a distance of less than 100 nm. The initial specification for the position accuracy is 1 μm. Furthermore, we plan to be able to move the head from track a to track b within 50 ms, if possible. Thus, we establish an initial system configuration as shown in Figure 9.30. This proposed closed-loop system uses a motor to actuate (move) the arm to the desired location on the disk.

9.11 SUMMARY

In this chapter, we discussed open- and closed-loop feedback control systems. Examples of control systems through the course of history were presented to motivate and connect the subject to the past. In terms of contemporary issues, key areas of application were discussed, including humanoid robots, unmanned aerial vehicles, wind energy, hybrid automobiles, and embedded control. The central role of controls in mechatronics was discussed. Mechatronics is the synergistic integration of mechanical, electrical, and computer systems. Finally, the design process was presented in a structured form and included the following steps: the establishment of goals and variables to be controlled, definition of specifications, system definition, modeling, and analysis. The iterative nature of design allows us to handle the design gap effectively while accomplishing necessary trade-offs in complexity, performance, and cost.

SKILLS CHECK

In this section, we provide three sets of problems to test your knowledge: True or False, Multiple Choice, and Word Match. To obtain direct feedback, check your answers with the answer key provided at the conclusion of the end-of-chapter problems.

In the following **True or False** and **Multiple Choice** problems, circle the correct answer.

1. The flyball governor is generally agreed to be the first automatic feedback controller used in an industrial process. *True or False*
2. A closed-loop control system uses a measurement of the output and feedback of the signal to compare it with the desired input. *True or False*
3. Engineering synthesis and engineering analysis are the same. *True or False*
4. The block diagram in Figure 9.31 is an example of a closed-loop feedback system. *True or False*

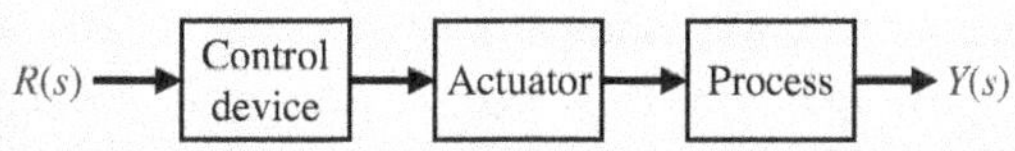

FIGURE 9.31 System with control device, actuator, and process.

5. A multivariable system is a system with more than one input and/or more than one output. *True or False*
6. Early applications of feedback control include which of the following?
   - **a.** Water clock of Ktesibios
   - **b.** Watt's flyball governor
   - **c.** Drebbel's temperature regulator
   - **d.** All of the above
7. Important modern applications of control systems include which of the following?
   - **a.** Fuel-efficient and safe automobiles
   - **b.** Autonomous robots
   - **c.** Automated manufacturing
   - **d.** All of the above
8. Complete the following sentence:

   Control of an industrial process by automatic rather than manual means is often called ________.
   - **a.** negative feedback
   - **b.** automation
   - **c.** a design gap
   - **d.** a specification
9. Complete the following sentence:

   ________ *are intrinsic in the progression from an initial concept to the final product.*
   - **a.** Closed-loop feedback systems
   - **b.** Flyball governors
   - **c.** Design gaps
   - **d.** Open-loop control systems

10. Complete the following sentence:

Control engineers are concerned with understanding and controlling segments of their environments, often called _______.

a. systems
b. design synthesis
c. trade-offs
d. risk

11. Early pioneers in the development of systems and control theory include:

a. H. Nyquist
b. H. W. Bode
c. H. S. Black
d. All of the above

12. Complete the following sentence:

An open-loop control system utilizes an actuating device to control a process _______.

a. without using feedback
b. using feedback
c. in engineering design
d. in engineering synthesis

13. A system with more than one input variable or more than one output variable is known by what name?

a. Closed-loop feedback system
b. Open-loop feedback system
c. Multivariable control system
d. Robust control system

14. Control engineering is applicable to which fields of engineering?

a. Mechanical and aerospace
b. Electrical and biomedical
c. Chemical and environmental
d. All of the above

15. Closed-loop control systems should have which of the following properties:

a. Good regulation against disturbances
b. Desirable responses to commands
c. Low sensitivity to changes in the plant parameters
d. All of the above

In the following **Word Match** problems, match the term with the definition by writing the correct letter in the space provided.

a. Optimization	The output signal is fed back so that it subtracts from the input signal.	_______
b. Risk	A system that uses a measurement of the output and compares it with the desired output.	_______
c. Complexity of design	A set of prescribed performance criteria.	_______
d. System	A measure of the output of the system used for feedback to control the system.	_______
e. Design	A system with more than one input variable or more than one output variable.	_______

f. Closed-loop feedback control system	The result of making a judgment about how much compromise must be made between conflicting criteria.	_______
g. Flyball governor	An interconnection of elements and devices for a desired purpose.	_______
h. Specifications	A reprogrammable, multifunctional manipulator used for a variety of tasks.	_______
i. Synthesis	A gap between the complex physical system and the design model intrinsic to the progression from the initial concept to the final product.	_______
j. Open-loop control system	The intricate pattern of interwoven parts and knowledge required.	_______
k. Feedback signal	The ratio of physical output to physical input of an industrial process.	_______
l. Robot	The process of designing a technical system.	_______
m. Multivariable control system	A system that utilizes a device to control the process without using feedback.	_______
n. Design gap	Uncertainties embodied in the unintended consequences of a design.	_______
o. Positive feedback	The process of conceiving or inventing the forms, parts, and details of a system to achieve a specified purpose.	_______
p. Negative feedback	The device, plant, or system under control.	_______
q. Trade-off	The output signal is fed back so that it adds to the input signal.	_______
r. Productivity	An interconnection of components forming a system configuration that will provide a desired response.	_______
s. Engineering design	The control of a process by automatic means.	_______
t. Process	The adjustment of the parameters to achieve the most favorable or advantageous design.	_______
u. Control system	The process by which new physical configurations are created.	_______
v. Automation	A mechanical device for controlling the speed of a steam engine.	_______

EXERCISES

Exercises are straightforward applications of the concepts of the chapter.

The following systems can be described by a block diagram showing the cause–effect relationship and the feedback (if present). Identify the function of each block and the desired input variable, output variable, and measured variable. Use Figure 9.3 as a model where appropriate.

E9.1 Describe typical sensors that can measure each of the following [93]:

a. Linear position
b. Velocity (or speed)
c. Nongravitational acceleration
d. Rotational position (or angle)
e. Rotational velocity
f. Temperature
g. Pressure
h. Liquid (or gas) flow rate
i. Torque
j. Force

E9.2 Describe typical actuators that can convert the following [93]:

a. Fluidic energy to mechanical energy
b. Electrical energy to mechanical energy

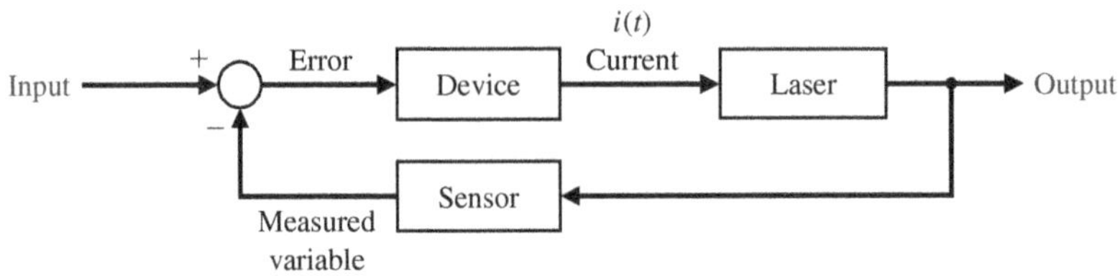

FIGURE E9.3 Partial block diagram of an optical source.

c. Mechanical deformation to electrical energy
d. Chemical energy to kinetic energy

E9.3 A precise optical signal source can control the output power level to within 1 percent [32]. A laser is controlled by an input current to yield the power output. A microprocessor controls the input current to the laser. The microprocessor compares the desired power level with a measured signal proportional to the laser power output obtained from a sensor. Complete the block diagram representing this closed-loop control system shown in Figure E9.3, identifying the output, input, and measured variables and the control device.

E9.4 An automobile driver uses a control system to maintain the speed of the car at a prescribed level. Sketch a block diagram to illustrate this feedback system.

E9.5 Fly-fishing is a sport that challenges the person to cast a small feathery fly using a light rod and line. The goal is to place the fly accurately and lightly on the distant surface of the stream [59]. Describe the fly-casting process and a model of this process.

E9.6 An autofocus camera will adjust the distance of the lens from the film by using a beam of infrared or ultrasound to determine the distance to the subject [42]. Sketch a block diagram of this open-loop control system, and briefly explain its operation.

E9.7 Because a sailboat cannot sail directly into the wind, and traveling straight downwind is usually slow, the shortest sailing distance is rarely a straight line. Thus sailboats tack upwind—the familiar zigzag course—and jibe downwind. A tactician's decision of when to tack and where to go can determine the outcome of a race.

Describe the process of tacking a sailboat as the wind shifts direction. Sketch a block diagram depicting this process.

E9.8 Modern automated highways are being implemented around the world. Consider two highway lanes merging into a single lane. Describe a feedback control system carried on the automobile trailing the lead automobile that ensures that the vehicles merge with a prescribed gap between the two vehicles.

E9.9 Describe the block diagram of the speed control system of a motorcycle with a human driver.

E9.10 Describe the process of human biofeedback used to regulate factors such as pain or body temperature. Biofeedback is a technique whereby a human can, with some success, consciously regulate pulse, reaction to pain, and body temperature.

E9.11 Future advanced commercial aircraft will be E-enabled. This will allow the aircraft to take advantage of continuing improvements in computer power and network growth. Aircraft can continuously communicate their location, speed, and critical health parameters to ground controllers, and gather and transmit local meteorological data. Sketch a block diagram showing how the meteorological data from multiple aircraft can be transmitted to the ground, combined using ground-based powerful networked computers to create an accurate weather situational awareness, and then transmitted back to the aircraft for optimal routing.

E9.12 Unmanned aerial vehicles (UAVs) are being developed to operate in the air autonomously for long periods of time (see Section 9.3). By autonomous, we mean that there is no interaction with human ground controllers. Sketch a block diagram of an autonomous UAV that is tasked for crop monitoring using aerial photography. The UAV must photograph and transmit the entire land area by flying a pre-specified trajectory as accurately as possible.

E9.13 Consider the inverted pendulum shown in Figure E9.13. Sketch the block diagram of a feedback control

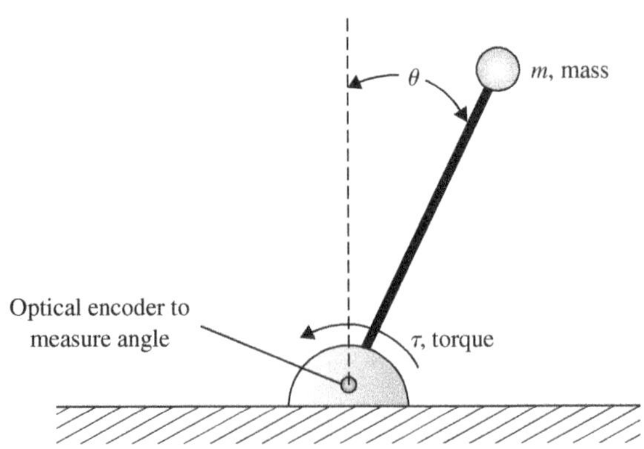

FIGURE E9.13 Inverted pendulum control.

system using Figure 9.3 as the model. Identify the process, sensor, actuator, and controller. The objective is keep the pendulum in the upright position, that is to keep $\theta = 0$ in the presence of disturbances.

E9.14 Describe the block diagram of a person playing a video game. Suppose that the input device is a joystick and the game is being played on a desktop computer. Use Figure 9.3 as a model of the block diagram.

PROBLEMS

Problems require extending the concepts of this chapter to new situations.

The following systems may be described by a block diagram showing the cause–effect relationship and the feedback (if present). Each block should describe its function. Use Figure 9.3 as a model where appropriate.

P9.1 Many luxury automobiles have thermostatically controlled air-conditioning systems for the comfort of the passengers. Sketch a block diagram of an air-conditioning system where the driver sets the desired interior temperature on a dashboard panel. Identify the function of each element of the thermostatically controlled cooling system.

P9.2 In the past, control systems used a human operator as part of a closed-loop control system. Sketch the block diagram of the valve control system shown in Figure P9.2.

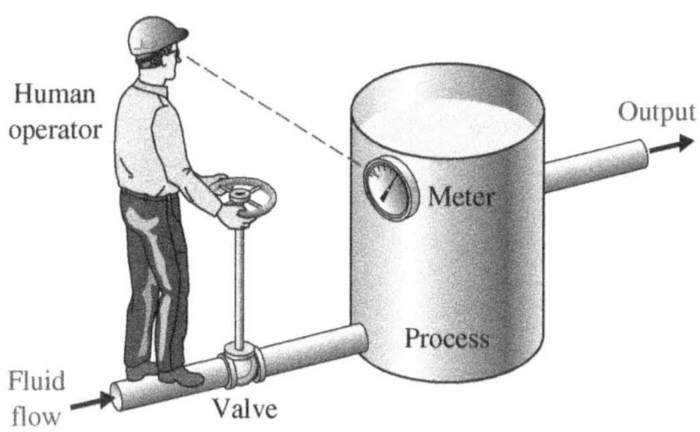

FIGURE P9.2 Fluid-flow control.

P9.3 In a chemical process control system, it is valuable to control the chemical composition of the product. To do so, a measurement of the composition can be obtained by using an infrared stream analyzer, as shown in Figure P9.3. The valve on the additive stream may be controlled. Complete the control feedback loop, and sketch a block diagram describing the operation of the control loop.

P9.4 The accurate control of a nuclear reactor is important for power system generators. Assuming the number of neutrons present is proportional to the power level, an ionization chamber is used to measure the power level. The current i_0 is proportional to the power level. The position of the graphite control rods moderates the power level. Complete the control system of the nuclear reactor shown in Figure P9.4 and sketch the block diagram describing the operation of the feedback control loop.

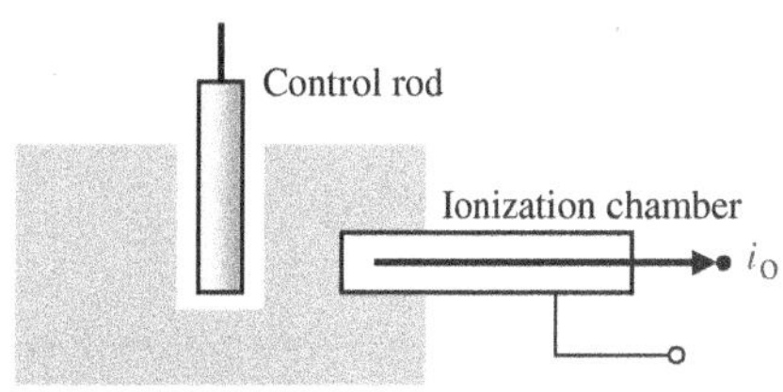

FIGURE P9.4 Nuclear reactor control.

P9.5 A light-seeking control system, used to track the sun, is shown in Figure P9.5. The output shaft, driven by the motor through a worm reduction gear, has a bracket attached on which are mounted two photocells. Complete the closed-loop system so that the system follows the light source.

P9.6 Feedback systems do not always involve negative feedback. Economic inflation, which is evidenced by continually rising prices, is a **positive feedback** system. A positive feedback control system, as shown in

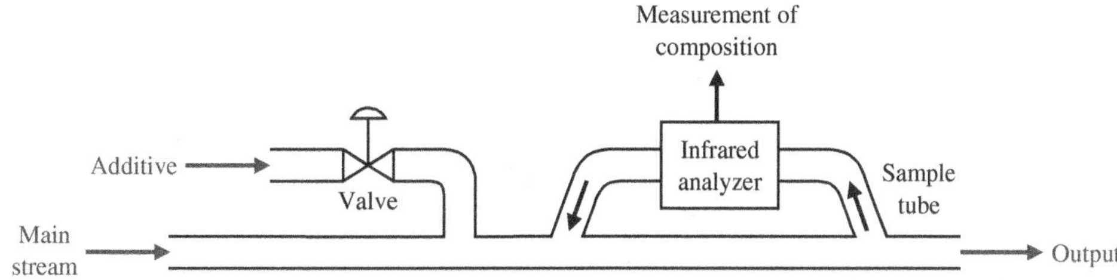

FIGURE P9.3 Chemical composition control.

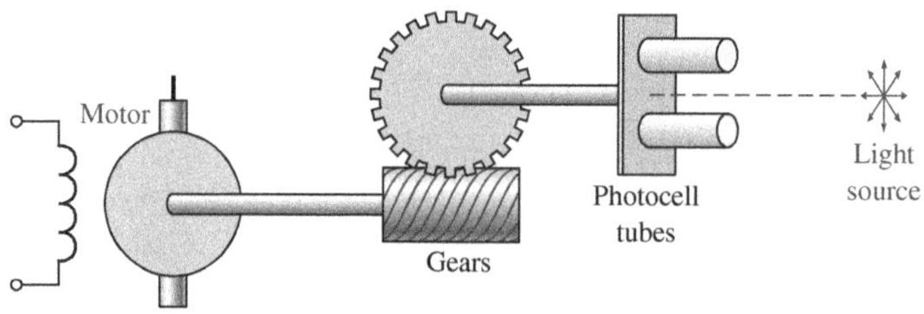

FIGURE P9.5 A photocell is mounted in each tube. The light reaching each cell is the same in both only when the light source is exactly in the middle as shown.

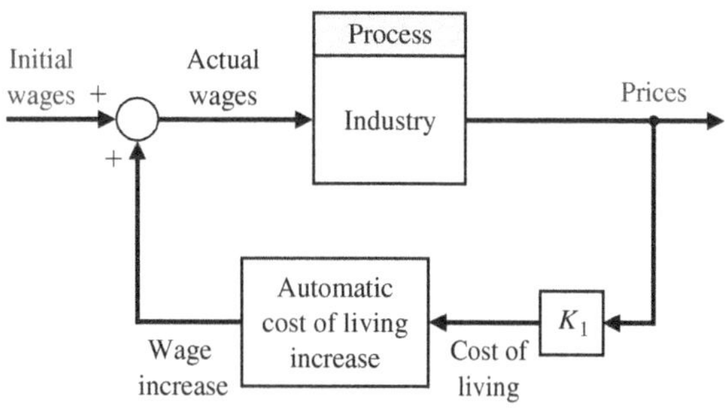

FIGURE P9.6 Positive feedback.

Figure P9.6, adds the feedback signal to the input signal, and the resulting signal is used as the input to the process. A simple model of the price–wage inflationary spiral is shown in Figure P9.6. Add additional feedback loops, such as legislative control or control of the tax rate, to stabilize the system. It is assumed that an increase in workers' salaries, after some time delay, results in an increase in prices. Under what conditions could prices be stabilized by falsifying or delaying the availability of cost-of-living data? How would a national wage and price economic guideline program affect the feedback system?

P9.7 The story is told about the sergeant who stopped at the jewelry store every morning at nine o'clock and compared and reset his watch with the chronometer in the window. Finally, one day the sergeant went into the store and complimented the owner on the accuracy of the chronometer.

"Is it set according to time signals from Arlington?" asked the sergeant.

"No," said the owner, "I set it by the five o'clock cannon fired from the fort each afternoon. Tell me, Sergeant, why do you stop every day and check your watch?"

The sergeant replied, "I'm the gunner at the fort!"

Is the feedback prevalent in this case positive or negative? The jeweler's chronometer loses two minutes each 24-hour period and the sergeant's watch loses three minutes during each eight hours. What is the net time error of the cannon at the fort after 12 days?

P9.8 The student–teacher learning process is inherently a feedback process intended to reduce the system error to a minimum. With the aid of Figure 9.3, construct a feedback model of the learning process and identify each block of the system.

P9.9 Models of physiological control systems are valuable aids to the medical profession. A model of the heart-rate control system is shown in Figure P9.9 [23, 24, 48]. This model includes the processing of the nerve signals by the brain. The heart-rate control

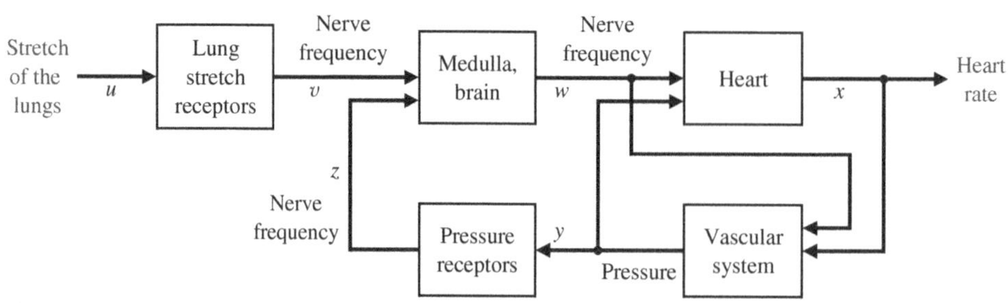

FIGURE P9.9 Heart-rate control.

system is, in fact, a multivariable system, and the variables x, y, w, v, z, and u are vector variables. In other words, the variable x represents many heart variables $x_1, x_2, \ldots, x_n$. Examine the model of the heart-rate control system and add or delete blocks, if necessary. Determine a control system model of one of the following physiological control systems:

1. Respiratory control system
2. Adrenaline control system
3. Human arm control system
4. Eye control system
5. Pancreas and the blood-sugar-level control system
6. Circulatory system

P9.10 The role of air traffic control systems is increasing as airplane traffic increases at busy airports. Engineers are developing air traffic control systems and collision avoidance systems using the Global Positioning System (GPS) navigation satellites [34, 55]. GPS allows each aircraft to know its position in the airspace landing corridor very precisely. Sketch a block diagram depicting how an air traffic controller might use GPS for aircraft collision avoidance.

P9.11 Automatic control of water level using a float level was used in the Middle East for a water clock [1, 11]. The water clock (Figure P9.11) was used from sometime before Christ until the 17th century. Discuss the operation of the water clock, and establish how the float provides a feedback control that maintains the accuracy of the clock. Sketch a block diagram of the feedback system.

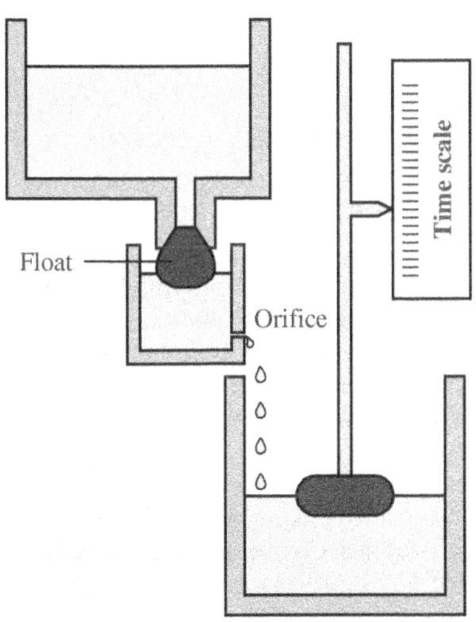

FIGURE P9.11 Water clock. (From Newton, Gould, and Kaiser, *Analytical Design of Linear Feedback Controls*. Wiley, New York, 1957, with permission.)

P9.12 An automatic turning gear for windmills was invented by Meikle in about 1750 [1, 11]. The fantail gear shown in Figure P9.12 automatically turns the windmill into the wind. The fantail windmill at right angle to the mainsail is used to turn the turret. The gear ratio is of the order of 3000 to 1. Discuss the operation of the windmill, and establish the feedback operation that maintains the main sails into the wind.

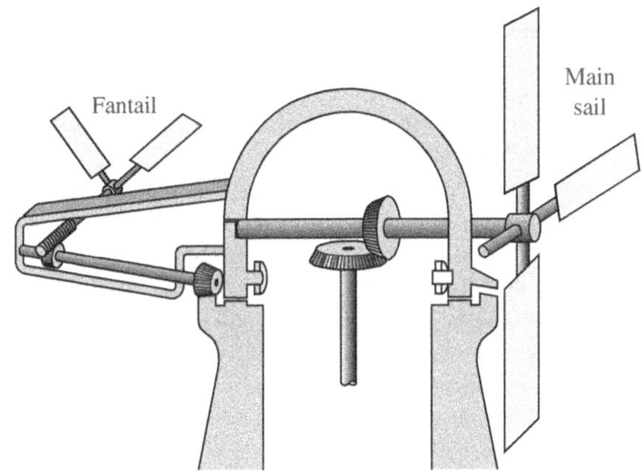

FIGURE P9.12 Automatic turning gear for windmills. (From Newton, Gould, and Kaiser, *Analytical Design of Linear Feedback Controls*. Wiley, New York, 1957, with permission.)

P9.13 A common example of a two-input control system is a home shower with separate valves for hot and cold water. The objective is to obtain (1) a desired temperature of the shower water and (2) a desired flow of water. Sketch a block diagram of the closed-loop control system.

P9.14 Adam Smith (1723–1790) discussed the issue of free competition between the participants of an economy in his book *Wealth of Nations*. It may be said that Smith employed social feedback mechanisms to explain his theories [41]. Smith suggests that (1) the available workers as a whole compare the various possible employments and enter that one offering the greatest rewards, and (2) in any employment the rewards diminish as the number of competing workers rises. Let r = total of rewards averaged over all trades, c = total of rewards in a particular trade, and q = influx of workers into the specific trade. Sketch a feedback system to represent this system.

P9.15 Small computers are used in automobiles to control emissions and obtain improved gas mileage. A computer-controlled fuel injection system that automatically adjusts the fuel–air mixture ratio could improve gas mileage and reduce unwanted polluting emissions significantly. Sketch a block diagram for such a system for an automobile.

P9.16 All humans have experienced a fever associated with an illness. A fever is related to the changing of the control input in the body's thermostat. This thermostat, within the brain, normally regulates temperature near 98°F in spite of external temperatures ranging from 0° to 100°F or more. For a fever, the input, or desired, temperature is increased. Even to many scientists, it often comes as a surprise to learn that fever does not indicate something wrong with body temperature control but rather well-contrived regulation at an elevated level of desired input. Sketch a block diagram of the temperature control system and explain how aspirin will lower a fever.

P9.17 Baseball players use feedback to judge a fly ball and to hit a pitch [35]. Describe a method used by a batter to judge the location of a pitch so that he can have the bat in the proper position to hit the ball.

P9.18 A cutaway view of a commonly used pressure regulator is shown in Figure P9.18. The desired pressure is set by turning a calibrated screw. This compresses the spring and sets up a force that opposes the upward motion of the diaphragm. The bottom side of the diaphragm is exposed to the water pressure that is to be controlled. Thus the motion of the diaphragm is an indication of the pressure difference between the desired and the actual pressures. It acts like a comparator. The valve is connected to the diaphragm and moves according to the pressure difference until it reaches a position in which the difference is zero. Sketch a block diagram showing the control system with the output pressure as the regulated variable.

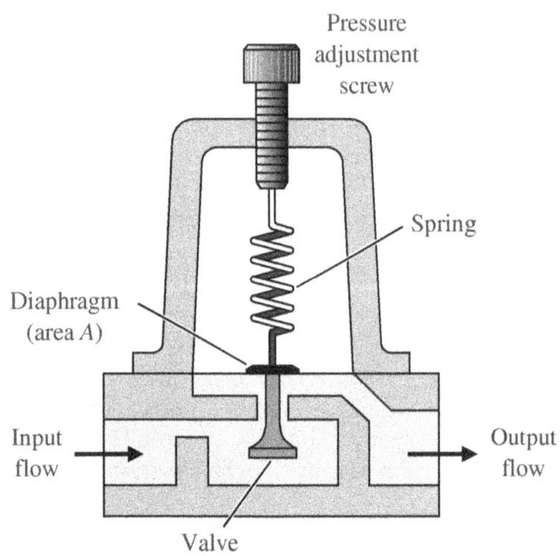

FIGURE P9.18 Pressure regulator.

P9.19 Ichiro Masaki of General Motors has patented a system that automatically adjusts a car's speed to keep a safe distance from vehicles in front. Using a video camera, the system detects and stores a reference image of the car in front. It then compares this image with a stream of incoming live images as the two cars move down the highway and calculates the distance. Masaki suggests that the system could control steering as well as speed, allowing drivers to lock on to the car ahead and get a "computerized tow." Sketch a block diagram for the control system.

P9.20 A high-performance race car with an adjustable wing (airfoil) is shown in Figure P9.20. Develop a block diagram describing the ability of the airfoil to keep a constant road adhesion between the car's tires and the race track surface. Why is it important to maintain good road adhesion?

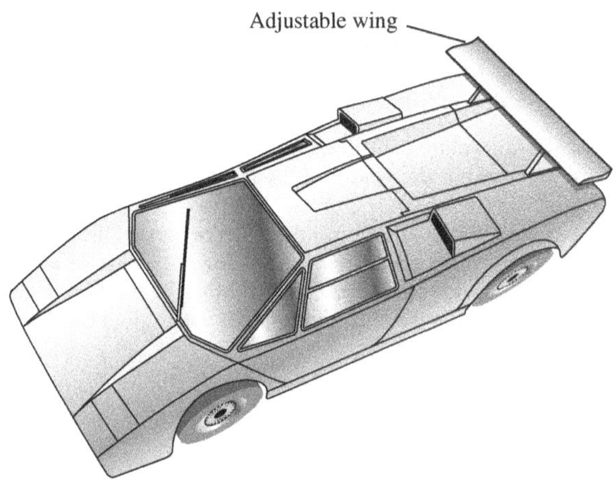

FIGURE P9.20 A high-performance race car with an adjustable wing.

P9.21 The potential of employing two or more helicopters for transporting payloads that are too heavy for a single helicopter is a well-addressed issue in the civil and military rotorcraft design arenas [37]. Overall requirements can be satisfied more efficiently with a smaller aircraft by using multilift for infrequent peak demands. Hence the principal motivation for using multilift can be attributed to the promise of obtaining increased productivity without having to manufacture larger and more expensive helicopters. A specific case of a multilift arrangement where two helicopters jointly transport payloads has been named **twin lift**. Figure P9.21

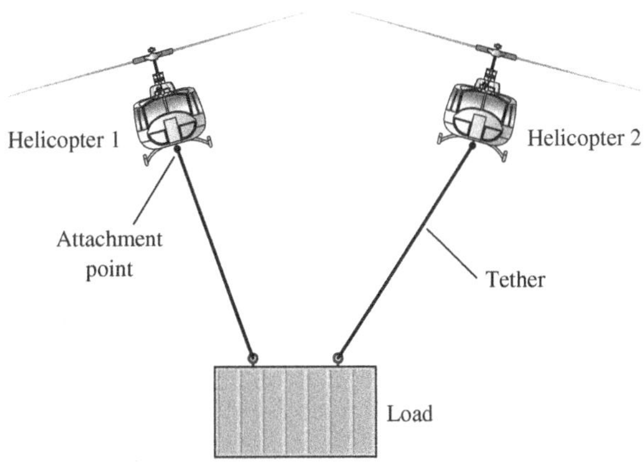

FIGURE P9.21 Two helicopters used to lift and move a large load.

shows a typical "two-point pendant" twin lift configuration in the lateral/vertical plane.

Develop the block diagram describing the pilots' action, the position of each helicopter, and the position of the load.

P9.22. Engineers want to design a control system that will allow a building or other structure to react to the force of an earthquake much as a human would. The structure would yield to the force, but only so much, before developing strength to push back [47]. Develop a block diagram of a control system to reduce the effect of an earthquake force.

P9.23 Engineers at the Science University of Tokyo are developing a robot with a humanlike face [52]. The robot can display facial expressions, so that it can work cooperatively with human workers. Sketch a block diagram for a facial expression control system of your own design.

P9.24 An innovation for an intermittent automobile windshield wiper is the concept of adjusting its wiping cycle according to the intensity of the rain [54]. Sketch a block diagram of the wiper control system.

P9.25 In the past 40 years, over 20,000 metric tons of hardware have been placed in Earth's orbit. During the same time span, over 15,000 metric tons of hardware returned to Earth. The objects remaining in Earth's orbit range in size from large operational spacecraft to tiny flecks of paint. There are about 150,000 objects in Earth's orbit 1 cm or larger in size. About 10,000 of the space objects are currently tracked from groundstations on the Earth. Space traffic control [61] is becoming an important issue, especially for commercial satellite companies that plan to "fly" their satellites through orbit altitudes where other satellites are operating, and through areas where high concentrations of space debris may exist. Sketch a block diagram of a space traffic control system that commercial companies might use to keep their satellites safe from collisions while operating in space.

P9.26 NASA is developing a compact rover designed to transmit data from the surface of an asteroid back to Earth, as illustrated in Figure P9.26. The rover will use a camera to take panoramic shots of the asteroid surface. The rover can position itself so that the camera can be pointed straight down at the surface or straight up at the sky. Sketch a block diagram illustrating how the microrover can be positioned to point the camera in the desired direction. Assume that the pointing commands are relayed from the Earth to the microrover and that the position of the camera is measured and relayed back to Earth.

FIGURE P9.26 Microrover designed to explore an asteroid. (Photo courtesy of NASA.)

P9.27 A direct methanol fuel cell is an electrochemical device that converts a methanol water solution to electricity [75]. Like rechargeable batteries, fuel cells directly convert chemicals to energy; they are very often compared to batteries, specifically rechargeable batteries. However, one significant difference between rechargeable batteries and direct methanol fuel cells is that, by adding more methanol water solution, the fuel cells recharge instantly. Sketch a block diagram of the direct methanol fuel cell recharging system that uses feedback (refer to Figure 9.3) to continuously monitor and recharge the fuel cell.

ADVANCED PROBLEMS

Advanced problems represent problems of increasing complexity.

AP9.1 The development of robotic microsurgery devices will have major implications on delicate eye and brain surgical procedures. The microsurgery devices employ feedback control to reduce the effects of the surgeon's muscle tremors. Precision movements by an articulated robotic arm can greatly help a surgeon by providing a carefully controlled hand. One such device is shown in Figure AP9.1. The microsurgical devices have been evaluated in clinical procedures and are now being commercialized. Sketch a block diagram of the surgical process with a microsurgical device in the loop being operated by a surgeon. Assume that the position of the end-effector on the microsurgical device can be measured and is available for feedback.

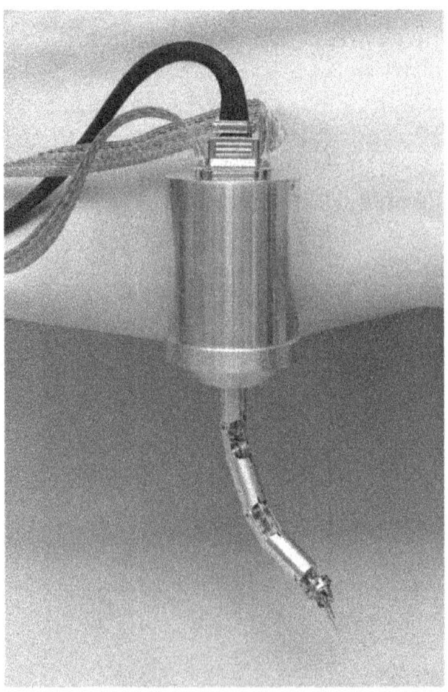

FIGURE AP9.1 Microsurgery robotic manipulator. (Photo courtesy of NASA.)

AP9.2 Advanced wind energy systems are being installed in many locations throughout the world as a way for nations to deal with rising fuel prices and energy shortages, and to reduce the negative effects of fossil fuel utilization on the quality of the air (refer to Example 9.2 in Section 9.6). The modern windmill can be viewed as a mechatronic system. Consider Figure 9.18, which illustrates the key elements of mechatronic systems. Using Figure 9.18 as a guide, think about how an advanced wind energy system would be designed as a mechatronic system. List the various components of the wind energy system and associate each component with one of the five elements of a mechatronic system: physical system modeling, signals and systems, computers and logic systems, software and data acquisition, and sensors and actuators.

AP9.3 Many modern luxury automobiles have an autopark option. This feature will parallel park an automobile without driver intervention. Figure AP9.3 illustrates the parallel parking scenario. Using Figure 9.3 as a model, sketch a block diagram of the automated parallel parking feedback control system. In your own words, describe the control problem and the challenges facing the designers of the control system.

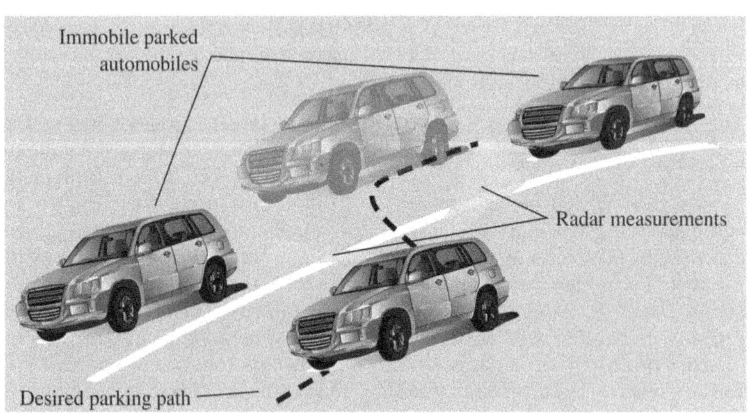

FIGURE AP9.3 Automated parallel parking of an automobile.

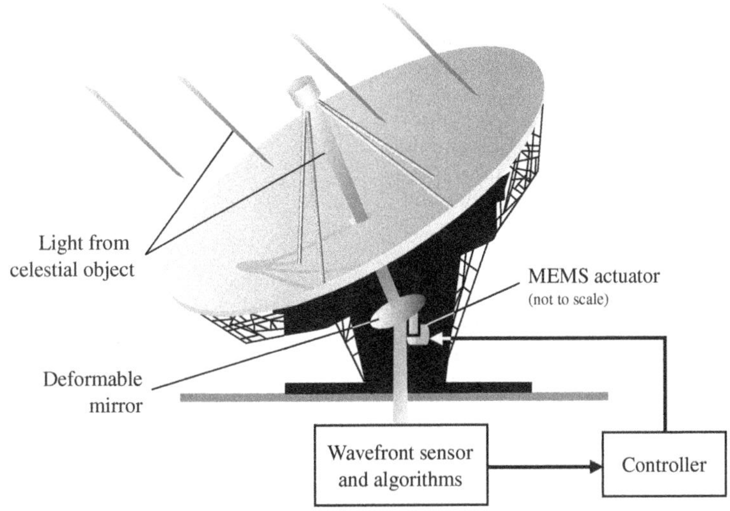

FIGURE AP9.4 Extremely large optical telescope with deformable mirrors for atmosphere compensation.

AP9.4 Adaptive optics has applications to a wide variety of key control problems, including imaging of the human retina and large-scale, ground-based astronomical observations [98]. In both cases, the approach is to use a wavefront sensor to measure distortions in the incoming light and to actively control and compensate to the errors induced by the distortions. Consider the case of an extremely large ground-based optical telescope, possibly an optical telescope up to 100 meters in diameter. The telescope components include deformable mirrors actuated by micro-electro-mechanical (MEMS) devices and sensors to measure the distortion of the incoming light as it passes through the turbulent and uncertain atmosphere of Earth.

There is at least one major technological barrier to constructing a 100-m optical telescope. The numerical computations associated with the control and compensation of the extremely large optical telescope can be on the order of 10^{10} calculations each 9.5 ms. To date, this computational power is unachievable. If we assume that the computational capability will ultimately be available, then one can consider the design of a feedback control system that uses the available computational power. We can consider many control issues associated with the large-scale optical telescope. Some of the controls problems that might be considered include controlling the pointing of the main dish, controlling the individual deformable mirrors, and attenuating the deformation of the dish due to changes in outside temperature.

Employing Figure 9.3 as a model for the block diagram, describe a closed-loop feedback control system to control one of the deformable mirrors to compensate for the distortions in the incoming light. Figure AP9.4 shows a diagram of the telescope with a single deformable mirror. Suppose that the mirror has an associated MEMS actuator that can be used to vary the orientation. Also, assume that the wavefront sensor and associated algorithms provide the desired configuration of the deformable mirror to the feedback control system.

AP9.5 The Burj Dubai is the tallest building in the world [94]. The building, shown in Figure AP9.5, stands at

FIGURE AP9.5 The world's tallest building in Dubai. (Photo courtesy of Alamy Images.)

over 800 m with more than 160 stories. There are 57 elevators servicing this tallest free-standing structure in the world. Traveling at up to 10 m/s, the elevators have the world's longest travel distance from lowest to highest stop. Describe a closed-loop feedback control system that guides an elevator of a high-rise building to a desired floor while maintaining a reasonable transit time [95]. Remember that high accelerations will make the passengers uncomfortable.

AP9.6 Control systems are aiding humans in maintaining their homes. The robotic vacuum cleaner depicted in Figure AP9.6 is an example of a mechatronic system under active control that relies on infrared sensors and microchip technology to navigate around furniture. Describe a closed-loop feedback control system that guides the robotic vacuum cleaner to avoid collisions with obstacles [96].

FIGURE AP9.6 A robotic vacuum cleaner communicates with the base station as it maneuvers around the room. (Photo courtesy of Alamy Images.)

DESIGN PROBLEMS

Design problems emphasize the design task. Continuous design problems (CDP) build upon a design problem from chapter to chapter.

CDP9.1 Increasingly stringent requirements of modern, high-precision machinery are placing increasing demands on slide systems [53]. The typical goal is to accurately control the desired path of the table shown in Figure CDP9.1. Sketch a block diagram model of a feedback system to achieve the desired goal. The table can move in the x direction as shown.

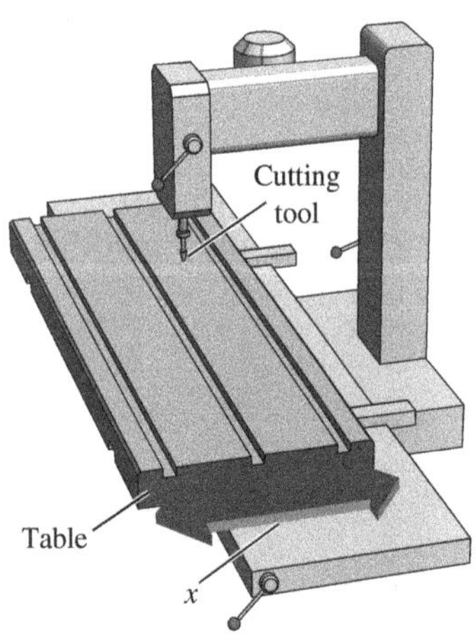

FIGURE CDP9.1 Machine tool with table.

DP9.1 The road and vehicle noise that invade an automobile's cabin hastens occupant fatigue [60]. Design the block diagram of an "antinoise" feedback system that will reduce the effect of unwanted noises. Indicate the device within each block.

DP9.2 Many cars are fitted with cruise control that, at the press of a button, automatically maintains a set speed. In this way, the driver can cruise at a speed limit or economic speed without continually checking the speedometer. Design a feedback-control in block diagram form for a cruise control system.

DP9.3 As part of the automation of a dairy farm, the automation of cow milking is under study [36]. Design a milking machine that can milk cows four or five times a day at the cow's demand. Sketch a block diagram and indicate the devices in each block.

DP9.4 A large, braced robot arm for welding large structures is shown in Figure DP9.4. Sketch the block diagram of a closed-loop feedback control system for accurately controlling the location of the weld tip.

DP9.5 Vehicle traction control, which includes antiskid braking and antispin acceleration, can enhance vehicle performance and handling. The objective of this control is to maximize tire traction by preventing locked brakes as well as tire spinning during acceleration. Wheel slip, the difference between the vehicle speed and the wheel speed, is chosen as the controlled variable because of its strong influence on the tractive force between the tire and the road [19]. The adhesion coefficient between the wheel and the road reaches a

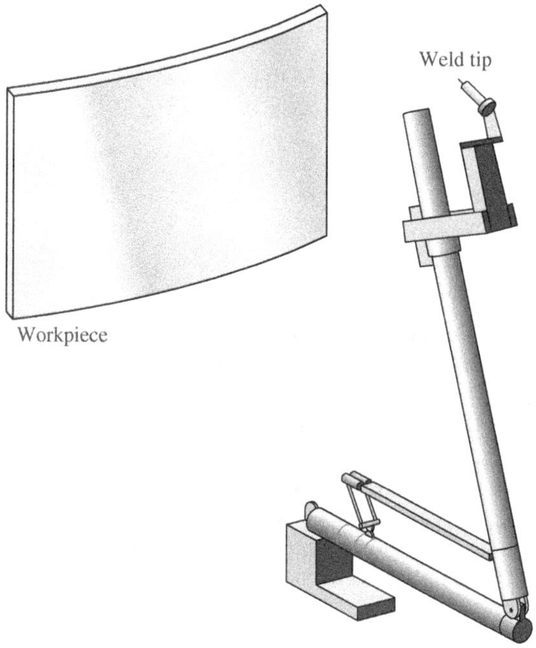

FIGURE DP9.4 Robot welder.

maximum at a low slip. Develop a block diagram model of one wheel of a traction control system.

DP9.6 The Hubble space telescope was repaired and modified in space on several occasions [44, 46, 49]. One challenging problem with controlling the Hubble is damping the jitter that vibrates the spacecraft each time it passes into or out of the Earth's shadow. The worst vibration has a period of about 20 seconds, or a frequency of 0.05 hertz. Design a feedback system that will reduce the vibrations of the Hubble space telescope.

DP9.7 A challenging application of control design is the use of nanorobots in medicine. Nanorobots will require onboard computing capability, and very tiny sensors and actuators. Fortunately, advances in biomolecular computing, bio-sensors, and actuators are promising to enable medical nanorobots to emerge within the next decade [99]. Many interesting medical applications will benefit from nanorobotics. For example, one use might be to use the robotic devices to precisely deliver anti-HIV drugs or to combat cancer by targeted delivering of chemotherapy.

At the present time, we cannot construct practical nanorobots, but we can consider the control design process that would enable the eventual development and installation of these tiny devices in the medical field. Consider the problem of designing a nanorobot to deliver a cancer drug to a specific location within the human body. The target site might be the location of a tumor, for example. Using the control design process illustrated in Figure 9.15, suggest one or more control goals that might guide the design process. Recommend the variables that should be controlled and provide a list of reasonable specifications for those variables.

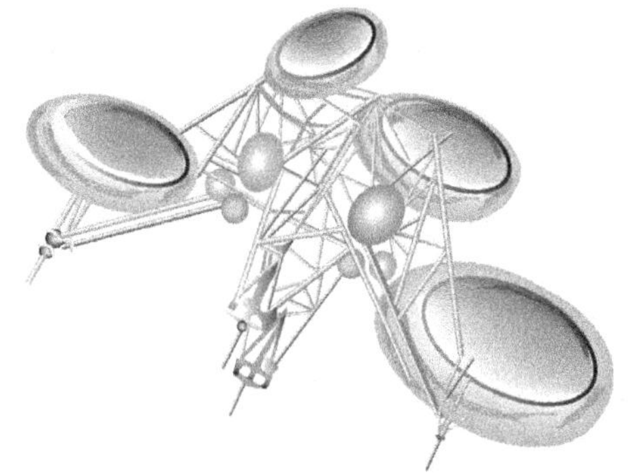

FIGURE DP9.7 An artist illustration of a nanorobot interacting with human blood cells.

DP9.8 Consider the human transportation vehicle (HTV) depicted in Figure DP9.8. The self-balancing HTV is actively controlled to allow safe and easy transportation of a single person [97]. Using Figure 9.3 as a model for the block diagram, describe a closed-loop feedback control system to assist the rider of the HTV in balancing and maneuvering the vehicle.

FIGURE DP9.8 Personal transportation vehicle. (Sergiy Kuzmin / Shutterstock)

ANSWERS TO SKILLS CHECK

True or False: (1) True; (2) True; (3) False; (4) False; (5) True

Multiple Choice: (6) d; (7) d; (8) b; (9) c; (10) a; (11) d; (12) a; (13) c; (14) d; (15) d

Word Match (in order, top to bottom): p, f, h, k, m, q, d, l, n, c, r, s, j, b, e, t, o, u, v, a, i, g

TERMS AND CONCEPTS

Automation The control of a process by automatic means.

Closed-loop feedback control system A system that uses a measurement of the output and compares it with the desired output to control the process.

Complexity of design The intricate pattern of interwoven parts and knowledge required.

Control system An interconnection of components forming a system configuration that will provide a desired response.

Design The process of conceiving or inventing the forms, parts, and details of a system to achieve a specified purpose.

Design gap A gap between the complex physical system and the design model intrinsic to the progression from the initial concept to the final product.

Disturbance An unwanted input signal that affects the output signal.

Embedded control Feedback control system that employs on-board special-purpose digital computers as integral components of the feedback loop.

Engineering design The process of designing a technical system.

Feedback signal A measure of the output of the system used for feedback to control the system.

Flyball governor A mechanical device for controlling the speed of a steam engine.

Hybrid fuel automobile An automobile that uses a conventional internal combustion engine in combination with an energy storage device to provide a propulsion system.

Measurement noise An unwanted input signal that affects the measured output signal.

Mechatronics The synergistic integration of mechanical, electrical, and computer systems.

Multiloop feedback control system A feedback control system with more than one feedback control loop.

Multivariable control system A system with more than one input variable or more than one output variable.

Negative feedback An output signal fed back so that it subtracts from the input signal.

Open-loop control system A system that uses a device to control the process without using feedback. Thus the output has no effect upon the signal to the process.

Optimization The adjustment of the parameters to achieve the most favorable or advantageous design.

Plant *See* Process.

Positive feedback An output signal fed back so that it adds to the input signal.

Process The device, plant, or system under control.

Productivity The ratio of physical output to physical input of an industrial process.

Risk Uncertainties embodied in the unintended consequences of a design.

Robot Programmable computers integrated with a manipulator. A reprogrammable, multifunctional manipulator used for a variety of tasks.

Specifications Statements that explicitly state what the device or product is to be and to do. A set of prescribed performance criteria.

Synthesis The process by which new physical configurations are created. The combining of separate elements or devices to form a coherent whole.

System An interconnection of elements and devices for a desired purpose.

Trade-off The result of making a judgment about how to compromise between conflicting criteria.

CHAPTER 10

Engineering Economics

The purpose of Chapter 10 is to present the concepts and principles of engineering economy.

Green Engineering in Action

Energy conservation comprises an important element in environmentally-conscious (green) engineering. In a Southeastern city, there are 310 traffic intersections that have been converted from incandescent lights to light-emitting diode (LED) lights. The study that led to this decision was conducted by the sustainability manager of the city. The wattage used at the intersections has been reduced from 150 watts to 15 watts at each traffic light. The resultant lighting bill has been lowered from $440,000 annually to $44,000 annually. When engineers went to check the traffic light meters for the first time, they were shocked by the low wattage numbers and the associated cost. One of them said, "We thought the meters were broken because the readings were so low." The annual savings of $396,000 per year from the traffic light conversion more than paid for the $150,000 cost of installing the LED lights. Chapter 10 introduces students to the decision-making process that accompanies "go/no go" evaluations of investments in engineering projects such as the one described above.

The best alternative may be the one you haven't yet discovered.

—Anonymous

Icons Used in This Book

Throughout this chapter, these two icons will appear in connection with numerous chapter opening materials, examples, and problems:

This icon identifies environmental (green) elements of the chapter. These elements pertain to engineering economy problems involving energy conservation, materials substitution, recycling, and other green situations.

VIDEO SOLUTIONS

This icon informs students of the availability of video tutorials for the examples and problems so marked. Students are encouraged to access the tutorials at `www.pearsonhighered.com/sullivan`. These icon-designated instances are intended to reinforce the learning of engineering economy through analogy with the marked problems and examples.

10.1 Introduction

The technological and social environments in which we live continue to change at a rapid rate. In recent decades, advances in science and engineering have transformed our transportation systems, revolutionized the practice of medicine, and miniaturized electronic circuits so that a computer can be placed on a semiconductor chip. The list of such achievements seems almost endless. In your science and engineering courses, you will learn about some of the physical laws that underlie these accomplishments.

The utilization of scientific and engineering knowledge for our benefit is achieved through the *design* of things we use, such as furnaces for vaporizing trash and structures for supporting magnetic railways. However, these achievements don't occur without a price, monetary or otherwise. Therefore, the purpose of this book is to develop and illustrate the principles and methodology required to answer the basic economic question of any design: Do its benefits exceed its costs?

The Accreditation Board for Engineering and Technology states that engineering "is the profession in which a knowledge of the mathematical and natural sciences gained by study, experience, and practice is applied with judgment to develop ways to utilize, economically, the materials and forces of nature for the benefit of mankind."* In this definition, the economic aspects of engineering are emphasized, as well as the physical aspects. Clearly, it is essential that the economic part of engineering practice be accomplished well. Thus, engineers use knowledge to find new ways of doing things economically.

* Accreditation Board of Engineering and Technology, *Criteria for Accrediting Programs in Engineering in the United States* (New York; Baltimore, MD: ABET, 1998).

Engineering economy involves the systematic evaluation of the economic merits of proposed solutions to engineering problems. To be economically acceptable (i.e., affordable), *solutions to engineering problems* must demonstrate a positive balance of long-term benefits over long-term costs, and they must also

- promote the well-being and survival of an organization,
- embody creative and innovative technology and ideas,
- permit identification and scrutiny of their estimated outcomes, and
- translate profitability to the "bottom line" through a valid and acceptable measure of merit.

Engineering economy is the dollars-and-cents side of the decisions that engineers make or recommend as they work to position a firm to be profitable in a highly competitive marketplace. Inherent to these decisions are trade-offs among different types of costs and the performance (response time, safety, weight, reliability, etc.) provided by the proposed design or problem solution. *The mission of engineering economy is to balance these trade-offs in the most economical manner.* For instance, if an engineer at Ford Motor Company invents a new transmission lubricant that increases fuel mileage by 10% and extends the life of the transmission by 30,000 miles, how much can the company afford to spend to implement this invention? Engineering economy can provide an answer.

A few more of the myriad situations in which engineering economy plays a crucial role in the analysis of project alternative come to mind:

1. Choosing the best design for a high-efficiency gas furnace
2. Selecting the most suitable robot for a welding operation on an automotive assembly line
3. Making a recommendation about whether jet airplanes for an overnight delivery service should be purchased or leased
4. Determining the optimal staffing plan for a computer help desk

From these illustrations, it should be obvious that engineering economy includes significant technical considerations. Thus, engineering economy involves technical analysis, with emphasis on the economic aspects, and has the objective of assisting decisions. This is true whether the decision maker is an engineer interactively analyzing alternatives at a computer-aided design workstation or the Chief Executive Officer (CEO) considering a new project. *An engineer who is unprepared to excel at engineering economy is not properly equipped for his or her job.*

10.2 The Principles of Engineering Economy

The development, study, and application of any discipline must begin with a basic foundation. We define the foundation for engineering economy to be a set of principles that provide a comprehensive doctrine for developing the methodology. These principles will be mastered by students as they progress. Once a problem or need has been clearly defined, the foundation of the discipline can be discussed in terms of seven principles.

PRINCIPLE 1

Develop the Alternatives

Carefully define the problem! Then the choice (decision) is among alternatives. The alternatives need to be identified and then defined for subsequent analysis.

A decision situation involves making a choice among two or more alternatives. Developing and defining the alternatives for detailed evaluation is important because of the resulting impact on the quality of the decision. Engineers and managers should place a high priority on this responsibility. Creativity and innovation are essential to the process.

One alternative that may be feasible in a decision situation is making no change to the current operation or set of conditions (i.e., doing nothing). If you judge this option feasible, make sure it is considered in the analysis. However, do not focus on the status quo to the detriment of innovative or necessary change.

PRINCIPLE 2

Focus on the Differences

Only the differences in expected future outcomes among the alternatives are relevant to their comparison and should be considered in the decision.

If all prospective outcomes of the feasible alternatives were exactly the same, there would be no basis or need for comparison. We would be indifferent among the alternatives and could make a decision using a random selection.

Obviously, only the differences in the future outcomes of the alternatives are important. Outcomes that are common to all alternatives can be disregarded in the comparison and decision. For example, if your feasible housing alternatives were two residences with the same purchase (or rental) price, price would be inconsequential to your final choice. Instead, the decision would depend on other factors, such as location and annual operating and maintenance expenses. This simple example illustrates Principle 2, which emphasizes the basic purpose of an engineering economic analysis: to recommend a future course of action based on the differences among feasible alternatives.

PRINCIPLE 3

Use a Consistent Viewpoint

The prospective outcomes of the alternatives, economic and other, should be consistently developed from a defined viewpoint (perspective).

The perspective of the decision maker, which is often that of the owners of the firm, would normally be used. However, it is important that the viewpoint for the particular decision be first defined and then used consistently in the description, analysis, and comparison of the alternatives.

As an example, consider a public organization operating for the purpose of developing a river basin, including the generation and wholesale distribution of electricity from dams on the river system. A program is being planned to upgrade and increase the capacity of the power generators at two sites. What perspective should be used in defining the technical alternatives for the program? The "owners of the firm" in this example means the segment of the public that will pay the cost of the program, and their viewpoint should be adopted in this situation.

Now let us look at an example where the viewpoint may not be that of the owners of the firm. Suppose that the company in this example is a private firm and that the problem deals with providing a flexible benefits package for the employees. Also, assume that the feasible alternatives for operating the plan all have the same future costs to the company. The alternatives, however, have differences from the perspective of the employees, and their satisfaction is an important decision criterion. The viewpoint for this analysis should be that of the employees of the company as a group, and the feasible alternatives should be defined from their perspective.

PRINCIPLE 4 Use a Common Unit of Measure

Using a common unit of measurement to enumerate as many of the prospective outcomes as possible will simplify the analysis of the alternatives.

It is desirable to make as many prospective outcomes as possible *commensurable* (directly comparable). For economic consequences, a monetary unit such as dollars is the common measure. You should also try to translate other outcomes (which do not initially appear to be economic) into the monetary unit. This translation, of course, will not be feasible with some of the outcomes, but the additional effort toward this goal will enhance commensurability and make the subsequent analysis of alternatives easier.

What should you do with the outcomes that are not economic (i.e., the expected consequences that cannot be translated (and estimated) using the monetary unit)? First, if possible, quantify the expected future results using an appropriate unit of measurement for each outcome. If this is not feasible for one or more outcomes, describe these consequences explicitly so that the information is useful to the decision maker in the comparison of the alternatives.

PRINCIPLE 5 Consider All Relevant Criteria

Selection of a preferred alternative (decision making) requires the use of a criterion (or several criteria). The decision process should consider both the outcomes enumerated in the monetary unit and those expressed in some other unit of measurement or made explicit in a descriptive manner.

The decision maker will normally select the alternative that will best serve the long-term interests of the owners of the organization. In engineering economic analysis, the primary criterion relates to the long-term financial interests of the owners. This is based on the assumption that available capital will be allocated to provide maximum monetary return to the owners. Often, though, there are other organizational objectives you would like to achieve with your decision, and these should be considered and given weight in the selection of an alternative. These nonmonetary attributes and multiple objectives become the basis for additional criteria in the decision-making process.

PRINCIPLE 6 Make Risk and Uncertainty Explicit

Risk and uncertainty are inherent in estimating the future outcomes of the alternatives and should be recognized in their analysis and comparison.

The analysis of the alternatives involves projecting or estimating the future consequences associated with each of them. The magnitude and the impact of future outcomes of any course of action are uncertain. Even if the alternative involves no change from current operations, the probability is high that today's estimates of, for example, future cash receipts and expenses will not be what eventually occurs. Thus, dealing with uncertainty is an important aspect of engineering economic analysis.

PRINCIPLE 7 Revisit Your Decisions

Improved decision making results from an adaptive process; to the extent practicable, the initial projected outcomes of the selected alternative should be subsequently compared with actual results achieved.

A good decision-making process can result in a decision that has an undesirable outcome. Other decisions, even though relatively successful, will have results significantly different from the initial estimates of the consequences. Learning from and adapting based on our experience are essential and are indicators of a good organization.

The evaluation of results versus the initial estimate of outcomes for the selected alternative is often considered impracticable or not worth the effort. Too often, no feedback to the decision-making process occurs. Organizational discipline is needed to ensure that implemented decisions are routinely postevaluated and that the results are used to improve future analyses and the quality of decision making. For example, a common mistake made in the comparison of alternatives is the failure to examine adequately the impact of uncertainty in the estimates for selected factors on the decision. Only postevaluations will highlight this type of weakness in the engineering economy studies being done in an organization.

10.3 Engineering Economy and the Design Process

An engineering economy study is accomplished using a structured procedure and mathematical modeling techniques. The economic results are then used in a decision situation that normally includes other engineering knowledge and input.

A sound engineering economic analysis procedure incorporates the basic principles discussed in Section 10.2 and involves several steps. We represent the procedure in terms of the *seven steps* listed in the left-hand column of Table 10-1. There are several feedback loops (not shown) within the procedure. For example, within Step 1, information developed in evaluating the problem will be used as feedback to refine the problem definition. As another example, information from the analysis of alternatives (Step 5) may indicate the need to change one or more of them or to develop additional alternatives.

The seven-step procedure is also used to assist decision making within the engineering design process, shown as the right-hand column in Table 10-1. In this case, activities in the design process contribute information to related steps in the economic analysis procedure. The general relationship between the activities in the design process and the steps of the economic analysis procedure is indicated in Table 10-1.

The engineering design process may be repeated in phases to accomplish a total design effort. For example, in the first phase, a full cycle of the process may be undertaken to select a conceptual or preliminary design alternative. Then, in the second phase, the activities are repeated to develop the preferred detailed design based on the selected preliminary design. The seven-step economic analysis

TABLE 10-1 The General Relationship between the Engineering Economic Analysis Procedure and the Engineering Design Process

Engineering Economic Analysis Procedure	Engineering Design Process (see Figure P10-15 on p. 350)
Step	*Activity*
1. Problem recognition, definition, and evaluation.	1. Problem/need definition.
	2. Problem/need formulation and evaluation.
2. Development of the feasible alternatives.	3. Synthesis of possible solutions (alternatives).
3. Development of the outcomes and cash flows for each alternative. 4. Selection of a criterion (or criteria). 5. Analysis and comparison of the alternatives.	4. Analysis, optimization, and evaluation.
6. Selection of the preferred alternative.	5. Specification of preferred alternative.
7. Performance monitoring and post-evaluation of results.	6. Communication.

procedure would be repeated as required to assist decision making in each phase of the total design effort. This procedure is discussed next.

10.3.1 Problem Definition

The first step of the engineering economic analysis procedure (problem definition) is particularly important, since it provides the basis for the rest of the analysis. A problem must be well understood and stated in an explicit form before the project team proceeds with the rest of the analysis.

The term *problem* is used here generically. It includes all decision situations for which an engineering economy analysis is required. Recognition of the problem is normally stimulated by internal or external organizational needs or requirements. An operating problem within a company (internal need) or a customer expectation about a product or service (external requirement) are examples.

Once the problem is recognized, its formulation should be viewed from a *systems perspective*. That is, the boundary or extent of the situation needs to be carefully defined, thus establishing the elements of the problem and what constitutes its environment.

Evaluation of the problem includes refinement of needs and requirements, and information from the evaluation phase may change the original formulation of the problem. In fact, redefining the problem until a consensus is reached may be the most important part of the problem-solving process!

10.3.2 Development of Alternatives*

The two primary actions in Step 2 of the procedure are (1) searching for potential alternatives and (2) screening them to select a smaller group of feasible alternatives for detailed analysis. The term *feasible* here means that each alternative selected for further analysis is judged, based on preliminary evaluation, to meet or exceed the requirements established for the situation.

10.3.2.1 Searching for Superior Alternatives In the discussion of Principle 1 (Section 10.2), creativity and resourcefulness were emphasized as being absolutely essential to the development of potential alternatives. The difference between good alternatives and great alternatives depends largely on an individual's or group's *problem-solving efficiency*. Such efficiency can be increased in the following ways:

1. Concentrate on redefining one problem at a time in Step 1.
2. Develop many redefinitions for the problem.
3. Avoid making judgments as new problem definitions are created.
4. Attempt to redefine a problem in terms that are dramatically different from the original Step 1 problem definition.

* This is sometimes called *option development*. This important step is described in detail in A. B. Van Gundy, *Techniques of Structured Problem Solving*, 2nd ed. (New York: Van Nostrand Reinhold Co., 1988). For additional reading, see E. Lumsdaine and M. Lumsdaine, *Creative Problem Solving—An Introductory Course for Engineering Students* (New York: McGraw-Hill Book Co., 1990) and J. L. Adams, *Conceptual Blockbusting—A Guide to Better Ideas* (Reading, MA: Addison-Wesley Publishing Co., 1986).

5. Make sure that the *true problem* is well researched and understood.

In searching for superior alternatives or identifying the true problem, several limitations invariably exist, including (1) lack of time and money, (2) preconceptions of what will and what will not work, and (3) lack of knowledge. Consequently, the engineer or project team will be working with less-than-perfect problem solutions in the practice of engineering.

EXAMPLE 10-1 Defining the Problem and Developing Alternatives

The management team of a small furniture-manufacturing company is under pressure to increase profitability to get a much-needed loan from the bank to purchase a more modern pattern-cutting machine. One proposed solution is to sell waste wood chips and shavings to a local charcoal manufacturer instead of using them to fuel space heaters for the company's office and factory areas.

(a) Define the company's problem. Next, reformulate the problem in a variety of creative ways.

(b) Develop at least one potential alternative for your reformulated problems in Part (a). (Don't concern yourself with feasibility at this point.)

Solution

(a) The company's problem appears to be that revenues are not sufficiently covering costs. Several reformulations can be posed:

1. The problem is to increase revenues while reducing costs.
2. The problem is to maintain revenues while reducing costs.
3. The problem is an accounting system that provides distorted cost information.
4. The problem is that the new machine is really not needed (and hence there is no need for a bank loan).

(b) Based only on reformulation 1, an alternative is to sell wood chips and shavings as long as increased revenue exceeds extra expenses that may be required to heat the buildings. Another alternative is to discontinue the manufacture of specialty items and concentrate on standardized, high-volume products. Yet another alternative is to pool purchasing, accounting, engineering, and other white-collar support services with other small firms in the area by contracting with a local company involved in providing these services.

10.3.2.2 Developing Investment Alternatives "It takes money to make money," as the old saying goes. Did you know that in the United States the average firm spends over $250,000 in capital on each of its employees? So, to make money, each firm must invest capital to support its important human resources—but in what else should an individual firm invest? There are usually hundreds of opportunities for a company to make money. Engineers are at the very heart of creating value for a firm by turning innovative and creative ideas into new or reengineered commercial products and services. Most of these ideas require investment of money, and only a few of all feasible ideas can be developed, due to lack of time, knowledge, or resources.

Consequently, most investment alternatives created by good engineering ideas are drawn from a larger population of equally good problem solutions. But how can this larger set of equally good solutions be tapped into? Interestingly, studies have concluded that designers and problem solvers tend to pursue a few ideas that involve "patching and repairing" an old idea.* Truly new ideas are often excluded from consideration! This section outlines two approaches that have found wide acceptance in industry for developing sound investment alternatives by removing some of the barriers to creative thinking: (1) classical brainstorming and (2) the Nominal Group Technique (NGT).

(1) Classical Brainstorming. Classical brainstorming is the most well-known and often-used technique for idea generation. It is based on the fundamental principles of *deferment of judgment* and that *quantity breeds quality*. There are four rules for successful brainstorming:

1. Criticism is ruled out.
2. Freewheeling is welcomed.
3. Quantity is wanted.
4. Combination and improvement are sought.

A. F. Osborn lays out a detailed procedure for successful brainstorming.† A classical brainstorming session has the following basic steps:

1. ***Preparation.*** The participants are selected, and a preliminary statement of the problem is circulated.
2. ***Brainstorming.*** A warm-up session with simple unrelated problems is conducted, the relevant problem and the four rules of brainstorming are presented, and ideas are generated and recorded using checklists and other techniques if necessary.
3. ***Evaluation.*** The ideas are evaluated relative to the problem.

Generally, a brainstorming group should consist of four to seven people, although some suggest larger groups.

(2) Nominal Group Technique. The NGT, developed by Andre P. Delbecq and Andrew H. Van de Ven,‡ involves a structured group meeting designed to incorporate individual ideas and judgments into a group consensus. By correctly applying the NGT, it is possible for groups of people (preferably, 5 to 10) to generate investment alternatives or other ideas for improving the competitiveness of the

* S. Finger and J. R. Dixon, "A Review of Research in Mechanical Engineering Design. Part I: Descriptive, Prescriptive, and Computer-Based Models of Design Processes," in *Research in Engineering Design* (New York: Springer-Verlag, 1990).

† A. F. Osborn, *Applied Imagination*, 3rd ed. (New York: Charles Scribner's Sons, 1963). Also refer to P. R. Scholtes, B. L. Joiner, and B. J. Streibel, *The Team Handbook*, 2nd ed. (Madison, WI: Oriel Inc., 1996).

‡ A. Van de Ven and A. Delbecq, "The Effectiveness of Nominal, Delphi, and Interactive Group Decision Making Processes," *Academy of Management Journal* 17, no. 4 (December 1974): 605–21.

firm. Indeed, the technique can be used to obtain group thinking (consensus) on a wide range of topics. For example, a question that might be given to the group is, "What are the most important problems or opportunities for improvement of . . .?"

The technique, when properly applied, draws on the creativity of the individual participants, while reducing two undesirable effects of most group meetings: (1) the dominance of one or more participants and (2) the suppression of conflicting ideas. The basic format of an NGT session is as follows:

1. Individual silent generation of ideas
2. Individual round-robin feedback and recording of ideas
3. Group clarification of each idea
4. Individual voting and ranking to prioritize ideas
5. Discussion of group consensus results

The NGT session begins with an explanation of the procedure and a statement of question(s), preferably written by the facilitator.* The group members are then asked to prepare individual listings of alternatives, such as investment ideas or issues that they feel are crucial for the survival and health of the organization. This is known as the silent-generation phase. After this phase has been completed, the facilitator calls on each participant, in round-robin fashion, to present one idea from his or her list (or further thoughts as the round-robin session is proceeding). Each idea (or opportunity) is then identified in turn and recorded on a flip chart or board by the NGT facilitator, leaving ample space between ideas for comments or clarification. This process continues until all the opportunities have been recorded, clarified, and displayed for all to see. At this point, a voting procedure is used to prioritize the ideas or opportunities. Finally, voting results lead to the development of group consensus on the topic being addressed.

10.3.3 Development of Prospective Outcomes

Step 3 of the engineering economic analysis procedure incorporates Principles 2, 3, and 4 from Section 10.2 and uses the basic *cash-flow approach* employed in engineering economy. A cash flow occurs when money is transferred from one organization or individual to another. Thus, a cash flow represents the economic effects of an alternative in terms of money spent and received.

Consider the concept of an organization having only one "window" to its external environment through which *all* monetary transactions occur—receipts of revenues and payments to suppliers, creditors, and employees. The key to developing the related cash flows for an alternative is estimating what would happen to the revenues and costs, as seen at this window, if the particular alternative were implemented. The *net cash flow* for an alternative is the difference between all cash inflows (receipts or savings) and cash outflows (costs or expenses) during each time period.

* A good example of the NGT is given in D. S. Sink, "Using the Nominal Group Technique Effectively," *National Productivity Review*, 2 (Spring 1983): 173–84.

In addition to the economic aspects of decision making, *nonmonetary factors (attributes)* often play a significant role in the final recommendation. Examples of objectives other than profit maximization or cost minimization that can be important to an organization include the following:

1. Meeting or exceeding customer expectations
2. Safety to employees and to the public
3. Improving employee satisfaction
4. Maintaining production flexibility to meet changing demands
5. Meeting or exceeding all environmental requirements
6. Achieving good public relations or being an exemplary member of the community

10.3.4 Selection of a Decision Criterion

The selection of a decision criterion (Step 4 of the analysis procedure) incorporates Principle 5 (consider all relevant criteria). The decision maker will normally select the alternative that will best serve the long-term interests of the owners of the organization. It is also true that the economic decision criterion should reflect a consistent and proper viewpoint (Principle 3) to be maintained throughout an engineering economy study.

10.3.5 Analysis and Comparison of Alternatives

Analysis of the economic aspects of an engineering problem (Step 5) is largely based on cash-flow estimates for the feasible alternatives selected for detailed study. A substantial effort is normally required to obtain reasonably accurate forecasts of cash flows and other factors in view of, for example, inflationary (or deflationary) pressures, exchange rate movements, and regulatory (legal) mandates that often occur. Clearly, the consideration of future uncertainties (Principle 6) is an essential part of an engineering economy study. When cash flow and other required estimates are eventually determined, alternatives can be compared based on their differences as called for by Principle 2. Usually, these differences will be quantified in terms of a monetary unit such as dollars.

10.3.6 Selection of the Preferred Alternative

When the first five steps of the engineering economic analysis procedure have been done properly, the preferred alternative (Step 6) is simply a result of the total effort. Thus, the soundness of the technical-economic modeling and analysis techniques dictates the quality of the results obtained and the recommended course of action. Step 6 is included in Activity 5 of the engineering design process (specification of the preferred alternative) when done as part of a design effort.

10.3.7 Performance Monitoring and Postevaluation of Results

This final step implements Principle 7 and is accomplished during and after the time that the results achieved from the selected alternative are collected. Monitoring project performance during its operational phase improves the achievement of related goals and objectives and reduces the variability in desired results. Step 7 is also the follow-up step to a previous analysis, comparing actual results achieved with the previously estimated outcomes. The aim is to learn how to do better analyses, and the feedback from postimplementation evaluation is important to the continuing improvement of operations in any organization. Unfortunately, like Step 1, this final step is often not done consistently or well in engineering practice; therefore, it needs particular attention to ensure feedback for use in ongoing and subsequent studies.

EXAMPLE 10-2 Application of the Engineering Economic Analysis Procedure

A friend of yours bought a small apartment building for \$100,000 in a college town. She spent \$10,000 of her own money for the building and obtained a mortgage from a local bank for the remaining \$90,000. The *annual* mortgage payment to the bank is \$10,500. Your friend also expects that annual maintenance on the building and grounds will be \$15,000. There are four apartments (two bedrooms each) in the building that can each be rented for \$360 per month.

Refer to the seven-step procedure in Table 10-1 (left-hand side) to answer these questions:

(a) Does your friend have a problem? If so, what is it?

(b) What are her alternatives? (Identify at least three.)

(c) Estimate the economic consequences and other required data for the alternatives in Part (b).

(d) Select a criterion for discriminating among alternatives, and use it to advise your friend on which course of action to pursue.

(e) Attempt to analyze and compare the alternatives in view of at least one criterion in addition to cost.

(f) What should your friend do based on the information you and she have generated?

Solution

(a) A quick set of calculations shows that your friend does indeed have a problem. A lot more money is being spent by your friend each year (\$10,500 + \$15,000 = \$25,500) than is being received (4 × \$360 × 12 = \$17,280). The problem could be that the monthly rent is too low. She's losing \$8,220 per year. (Now, that's a problem!)

(b) Option (1). Raise the rent. (Will the market bear an increase?)
Option (2). Lower maintenance expenses (but not so far as to cause safety problems).
Option (3). Sell the apartment building. (What about a loss?)
Option (4). Abandon the building (bad for your friend's reputation).

(c) Option (1). Raise total monthly rent to \$1,440 + \$*R* for the four apartments to cover monthly expenses of \$2,125. Note that the minimum increase in rent would be (\$2,125 – \$1,440)/4 = \$171.25 per apartment per month (almost a 50% increase!).

Option (2). Lower monthly expenses to \$2,125 – \$C so that these expenses are covered by the monthly revenue of \$1,440 per month. This would have to be accomplished primarily by lowering the maintenance cost. (There's not much to be done about the annual mortgage cost unless a favorable refinancing opportunity presents itself.) Monthly maintenance expenses would have to be reduced to (\$1,440 – \$10,500/12) = \$565. This represents more than a 50% decrease in maintenance expenses.

Option (3). Try to sell the apartment building for \$X, which recovers the original \$10,000 investment and (ideally) recovers the \$685 per month loss (\$8,220 ÷ 12) on the venture during the time it was owned.

Option (4). Walk away from the venture and kiss your investment good-bye. The bank would likely assume possession through foreclosure and may try to collect fees from your friend. This option would also be very bad for your friend's credit rating.

(d) One criterion could be to minimize the expected loss of money. In this case, you might advise your friend to pursue Option (1) or (3).

(e) For example, let's use "credit worthiness" as an additional criterion. Option (4) is immediately ruled out. Exercising Option (3) could also harm your friend's credit rating. Thus, Options (1) and (2) may be her only realistic and acceptable alternatives.

(f) Your friend should probably do a market analysis of comparable housing in the area to see if the rent could be raised (Option 1). Maybe a fresh coat of paint and new carpeting would make the apartments more appealing to prospective renters. If so, the rent can probably be raised while keeping 100% occupancy of the four apartments.

A tip to the wise—as an aside to Example 10-2, your friend would need a good credit report to get her mortgage approved. In this regard, there are three major credit bureaus in the United States: Equifax, Experian, and TransUnion. It's a good idea to regularly review your own credit report for unauthorized activity. You are entitled to a free copy of your report once per year from each bureau. Consider getting a report every four months from `www.annualcreditreport.com`.

EXAMPLE 10-3 Get Rid of the Old Clunker?

Engineering economy is all about deciding among competing alternatives.

Consider this situation: Linda and Jerry are faced with a car replacement opportunity where an interest rate can be ignored. Jerry's old clunker that averages 10 miles per gallon (mpg) of gasoline can be traded in toward a vehicle that gets 15 mpg. Or, as an alternative, Linda's 25 mpg car can be traded in toward a new hybrid vehicle that averages 50 mpg. If they drive both cars 12,000 miles per year and their goal is to minimize annual gas consumption, which car should be replaced—Jerry's or Linda's? They can only afford to upgrade one car at this time.

Solution

Jerry's trade-in will save (12,000 miles/year)/10 mpg − (12,000 miles/year)/15 mpg = 1,200 gallons/year − 800 gallons/year = 400 gallons/year.

Linda's trade-in will save (12,000 miles/year)/25 mpg − (12,000 miles/year)/50 mpg = 480 gallons/year − 240 gallons/year = 240 gallons/year. Therefore, Jerry should trade in his vehicle to save more gasoline.

10.4 Using Spreadsheets in Engineering Economic Analysis

Spreadsheets are a useful tool for solving engineering economy problems. Most engineering economy problems are amenable to spreadsheet solution for the following reasons:

1. They consist of structured, repetitive calculations that can be expressed as formulas that rely on a few functional relationships.
2. The parameters of the problem are subject to change.
3. The results and the underlying calculations must be documented.
4. Graphical output is often required, as well as control over the format of the graphs.

Spreadsheets allow the analyst to develop an application rapidly, without being inundated by the housekeeping details of programming languages. They relieve the analyst of the drudgery of number crunching but still focus on problem formulation. Computer spreadsheets created in Excel are integrated throughout all chapters in this book.

10.5 Try Your Skills

The number in parentheses that follows each problem refers to the section from which the problem is taken. Solutions to these problems can be found in the Appendix.

1-A. For every penny that the price of gasoline goes up, the U.S. Postal Service (USPS) experiences a monthly fuel cost increase of $8 million. State what assumptions you need to make to answer this question: "How many mail delivery vehicles does the USPS have in the United States?"

1-B. Assume that your employer is a manufacturing firm that produces several different electronic consumer products. What are five nonmonetary factors (attributes) that may be important when a significant change is considered in the design of the current bestselling product? **(10.2, 10.3)**

10.6 Summary

In this chapter, we defined engineering economy and presented the fundamental concepts in terms of seven basic principles (see pp. 335–338). Experience has shown that most errors in engineering economic analyses can be traced to some violation of these principles. We will continue to stress these principles in the chapters that follow.

The seven-step engineering economic analysis procedure described in this chapter (see p. 339) has direct ties to the engineering design process. Following this systematic approach will assist engineers in designing products and systems and in providing technical services that promote the economic welfare of the company they work for. This same approach will also help you as an individual make sound financial decisions in your personal life.

In summary, engineering economy is a collection of problem-solving tools and techniques that are applied to engineering, business, and environmental issues. Common, yet often complex, problems involving money are easier to understand and solve when you have a good grasp on the engineering economy approach to problem solving and decision making. The problem-solving focus of this text will enable you to master the theoretical and applied principles of engineering economy.

Problems

The number in parentheses that follows each problem refers to the section from which the problem is taken.

10-1. Stan Moneymaker needs 15 gallons of gasoline to top off his automobile's gas tank. If he drives an extra eight miles (round trip) to a gas station on the outskirts of town, Stan can save \$0.10 per gallon on the price of gasoline. Suppose gasoline costs \$3.90 per gallon and Stan's car gets 25 mpg for in-town driving. Should Stan make the trip to get less expensive gasoline? Each mile that Stan drives creates one pound of carbon dioxide. Each pound of CO_2 has a cost impact of \$0.02 on the environment. What other factors (cost and otherwise) should Stan consider in his decision making? **(10.2)**

10-2. The decision was made by NASA to abandon rocket-launched payloads into orbit around the earth. We must now rely on the Russians for this capability. Use the principles of engineering economy to examine this decision. **(10.2)**

10-3. A typical discounted price of a AAA battery is \$0.75. It is designed to provide 1.5 volts and 1.0 amps for about an hour. Now we multiply volts and amps to obtain power of 1.5 watts from the battery. Thus, it costs \$0.75 for 1.5 Watt-hours of energy. How much would it cost to deliver one kilo Watt-hour? How does this compare with the cost of energy from your local electric utility at \$0.10 per kilo Watt-hour? **(10.2, 10.3)**

10-4. Tyler just wrecked his new Nissan, and the accident was his fault. The owner of the other vehicle got two estimates for the repairs: one was for \$803 and the other was for \$852. Tyler is thinking of keeping the insurance companies out of the incident to keep his driving record "clean." Tyler's deductible on his comprehensive coverage insurance is \$500, and he does not want his premium to increase because of the accident. In this regard, Tyler estimates that his semiannual premium will rise by \$60 if he files a claim against his insurance company. In view of the above information, Tyler's initial decision is to write a personal check for \$803 payable to the owner of the other vehicle.

Did Tyler make the most economical decision? What other options should Tyler have explored? In your answer, be sure to state your assumptions and quantify your thinking. **(10.3)**

10-5. Henry Ford's Model T was originally designed and built to run on ethanol. Today, ethanol (190-proof alcohol) can be produced with domestic stills for about \$0.85 per gallon. When blended with gasoline costing \$4.00 per gallon, a 20% ethanol and 80% gasoline mixture costs \$3.37 per gallon. Assume fuel consumption at 25 mpg and engine performance in general are not adversely affected with this 20–80 blend (called E20). **(10.3)**

a. How much money can be saved for 15,000 miles of driving per year?

b. How much gasoline per year is being converted if one million people use the E20 fuel?

10-6. The Russian air force is being called on this year to intercept storms advancing on Moscow and to seed them with dry ice and silver iodine particles. The idea is to make the snow drop on villages in the countryside instead of piling up in Moscow. The cost of this initiative will be 180 million rubles, and the savings in snow removal will be in the neighborhood of 300 million rubles. The exchange rate is 30 rubles per dollar. Comment on the hidden costs and benefits of such a plan from the viewpoint of the villagers in terms of dollars. **(10.2)**

10-7. A large electric utility company has proposed building an \$820 million combined cycle, gas-powered plant to replace the electric generation capacity at one of its coal-fired facilities. Develop three other alternatives for replacing this electric generation capacity.

10-8. Studies have concluded that a college degree is a very good investment. Suppose that a college graduate earns about 75% more money per hour than a high-school graduate. If the lifetime earnings of a high-school graduate average \$1,200,000, what is the expected value of earning a college degree? **(10.3)**

10-9. Automobile repair shops typically recommend that their customers change their oil and oil filter every 3,000 miles. Your automobile user's manual suggests changing your oil every 5,000–7,000 miles. If you drive your car 15,000 miles each year and an oil and filter change costs \$30, how much money would you save each year if you had this service performed every 5,000 miles? **(10.3)**

10-10. Often it makes a lot of sense to spend some money now so you can save more money in the future. Consider filtered water. A high-tech water filter cost about \$60 and can filter 7,200 ounces of water. This will save you purchasing two 20-ounce bottle of filtered water every day, each costing \$1.15. The filter will need replacing every 6 months. How much will this filter save you in a year's time?

10-11. The manufacturer of Brand A automobile tires claims that its tire can save 110 gallons of fuel over 55,000 miles of driving, as compared to a popular competitor (Brand B). If gasoline costs \$4.00 per gallon, how much per mile driven does this tire save the customer (Brand A versus Brand B)?

10-12. During your first month as an employee at Greenfield Industries (a large drill-bit manufacturer), you are asked to evaluate alternatives for producing a newly designed drill bit on a turning machine. Your boss' memorandum to you has practically no information about what the alternatives are and what criteria should be used. The same task was posed to a previous employee who could not finish the analysis, but she has given you the following information: An old turning machine valued at \$350,000 exists (in the warehouse) that can be modified for the new drill bit. The in-house technicians have given an estimate of \$40,000 to modify this machine, and they assure you that they will have the machine ready before the projected start date (although they have never done any modifications of this type). It is hoped that the old turning machine will be able to meet production requirements at full capacity. An outside company, McDonald Inc., made the machine seven years ago and can easily do the same modifications for \$60,000. The cooling system used for this machine is not environmentally safe and would require some disposal costs. McDonald Inc. has offered to build a new turning machine with more environmental safeguards and higher capacity for a price of \$450,000. McDonald Inc. has promised this machine before the startup date and is willing to pay any late costs. Your company has \$100,000 set aside for the start-up of the new product line of drill bits. For this situation,

a. Define the problem.

b. List key assumptions.

c. List alternatives facing Greenfield Industries.

d. Select a criterion for evaluation of alternatives.

e. Introduce risk into this situation.
f. Discuss how nonmonetary considerations may impact the selection.
g. Describe how a postaudit could be performed.

10-13. *The Almost-Graduating Senior Problem.* Consider this situation faced by a first-semester senior in civil engineering who is exhausted from extensive job interviewing and penniless from excessive partying. Mary's impulse is to accept immediately a highly attractive job offer to work in her brother's successful manufacturing company. She would then be able to relax for a year or two, save some money, and then return to college to complete her senior year and graduate. Mary is cautious about this impulsive desire, because it may lead to no college degree at all!

a. Develop at least two formulations for Mary's problem.
b. Identify feasible solutions for each problem formulation in Part (a). *Be creative!*

10-14. While studying for the engineering economy final exam, you and two friends find yourselves craving a fresh pizza. You can't spare the time to pick up the pizza and must have it delivered. "Pick-Up-Sticks" offers a 1-1/4-inch-thick (including toppings), 20-inch square pizza with your choice of two toppings for $15 plus 5% sales tax and a $1.50 delivery charge (no sales tax on delivery charge). "Fred's" offers the round, deep-dish Sasquatch, which is 20 inches in diameter. It is 1-3/4 inches thick, includes two toppings, and costs $17.25 plus 5% sales tax and free delivery.

a. What is the problem in this situation? Please state it in an explicit and precise manner.
b. Systematically apply the seven principles of engineering economy (pp. 335–338) to the problem you have defined in Part (a).
c. Assuming that your common unit of measure is dollars (i.e., cost), what is the better value for getting a pizza based on the criterion of *minimizing cost per unit of volume?*
d. What other criteria might be used to select which pizza to purchase?

10-15. Storm doors have been installed on 50% of all homes in Anytown, USA. The remaining 50% of homeowners without storm doors think they may have a problem that a storm door could solve, but they're not sure. Use Activities 1, 2, and 3 in the engineering design process (Table 10-1) to help these homeowners systematically think through the definition of their need (Activity 1), a formal statement of their problem (Activity 2), and the generation of alternatives (Activity 3).

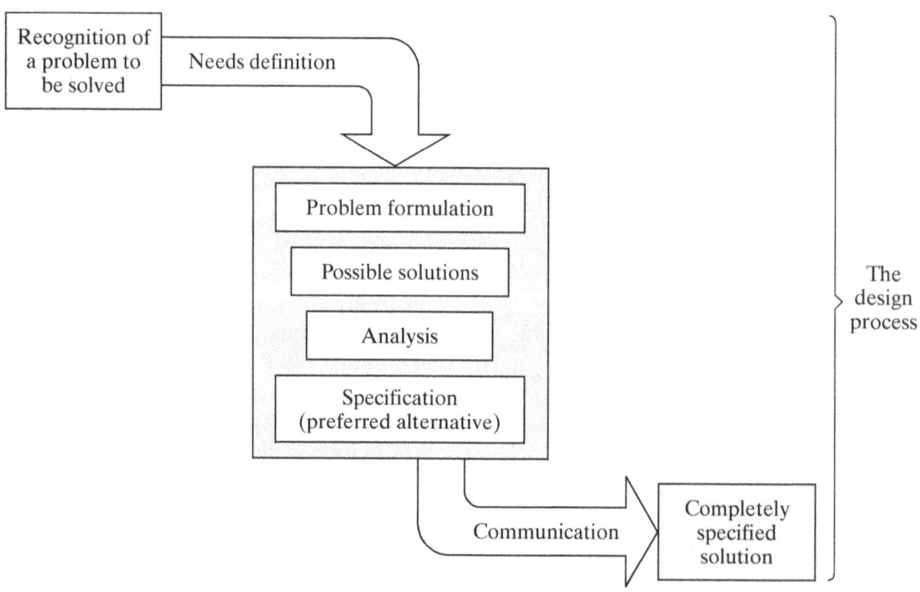

Figure P10-15 Figure for Problem 10-15

The design process begins in Figure P10-15 with a statement of need and terminates with the specifications for a means of fulfilling that need.

10-16. *Extended Learning Exercise.*
Bad news: You have just wrecked your car! You need another car immediately because you have decided that walking, riding a bike, and taking a bus are not acceptable. An automobile wholesaler offers you \$2,000 for your wrecked car "as is." Also, your insurance company's claims adjuster estimates that there is \$2,000 in damages to your car. Because you have collision insurance with a \$1,000 deductibility provision, the insurance company mails you a check for \$1,000. The odometer reading on your wrecked car is 58,000 miles.

What should you do? Use the seven-step procedure from Table 10-1 to analyze your situation. Also, identify which principles accompany each step.

10-17. "What you do at work is your boss' business" is a timely warning for all employees to heed. Last year, your company installed a new computer surveillance program in an effort to improve office productivity. As a courtesy, all employees were informed of this change. The license for the software costs \$30,000 per year. After a year of use, productivity has risen 10%, which translates into a savings of \$30,000. Discuss other factors, in addition to productivity, that could have been used to justify the surveillance software.

10-18. Owing to the rising cost of copper, in 1982 the U.S. Mint changed the composition of pennies from 95% copper (and 5% zinc) to 2.5% copper (and 97.5% zinc) to save money. Your favorite aunt has a collection of 5,000 pennies minted before 1982, and she intends on gifting the collection to you.

a. What is the collection's value based on metal content alone? Copper sells for \$3.50 per pound and zinc for \$1 per pound. It takes approximately 130 pre-1982 pennies to add up to one pound of total weight.

b. If it cost the U.S. Mint \$0.017 to produce a penny in 2012, is it time to eliminate pennies and round off all financial transactions to the nearest 5 cents (nickel)? As a matter of interest, it cost the government almost 10 cents to produce a nickel in 2012.

10-19. A home mortgage is "under water" when the amount of money owed on it is much greater than (say, twice) the market value of the home. Discuss the economic and ethical issues of walking away from (i.e., defaulting on) an underwater loan. Assume you have \$10,000 equity in the home and your monthly payments are \$938. **(10.3)**

10-20. A deep-water oil rig has just collapsed into the Gulf of Mexico. Its blowout-preventer system has failed, so thousands of barrels of crude oil each day are gushing into the ocean. List some alternatives for stopping the unchecked flow of oil into the Gulf. **(10.3)**

10-21. Energy can be conserved when your home heating/cooling system works less during the heating and cooling seasons. In fact a one degree Fahrenheit difference in your thermostat setting can reduce energy consumption by up to 5%. Identify the assumptions necessary to make this statement valid for heating and cooling a 2,000 square foot home. **(10.3)**

11 OPERATIONS RESEARCH

11.1 INTRODUCTION—DEFINITION

Operations Research has been defined by the Operational Research Society of the United Kingdom as follows:

> The attack of modem science on complex problems arising in the direction and *management of large systems of men, machines, materials, and money in industry, business, government, and defense*. The distinctive approach is to develop a *scientific model* of the system, incorporating measurement of factors such as chance and risk, with which *to predict and compare outcomes of alternative decisions, strategies, or controls*. The purpose is to *help management determine its policies and actions scientifically*.

Let's examine the italicized terms more closely. First, the definition says that for operations research to be applicable, there must be systems of men, machines, materials, and money and that the techniques of O.R. are applicable to systems in industry, business, government, and defense. We can say, then, that O.R. is applicable just about anywhere there are systems that need to be managed.

The next italicized phrase, a "scientific model," is perhaps the key phrase in the definition. This implies that unless a scientific model (usually mathematical) is developed, it cannot be called operations research. Notice that many of the techniques we have discussed in previous chapters definitely use scientific models.

The next italicized phrase says that the objectives are to "predict and compare the outcomes of alternative decisions, strategies, or controls." This implies that any scientific model that predicts or evaluates the results of decisions or policies is operations research.

Finally, the overall objective of operations research is well stated by the last italicized phrase, "help management determine its policies and actions scientifically." It is significant that we say to *help* management. No O.R. tool "makes" decisions. It is only an aid to the decision maker.

11.2 SIMILARITY TO INDUSTRIAL ENGINEERING

The definition of industrial engineering, as adopted by the Institute of Industrial Engineers, is as follows:

> Industrial engineering is concerned with the design, improvement, and installation of *integrated systems of people, materials, information, equipment, and energy.* It draws upon specialized knowledge and skill in the *mathematical, physical, and social sciences* together with the principles and methods of engineering analysis and design to *specify, predict, and evaluate* the results to be obtained from such systems.

It is obvious that I.E. and O.R. have commonalities, as signified by the similarity in the definitions (note especially the italicized terms). It is the authors' opinion that operations research and industrial engineering indeed do have many of the same objectives and work on many of the same problems. The primary difference is that operations research has a higher level of theoretical and mathematical orientation, providing a major portion of the science base of industrial engineering.

Operations research carries a connotation of mathematical orientation, but industrial engineering does not restrict itself to any specific approach. Figure 11.1 illustrates the relationship between the two disciplines in terms of mathematical sophistication.

Many industrial engineers work in the area of operations research, as do mathematicians, statisticians, physicists, sociologists, and others. It is significant to observe that many O.R. programs in universities are taught by industrial engineering faculty and many O.R. programs are taught by mathematics and statistics faculty. Some universities have individual programs called Operations Research that are taught by Operations Research faculty.

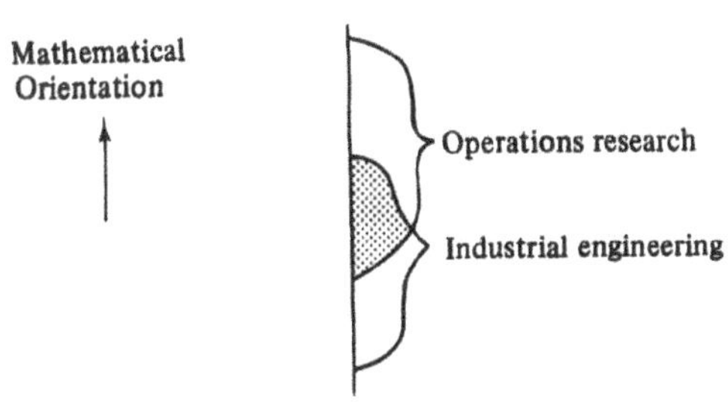

Fig. 11.1. Industrial engineering and operations research.

11.3 NATURE OF OPERATIONS RESEARCH

Since we have already touched upon this subject, let's explore the concept of mathematical orientation in some depth. We have said that a chief distinguishing characteristic of O.R. is the mathematical involvement. Let's look at some of the models we have already studied under industrial engineering.

11.3.1 Economic Order Quantity

We developed a mathematical model depicting the total variable cost of an inventory situation. That equation is repeated below:

$$TC = PC\frac{D}{Q} + CC\frac{Q}{2} \tag{11.1}$$

We took a first derivative, set it equal to 0, and solved for the optimum order quantity Q_0

$$Q_0 = \sqrt{\frac{(2)(PC)(D)}{CC}} \tag{11.2}$$

11.3.2 Plant Location

We developed a mathematical programming formulation of a plant location problem and stated that it could be solved by using a technique consisting of linear programming and branch and bound search (although we did not study them). That formulation is repeated below:

$$\text{Minimize } TC = \sum_{i=1}^{m}\sum_{j=1}^{n} c_{ij}\, x_{ij} + \sum_{j=1}^{n} k_j y_j \tag{11.3}$$

subject to

$$\sum_{i=1}^{m} x_{ij} \le my_j \qquad (j = 1,\ldots,n) \tag{11.4}$$

$$\sum_{j=1}^{n} x_{ij} = 1 \qquad (i = 1),\ldots,m) \tag{11.5}$$

$$y_j = 0,1 \qquad (j = 1,\ldots,n) \tag{11.6}$$

$$x_{ij} \ge 0 \qquad (i = 1,\ldots,m \text{ and } j = 1,\ldots,n) \tag{11.7}$$

11.3.3 Job Evaluation

In both point rating and factor comparison methods of job evaluation we assumed an underlying mathematical model that says that a measure of a job's total complexity can be obtained by *summing* the measures of complexity for each subfactor.

11.3.4 Quality Control

In the construction of $\overline{x}$ charts we assumed the probability distribution of sample means can be represented by the normal distribution where the normal distribution is defined (in Appendix A) as

$$f(x) = \frac{1}{\sqrt{2\pi}\,\sigma} e^{-(x-\mu)^2/2\sigma^2}$$

11.3.5 Others

This discussion could be extensively continued. The transportation programming problem, the assignment problem, the computerized plant layout packages (scoring), and the traveling salesman problem and routing problems are just a few other examples of industrial engineering techniques with mathematical orientations.

Given all these formulations, would you say that they are industrial engineering or that they are operations research? Actually, they are both and would appear in the shaded area of Figure 11.1. A flow process chart might be an example of a technique that could be called industrial engineering, but not operations research (although we could argue that a flow process chart is a *scientific model* of a real-world situation). Research into new mathematical techniques could be called O.R. but not I.E., although the research is often done by industrial engineers.

11.4 CATEGORIZATION OF OPERATIONS RESEARCH

Let's explore the nature of operations research in an attempt to categorize the techniques. In every situation that we wish to model using an O.R. approach we encounter the problem of estimating the values of the parameters (factors such as the price of raw material or time to produce a part). Those parameters may not be constant over time. That is, they may behave as random variables or perhaps change in some predictable fashion. One approach is to forget that they are random variables and use a mathematical model that does not recognize variation. This approach is called the *deterministic* approach and is often used. Actually, all models we have seen thus far have been deterministic. If the model recognizes this random variation, the approach is called the *probabilistic* approach. We can categorize O.R. techniques then as deterministic or probabilistic, and this is done in many I.E. and O.R. curricula in which the first course(s) explores deterministic models.

As an example (grossly oversimplified) of these two approaches, consider the inventory problem. We used a deterministic model to determine the optimum order quantity Q_0. Now let's determine when to order if we assume that there is a lead time (delay from the time the order is issued until the material is received) involved. We shall use both a deterministic and a probabilistic approach.

11.4.1 Deterministic Approach

The model represented the inventory situation when there was zero lead time and no variation in demand. If we assume that the lead time can be specified by l and that it is deterministic (does not change), we can draw the figure for stock on hand as shown in Figure 11.2. The figure simply shows that the time to issue an order must be exactly l time units before the stock on hand runs out. This also specifies the inventory at that reorder time as

$$\text{RQ} = lD$$

where l is the lead time in units of time and D is the demand per unit of time.

11.4.2 Probabilistic Approach

Now let's assume that the demand is still constant and deterministic but that the lead time is a random variable with a mean of l. This means that we can't be sure how long after we issue an order the order will be received, but we do know that it averages l time units.

Figure 11.3 depicts the situation that might occur if we calculated the reorder quantity and time as before. We can see from Figure 11.3 that many things can occur when random variation is allowed. In the inventory situation the results usually include excess inventory at times and stockouts at other times (Figure 11.3 assumes that the customer waits and his order is filled as soon as the material arrives).

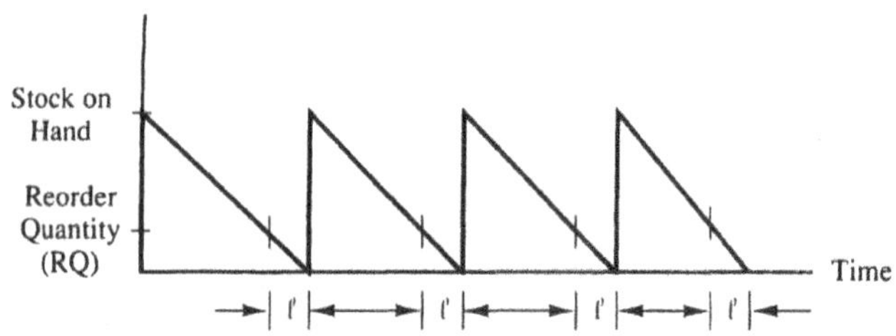

Fig. 11.2. Deterministic lead time.

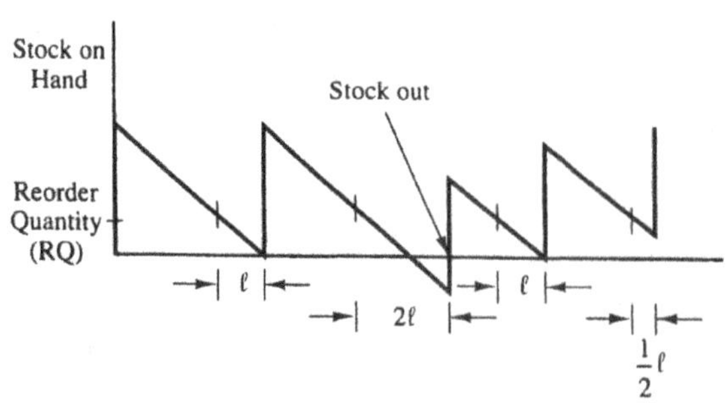

Fig. 11.3. Probabilistic lead time.

Any time random variation is allowed, the model is probabilistic; any time random variation is not allowed, the model is deterministic. This chapter discusses some elementary and basic deterministic models.

11.5 DETERMINISTIC OPERATIONS RESEARCH

In this section and the rest of the chapter we shall cover some of the deterministic tools of operations research. We certainly cannot even mention all the areas of deterministic O.R., but we can provide a brief contact with almost all of the important ones and maybe a little insight into a few of the simpler ones. Almost all industrial engineering curricula contain several courses that cover the material presented.

11.6 MATHEMATICAL PROGRAMMING

It is possible to present one general problem formulation that generalizes all deterministic operations research techniques. That problem formulation is called the *mathematical programming problem* and is presented below:

$$\text{Optimize } f(x) \tag{11.8}$$

subject to

$$g_i(x) \le b_i \qquad (1 = 1, \ldots, m) \tag{11.9}$$

Here, we have a mathematical function [$f(x)$] that is to be optimized in some manner (probably minimized or maximized). The function involves some decision variables x, where x is a vector of N components $(x_1, x_2, \ldots, x_N)$. There are m constraints that restrict the choice or combination of variables so that the solution is nontrivial. (Of course, m could be zero and no constraints could be present such that an unconstrained problem like the inventory control order quantity problem results.)

There could also be other constraints. For example, some or all of the x's could be restricted to integer quantities or just to values of 0 or 1. The decision variables could be restricted to only positive values (as is usually the case), or other special restrictions could be present. Finally, the constraints are not limited to less than or equal to inequalities.

For example,

$$x_1 + x_2 \ge 5 \tag{11.10}$$

can be shown as

$$-x_1 - x_2 \le -5 \tag{11.11}$$

and still meet the definition.

11.7 UNCONSTRAINED OPTIMIZATION

The unconstrained optimization problem is one studied in calculus and is simply a mathematical programming problem where $m = 0$ (no constraints are present). The inventory control problem was just that, since the objective was to minimize the total cost equation given by

$$TC = P\frac{D}{Q} + CC\frac{Q}{2} \tag{11.12}$$

The only constraint present is that positive quantities of Q are required, but we solved the problem without recognizing the constraint and found that there would always be a positive answer.

The unconstrained optimization problem is

$$\text{Optimize } f(x) \tag{11.13}$$

with no constraints. Calculus says that a *necessary condition* for x^* to be an optimum point of $f(x)$ is that the first derivative must be equal to zero.

$$\left.\frac{df(x)}{dx}\right|_{x = x^*} = 0 \tag{11.14}$$

A point x^* satisfying Equation (11.14) is called a *stationary point*. Since we know that it is possible for the derivative to be equal to zero and the point to be a point of inflection, the first derivative being equal to zero is not enough. We then say it is *sufficient* for x^* to be an optimum point if the second derivative at x^* is positive or negative. The point x^* is a minimum point if the second derivative is positive and a maximum point if the second derivative is negative. If the second derivative is zero, then higher-order derivatives must be taken until the first nonzero one is found at, say, the nth derivative.

$$\left.\frac{d^n f}{dx^n}\right|_{x = x^*} \neq 0 \tag{11.15}$$

(1) If n is odd, x^* is a point of inflection.
(2) If n is even and

$$\left.\frac{d^n f}{dx^n}\right|_{x = x^*} < 0 \tag{11.16}$$

then x^* is a local maximum. If

$$\left.\frac{d^n f}{dx^n}\right|_{x = x^*} > 0 \tag{11.17}$$

then x^* is a local minimum.

As an example, consider the inventory control problem of Equation (11.12). The *necessary* condition is that

$$\left.\frac{dTC}{dQ}\right|_{Q=Q^*} = \frac{-(PC)(D)}{Q^2} + \frac{CC}{2} = 0 \tag{11.18}$$

and the *sufficient* condition for Q^* to be a minimum point is that

$$\left.\frac{d^2TC}{dQ^2}\right|_{Q=Q^*} = \frac{2(PC)(D)}{Q^3} > 0 \tag{11.19}$$

which is greater than zero for all positive quantities of Q. Therefore, any positive Q^* that satisfies the necessary conditions of Equation (11.18) is a minimum point.

As another example, suppose that the percent of undesirable by-product in a chemical reaction is a function of time as represented by[1]

$$f(t) = 3t^4 - 4t^3 + 2 \tag{11.20}$$

Suppose further that we wish to minimize this percent. The necessary condition is that

$$\frac{df(t)}{dt} = 12t^3 - 12t^2 = 0 \tag{11.21}$$

There are two points that satisfy this relationship:

$$t = 0 \qquad \text{and} \qquad t = 1$$

The second derivative is given by

$$\frac{d^2f(t)}{dt^2} = 36t^2 - 24t \tag{11.22}$$

At $t = 1$,

$$\frac{d^2f(t)}{dt^2} = 12 \tag{11.23}$$

Therefore, $t = 1$ is a *minimum point.*

At $t = 0$,

$$\frac{d^2f(t)}{dt^2} = 0 \tag{11.24}$$

[1] Extracted from William W. Claycombe and William G. Sullivan, *Foundations of Mathematical Programming* (Reston, Va.: Reston Publishing Co., Inc., 1975).

Therefore we don't know what it is and we should take higher-order derivatives until one is nonzero.

At $n = 3$ and $t = 0$,

$$\frac{d^3 f(t)}{dt^3} = 72t - 24 = -24 \tag{11.25}$$

Now, since $n = 3$ is odd, the rule states that $t = 0$ is a point of inflection.

For a geometrical interpretation of the above, consider Figures 11.4, 11.5 and 11.6. In all the figures, x^* satisfies the necessary condition that the first derivative equals zero. In Figure 11.4 the second derivative is positive at x^*; therefore, x^* is a minimum point. Note also that the second derivative would be positive at all points. A function like this is said to be *convex* and any stationary point is a minimum point for a convex function.

In Figure 11.5 the second derivative is negative for all values of x; therefore, any stationary point should be a maximum. There is only one stationary point and that is x^* which is a maximum point. A function whose second derivative is negative for all values of x is said to be *concave*.

Figure 11.6 is slightly different. We can see that for any $x < x^*$, the second derivative is negative; for any $x > x^*$, the second derivative is positive; and at $x = x^*$, the second derivative is zero. If continued, we would find that some higher-order nth derivative, where n is odd, evaluated at x^* would be nonzero, signifying that x^* is a point of inflection, which it is.

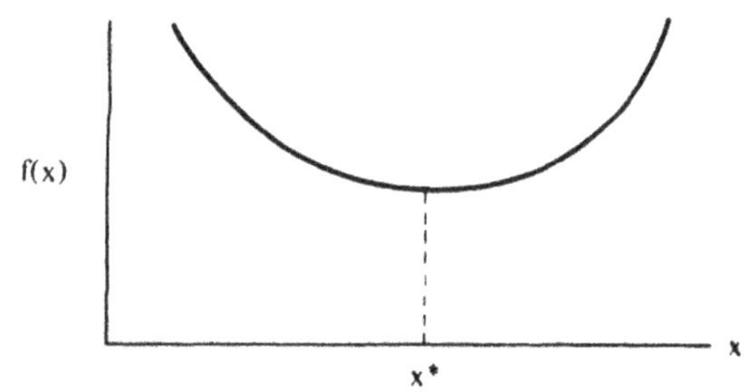

Fig. 11.4. Minimum point-single variable.

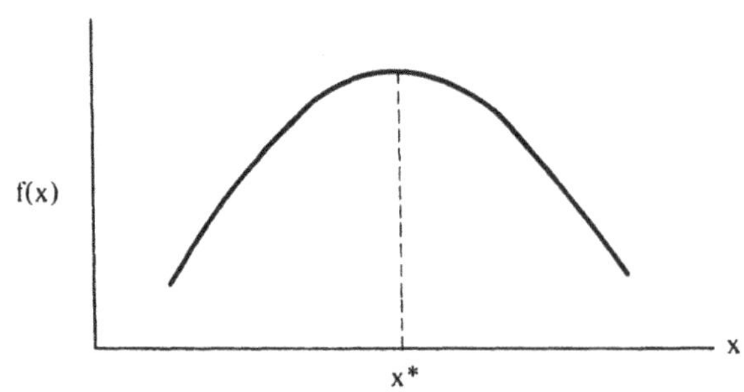

Fig. 11.5. Maximum point-single variable.

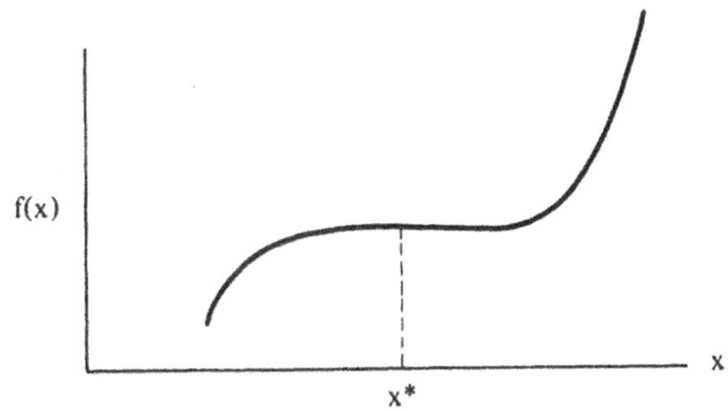

Fig. 11.6. Point of inflection—single variable.

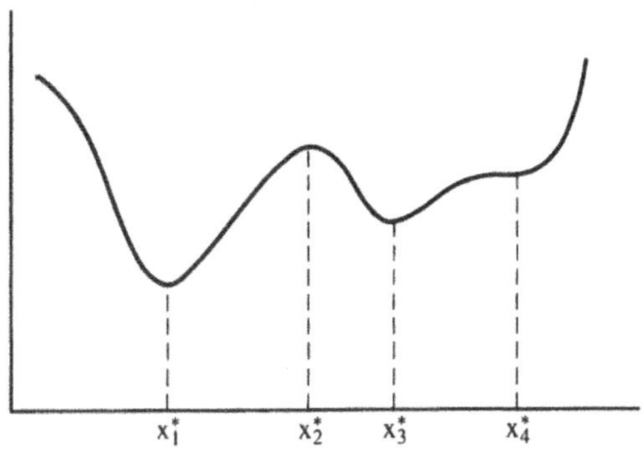

Fig. 11.7. Function with multiple stationary points.

Finally, it is possible for a function to have several minima, maxima, and points of inflection. Figure 11.7 demonstrates this possibility.

In this figure x_1^*, x_2^*, x_3^*, and x_4^* are all stationary points (first derivative = 0). Points x_1^* and x_3^* are minimum points (second derivative positive), x_2^* is a maximum point (second derivative negative), and x_4^* is a point of inflection (second derivative equal to zero and some higher-odd order derivative not equal to zero). Of course, if the function is convex (concave) for all values of x, there can only be one stationary point and that point is a unique minimum (maximum). When there is only one minimum (maximum), that minimum (maximum) is called the *global* minimum (maximum). When there is more than one minimum (maximum), the smallest of all the minima (largest of all the maxima) is also called the *global* minimum (maximum).

For functions of more than one variable, the same essential relationships hold. For example, the necessary condition for a vector

$$x^* = (x_1, x_2, \ldots, x_n) \tag{11.26}$$

to be a stationary point of a function $F(x)$, all elements of the vector of first partial derivatives (called the *gradient*) evaluated at x^* must be zero.

$$\nabla F(x) = \left(\frac{\partial F(x)}{\partial x_1}, \frac{\partial F(x)}{\partial x_2}, \ldots, \frac{\partial F(x)}{\partial x_n} \right)^T = 0 \qquad (11.27)$$

As an example, suppose that the cost of producing a product in a continuous process is a function of the amount of material added (x_1) and the temperature (x_2) as shown in

$$TC = 2x_1^2 - 2x_1x_2 + x_2^2 - 4x_2 + 5 \qquad (11.28)$$

The necessary condition [Equation (11.27)] states that

$$\frac{\partial TC}{\partial x_1} = 4x_1 - 2x_2 = 0 \qquad (11.29)$$

and

$$\frac{\partial TC}{\partial x_2} = -2x_1 + 2x_2 - 4 = 0 \qquad (11.30)$$

Solving these two equations simultaneously yields

$$x_1^* = 2 \quad \text{and} \quad x_2^* = 4 \qquad (11.31)$$

or

$$x^* = (2, 4) \qquad (11.32)$$

Now, we know that there is only one stationary point, but we do not know whether it is a minimum, maximum, or point of inflection (a point of inflection in more than one dimension is often called a *saddle point*). To find this, we need to evaluate the Hessian matrix, which is the matrix of second partials as shown in Figure 11.8.

The sufficiency test involves testing this Hessian matrix for something called positive or negative definiteness (note the similarity to the one-dimensional case). If the matrix is positive definite at a stationary point x^*, then x^* is a minimum point; if the matrix is negative definite, then x^* is a maximum point. Testing for this definiteness is beyond the scope of this book, but it involves a check of the eigenvalues of the matrix or calculation of determinants of the principle minor submatrices. As for the single variable case, we can show that if the Hessian matrix is positive (negative)

$$\begin{bmatrix} \frac{\partial^2 f}{\partial x_1^2} & \frac{\partial^2 f}{\partial x_1\, \partial x_2} & \cdots & \frac{\partial^2 f}{\partial x_1\, \partial x_n} \\ \frac{\partial^2 f}{\partial x_2\, \partial x_1} & \frac{\partial^2 f}{\partial x_2^2} & \cdots & \frac{\partial^2 f}{\partial x_2\, \partial x_n} \\ \vdots & \vdots & \cdots & \vdots \\ \frac{\partial^2 f}{\partial x_n\, \partial x_1} & \frac{\partial^2 f}{\partial x_n\, \partial x_2} & \cdots & \frac{\partial^2 f}{\partial x_n^2} \end{bmatrix}$$

Fig. 11.8. Hessian matrix

definite for all values of x, then any stationary point is a minimum (maximum). The function is then said to be convex (concave).

Going back to the example problem of Equation (11.28), we can see that the Hessian matrix is

$$\begin{bmatrix} 4 & -2 \\ -2 & 2 \end{bmatrix}$$

We can show that this matrix is positive definite for all values of x. Therefore, the function is convex and any stationary point is a minimum point. The point

$$x^* = (2, 4) \qquad (11.33)$$

is a minimum point.

Sometimes the methods of calculus covered above are not applicable for one of several possible reasons. The most frequently occurring is that the mathematical model of the system is not known. For example, a new chemical process may have several variables, including items such as composition, temperature, and humidity and to run the process may be extremely costly. Therefore, it is virtually impossible to run the system often enough to be able to *fit* a mathematical function to the results. It is necessary, however, to be able to determine *good* if not *optimum* values of the variables. This also occurs when a system is being *simulated* on a computer. Another problem may occur when the mathematical function can be stated, but taking derivatives, etc., would be very difficult. In situations like these the techniques of *search theory* are often used.

Search theory, as the name implies, is a *search* over the feasible domain of the variables in an effort to determine *good* values for the variables. There are two essential search procedures—those that use derivatives and those that don't. Search procedures that use derivatives are limited in scope because derivatives aren't always available and they are more difficult to explain than the procedures that don't require derivatives. Consequently, we shall cover only that class of search procedures that does not require derivatives.

For one-dimensional problems, there are several very efficient search procedures available, for example, Bolzano, Fibonacci, paired block, and golden section. We shall cover here only the paired block procedure. Suppose that we have a production process whose total cost is completely dependent on the temperature of the process and suppose that we wish to minimize this total cost by choosing a good value for the temperature. Suppose also that it is possible to say that the temperature must lie between 50 and 150°F and our controls are capable of distinguishing only full degree differences, not fractional ones. (Minimum spacing for distinguishing between values is called *resolution*.) The paired block procedure places two observations as close together as possible (i.e., one resolution apart) in the middle of the range. In our example the process should be run at 99.5 and 100.5°F for the first two trials. Since we can only run at full degrees, we arbitrarily choose to run at 100 and 101°F, yielding the results shown in Figure 11.9 in which the height of the vertical line represents the total cost of the process at those points. Obviously, 101°F is cheaper and, therefore, most likely (definitely if the function is convex) the best value will lie between 100 and 150°F. We should then try the process at two points close to the middle of the remaining range or

$$100°\text{F} + \left(\frac{1}{2}\right)(150°\text{F} - 100°\text{F}) = 125°\text{F}$$

Arbitrarily, we choose to try 124 and 125°F (125 and 126°F would have worked just as well). The results are shown in Figure 11.10. From these results we can now say that the optimum value should lie between 100 and 125°F. We repeat this process until we can do it no more or we are satisfied with the results.

The simplest of the nonderivative oriented search procedures for multivariate functions is called *sectioning* or *one-at-a-time*. In sectioning, the searcher starts at a

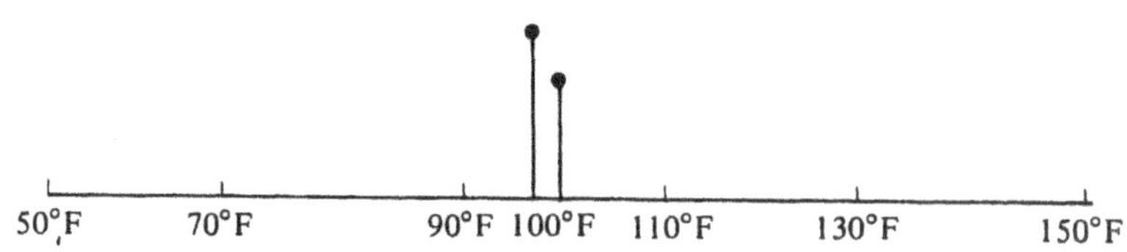

Fig. 11.9. First results of paired block search.

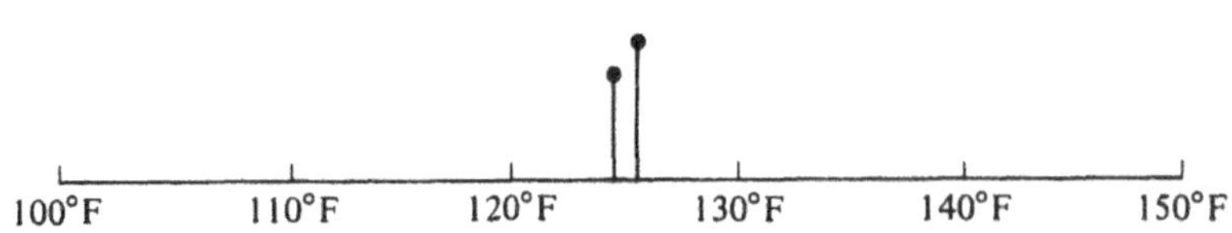

Fig. 11.10. Second results of paired block search.

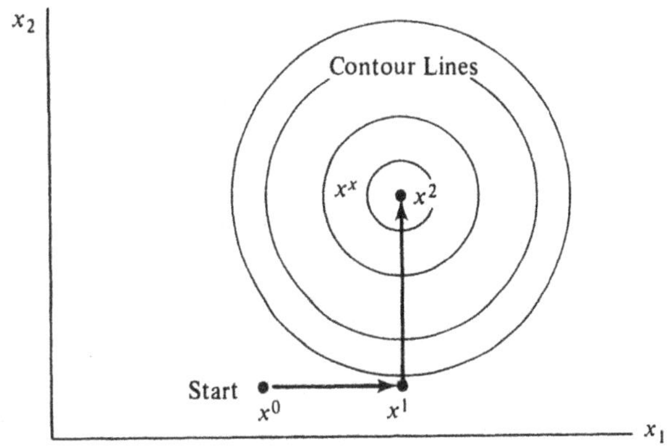

Fig. 11.11. Sectioning procedure.

point and proceeds in one coordinate direction[2] (holding all other variables constant) in iterative steps until the value observed stops improving. The searcher then chooses another coordinate direction and repeats the process, again holding all other variables constant. This is repeated until successive searches on all N variables do not result in any improvements or the searcher is satisfied with the results. As an example, consider the two variable functions shown in Figure 11.11 in which the starting point is shown and the circular contour lines are various levels of the function. (Imagine a top view of a bowl "cut" at various levels.)

In this case, the searcher started at x^0, held x_2 constant, proceeded in the positive x_1 direction until the function stopped improving at point x^1. Then the searcher proceeded in the positive x_2 direction holding x_2 constant at x^1 and found the optimum point. Sectioning will always yield the optimum answer for perfect functions such as the one shown in Figure 11.11 in N moves, where N is the number of variables (two in this case). Usually, the function is not so well-behaved and many moves are necessary before the searcher is satisfied. Also, each move entails the evaluation of many points as small steps are taken.

As a matter of interest, the function shown in Figure 11.11 is a quadratic sum of squares with no interactive terms and equal multipliers such as

$$x_1^2 + x_2^2 \quad \text{or} \quad 2x_1^2 + 2x_2^2$$

11.8 LINEAR PROGRAMMING

One of the first operations research methods was that of linear programming invented by George Dantzig in 1947. Linear programming is still a widely used operations research technique. Many packaged programs are available for solving linear programming problems and the basic theory is very simple, which probably accounts for its popularity.

To define linear programming (L.P.), we must go back to Equations (11.8) and (11.9) that define the general mathematical programming problem. If the objective

[2]One step in the positive direction can be tried first; if it does not improve the value, then one step in the negative direction can be attempted.

function $f(x)$ and the constraint set functions $g_i(x)$ are all linear or first-order (the exponent of all variables equals one), then the problem is L.P. Also, since we normally deal with positive values of variables, we can add another constraint set

$$x_i \geq 0 \quad (i = 1, \ldots, N) \tag{11.34}$$

Restating (to show that all functions are linear), we have the following as the L.P. formulation:

$$\text{Optimize } C'X \tag{11.35}$$

subject to

$$GX \leq B \tag{11.36}$$

$$x \geq 0 \tag{11.37}$$

As illustrations, Figure 11.12 presents two linear programming problems and two problems that cannot be classified as L.P. because of the underlined elements that are not linear.

Example 11.1[3]

Gadgets, Inc., is trying to decide how many of each of two products are to be manufactured. A single unit of Gadgets model I requires 2.4 minutes of punch press time and 5.0 minutes of assembly time and yields a profit of \$0.80 per unit. A single unit of model II requires 3.0 minutes of punch press time and 2.5 minutes of welding time and yields a profit of \$0.70 per unit. If the punch press department has 1,200 minutes available per week, the welding department 600 minutes, and the assembly department 1,500 minutes per week, what is the product mix (quantity of each to be produced) that maximizes profit?

Linear Programming Problems	Nonlinear Programming Problems
Min. $5x_1 + 3x_2$	Min. $5x_1 + 3x_2$
S.T. $x_1 + x_2 \leq 5$	S.T. $\underline{x_1 x_2} \leq 5$
$x_1 - x_2 \leq 7$	$x_1 - x_2 \leq 7$
$x_1 \geq 0 \; x_2 \geq 0$	$x_1 \geq 0 \; x_2 \geq 0$
Max. $3x_1 - x_2$	Max. $3x_1 - x_2^2$
S.T. $x_1 + x_2 \leq 4$	S.T. $x_1 + x_2 \leq 4$
$x_1 - 2x_2 \leq 1$	$\underline{\text{Tan } x_1} - 2x_2 \leq 1$
$x_1 \geq 0 \; x_2 \geq 0$	$x_1 \geq 0 \; x_2 \geq 0$

Fig. 11.12. Sample mathematical programming problems.

[3] Partially extracted from W. J. Fabrycky, P. M. Ghare, and P.E. Torgersen, *Applied Operations Research and Management Science* (Englewood Cliffs, N.J.: Prentice Hall, 1983).

Following the model of Equations (11.35), (11.36), and (11.37), we see that the linear programming formulation of the problem is as follows:

$$\text{Maximize } 0.8x_1 + 0.7x_2 \tag{11.38}$$

$$2.4x_1 + 3.0x_2 \leq 1{,}200 \tag{11.39}$$

$$5.0x_1 \qquad \leq 1{,}500 \tag{11.40}$$

$$2.5x_2 \leq 600 \tag{11.41}$$

$$x_1 \geq 0 \quad x_2 \geq 0 \tag{11.42}$$

Here x_1 is the number of units of model I to produce and x_2 is the number of units of model II. The objective function [Equation (11.38)] says that the total profit is to be maximized. Constraints 1, 2, and 3 [Equations (11.39), (11.40), and (11.41)] state that the punch press, assembly, and welding departments have only so many hours available per week. The last constraints [Equation (11.42)] say that only positive quantities of products can be produced.

It is a relatively simple matter to solve this problem because it is linear and there are only two variables. A two-variable problem can be drawn as shown in Figure 11.13. Each constraint is first drawn as an equation (straight line); then the appropriate side ("less than" in this case) is dashed as shown. Finally, the intersection of these areas is the "feasible region*' for the problem and is shaded in Figure 11.13. For any point to be a possible solution to this problem, that point must lie within the feasible region (shaded area in the figure).

The profit for the problem is given by

$$TP = 0.8x_1 + 0.7x_2 \tag{11.43}$$

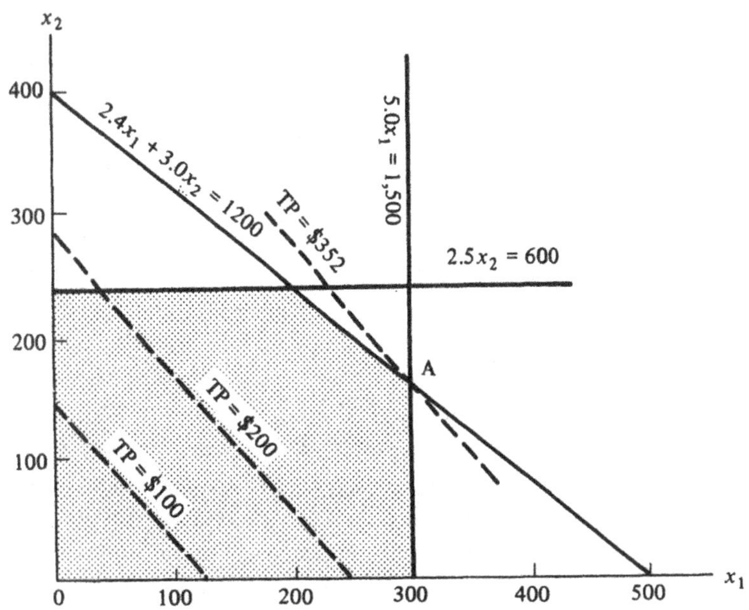

Fig. 11.13. Linear programming formulation of production problem.

This defines a straight line for any value of profit reflecting the different combinations of x_1 and x_2 that yield that value of profit. The straight lines for profits of $100 and $200 are shown in Figure 11.13 by dashed lines.

It should be obvious that the total profit increases as x_1 and x_2 either move away from the origin or increase in value. Also, the *TP* lines are parallel. Since the solution must lie within the shaded region, the optimum point should be the point where the maximum *TP* line just touches the feasible region. With the slope of these *TP* lines, the maximum point should be point *A* where two of the constraints meet. Those two constraints are

$$2.4x_1 + 3.0x_2 = 1{,}200 \qquad (11.44)$$

$$5.0x_1 \qquad = 1{,}200 \qquad (11.45)$$

Solving these simultaneously shows that point *A* is located at (300, 160) and that the total profit at that point is

$$0.8(300) + 0.7(160) = \$352 \qquad (11.46)$$

That *TP* line is shown by the dashed line segment in Figure 11.13.

A minimization problem is very similar to a maximization problem except that the constraints are usually reversed (greater than or equal to) and it is desirable to bring the total profit line as *close* to the origin as possible. For example, Figure 11.14 represents the following linear programming problem:

$$\text{minimize total cost } TC = 5x_1 + 4x_2 \qquad (11.47)$$

$$\text{subject to} \qquad 5x_1 + x_2 \geq 5 \qquad (11.48)$$

$$x_1 + 5x_2 \geq 5 \qquad (11.49)$$

The optimum point is at (5/6, 5/6), which yields a total cost of 7.5.

It is possible, but very difficult, to draw a linear programming problem of three variables (we shall not attempt one here). With three variables, the objective function is a *plane* instead of a straight line and the plane must be moved until it strikes an extreme point (edge or vertice) of the feasible region that yields an optimal answer.

Obviously, since almost all linear programming problems have more than two or three variables, how can we solve them? Since few, if any, L.P. problems are ever solved graphically, the graphical analysis of Figures 11.13 and 11.14 is good only for demonstration. Actually, almost all L.P. problems are solved by using the simplex method.

The simplex method is a mathematical procedure that systematically examines the extreme points of the feasible region and locates an optimum value. We can prove that at least one extreme point is an optimum for any linear programming problem; hence, the simplex method can concentrate on extreme points only. The examples of Figures 11.13 and 11.14 also *show* that the optimum must lie on a boundary, meaning that at least one optimum point is an extreme point.

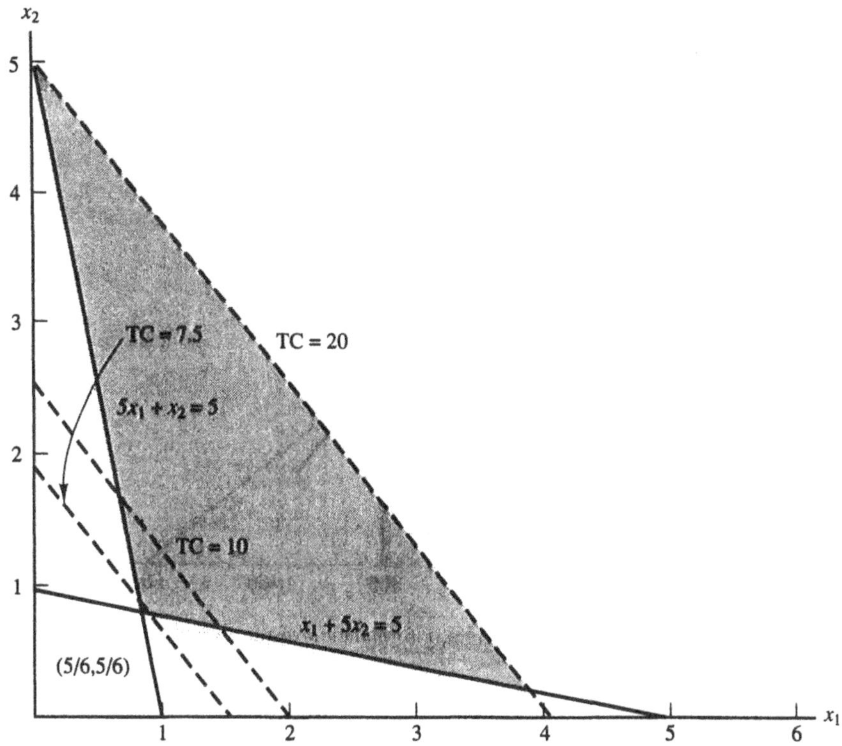

Fig. 11.14. Graphical minimization of a two-variable linear programming problem.

Although there are many programs available for the simplex method, some people write their own programs because the simplex method is very simple to learn. The simplex method is covered in depth in industrial engineering courses and will not be covered here. Instead, let's look at some specialized linear programming problems that can be solved more easily.

11.8.1 Assignment Problem

The assignment problem is perhaps the simplest of all linear programming problems both to understand and to solve.

The assignment problem is encountered any time there are N items or services available at N locations and N other locations (some may be similar) that require one and only one of these N items or services. As an example, suppose that we have four products that can be produced by any one of four machines, and that any machine can produce only one product. Furthermore, because of relative efficiencies on the different products, the cost varies between the assignments as shown in the cost matrix below. The objective is *to assign* the jobs to the machines in order to minimize total cost.

It would be possible to solve the problem above by using a mathematical programming package, but a more efficient method, the *Hungarian method*, has been developed. The first part of the Hungarian method is repeated here.

(1) Subtract the smallest number in each row from every entry in that row.

(2) In the matrix resulting from step 1, subtract the smallest number in each column from every entry in that column.

(3) In the matrix resulting from step 2, a zero cost assignment (if it can be made) is optimal. If not, complete the Hungarian method.

	Machines			
	A	B	C	D
Job 1	16	14	15	18
2	12	13	16	14
3	14	13	11	12
4	16	18	15	17

As step 3 implies, the procedure above frequently is all that is necessary; it is impossible, however, to find a zero cost assignment for the reduced matrix shown in Figure 11.15. Here the assignment matrix has been reduced as far as possible with no zero cost assignment available.

To solve this, it is necessary to enter the next phase of the Hungarian method, which attempts to create more zeros in the matrix so that a zero cost assignment can be made. The steps of the second phase of the Hungarian method are as follows:

(4) Draw the minimum number of horizontal and vertical lines (no diagonals) that intersect all the zeros (four in this case as shown below).

	1	2	3	4	5
1	7	3	2	0	0
2	4	0	1	1	1
3	0	0	0	1	1
4	1	1	1	3	0
5	1	1	1	6	0

	1	2	3	4	5
1	7	3	2	0	0
2	4	0	1	1	1
3	0	0	0	1	1
4	1	1	1	3	0
5	1	1	1	6	0

Fig. 11.15. Reduced assignment matrix with no zero cost assignment.

(5) Subtract the smallest uncrossed number from every other uncrossed number and add it to all elements where the vertical and horizontal lines intersect. (The smallest uncrossed number is 1.)

	1	2	3	4	5
1	7	3	2	0	1
2	4	0	1	1	2
3	0	0	0	1	2
4	0	0	0	2	0
5	0	0	0	5	0

(6) If a zero cost assignment can be made, it is optimal. If not, repeat steps 4 and 5 until one can be made. (In this case, a zero cost assignment can be made as shown by the circled items below. That solution is optimal.)

	1	2	3	4	5
1	7	3	2	⓪	1
2	4	⓪	1	1	2
3	0	0	⓪	1	2
4	⓪	0	0	2	0
5	0	0	0	5	⓪

$\Rightarrow$

	1	2	3	4	5
1				1	
2		1			
3			1		
4	1				
5					1

Solution

The assignment problem is not a difficult problem to solve manually even when the second phase of the Hungarian method is required. It is extremely difficult, however, to program a computer to follow this procedure and do it efficiently.

We have mentioned previously that unbalanced assignment problems (number of sources does not equal number of destinations) can be solved by adding dummy rows or columns. The procedure is very simple, for all we need do is add enough dummy sources or destinations to "balance" the problem. The costs associated with these rows or columns are all zeros to ensure that the solution is not affected.

Figure 11.16 presents two unbalanced assignment problems that have been balanced so that they can be solved by using the standard procedure.

Finally, the objective is sometimes maximizing as opposed to minimizing. In the previous example since max $f(x)$ yields the same *solution point* as min $-f(x)$, the change for maximization is very simple. We replace all the cost elements by their negative and continue with the minimization procedure. As an example, consider the problem regarding the assignment of salesmen to territories, shown in Figure 11.17, in which each element in the matrix shows profits and the objective is to assign the salesmen to maximize profits.

Unbalanced

	1	2	3	4	5
1	5	3	4	1	6
2	2	5	6	7	3
3	5	2	1	6	4
4	4	5	1	2	1

Add Dummy Row

Balanced

	1	2	3	4	5
1	5	3	4	1	6
2	2	5	6	7	3
3	5	2	1	6	4
4	4	5	1	2	1
5	0	0	0	0	0

	1	2	3
1	5	3	1
2	1	4	2
3	3	1	4
4	6	5	6
5	3	6	2

Add Two Dummy Columns

	1	2	3	4	5
1	5	3	1	0	0
2	1	4	2	0	0
3	3	1	4	0	0
4	6	5	6	0	0
5	3	6	2	0	0

Fig. 11.16. Unbalanced assignment problems.

Territory

Salesmen	1	2	3	4
1	5	3	5	4
2	2	1	6	3
3	5	4	6	2
4	6	3	1	4

Entries Are In 100,000's of Dollars

Fig. 11.17. Maximization assignment problem.

	1	2	3	4	Smallest Number
1	−5	−3	−5	−4	−5
2	−2	−1	−6	−3	−6
3	−5	−4	−6	−2	−6
4	−6	−3	−1	−4	−6

⇒

	1	2	3	4
1	0	2	0	1
2	4	5	0	3
3	1	2	0	4
4	0	3	5	2
Smallest Number	0	2	0	1

⇒

	1	2	3	4
1	0	0	0	(0)
2	4	3	(0)	2
3	1	(0)	0	3
4	(0)	1	5	1

Fig. 11.18. Maximization assignment problem solution.

First, we must replace all elements with their negative. Then we must solve the problem by using the Hungarian method as shown in Figure 11.18. Note that only the first phase of the Hungarian method is necessary and that the solution involves assigning salesman 4 to territory 1, salesman 3 to territory 2, salesman 2 to territory 3, and salesman 1 to territory 4.

11.8.2 Transportation Problem

The transportation problem is actually similar to the assignment problem in that there are M sources with something available and N destinations needing something. The difference is that under the transportation formulation, each source may have more than one unit available and each destination may need more than one unit.

The mathematical formulation is presented below for a typical minimization transportation problem:

$$\text{Minimize } E = \sum_{i=1}^{M} \sum_{j=1}^{N} e_{ij} x_{ij} \tag{11.50}$$

subject to

$$\sum_{i=1}^{M} x_{ij} = b_j \quad (j = 1, \ldots, N) \tag{11.51}$$

$$\sum_{j=1}^{N} x_{ij} = a_i \quad (i = 1, \ldots, M) \tag{11.52}$$

where

$$\sum_{i=1}^{M} a_i = \sum_{j=1}^{N} b_j \tag{11.53}$$

e_{ij} = cost to ship 1 unit from i to j
x_{ij} = number of units shipped from source i to destination j (integers only)
a_i = number units available at i
b_j = number units needed at j

The objective equation (11.50) states that the cost of transporting the units increases linearly with the number of units such that the total cost is obtained by multiplying the cost per unit (e_{ij}) by the number of units (x_{ij}). The first set of constraints [Equation (11.51)] states that each destination needs exactly b_j units so that the sum of units shipped from all sources to the destination must be bj. Similarly, since each source has exactly a_i units available, the total shipped out must be exactly a_i [Equation (11.52)]. Finally, the algorithm for optimally solving the transportation problem requires that the total number of units available must equal the total number needed [Equation (11.53)]. If this is not true, it can be forced by adding a dummy row or column with zero cost elements and required number of units to balance.

The only solution procedure presented thus far is the least cost assignment whereby the analyst chooses the least available cost and assigns as much as possible. Next, the analyst corrects the source availabilities and destination requirements and repeats the procedure until all allocations have been made. It should be intuitively obvious that this procedure does not yield an optimum answer but that it does present a very good starting solution that may be improved.

As an example, let us use the transportation problem and solution demonstrated in Figure 11.19. Although there are other shorter ways, the easiest procedure to understand is to check each combination that is *not* currently in the solution to see if the solution is improved by including it in the solution.

For example, the Tulsa–Washington (1,1) combination is not in the solution. If one unit were shipped from Tulsa to Washington, then a unit would have to be subtracted from the Tulsa–Lincoln (1,3) combination or the Tulsa–San Francisco (1, 4) combination to balance the load. If one unit were subtracted from the Tulsa-Lincoln (1,3) combination, then it would have to be added somewhere in column 3. There is no way of doing this because there are no more entries in column 3. (We cannot *add* a new one since we can prove that the transportation solution can have at *most* $M + N - 1$ combinations in the solution, 7 in this case. Therefore, the only elements that can be added to or subtracted from, *except for the one being examined*, must already be in the solution.) For example, considering the shipment of one additional unit from Tulsa to either Washington, Cleveland, or Phoenix results in cost dy as follows:

$$\begin{aligned} d_{11} = c_{11} - c_{14} + c_{34} - c_{31} &= 80.00 - 78.00 + 74.50 - 70.80 \\ &= \$5.70 \\ d_{12} = c_{12} - c_{14} + c_{34} - c_{32} &= 78.00 - 78.00 + 74.50 - 72.00 \\ &= \$2.50 \\ d_{15} = c_{15} - c_{14} + c_{24} - c_{25} &= 77.00 - 78.00 + 71.50 - 70.70 \\ &= -\$0.20 \end{aligned}$$

From \ To	1. Washington	2. Cleveland	3. Lincoln	4. San Francisco	5. Phoenix	Plant Capacity
1. Tulsa	80.00	78.00	77.00 4,000	78.00 3,000	77.00	7,000
2. Tempe	76.50	75.00	73.50	71.50 2,500	70.20 3,000	5,500
3. Sparta	70.80 5,000	72.00 6,000	75.00	74.50 1,500	75.00	12,500
	5,000	6,000	4,000	7,000	3,000	

Fig. 11.19. Solution to transportation problem by using least cost assignment.

To / From	1. Washington	2. Cleveland	3. Lincoln	4. San Francisco	5. Phoenix	Plant Capacity
1. Tulsa	80.00	78.00	77.00 4,000	78.00	77.00 3,000	7,000
2. Tempe	76.50	75.00	73.50	71.50 5,500	70.70	5,500
3. Sparta	70.80 5,000	72.00 6,000	75.00	74.50 1,500	75.00	12,500
	5,000	6,000	4,000	7,000	3,000	

Fig. 11.20. Revised solution to transportation problem.

Shipping a unit from Tulsa to Phoenix (1, 5) would improve the solution by $0.20. Since it is logical to ship as many as possible, we would have to look at the other elements affected. To add to (1, 5), we would have to subtract from (2, 5), add to (2, 4), and subtract from (1, 4). We can see, then, that we can ship 3,000 units from 1 to 5, none from 2 to 5, 5,500 from 2 to 4 (2,500 + 3,000), and none from 1 to 4, yielding the solution shown in Figure 11.20.

To check for optimality, the calculations for all the empty combinations would have to be done again. If all are positive, the solution is optimal. In this case, they are all positive. Therefore, the solution shown in Figure 11.20 is optimal.

We mentioned earlier that the transportation problem cannot have more than $M + N - 1$ positive entries in the solution matrix. It can, however, have fewer. In Figure 11.20 the solution has only $M + N - 2$ positive entries. This is okay, but it does cause some problems in trying to calculate optimality. We shall not discuss that here except to note that we might have difficulty showing that the solution in Figure 11.20 is optimal. It can be done, however.

11.9 OTHER TECHNIQUES

In this chapter we have only begun to skim the surface of deterministic operations research techniques. Only unconstrained optimization and linear programming have been discussed and they haven't been covered in any depth. Some of the other areas are mentioned below in just enough depth to describe them. Curricula in industrial engineering cover almost all of these techniques in detail.

11.9.1 Nonlinear Programming

We spent a large amount of time discussing linear programming because it is the simplest to understand. When the objective function and/or the constraints are nonlinear, the problem is much more difficult to solve and is called nonlinear programming. This terminology covers a large area that actually includes some of the other techniques described below. Basically, nonlinear programming includes any mathematical programming problem that does not have linear constraints and objective function.

11.9.2 Integer Programming

Sometimes a mathematical programming problem has the decision variables limited to integer values. These problems, whether linear or not, are called integer programming problems. The transportation problem is a linear integer programming problem since fractional units cannot be shipped.

11.9.3 Zero-One Programming

When an integer programming problem has the integer variables limited to the values of zero or one, it is called a zero–one programming problem. The assignment problem is an example of zero–one programming.

11.9.4 Quadratic Programming

Sometimes a problem has linear constraints but a quadratic (second-order) objective function. This is called quadratic programming.

11.9.5 Geometric Programming

Often, a mathematical programming problem has a structure such that it can be solved by using a technique called geometric programming. The structure is too complex to explain here.

11.9.6 Other Programming

Some of the other techniques encountered in the study of deterministic operations research methods are *dynamic programming, separable programming, branch and bound search*, and *mixed integer programming*. Although this list is not exhaustive, it does give an idea of the extensive field covered by deterministic operations research.

11.10 IMPACT OF COMPUTERS

It should be obvious that almost all tools of operations research require some mathematical sophistication and a large amount of computation capability and storage space. Without the speed of computers *almost* all operations research techniques would not be applicable to real-world problems. With the storage available on modern computers and the rapid computation time, many techniques can be used today that were only good for textbook discussion just a few years ago.

As an example, linear programming problems have been solved with many hundreds of variables and almost that many constraints. This would be completely impossible without computers. Software programs also exist for all the areas mentioned in this chapter.

Also, the unbelievable growth in power and affordability of personal computers have made larger O.R. models more feasible for the average industrial engineer. It would be rare indeed for an industrial engineer today not to have access to sufficient computing power to enable him or her to use most O.R. models in day-to-day operations. The future is practically unlimited.

DISCUSSION QUESTIONS

1. Define operations research and state how it is similar to and different from industrial engineering.
2. Distinguish between deterministic and probabilistic models of operations research.
3. It has been said that *everything* varies. Since deterministic models do not allow for random variation, why are they still very useful?
4. Why is the large-scale or systems approach so important in industrial engineering and operations research?
5. Define the mathematical programming problem.
6. State the rules for finding the extreme points of a function of one variable and for telling whether the extreme points are minima, maxima, or points of inflection.
7. Graph the general shape of a quadratic function of one variable that has a positive second derivative for all values. Repeat for a quadratic function that has a negative second derivative for all values.
8. Draw a function that has stationary points at which the second derivative is positive, negative, and zero. Clearly label these points.
9. Distinguish between local and global optima.
10. Draw a function for which a search procedure would probably miss finding the global maximum or minimum.
11. Describe *paired block* and *sectioning* search procedures. State when a search procedure might be appropriate.
12. Define a linear programming problem and give a real example. Discuss the objective function and constraints.
13. Define an assignment problem and give a real example.
14. Define a transportation problem and give a real example.
15. Give three examples of tools for decision making that could be called "operations research" or "industrial engineering."

PROBLEMS

1. Assume a situation in which the demand per year for an item is 980 units. The carrying costs are \$20 per unit per year and procurement costs are \$50 each time an order is issued. If lead time is 15 days, what amount should be ordered each time and what is the reorder point?
2. For the facts given in Problem 1, determine how many short or over the inventory will be when the new order is received if lead time is twice as long as usual. Repeat this for half as long as usual.

3. Find the extreme points of $2x^3 - 7x + 3$ and state whether they are minima, maxima, or points of inflection.
4. Find the extreme points of $2t^4 + 6t^3 - 5$ and state whether they are minima, maxima, or points of inflection.
5. Graph Problems 3 and 4 and show where the extreme points are. (This would be a good opportunity to use a spreadsheet.)
6. Calculate the Hessian matrix for $x_1^3 + x_2^3 + x_2 x_3^2 + 3x_3 - 4$. Evaluate it at the point (0, 1, 1).
7. Use a paired block search to determine the minimum of the equation TC(x) = 3x2 – 40x + 150 in the range of $0 \le x \le 10$ using increments of 0.5.

Problems 8, 9, 11, and 13 adapted, with permission, from W. J. Fabrycky, P. M. Ghare, and P. E. Torgersen, Industrial Operations Research (Englewood Cliffs, N.J.: Prentice Hall, 1972).

8. Solve graphically for the values of x_1 and x_2 that

$$\text{minimize } TC = 7x_1 + 9x_2$$

subject to:

$$18x_1 + 7x_2 \ge 250$$

$$3x_1 + 5x_2 \ge 84$$

$$9x_1 + 2x_2 \ge 60$$

$$x_1 \ge 0 \quad x_2 \ge 0$$

9. In the linear programming example given by Equations (11.38) through (11.42) determine the optimum point if the following constraint is added:

$$5x_1 - 2x_2 \le 1,000$$

10. A small machine shop has capacity to do turning, milling, drilling, and welding. Two products are under consideration. Each will yield a net profit of $3.50 per unit and will require the following machine time:

	Product		*Capacity*
Operation	*1*	*2*	*(hours)*
Turning	0.092	0.248	32
Milling	0.224	0.096	16
Drilling	0.080	0.000	32
Welding	0.000	0.240	16

(a) Formulate this as a linear programming problem similar to Equations (11.38) to (11.42).

(b) Solve graphically for the number of units of each product that should be scheduled to maximize profit.

11. Six service technicians are available to assist five customers. The time in hours necessary to travel roundtrip is shown in the assignment matrix. Optimally solve the assignment problem.

Service Technician	*Customer*				
	1	*2*	*3*	*4*	*5*
1	5	4	3	6	7
2	2	1	5	6	6
3	3	8	7	2	5
4	5	1	1	7	8
5	2	4	3	7	8
6	8	7	3	2	3

12. The foreman of a machine shop has five machinists available to assign to jobs for the day. Five jobs are offered with the expected profit in dollars for each machinist on each job being as follows:

Machinist	*Task*				
	A	*B*	C	*D*	*E*
1	65	45	100	75	90
2	70	55	105	90	93
3	60	50	100	85	95
4	60	40	120	85	97
5	70	60	110	97	85

Determine the assignment of machinists to jobs that will result in maximum profit.

13. Optimally solve the following transportation problem to allocate parts available to locations needed:

From \ *To*	*1*	*2*	*3*	*4*	*5*	*6*	*Available*
1	2	7	3	5	3	2	11
2	5	7	4	6	3	1	21
3	1	2	5	6	7	4	16
4	3	6	7	5	4	6	31
Needed	9	4	14	9	19	24	

14. Three plants produce parts that are to be shipped to four distribution centers. Plants 1, 2, and 3 produce 24, 34, and 22 shipments worth of parts per month, respectively.

Each distribution center needs to receive 20 shipments per month. The distance from each plant to each distribution center is given as follows:

	Distribution Center			
Plant	*1*	2	*3*	*4*
1	400	650	200	350
2	550	700	300	500
3	300	600	400	450

The transportation cost for each shipment is \$300 + \$0.75 per mile. How many shipments should be sent from each plant to each of the distribution centers to minimize total transportation costs? Before trying to solve this, be sure to formulate the problem as a transportation problem by constructing the appropriate cost and requirements table.

12 ERGONOMICS AND HUMAN FACTORS[1]

12.1 INTRODUCTION

"Managers at all levels and in all organizations are striving to achieve results through people."[2] To do this all sorts of control systems have been designed to measure and enhance organizational performance. Unfortunately, many of the popular control systems monitor only a small part of total system performance. Recently, however, great strides have been made toward the design of effective and efficient control systems. Industrial engineers are leading the way.

In the first edition, this chapter focused on work measurement and design at the workplace. That was an accurate portrayal of industrial engineering involvement at the time but was microscopic or myopic in that it did not measure organizational performance. It did measure individual efficiency. In the second edition we modernized that a bit by talking more about true "productivity," as explained below. Thus, the area is becoming more system or total system performance oriented.

Now, the third edition recognizes that this trend is continuing and perhaps accelerating. In fact, many industrial engineering curricula are dropping the more traditional areas of workplace measurement and improvement. We feel, however, that it is important to keep this in perspective. Thus, this chapter will briefly cover traditional areas and move on to the more company-wide techniques.

Perhaps Figure 12.1 does the best job of placing all this in perspective. Organizational productivity is simply a measurement of an organization's ability to turn inputs into outputs. Equation (12.1) states this:

$$\text{productivity} = \frac{\text{output}}{\text{input}} \tag{12.1}$$

Outputs and inputs can be partial or composite. For example, units per person-hour is a measure of labor productivity only and is, therefore, partial. Assuming that we desire quality outputs which meet the consumer demand (effectiveness), we can see that productivity is affected by efficiency, effectiveness, and quality, as shown in

[1]"The authors are deeply indebted to Dr. D. Scott Sink and Ms. Leva K. Swim for their guidance and help with this chapter.

[2]D. Scott Sink, "Organizational System Performance: Is Productivity a Critical Component?" *Proceedings* 1983 *Annual Industrial Engineering Conference*, Louisville, Ky., May 1983.

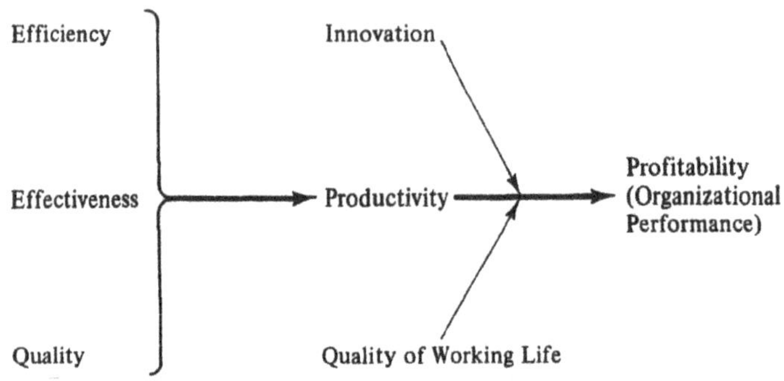

Fig. 12.1. Total organization performance. [Adapted from D.S. Sink, *Productivity Management: Planning Measurement, Evaluation, Control, and Improvement* (New York: John Wiley & Sons, Inc., 1984)].

Figure 12.1. Productivity, together with innovation and quality of working life, determine the total organizational performance, which is usually measured by profitability in the free-enterprise system.

Now, where does work design and measurement (timing of individual operations) fit? Referring to Figure 12.1, we see that *work measurement is the standard by which we measure efficiency. Similarly, work design helps determine efficiency.* Work measurement and design can be distinguished from total organizational performance through considerations of *focus* and *scope.*

Total organizational system performance has a much broader *focus* in that it addresses more performance criteria than just efficiency (i.e., quality, quality of working life, effectiveness, innovation, profitability, productivity). Work measurement/design in the workplace has a narrow focus in that it addresses efficiency, primarily. In regard to *scope*, total organizational system performance is concerned with the collection of individuals organized to accomplish a purpose. Work measurement/design concerns itself with the individual person and the person's interaction with his or her environment.

This chapter examines the measurement of organizational performance and the design of individual workstations. It does that by first discussing work simplification (probably the birthplace of industrial engineering) and then work measurement. Then the chapter discusses organizational performance with an emphasis on productivity measurement.

Some of you may never do work methods simplification or work measurement, as many industrial engineers are branching out into different areas. However, you should understand that industrial engineering started here and that more I.E.s are employed in this area today than in any other. It is sometimes very helpful to know your discipline's birthplace.

The following example is presented to demonstrate the need for the topics of this chapter.

Example 12.1

You have been contacted by the president of Gadgets, Inc., in Corvallis, Oregon, and asked to design and lay out the manufacturing processes for production of their new "super- gadget." Furthermore, he wants you to stay on and manage the plant once it is completed.

Since Gadgets, Inc., has done a great deal of research on processing methods, you find that there is little work needed there. Your goal is to produce 10,000 units per year. You must gear up to produce that, *but* to be able to determine how much equipment is necessary, you must know how long each operation takes. Since Gadgets, Inc., does not have that kind of information, you hire a time study analyst to do work measurement. Furthermore, since the equipment is expensive, you want the most efficient methods possible. The analyst must also design the most efficient work procedures for the workers to follow.

With the results of the time study analyst's work, you have the information necessary for determining manpower needs. Again, you find you must know how many super-gadgets each person can produce. You go back to the previously taken time studies to determine your labor needs.

Of course, the people must be trained. Who is going to train them, and what methods should they use? You decide to let your foremen do the training, *and* you find that the time study analyst specified the proper method for each job. The training is then simplified.

Because you are human relations oriented, you want to pay your people on the basis of their work quantity and quality. One way of doing this is to have an *incentive system* in which the people are paid according to how much they do. The previously taken time studies provide you with the time information you need for this incentive system.

Next, it is necessary to start production (construction and layout are complete). One of the first things you must do is schedule the orders for super-gadgets (each order has a unique requirement or design change) throughout the shop. To do this, you must figure how long a production order takes for each operation and the total time necessary in the plant. Since your time studies provide you with this information, you can do the scheduling. Production now begins.

As you sit back in your chair, you reflect, "How can I know if the plant is performing as it should?" Once again, the time studies give you a bench mark against which you can compare performances since you know how long the operations should take.

Finally, you need to decide how much to charge for super-gadgets. There are many considerations, for example, the market supply, but one consideration is the cost of production. Since the plant is just starting, you do not know the cost of production. The time studies tell you what the labor cost *should* be. You then add on material and overhead cost and you are ready.

Or, at least you thought you were. The president recently attended a productivity improvement seminar; so he tells you to design a productivity measurement model for the company. Using the material of this chapter, you find you can, indeed, measure productivity. In fact, you feel the need to inform the president that he also needs to worry about total organizational performance (hopefully, profitability) and that productivity is only a part of total performance.

Although the above example is grossly oversimplified, it does point out *some* of the uses of work simplification and measurement. Those uses were in:

(1) Determining labor requirements.
(2) Designing training methods.
(3) Designing scheduling procedures.
(4) Designing incentive systems.
(5) Gauging work performance.
(6) Estimating cost.

This is not an exhaustive list by any means, but it should be sufficient to demonstrate the importance of work simplification and measurement.

We have demonstrated that the correct design of work methods and timing of these methods is very important to an organization, but how can these be done? For obvious reasons, work simplification should be performed on a task before that task is timed. Therefore, it is logical to examine some procedures for work simplification first.

12.2 METHODS IMPROVEMENT

Over the evolutionary sequence of industrial engineering, work methods improvement has been called many things, including work simplification, methods improvement, and work design. Whatever it is called (we chose the term *work methods improvement)* a definition might be: *A systematic approach to finding easier and better ways to accomplish a task.*

The basic objective is to avoid waste of any kind (time, human effort, materials, capital, etc.) or, as often quoted

> Work smart, not hard.[3]

You have been making attempts at work methods improvement all your life whenever you try to find better ways to cut grass, wash cars, travel to work, and so on, but have you used a systematic approach? In some cases, probably yes, but in others, no. To do work methods improvement correctly, a systematic approach is necessary. We will discuss some systematic approaches in this section. A five- step procedure might be:

(1) *Selecting a Job*—One must select a job to study. This is often done because the job is the most expensive or obviously in need of such a study.
(2) *Getting and Recording the Facts*—The present method of doing the job must be recorded along with other facts such as the frequency of the job.

[3]Stephan Konz, *Work Design: Industrial Ergonomics* (Columbus, Ohio: Grid Publishing, Inc., 1983), p. 21.

(3) *Questioning Every Detail*—The present method must be questioned. Some elements may be deleted; most can be changed.
(4) *Developing and Testing a Better Method*—From step 3, better efforts may be obvious. These ideas should be developed and tried.
(5) *Installing and Maintaining Improvements*—For the study to be worthwhile an improved procedure must be installed and maintained.

Charts are a convenient way of applying parts of the foregoing procedure, especially in steps 2 and 3. The operations process chart already covered is one. Some of the other charts will be discussed next in their approximate order of application.

12.2.1 Flow Process Charts

A flow process chart is a chart of all the activities involved in a process. It is similar to an operations process chart, except that more detail is shown by including transportations and delays as well as operations, inspections, and storages. As in operations process charts, the objective is to list every detail and to examine each detail closely. The following questions should be asked about each activity:

(1) Is this activity necessary, or can it be eliminated?
(2) Can this activity be combined with another or others?
(3) Is this the proper sequence of activities, or should the sequence be changed?
(4) Can this activity be improved?
(5) Is this the proper person to be doing this activity?

The construction is similar to that of an operations process chart in that flow is from the top to the bottom. Only one item can be charted per form, however, and that item can be either a person or an object. A blank flow process chart is shown in Figure 12.2.[4] An example of its use is demonstrated below.

Example 12.2

Mr. Cafe, a school food cafeteria supervisor, has asked you to prepare a flow process chart for the preparation of spaghetti. You have decided to make a chart of the noodles going through the cafeteria and see how the process can be improved.

The noodles start on the storage shelf. They are picked up and carried to the cooker, which is a large vat in which the noodles are cooked. Next they are placed in a large bowl, which is carried to a sink where they are washed. Then the bowl is carried to a work area where the noodles are placed in six serving pans. The meat and sauce (a separate flow process chart covers their production) are added, and the pans are carried to a heating unit where they are kept warm unitl ready for serving.

[4]This is a simplified form in that much more information usually is presented. But either form or another can be used, depending on the preference of the user.

You prepare a flow process chart (shown in Figure 12.3) After consultation with Mr. Cafe and the operators and examining the chart carefully, you decide that by adding a water tap adjacent to the cooker (it already has a drain) you could wash the noodles right in the cooker, add the meat and sauce directly, and keep them warm in the cooker. This saves several transportations. The revised procedure is shown in Figure 12.4.

The advantages of the revised procedure over the existing one are obvious and numerous. No heavy lifting is required, only three transportations are required—a saving of four (57%)—and the same number of operations and storages are required.

In the existing procedure, steps 4 and 8 are called *transportations,* but they could have been called *operations.* Either term would have worked very well since the term is not important *in cases like this.* Many times an activity can be called several things; usually all are shaded on the chart, *or* only the most predominant ones are shown, as was done here.

12.2.2 Left-Hand–Right-Hand Charts

Left-hand–right-hand charts are useful in analyzing the work performed by one person at one specific workstation. As the name implies, the chart follows the motion of the left and right hands of one operator. Consequently, it is applicable only when one operator is performing repetitive work—usually at one station. An example is given below.

Process Begins Charter Present ______ Proposed ______		Man ______ or Material ______ Ends Date
Steps	**Symbol**	**Notes**
1.		
2.		
3.		
4.		
5.		
⋮		

Fig. 12.2. Flow process chart.

Process: Prepare Spaghetti	Man ___ or Material _X_ Noodles
Begins: In Storage	Ends: Ready to Serve
Charter: Ind. Eng.	Date: 7/24
Present _X_ Proposed _____	

Steps	Symbol	Notes
1. Noodles on Shelf	▽	
2. Carried to Cooker	⇨	
3. Cooked	○	
4. Placed in Bowl	⇨	Wash in Cooker?
5. Carried to Sink	⇨	Very Heavy
6. Washed	○	
7. Bowl Carried to Work Area	⇨	Very Heavy
8. Placed in Serving Pans	⇨	Six Pans-Full
9. Meat and Sauce Added to Pans	○	Repeated Six Times
10. Pans Carried to Heater	⇨	Repeated Six Times
11. Warmed	▽	
12. Carried to Serving Line	⇨	Repeated Six Times

Fig. 12.3. Flow process chart—present method.

Process: Prepare Spaghetti	Man ___ or Material _X_ Noodles
Begins: In Storage	Ends: Ready to Serve
Charter: Ind. Eng.	Date: 7/26
Present ____ Proposed X____	

Steps	Symbol	Notes
1. Noodles on Shelf	▽	
2. Carried to Cooker	⇨	
3. Cooked	○	
4. Washed in Cooker	○	Add a Tap Adjacent to Cooker
5. Meat and Sauce Added	○	
6. Warmed	▽	
7. Placed in Serving Pans	⇨	Repeat Six Times
8. Carried to Serving Line	⇨	Repeat Six Times

Fig. 12.4. Flow process chart—proposed method.

Example 12.3

In this example a pot handle is formed by plastic injection molding. This process requires a sophisticated die, and consequently, much time is devoted to the design of the die. Regardless of the amount of time spent on the design, a small burr is left on the part after it is removed from the die. This burr is ground in a small grinder; then it is polished by a buffing wheel to eliminate any grinding marks. The process used is presented below. Plastics, Inc., has asked you to improve the process and you have decided to use a left-hand-right- hand chart. First, you ask for a drawing of a typical burr. That is presented in Figure 12.5. Then you ask for a sketch of the workplace and a description of the process. The workplace sketch is shown in Figure 12.6.

All pot handles are placed on the gravity chute by a material handler. The chute is inclined so that parts are always available to the operator. The operator reaches to the chute with her left hand and picks up three pot handles. In the center of the work station she holds the handles with her left hand. Her right hand picks one pot handle out of the left hand, grinds the pot handle (takes about 2 seconds), and drops it in a cardboard box on the floor marked for completed units. The right hand does all three parts and the left hand reaches for three more pot handles. This is continued approximately 56 times until the cardboard box on the floor is full. The operator stops, shoves the box over to the area marked full boxes, and slides another empty box into place. The process is repeated. (The material handler keeps a supply of empty boxes in place.)

Figure 12.7 shows a simplified left-hand-right-hand chart. First, the information is filled in on the part number, and so on; then a sketch of the workplace is

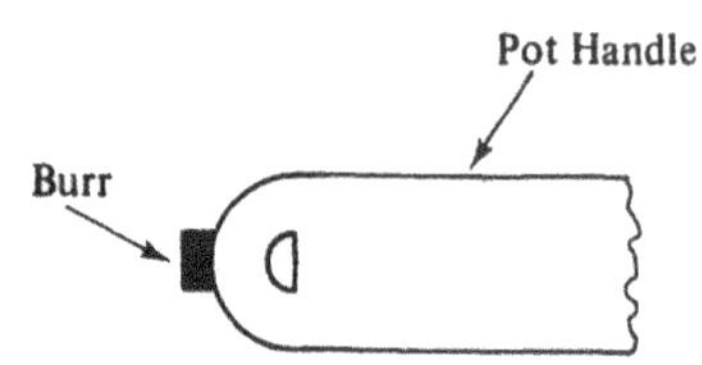

Fig. 12.5. Typical burr.

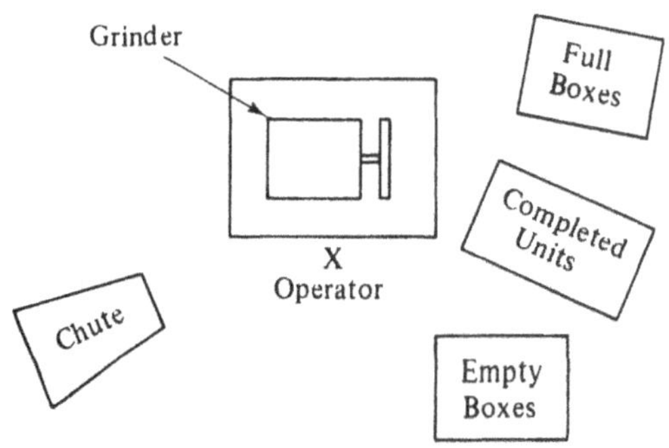

Fig. 12.6. Workstation layout.

Operation ______ Name ______ Date ______

Existing ______ Proposed ______

(Sketch of workplace and part if necessary)

Left-hand Activity	Symbol	Symbol	Right-hand Activity

Fig. 12.7. Left-hand–right-hand chart.

made. After that, the left and right hands are simultaneously charted so that at any time the movement of the left hand should match that of the right. Accordingly, a description of activities being on the same horizontal line implies that the activities are done at the same time.

The symbols are similar to those of a flow process chart, for example:

O Operation: A hand is doing some productive work

⇨ Transport: A hand is moving from one place to another

D Delay: A hand is not accomplishing anything but is waiting

∇ Hold: A hand is holding an object.

Example 12.3 (Continued)

You construct a left-hand-right-hand chart for the operation. That chart is shown in Figure 12.8. You notice that the right hand is delayed a large portion of the time while it is waiting for the left hand to obtain the parts. Also, the left hand holds the parts while the right hand does the work. (Holding is really not doing any productive work.) Productive work then really only occupies a small percentage of the total cycle.

After working with the operator and the foreman and after trying several new methods, you come up with a proposed procedure. (You also consider the principles of motion economy that will be discussed in Section 12.2.4.) That proposed procedure is shown in Figure 12.9 with a sketch of the new work plan. You time the two procedures and find that you save 37% of the time formerly used, not counting the substantial material handling savings.

Operation Grinding Burrs Name Bear Brian Date 7/30

Existing X Proposed ______

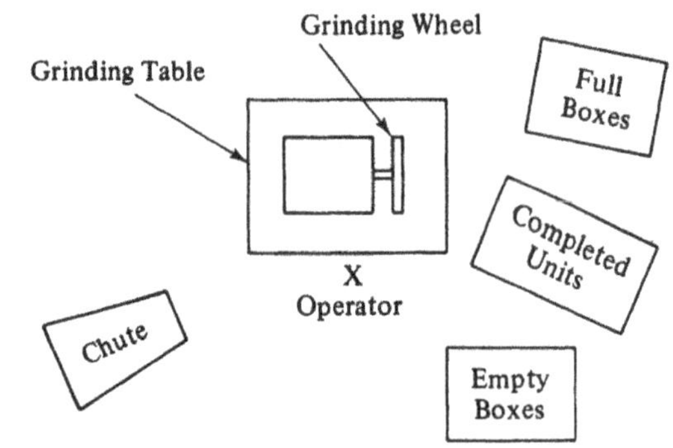

Left-hand Activity	Symbol	Symbol	Right-hand Activity
1. Reach to Chute	⇨	D	1. Wait
2. Pick up 3 Parts	○	D	2. Wait
3. Move to Center	⇨	D	3. Wait
4. Hold Parts	▽	○	4. Position Part and Grind
5. Hold Parts	▽	⇨	5. Move to Right and Drop in Box
6. Hold Parts	▽	⇨	6. Move Back to Center
7. Hold Parts	▽	○	7. Repeat 4
8. Hold Parts	▽	⇨	8. Repeat 5
9. Hold Parts	▽	⇨	9. Repeat 6
10. Hold Parts	▽	○	10. Repeat 4
11. Hold Parts	▽	⇨	11. Repeat 5
Repeat Steps 1 through 8 56 times.			
9. Walk to Box	D	D	9. Same as left hand
10. Slide Full Box Out	⇨	⇨	10. Same as left hand
11. Move Empty Box In	⇨	⇨	11. Same as left hand
12. Walk Back	D	D	12. Same as left hand

Fig. 12.8. Left-hand–right-hand chart—existing procedure.

Now there are two chutes feeding the operator from the left and the right. There is also a work shelf added to the grinding table. The work shelf has a drop hole in it directly in front of the wheel. The operator reaches simultaneously left and right and slides two parts on the work shelf (one per hand) to the two grinding wheels. After the grinding occurs, the operator slides the parts to the middle where

Operation Grinding Burrs Name Ara Smith Date 7/30

Existing ____ Proposed X

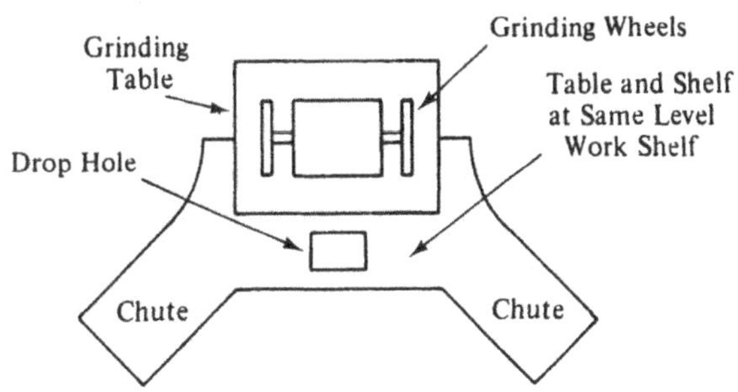

Left-hand Activity			Right-hand Activity
1. Reach to and Preposition Handle	⇨	⇨	1. Reach to and Preposition Handle
2. Slide to Grinder	⇨	⇨	2. Slide to Grinder
3. Grind	○	○	3. Grind
4. Slide to Center and Drop	⇨	⇨	4. Slide to Center and Drop

Fig. 12.9. Left-hand–right-hand chart—proposed procedure.

they drop through to a conveyor underneath. The conveyor carries the handles to the polishing operation. The cycle is repeated continuously because boxes never have to be moved.

12.2.3 Other Charts

This discussion only skims the surface of charting for work simplification. There are many other charts that have been used in designing work for many different types of jobs. Some of the other popular charts are described briefly below.

Flow Diagram. A flow diagram is essentially a flow process chart drawn to show the layout of a facility. Only the symbols and the step numbers are shown. Usually, the objective is to look for spatial relationships.

Multiple Activity Chart. A multiple activity chart is similar to a left-hand–right-hand chart in that the simultaneous activities of two or more objects or persons are shown. A chart simultaneously showing the activities of a worker and those of one or more machines he is operating is an example. The simultaneous activity of several workers on an assembly line is another. The objective is to obtain good balance—minimizing the total cycle time and subsequently the waiting time of any of the components.

Work Distribution Chart. A work distribution chart is a listing of all activities or responsibilities of every person in a department or group. It is similar to a multiple activity chart except that the activities are not necessarily simultaneously performed. The objective is to obtain proper balance of assignments and to be certain that the proper level of employment is performing the function (e.g., a department head should not deliver mail or answer telephone calls).

Operation Process Chart (already discussed)—The objective is to eliminate unnecessary operations, inspections, or storages and to obtain the proper sequence of them.

Gantt Chart. A Gantt chart is a horizontal time bar chart that shows relative timing of various activities. It is not often used in work simplification but may be used in scheduling. The objective is usually to perform all activities in the least amount of total time—often subject to some resource availability constraints.

12.2.4 Principles of Motion Economy

The *principles of motion economy* is a list of thoughts or concepts that have proven helpful in designing efficient work methods. There have been many lists generated by different authors; but the lists always carry the same essential concepts.

Many industrial engineers use motion economy in conjunction with left-hand–right-hand charts, since the list is most useful for single workstations. There are other uses, however, and one of them is to use the principles of motion economy as a list of commandments in designing and performing work. This last is perhaps the most rewarding but least utilized approach.

The following list of motion economy principles is taken from Mundel.[5]

A. Elimination:
   1. Eliminate all possible jobs, steps, or motions. (This applies to body, leg, arm, hand, or eye.)
   2. Eliminate irregularities in a job so as to facilitate automaticity. Provide fixed places for things.
   3. Eliminate the use of the hand as a holding device.
   4. Eliminate awkward or abnormal motions.
   5. Eliminate the use of muscles to maintain a fixed position.
   6. Eliminate muscular force by using power tools, power feeds, etc.
   7. Eliminate the overcoming of momentum.
   8. Eliminate danger.
   9. Eliminate idle time unless needed for rest.

B. Combination:
   1. Replace with one continuous curved motion short motions which are connected with sudden changes in direction.
   2. With fixed machine cycles, make a maximum of work internal to the machine cycle.
   3. Combine tools.
   4. Combine controls.
   5. Combine motions.

C. Rearrangement:
   1. Distribute the work evenly between the two hands. A simultaneous symmetrical motion pattern is most effective. (This frequently involves working on two parts at the same time.) With crew work, distribute the work evenly among members of the crew.
   2. Shift work from the hands to the eyes.
   3. Arrange for a straightforward order of work.

D. Simplification:
   1. Use the smallest muscle group capable of doing the work, providing for intermittent use of muscle groups as needed.
   2. Reduce eye travel and the number of fixations.
   3. Keep work in the normal work area.
   4. Shorten motions.
   5. Adapt handles, levers, pedals, buttons, etc., to human dimensions and musculatures.
   6. Use momentum to build up energy in place of the intense application of muscular force.

---

[5]Reproduced, with permission, from Marvin E. Mundel, *Motion and Time Study'. Improving Productivity*, 5th ed. (Englewood Cliffs, N.J.: Prentice Hall, 1978), pp. 173–174.

7. Use the simplest possible combination of therbligs. (A *therblig* is an elementary activity such as reach, grasp, pick up.)
8. Reduce the complexity of each therblig, particularly the "terminal" therbligs.

12.2.5 Human Engineering

Human engineering (or ergonomics) may be defined as the design of the human-machine interface so that workers and machines may function more effectively and efficiently as integrated systems. Much of what has been covered in this chapter can be called human engineering, but there is much more to human engineering than work simplification and principles of motion economy.

Formerly, machines were designed with an emphasis on their capabilities for production but with little or no emphasis on the human element. Now, with human engineering, the machine may be lowered, tilted, or completely redesigned so that the worker and the machinc may function better as a complete system. Much work in human engineering is done in design and location of dials for maximum effectiveness and efficiency. The design of a cockpit for a large modem jet is an excellent example of human engineering.

Human engineering is also useful when applied to day-to-day operations. For example, at an individual workstation it is important that all work be located close to and in front of the worker, as implied by the principles of motion economy. Using human engineering, we can design the workstation for the "average" person. Human engineers have calculated the reach of the average worker for both easy and strained moves. The workstation can be designed so that as many moves as possible are within easy reach. Figure 12.10 is an example of such measurements.

Another example of human engineering is the design of tools for more effective use. Pliers and scissors with curved handles are results of human engineering studies in which it was found that correctly designed curved handles are easier for human beings to use.

12.3 WORK MEASUREMENT

The objective of any work measurement system is to determine the time it should take an average, trained person to perform a task if he or she were doing that over an 8-hour day under usual working conditions and working at a normal pace. This time is called *standard time*. Historically, there have been two basic approaches to defining standard time. We shall call these two the bottom-up and top-down approaches. The *bottom-up approach* starts with a basic measurement of time, adjusts for operator pace, and then allows for fatigue, personal needs, and delays. The *top-down approach* is used in many labor contracts, and it normally defines standard time as that time under which a qualified employee working under usual conditions *can* make an incentive pay (specified) percent above base pay. Whichever definition is used, the approach taken to calculate standard time is *almost* always the bottom- up approach. Consequently, we shall explore that approach. First, some definitions are in order.

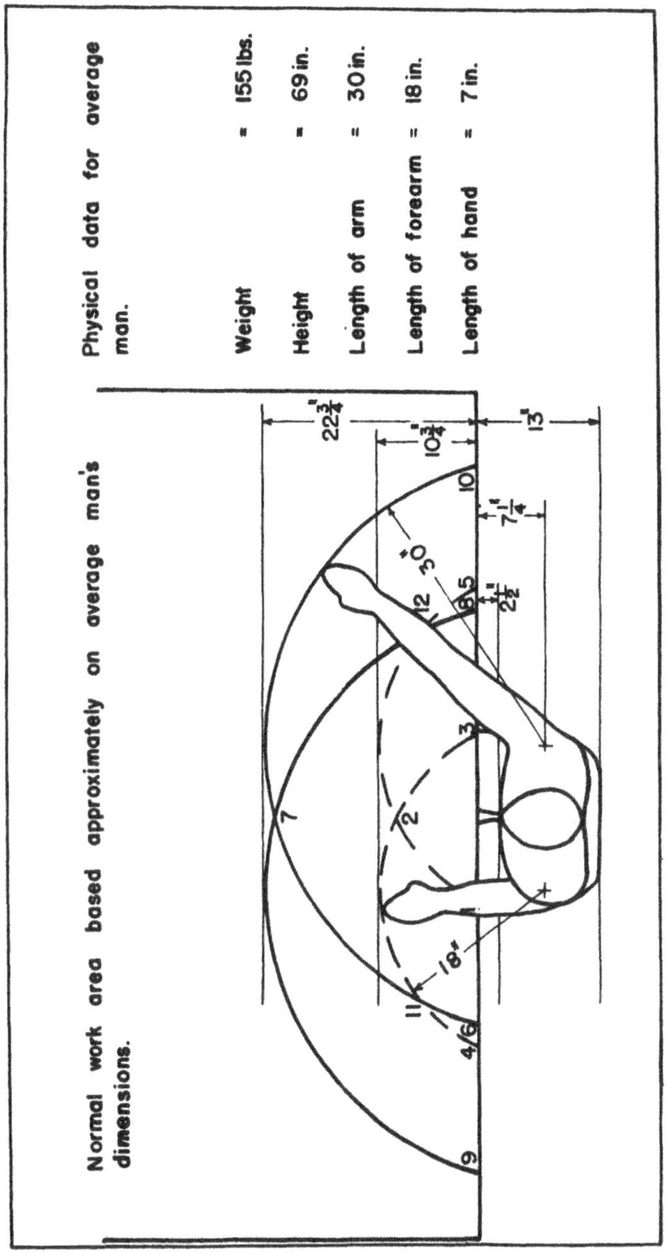

Fig. 12.10. Human work areas.

Normal Time. The time required for an average, trained operator to perform a task under usual working conditions and working at a normal pace. (It does not include allowances for personal needs and delays that would be necessary if the task were done all 8 hours.)

Normal Pace. The pace of an average, trained, and conscientious operator working over an 8-hour day.

Actual Time. The observed time required for an operator to perform a task.

Allowances. The amount of time added to the *normal time* to provide for personal needs, unavoidable delays, and fatigue.

We can now examine some of the possible methods of determining standard times. Following are the methods to be discussed:
(1) Direct time study.
(2) Time study standard data.
(3) Predetermined time systems.
(4) Predetermined time systems standard data.
(5) Work sampling.

12.3.1 Direct Time Study

In this context, direct time study is a work measurement technique in which a physical measurement is made of the actual time required to do a task by using a watch or some other timing device. This measured time is then modified by considering the operator's pace, and finally, allowances are added.

The form used to record task times varies greatly among companies. A general rule of thumb is to present as much information as might be required to duplicate the operation from the form alone. Therefore, the form should include a good description of the task involved, a sketch of the workplace, perhaps a left-hand–right-hand chart of the operation, and a clear description of the elemental breakdown.

Figure 12.11 shows a typical watch used in direct time studies and Figure 12.12 shows a typical time study board. A time study board is a device for holding the forms so that the analyst can make a recording while standing or even moving.

Figure 12.13 shows a simplified sample time study sheet for a continuous reading time study (the watch is not stopped). The front of the sheet is for recording the task being studied, the existing procedure in detail, and any other pertinent information. The back of the sheet is for recording the actual time study. The task is broken down into easily discernible elements, the elements are recorded, and then the timing is done. The watch reading is recorded in column R, and the time for each element is entered in column T later. For example, the following might be sample results:

	1		2		3		4	
Element	*R*	*T*	*R*	*T*	*R*	*T*	*R*	*T*
1	5		18		27		40	
2	10		22		32		45	
3	12		23		35		47	

	1		2		3		4	
Element	*R*	*T*	*R*	*T*	*R*	*T*	*R*	*T*
1	5	5	18	6	27	4	40	5
2	10	5	22	4	32	5	45	5
3	12	2	23	1	35	3	47	2

Fig. 12.11. Time study watch.

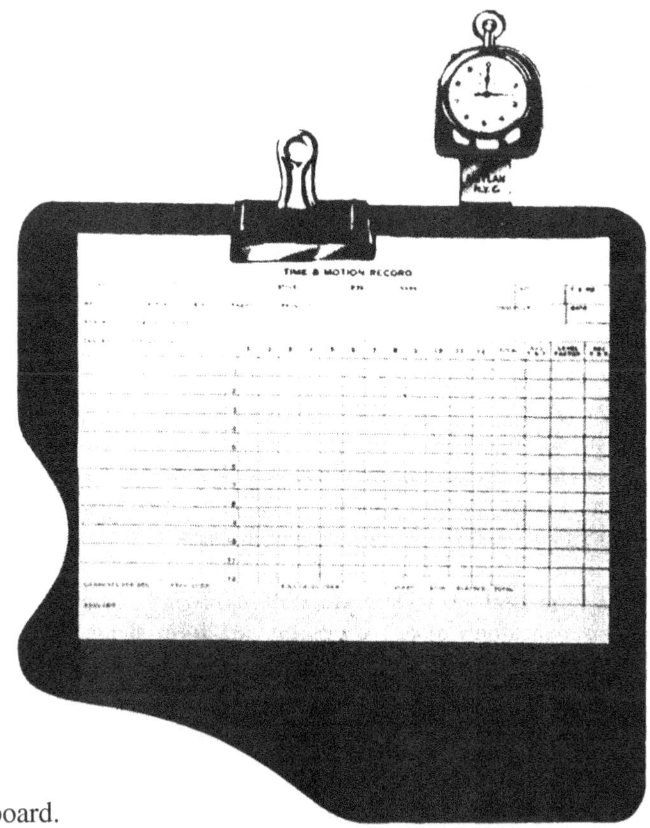

Fig. 12.12. Time study board.

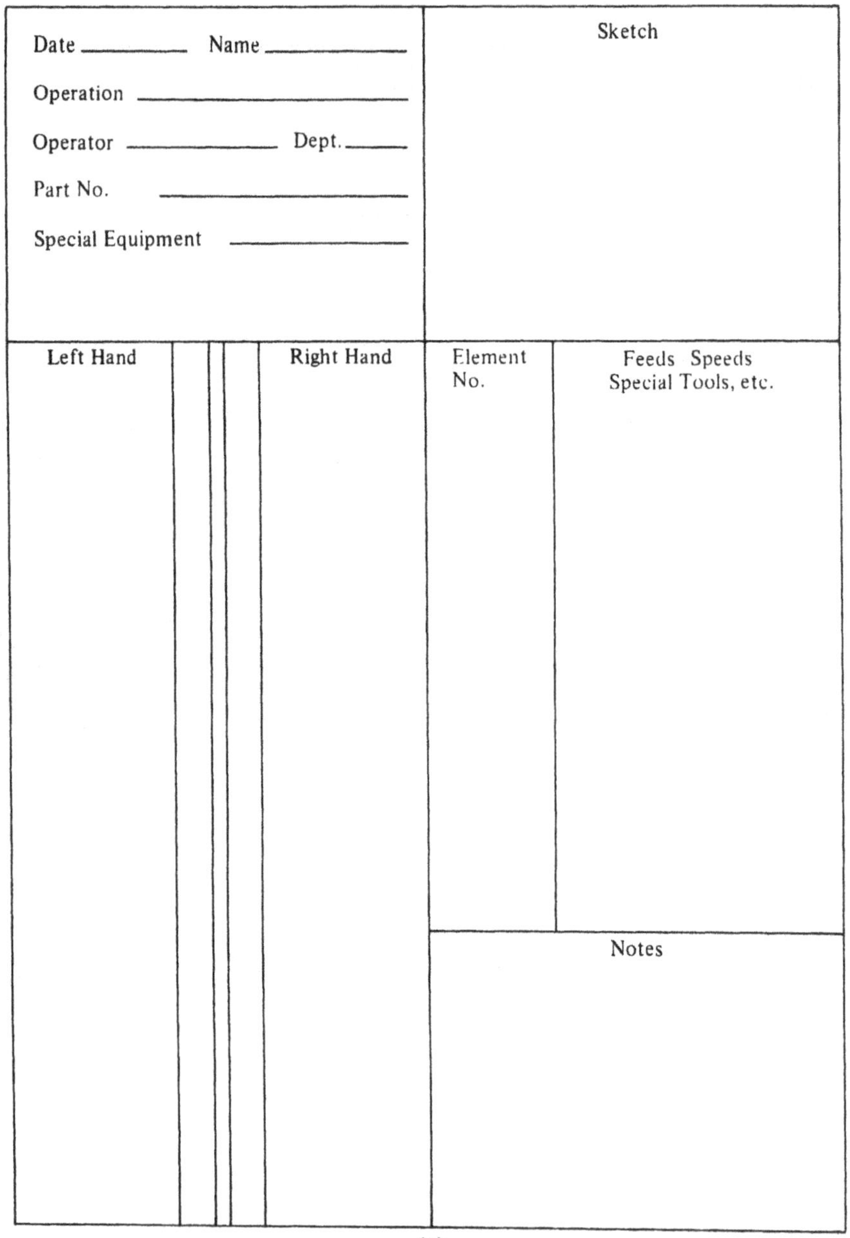

Date ________ Name ________

Operation ________

Operator ________ Dept. ________

Part No. ________

Special Equipment ________

Sketch

Left Hand

Right Hand

Element No.

Feeds Speeds Special Tools, etc.

Notes

(a)

Fig. 12.13. Time study form.

No.	Element Description	Cycle 1	2	3	4	5	6	7	8	9	10	11	12	13	Average
		R T	R T	R T	R T	R T	R T	R T	R T	R T	R T	R T	R T	R T	

Symbols Explanation	Calculations
	Performance Rating ____________

(b)

Fig. 12.13. *(con'd)*

Next, each element is averaged. Sometimes a performance rating is applied to each element, but in this case, we shall assume that the entire job (not each element) is rated. The calculations for a *different* example would be as follows (the average time for each element is obtained by averaging all the good readings for that element):

Element	*Average*
1	15
2	7
3	5
4	8

Total actual time = 15 + 7 + 5 + 8
= 35
Performance rating = 115%
Normal time = (35)(1.15) = 40.25
Allowances = 12%
Standard time = (40.25)(1.12) = **45.08**

In this case, the analyst believes that the operator is performing at 115% or 15% faster than would normally be expected. Consequently, the normal time should be adjusted to show that *more time is allowed.* To obtain the normal time, the actual time is multiplied by the performance rating. If the operator has been performing at 90%, or 10% less than what normally would be expected, the normal time would be

$$(35)(0.90) = 28.50$$

Now, realizing that there will be other problems, the problems must be "allowed" for. For example, the operator will have personal needs (rest room, water, etc.) and will become fatigued as the day progresses. Finally, there will be unavoidable delays (machine breakdowns, shortage of parts). We make a study and find that personal needs occupy 4% of the workday, fatigue accounts for 5% of the performance, and unavoidable delays occur 3% of the time. The total allowance then is

$$4\% + 5\% + 3\% = 12\%$$

To calculate the standard time, we add the allowances to the normal time and find

$$\text{Standard time} = (40.25)(1.12) = 45.08$$

The form and example above are highly simplified. No foreign elements (unnecessary moves, for example) were encountered, and performance rating was on the entire job, not on each element. Consequently, real-world time studies can often be very complicated. A competent time study analyst is a highly trained technical person who, since he or she is dealing with other employees almost all the time, must also be a human relations expert.

12.3.2 Time Study Standard Data

Often, elements of a task are repeated many times throughout an organization. Instead of a detailed study of each time a task is performed, the task could be studied in depth once and a standard data file developed. Then, whenever that element reappears, the time could be directly applied from the standard data file. *Time study standard data* may be defined as normal time values not obtained from direct time measurement of the *particular* element but obtained from direct time measurement of the element in a similar operation earlier.

For example, in a machine shop that produces many small parts all of which are the same approximate size, the element "obtain two parts and place in fixture, *X* in. to parts and *Y* in. to fixture" would probably be repeated very often. A left-hand-right-hand chart of the element is shown in Figure 12.14.

After they have been studied, the elements could be placed in a standard data file, such as that shown in Table 12.1. (The values in this table are fictitious.) Then, any time a time study analyst encounters this element, he or she can obtain the normal time directly without having to record stopwatch readings or performance ratings.

The advantages in using standard data are numerous. It must be realized, however, that standard data are not always applicable and that even when they are, some other measurement must be used for unusual elements. A list of advantages given below.

(1) Individual time studies are quicker and less costly.
(2) Consistency exists between time studies.
(3) There is less chance for error on a study.
(4) Fewer analysts may be needed.
(5) Some work simplification is done automatically as the method is prescribed.
(6) It is good for cost estimating and production planning before a job is run.
(7) There is less disturbance in the shop while the time study is being taken.

The chief disadvantage is that there is a high developmental cost to obtain the necessary data. In some organizations the cost can be completely prohibitive. In general, however, whenever standard data are applicable and the developmental cost is not prohibitive, the data should be used.

Left-hand Activity	Symbol	Symbol	Right-hand Activity
Reach X "	⇨	⇨	Reach X "
Grasp	○	○	Grasp
Move Y "	⇨	⇨	Move Y "
Position	○	○	Position
Release	○	○	Release

Fig. 12.14. Left-hand-right–hand chart of the element, "Obtain two parts and place in fixture, X in. to parts and Y in. to fixture" (time in minutes).

Table 12.1. STANDARD DATA FILE FOR "OBTAIN TWO PARTS AND PLACE IN FIXTURE, X IN. TO PARTS AND Y IN. TO FIXTURE"
(Time in minutes)

	Y			
X	*6 in.*	*8 in.*	*10 in.*	*12 in.*
6 in.	0.025	0.026	0.026	0.027
8 in.	0.026	0.026	0.027	0.027
10 in.	0.027	0.028	0.028	0.029
12 in.	0.027	0.028	0.029	0.030

12.3.3 Predetermined Times

If jobs or tasks are broken down into finer and finer elements, there will be a point reached where all tasks are made up of the same set of elements. At this point, time values can be assigned to these elements, and the total time required to perform a task can be determined by adding the appropriate elements. This approach is called *predetermined times.*

Predetermined time systems are very similar to standard data in that the time values can be read directly from charts. They are not unique in their application, however, and can be used by any company.

There are several systems in use today. Among the better known are (1) work factor and (2) methods time measurement (MTM).

The time values applied to the various elements can be obtained in several ways. One approach is to use highly accurate movie cameras and use the films (slowed down) to obtain the times (micromotion analysis). Another approach is to use electronic timing devices. Finally, time studies can be used when appropriate.

As an example, one such element might be "move object to other hand," and the time might be recorded as

Weight > 2 lb

Distance	*Time*
4 in.	0.00412 min
8 in.	0.00591 min

Of course, since the data must be extremely accurate, more decimal places are necessary.

12.3.4 Predetermined Time Standard Data

Predetermined time standard data are standard data exactly as in time study standard data *except* for the source of the information. Here, the repeated elements are timed by using predetermined time systems. For example, Figure 12.14 and Table would appear exactly the same, except that now times would be applied to the sub-elements in the chart of Figure 12.14 from predetermined time systems. Often this distinction is made by saying predetermined systems time elements at a level 1. The level 2 elements (standard data) may be made up of several level 1 elements.

The advantages and disadvantages are exactly the same as those for time study standard data. The only possible difference is that predetermined time standard data may be developed more quickly since all times are already available.

12.3.5 Work Sampling

Work sampling is the estimating of the proportion of time devoted to a given type of activity by the relative frequency occurrence of intermittent randomly spaced instantaneous observations. Suppose, for example, that it is desired to estimate the amount of time that should be set aside for personal needs during a workday or that it is desired to know the amount of time taken up by unavoidable delays such as material shortages or machine breakdowns. This can be done by sampling over an extended period of time. Basically, a series of observations is made randomly (probably through the use of random number tables such as those available in almost all texts on work sampling), and the results are recorded. Sample random numbers are shown below with possible interpretations.

Random Number	*Interpretation (When to Observe)*	
915	5:15 P.M.	(starting with 100 as 8:00 A.M. and skipping lunch)
723	3:23 P.M.	
019	8:19 A.M.	

Once a series of numbers has been generated, the observations can be made and the percent of time occupied by one activity can be estimated as follows:

$$\text{Percent occurrence for activity A} = \frac{\text{number times A occurred}}{\text{total number of observations}}$$

The number of observations necessary can be determined scientifically, but in general, enough observations are taken so that the accumulated data stabilize. For example, Figure 12.15 shows the typical response and the approximate total number of observations that should be taken—N^*.

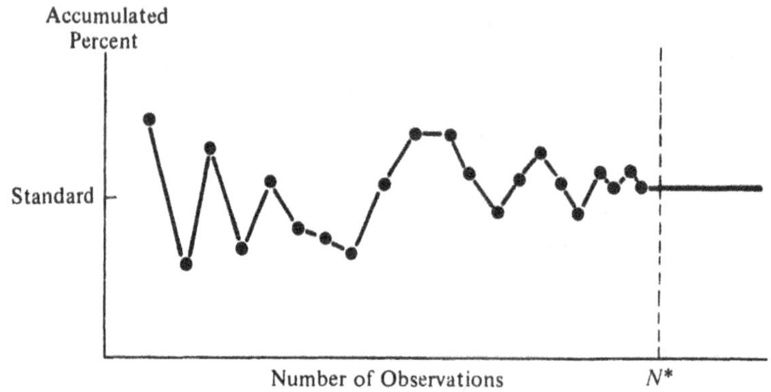

Fig. 12.15. Typical work sampling response.

12.4 ORGANIZATIONAL SYSTEM PERFORMANCE MEASUREMENT

As discussed in Section 12.1, the goal of an organization is usually some composite function of several measures. Profitability, for example, is a composite function bringing together quality, customer satisfaction, sales, cost of labor, material, energy, and others. Naturally, there are other objectives that may not be measured by profitability, but they are usually less important and we will not discuss them here.

A control system designed to optimize profitability is really a series of control subsystems—each with its own sub objectives. For example, a measured day work system controls direct labor by comparing its performance to that determined by standard times (work measurement). This is efficiency measurement in Figure 12.1. Another popular control scheme is quality control.

Recently, the industrial engineering profession has recognized that another control system is needed. *Productivity measurement and control* is the newest control system and it is receiving a tremendous amount of attention in the literature, at conferences, and in day-to-day discussions. The rest of this chapter will discuss productivity measurement (control is exercised as a result of the measurement system, hopefully).

One must question whether or not productivity measurement is necessary. Are our time-proven systems of quality control and work measurement enough, or do we need more? There is increasing evidence that more is directly needed. Once the leading nation in productivity improvement, we (the United States) are finding our rate of productivity improvement very low—so low, in fact, that we are trailing most industrialized nations in the rate of productivity improvement. Perhaps by measuring and improving productivity, we can reverse this trend.

12.4.1 Productivity Measurement Basics

Productivity measurement is simply a ratio of organizational outputs to organizational inputs [Equation (12.1)]. The ratios can be total or partial. For example, consider Figure 12.16. Given this introduction, let's look more deeply into productivity measurement. First, productivity measurement seems exceedingly simple, and, in fact, it can be. However, the following often make this measurement more complex:

(1) The relationship between input and output may be obscure.
(2) It may be difficult to measure outputs and/or inputs (e.g., hospitals and banks).
(3) Product mixes may vary considerably, affecting the ratio in a fashion difficult to understand.
(4) Input and/or output quality may vary.
(5) Data may be hard to obtain.

In addition to being total or partial measures[6] (Figure 12.16), productivity indices can be "static" or "dynamic." *Static measures* are ratios taken as a snapshot with no base year comparison. All the ratios in Figure 12.16 are static. *Dynamic measures* are actually ratios of static productivity indices, as shown in the equation

$$\text{dynamic productivity index} = \frac{\dfrac{\text{outputs this year}}{\text{inputs this year}}}{\dfrac{\text{outputs base year}}{\text{inputs base year}}} \tag{12.2}$$

Therefore, indices greater than 1 show productivity improvement, while indices less than 1 show productivity deterioration. As with static measures, dynamic measures can be partial or total.

There are two popular productivity measurement approaches. The first uses a group-generated model and is called the *normative productivity measurement methodology* (NPMM). The second is less participative in that one model can be modified to fit any organizational scheme. It is called the *multifactor productivity measurement model* (MFPMM). Each will be discussed in some detail.

12.4.2 Normative Productivity Measurement Model

The NPMM relies on experienced managers to develop the productivity measurement ratios or indices applicable to a specific company. It does this through a structured group process known as the *nominal group technique* (NGT). Although not a panacea to productivity measurement, the NPMM can usually lead to group consensus on what is important to monitor.

The procedure is first to develop the group. Normally, 5 to 12 people are chosen. They should all be familiar with the organization and should represent various interest groups. Give the group a charge. For productivity measurement, the charge might be something like: "Develop a list of productivity measures, ratios, or indices important to our organization." Then the group is immersed into the NGT. The process (which must be carefully adhered to by a chairperson) is as follows:

[6] Sink, "Organizational System Performance."

Partial Measures:	Composite:
Units/Person Hour Units/Pound of Material Units/Million BTU of Energy	$\frac{\text{Units}}{\text{\$ Labor Cost + \$ Material Cost + \$ Energy Cost}}$

Fig. 12.16. Productivity indices.

(1) *Silent Generation*—Each person is asked to generate a list of measures.
(2) *Round Robin*—After 5 to 15 minutes, the group members are asked to present their ideas. Each person gives a succinct description of one measure. The next person does the same; and so on. This continues around the group as many times as necessary to exhaust everyone's list.
(3) *Group Clarification*—First, all ideas are clarified. Then overlap or duplication of ideas is identified and remedied. (Normally, several ratios are the same or very similar and can be combined.)
(4) *Voting*—Each person is given N cards (N is determined by the leader) and asked to list the N best ideas or ratios. After this is done, each person chooses:
    (a) The best one and gives it N points.
    (b) The lowest of their list and gives it 1 point.
    (c) The second best and gives it $N - 1$ points.
    (d) The second worst and gives it 2 points.
    Etc.

The results might be as shown in Figure 12.17. From these results, we can see two things. First and most important, there is a strong consensus on what is important and what should be monitored. Second, there is some confusion on productivity. Measures 1, 4, and 5 are productivity measures, but 2 and 3 really are not productivity measures (they are more measures of effectiveness than ratios of output to input). This confusion is normal and should be expected.

	Individual Votes (5 = most impt., 1 = least)	Number of Votes/Vote Score (Total)
1. Units per Direct Labor Hour	5–4–4–3–4–5	6/25
2. Number Orders Completed on Time	3–2–2–1–1	5/9
3. Number Customers with Repeat Orders	1–1–3–5	4/10
4. Units per Million BTU (Energy)	4–5–5–4–5	5/23
5. Units per Ton of Purchased Material	2–3–1	3/6

Fig. 12.17. Results from NGT on productivity rations.

At any rate, there is a strong consensus on units per direct labor hour as being the most important; units per million Btu comes in a close second (perhaps the company is very energy intensive); and so on. From this the list of final measures can be developed and the measurement process started.

Given the list of ranked measures, it is possible to develop a composite model that actually combines these identified ratios through weighting techniques. One such process is known as the *multicriteria performance/productivity measurement technique*.[7] Details will not be presented here.

12.4.3 Multifactor Productivity Measurement Model

MFPMM is a more comprehensive and analytical approach to measuring changes in productivity. Basically, MFPMM recognizes that there are two ways profitability can be affected in an organization:

(1) Price recovery.
(2) Productivity changes.

Under price recovery, prices can be raised or lowered as the market allows or demands. Both can dramatically affect profitability. Price increases can usually be tolerated only as long as competition and consumer reaction eventually occur. The desired change is through productivity improvement(s).

The MFPMM uses accounting techniques to break the total variation (change in profit) into price effects and productivity effects, resulting in a valid productivity measurement model. This analysis can be done on both inputs and outputs. For example, profitability is affected by sales price and sales quantity (outputs). It is also, however, affected by price and quantity changes in labor, energy, materials, and so on (inputs). Careful manipulation of the data will demonstrate what effects each change has.[8]

12.5 COMPUTERS AND WORK MEASUREMENT AND DESIGN

Computers have had a strong impact on the field of work measurement and design, but particularly on work measurement. For example, several firms use microprocessors directly wired to workstations. This way, a long-range time study can be conducted yielding very accurate measurements.

There has been a great deal of work toward computerizing predetermined time systems. For example, the computer can take simple coded activities, look up the predetermined times, and do all the calculations. Another more sophisticated approach is to store some previously determined times for higher levels of activities (several predetermined elements as in standard data) and to use these whenever appropriate.

[7]D. Scott Sink, "Multi-criteria Performance/Productivity Measurement Technique," *Productivity Management*, Vol. II, No. IV, Stillwater, Okla., Fall 1983.

[8]The details of MFPMM techniques are beyond the scope of this chapter and will not be presented. For more, see J. C. Swaim et al., "Current Developments in Firm or Corporate Level Productivity Measurement and Evaluation," *Proceedings 1983 Fall Industrial Engineering Conference*, Toronto, Canada, Nov. 1983.

When similar products are produced, it may be possible to store standard times by product descriptions, such as weight or material used. By adequately describing the product to the computer, the appropriate standard time could be obtained. Interpolation could be used for given characteristics that have not previously been calculated.

Finally, there are numerous routines available on computers that generate random numbers. These random numbers can then be used in the design of a work sampling procedure.

12.6 PERSPECTIVE

In designing productive work systems, I.E.s put together various components that will operate in such a way as to achieve the outputs desired from the system. The components making up a productive work system include:

- *Equipment*—Machines, material handling devices, gauges, instruments, etc.
- *Processes*—Chemical treatments, thermal treatments, laser hardening, etc.
- *Materials*—Raw materials, purchased parts, lubricants, chemicals, supplies, etc.
- *Capital*—Cash, credits, debits, investments, etc.
- *Energy*—Electrical, steam, mechanical.
- *Facilities*—Buildings, other structures.
- *Information*—Forecasts, sales records, costs, production orders, vendor histories, shop floor status, etc.
- *People*—Workers, supervisors, design engineers, quality control engineers, salespeople, managers, etc.

When the I.E. specifies a specific machine to include in the system, he does so knowing the performance characteristics of that machine. Among the characteristics known about a typical machine are:

- *Range of Operations*—Operations it is capable of performing.
- *Size of Workpiece*—The machine can accommodate a workpiece up to a certain size and shape.
- *Speed of Operation*—Spindle speed, etc.
- *Tolerance Capabilities*—The tolerances the machine is capable of holding.
- *Energy Consumption*—Energy consumed in operation.
- *Maintenance Requirements*—Frequency of overhaul, other maintenance operations.

When the I.E. specifies that a human being should be included as a specific component in the system, it is equally important that proper consideration be given to the performance characteristics of that person. People have performance characteristics just as machines do.

Figure 12.17 illustrates some of the performance characteristics of a "typical" human being compared to a "typical" machine tool. It should be clear from these comparisons that human beings are ideally suited for many functions in a system, but woefully inadequate for others.

The field of study that is concerned with quantifying and measuring the performance characteristics of human beings is called *human factors engineering* or simply *human factors.* Most of the research and teaching associated with human factors is conducted within industrial engineering departments, usually at the graduate level. Human factors is an interdisciplinary field that includes professionals from the behavioral sciences (e.g., psychologists and sociologists) and the biological sciences (e.g., physiologists and bioengineers) as well as industrial engineers.

Human performance characteristics fall into two broad categories, physiological and psychological. *Physiological performance characteristics* have to do with the physical aspects of human activities (we are concerned primarily with work activities), such as lifting, reaching, carrying, hearing, seeing, and speaking. *Psychological performance characteristics* are those dealing with the mental aspects of human activities, such as stress, boredom, and motivation.

12.7 PHYSIOLOGICAL ASPECTS OF HUMAN PERFORMANCE[9]

Human beings, by virtue of the biological characteristics of their bodies, are limited in what they can do in a work system. Specifically, the abilities of people to perform various types of motor activities depend on the physical structure of the skeleton, the skeletal muscles, the nervous system, and the metabolic processes.

We are concerned primarily with the bones comprising the upper and lower extremities and the bones of the spine, since these are concerned with the execution of physical activities. These skeletal bones are connected at body joints, such as the fingers, wrists, elbows, shoulders, hips, knees, ankles, and feet. The manner in which they are shaped and connected puts limitations on our ability to perform physical activities. The wrist will rotate only so far. Our knees bend in one direction but not the other. Our thumbs oppose the four fingers, giving us considerably more dexterity in our hands than any other animal. (Some anthropologists claim that this characteristic is at the root of human mental superiority in the animal world.)

The bones are held together at their joints by ligaments. The skeletal muscles are connected to different bones through tendons, such that when the muscles are activated, the bones apply mechanical leverage. The fibers of the muscles convert

[9]Our coverage of human factors will necessarily be brief. For more complete coverage, see, for example, M. S. Sanders and E. J. McCormick, *Human Factors in Engineering and Design*, 6th ed. (New York: McGraw-Hill Book Company, 1987).

Performance Characteristic	*"Typical" Machine Tool*	*"Typical" Human Being*
Range of Operations	Most M.T.'s perform one basic operation; some perform up to 10.	Extremely broad range.
Size of Workpiece	Some machines perform operations on microscopic-sized workpieces; others machine workpieces weighing serveral tons.	Very limited range. Cannot perform operations on extremely small items. Can perform repetitive operations on items weighing up to 50 pounds, where physical handling is required.
Speed of Operation	Can be extremely fast.	Usually very slow.
Tolerance Capabilities	Machines routinely hold tolerances to 0.001 inch.	Very poor at accuracy and repeatability.
Energy Consumption	Very efficient in converting energy into useful output.	Very low level of energy output; cannot sustain output on even level.
Maintenance Requirements	Regular maintenance required; complete overhaul needed periodically.	Needs minor "maintenance" frequently, such as water, food, oxygen, rest room breaks, bandaids, etc. Major overhauls are performed continuously, as the body mends itself.
Response to Unexpected Occurrences	Very limited capability.	Extremely resourceful and creative.

Fig. 12.17. Performance characteristics of a typical machine tool vs. a human being.

chemical energy (from food) into mechanical work, such as lifting, walking, and so on.

The actions of the muscles are controlled by motor nerves. The nerves convey conscious and subconscious messages from the brain to the muscles, causing physical activities to occur. A machine operator reads a work instruction from a blueprint, then executes that instruction by sending the appropriate message over the nervous system to the appropriate muscles, which move bones so as to accomplish the specified task.

Muscles contract to move bones. Contraction requires energy. As it is consumed, it must be replenished through the circulatory system. The heart continuously pumps blood through the circulatory system. The blood picks up nutrients from the digestive process and it replenishes its supply of oxygen from the lungs. The nutrients and the oxygen are carried by the blood to the muscles throughout the body.

Researchers in human factors have carefully studied the process described above and have determined the energy requirements for various types of jobs. Based on these data, jobs can be designed which can be performed by a large percentage of people likely to be involved in such work. Also, optimal rest periods can be specified.

Substantial amounts of detailed information have been generated concerning the biomechanics of motion of the human body. For example, tables of data exist which characterize the movements of the arms, legs, and other body members. As a specific example, the average person's elbow can bend 142 degrees. People at the 5th percentile can bend their elbows only 126 degrees, while people at the 95th percentile can bend their elbows 158 degrees.

Data such as these are extremely useful in designing workplaces for optimum human performance. They are also very useful in designing tools. Better work methods can also be determined, such as having people lift with the large muscles of their legs rather than with the much weaker muscles of their back.

12.8 PSYCHOLOGICAL ASPECTS OF HUMAN ACTIVITIES

The physiological aspects of human activities discussed in the preceding section view the human being as a mechanism, much as one would view a machine. While the human body is, indeed, a mechanism governed by physical laws, it also contains a thinking, reasoning, feeling brain. Human beings experience pain, joy, sadness, depression, anger, boredom, frustration, fear, outrage, jealousy, love, hate, and (occasionally) schizophrenia.

We have not made very much progress in quantifying, measuring, or understanding the psychological aspects of human performance. We can predict with much greater accuracy the limit of how much a person can lift than the limit of how much boredom that same person can withstand.

Do people respond better in large open offices or smaller walled ones? With windows or without? Does soft music stimulate thinking, or slow down the work pace?

Does fast music result in more output per hour? Unfortunately, the research that has been performed on these questions is inconclusive. Individual differences on such questions are great. Also, people tend to be very adaptable and seem to quickly become oblivious to many such factors.

We can only make some broad, sweeping statements about the psychological aspects of human activity:

- *People are still motivated by money.* Even among people in higher income brackets, the size of their salary is a quantitative measure of their relative worth to the organization.
- *People would rather be doing productive work than be idle.* This has less to do with the "work ethic" than with basic human nature. People are bored when they have nothing to do.
- *Many American workers do not necessarily want "job enrichment."* Behavioral scientists have long proclaimed that workers want more variety and more challenge in their work. Recent attempts to implement "job enrichment" programs have met with surprising resistance.
- *Workers themselves often have the best ideas on how to increase productivity, if they were only asked.* It is astounding that most American managers have not yet learned that their employees have brains as well as hands.
- *Most people respond positively when they know the objectives of their organization, what their own jobs are, and that they will be treated fairly, consistently, and with respect.*

12.9 HUMAN INTERFACE WITH THE WORLD OF WORK

An important dimension of human performance is the various ways in which human beings interact with their world. Some of the important interfaces are with the following entities:

- *Work Environment*—The physical environment in which the productive work system exists.
- *Social Structures*—Basic values, creeds, laws, traditions, community arrangements, legal structures, educational systems.
- *Machines*—The physical components of equipment and processes in the workplace.
- *Information/Communication Systems*—The total set of systems and procedures by which work system components exchange information.
- *Organizational/Supervisory Structure*—The total set of authority/responsibility/accountability relationships within the work system.
- *Robots/"Intelligent" Machines*—The rapidly emerging set of electromechanical devices which, via embedded computer-based "intelligence," are assuming certain humanlike characteristics.

As industrial engineers, we are particularly interested in human interfaces with the work environment, machines, information/communication system, organizational/supervisory structure, and robot/"intelligent" machines. Each of these will be explored briefly in following sections.

12.9.1 Human Interface with the Work Environment

Industrial engineers design workplaces. These may be some combination of tables, desks, seats, chairs, and so on. Machines, equipment, telephones, computer terminals, and other devices are incorporated into the workplace according to the needs of the job.

Most workplaces are such that several different people will work there over a time interval. If the company is working multiple shifts, two or three different people may use the same workplace in one day. These factors must be considered when the I.E. specifies work-space dimensions. Where possible, adjustable chair heights and other dimensions are specified.

Considerable effort is also required to arrange various components of the workplace. Consider the workplace of an airline pilot. Where is the optimal location of the dozens of dials, gauges, levers, and buttons? Some of the components that must be arranged are visual displays, hand controls, foot controls, audio inputs, and audio outputs.

How much light (illumination) is needed for specific jobs? Considerable data exist that provide illumination standards for a variety of jobs. The widespread use of computer terminals has required modification of many lighting systems to reduce glare on the screens. Safety relative to lengthy exposure to CRTs is a growing concern.

Atmospheric conditions are also important in designing work environments. Temperature, humidity, and air cleanliness are the factors considered most frequently. If, for example, dust cannot be controlled adequately in the work environment, respirators should be furnished to all employees affected. Noise is still another atmospheric factor that must be considered when designing work environments. Excessive noise is not only annoying and detrimental to good work, but can be dangerous to the health of the employees.

Finally, the work environment should be safe and attractive. It should be designed with good maintenance in mind. Good housekeeping practices are essential. Guards, carefully marked walkways, warning lights, and so on, are all important in designing work areas with employee safety in mind.

12.9.2 Human Interface with Machines

Virtually all human beings interface with machines of one type or another. The human/machine interface is obvious for truck drivers, lathe operators, crane operators, typists, and computer operators. In most productive work systems involving people and machines, the human being serves a control function. The person adjusts the workpiece in the machine, pushes the start button, watches as the machine operates, stops the machine in emergencies, decides when to shut the machine off, and so on.

Human factors data are extremely helpful in selecting the size, shape, and location of various controls. Consider the human-machine interface involved in driving an automobile. All the displays need to be designed carefully to be immediately recognizable. They also need to be located such that they are easily visible, even when the auto is in motion. The steering wheel needs to be correctly sized so that the driver has the proper "feel" of the road as he makes steering corrections. The brake pedal must be located properly and the correct amount of resistance built in. Warning lights and signals indicate "out-of-norm" conditions. If the oil in the crankcase is too low, a red light comes on. If the cooling system loses its coolant, the temperature light comes on. Recent automobiles provide verbal warnings.

One of the most important tasks of the I.E. is to allocate functions among machines and human beings in the system. In making this allocation, some general guidelines can serve to delineate between people and machines.

In general, human beings are better in performing the following types of tasks:

(1) Recognizing complex patterns, which may vary considerably over time.
(2) Recognizing unusual and unexpected events.
(3) Storing large amounts of information over long time periods, then assimilating this information into general principles and strategies. (This is the basis of intuition.)
(4) Quickly retrieving pertinent information; using associative abilities to retrieve many related items of information and to project these to inferences involving complex associations between related information.
(5) Adapting the decision process to differing situations.
(6) Reasoning inductively, generalizing from specific observations.
(7) Making subjective evaluations and decisions.
(8) Discerning the most important activities to concentrate on when overload exists.

In general, machines are better in performing the following types of tasks:

(1) Monitoring for prespecified events.
(2) Storing quantitative data quickly and in great volume.
(3) Retrieving and processing quantitative data quickly.
(4) Performing repetitive activities reliably over extended periods of time.
(5) Responding to inputs rapidly and consistently.
(6) Ignoring distractions.

These and other factors are considered when allocating functions among machines and human beings in a complex work system.

12.9.3 Human Interface with Information/Communication Systems

The human being is an essential component in essentially all information/communication systems. These systems are becoming very complex in most modem organizations.

People receive information person to person and from written reports, information displays, telephone calls, buzzers, lights, signals, dials, sounds, and so on. People transmit information in writing, through keyboards, verbally, and by actuating certain input devices.

This is a complex subject. For our purposes, we will examine one small portion of this topic by considering when to use auditory versus visual presentation of information. This is summarized in Figure 12.18.

12.9.4 Human Interface with Organizational/Supervisory Structure

This particular interface has to do with the entire set of issues involving job design, reporting relationships, performance appraisal, evaluation, and training. These issues are closely interwoven with the concepts discussed in Section 12.8.

We made the point in Section 12.8 that most people respond positively when they know the objectives of their organization, what their own jobs are, and that they will be treated fairly, consistently, and with respect. People want to be treated as individuals and not as numbers or as blocks on an organization chart. They also want to have some input into decisions that affect them directly, such as the arrangement of the workplace and the inspection of the work they perform.

During the early days of the American industrial revolution (roughly 1850-1920), the work force in our factories came primarily from farms and from the masses of people who were immigrating to the United States. These people were

Use Auditory Presentation if:	*Use Visual Presentation if:*
• Message is simple	• Message is complex
• Message is short	• Message is long
• Message not referred to later	• Message will be referred to later
• Message deals with events in time	• Message deals with location in space
• Message calls for immediate action	• Message does not call for immediate action
• Visual systems of the person is overburdened	• Auditory system of the person is overburdened
• Receiving location is too bright	• Receiving location is too noisy
• Person's job requires worker to move about continually	• Person's job allows worker to remain in one position

Fig. 12.18. When to use auditory or visual presentation [From B. H. Deatherage, "Auditory and Other Sensory Forms of Information Presentation," in H. P. Van Cott and R. G. Kinkade (eds.), *Human Engineering Guide to Equipment Design,* rev. ed. (Washington, D.C., U.S. Government Printing Office, 1972).]

largely uneducated and unskilled. As the industries grew from small numbers of workers (say 30) to very large work forces (several thousand) the company managers were faced with the problem of how to group (organize) the workers for maximum efficiency. The early efforts at organizing were largely trial and error.

Organizing and managing large work forces was a new endeavor. The only large organizations in the past had been military organizations. Consequently, company managers adopted many of the practices of military organizations, this being the only experience base from which to draw.

The result has been the rigid, formal-looking organization charts that we see on company walls that are supposed to represent authority/responsibility relationships. We hear a lot about "the chain of command," or "we must go through channels." Yet that is not how real organizations actually work.

The military model also gave managers their supervisory style, which basically amounts to "do what you are told, don't ask questions, and have unswerving loyalty to the company." Unfortunately, this *is* the way the supervisory style actually works in many organizations.

There are no clear-cut solutions to these problems at the present time, but we are beginning to see some alternative styles of organization and supervision. These are largely experimental, and we have a long way to go.

A great amount has been written in recent years about "quality circles," "Japanese management style," "participatory management," and so on. The industrial engineer must play a central role in finding more effective means of organizing and supervising the work force of organizations.

12.9.5 Human Interface with Robots and Intelligent Machines

If any one thing symbolizes the second industrial revolution that is now under way, it is the robot. Although robots are certainly an important aspect of the new manufacturing technology, they are by no means the dominant driving force.

Robots and robotic devices have been sensationalized by the popular press and by science fiction movies. In addition, there is a concern among many blue-collar workers that robots will displace them. Thus, to many uninformed people, robots are mysterious things that appear to be threatening and to have humanlike qualities.

One thing we will *not* see is a bunch of two-legged robots marching into a factory having today's typical equipment, and operating the system in the same manner as would human operators. In those situations where robots can play an important role, the total set of manufacturing equipment with which the robot must interact is specifically designed with the robot in mind. It would be foolish (or at best poor design) to consider robots as one-to-one replacements of human beings with all other equipment unchanged. Such an approach would not take advantage of the robot's superiority to the human being in certain areas, nor would it consider the inferiority of the robot in other areas.

Many early applications of robots did not take advantage of their capabilities, leading one manager to remark that his industrial engineers should design a job enrichment program for the robot! In this particular case, a large, expensive robot was simply grasping items coming from a conveyor, raising them 2 feet, and then loading them onto a take-away conveyor that ran at right angles to the first conveyor. The robot possessed accuracy and repeatability capabilities on the order of 10 times greater than were needed. Furthermore, a fairly standard transfer mechanism using hard automation concepts would have been far superior and much less expensive.

Many factories of the future will contain numerous robots. Within the foreseeable future, there will also be many human beings in these factories. Industrial engineers must develop methods for designing productive work systems which include a varied mix of human beings, robots, and other "intelligent" machines.[10]

There are at least three sets of issues involving the interface between human beings and intelligent machines. The first set of issues is concerned with how to allocate work functions and control functions optimally among the human and machine elements comprising the system. Figure 12.17 compared certain performance characteristics of human beings and "unintelligent" machines. A comparable analysis would have to be conducted for each different type of intelligent machine.

The second set of issues concerning the human being and intelligent machine interface is that of safety. A robot or any machine will activate its components (such as swinging an arm) in response to specific commands. It will not know to stop its movement if a human being happens to be in its path. There have been several confirmed deaths of human beings caused by robot arm movement. Considerable effort is being directed toward the provision of safety features on robotic systems. However, this is expected to remain a critical issue for many years to come.

The third set of issues concerning the interface between human beings and intelligent machines has to do with the question of "who has the final say" when an intelligent machine wants to do one thing and the person wants to do another.

At what point, and under what conditions, should a human being be permitted to override a decision made by an intelligent machine? What about the reverse situation? Should a machine ever be permitted to override the decision of a human being?

Research into these issues will be pursued for many years. Industrial and systems engineers have an opportunity to contribute to this important effort.

---

[10]The term *machine intelligence* is frequently used to characterize such capabilities as machine vision (image processing), pattern recognition, audio input/output, tactile sensing, and adaptive control (learning, or modifying its own behavior).

DISCUSSION QUESTIONS

1. Define work measurement, work design, and productivity measurement. Discuss the relationship between work measurement and productivity measurement.
2. Discuss organizational systems performance measures other than "productivity" that might be important.
3. Ajax Manufacturing produces wooden kegs utilized in plants in Kentucky and Tennessee. The only product is wooden kegs, but main inputs include labor, energy, and wood. List potential productivity measures for the plant and state whether they are partial or total.
4. You are asked to study the procedure for issuing and executing back orders. (Back orders occur when inventory is temporarily depleted. An order is issued so that the product will be shipped as soon as it is available.)

   In another situation, you are asked to study the procedure for assemblying fountain pens. Basically, one operator assembles all fountain pens and all parts are brought to her workstation.

   State what tools of work methods simplification you might utilize.
5. Define normal times, normal pace, allowances, standard times, direct time studies, standard data, predetermined time systems, and work sampling. Discuss the uses of the last four.
6. Give the five-step approach to work simplification and show the use of charting in the five steps.
7. Discuss the advantages and disadvantages of predetermined time studies over direct time studies.
8. Discuss the primary difference between a static productivity ratio and a dynamic productivity index.
9. List several performance measures that a university might use.
10. Discuss why you think the nominal group technique (NGT) might work well for group consensus on productivity measurement.
11. Eskimo Mike's is a tremendously popular college town pub. It produces and sells hamburgers, beer, mixed drinks, and other fast foods. Most of the employees are older students and recent graduates. You are the manager. List one example each of the measures shown in Fig. 12.1 (efficiency, effectiveness, quality, innovation, productivity, quality of working life, profitability).
12. Is it proper to compare "performance characteristics" of human beings and machines? Discuss thoroughly and justify your answer.
13. Distinguish between physiological and psychological human performance characteristics.
14. Study carefully the furniture in your apartment or dormitory room. Pay especially careful attention to the physical dimensions of the furniture. Speculate on how each major piece of furniture was given its dimensions. Do you believe that these dimensions are proper for the 1990s and the 2000s?
15. Apply the general concepts discussed in this chapter to the design of an ideal engineering student workstation for a dormitory room.
16. Conduct a library research project on the topic "quality circles." Trace the development of this movement and present your conclusions and recommendations.

17. Conduct a library research project on "human-machine interface" in industrial organizations. State your conclusions.
18. From literature in your library, speculate on the degree to which robots will have humanlike characteristics by 2000.
19. With human beings having less and less direct involvement with the manufacturing process, per se, do you believe that the study of human factors will diminish in importance in the future?
20. Conduct a library research project on recent innovations in organizational structures. Is it necessarily true that hierarchical organizations are a thing of he past? Justify your response with an appropriate comparative analysis.

PROBLEMS

1. Given the following data, compute the production standard (pieces per hour), the standard labor cost per 100 pieces, and the standard hours per 100 pieces. The allowances are 4% for fatigue, 8% for unavoidable delays, and 8% for personal needs. The operator is paid $5.00 per hour. Performance rating = 105%. (Instructor: Look also at Problem 12.)

	Reading (min)			
Element	*1*	*2*	*3*	*4*
1	*0.020*	*0.010*	*0.009*	*0.012*
2	*0.002*	*0.003*	*0.004*	*0.004*
3	*0.100*	*0.110*	*0.009*	*0.050*
4	*0.205*	*0.195*	*0.200*	*0.210*

2. Take an operation that is performed at one workstation with which you are very familiar and make a left-hand–right-hand chart of it. List the principles of motion economy and show which areas are applicable to that operation. Finally, show a revised left-hand–right-hand chart and calculate labor savings for an appropriate rate.
3. Make a flow process chart for a process with which you are familiar. Examine it and see which activities can be eliminated, combined, changed in sequence, or improved (in that order of preference). Show the revised flow process chart.
4. You are working with Giziwizets, Inc., on productivity measurement. They give you the data below. For those data, calculate:
   (a) Partial productivity ratios for this year and the base year.
   (b) Productivity ratios for this year.
   What can you say about the company's productivity?

	Base Year	*This Year*
Output (number of giziwizits)	110	160
Labor hours	250	275
Energy (10^6 Btu)	10,000	60,000
Material (lb)	500	675

5. You have made time studies of a task that occures frequently within your plant. You find that standard time varies directly with the weight of the product. For the given values, graph time vs. weight and draw a straight line. Fill in your "estimate" for the line (i.e., visually estimate ***a*** and ***b*** below).

$$y = a + bx$$
$a = y$ intersect
$b =$ slope
$x =$ weight
$y =$ time

Weight (lb)	*Time (min)*
1.00	0.190
5.75	0.300
3.50	0.230
4.00	0.260

Why is a not a zero value?

6. You are to design the workstation for installing shoelaces in shoes. The shoes are delivered to the workstation in a cardboard box and are taken to the next station in the same box. (Assume that you must use the box even though a conveyor would be better and assume that the operator does not have to move the boxes except to move the full one out of the way and an empty one in its place.) Using the principles of motion economy, design the best method and record it on a left-hand-right-hand chart. Time the method by using a stopwatch obtained from your instructor (ignore any fumbles by dropping elements where fumbles occur). Using allowances of 15%, determine the standard time per pair of shoes. Assume a standard pair of shoes with four holes on each side and 18-in. laces.
7. Repeat Problem 6 for an operation with which you are familiar, for example, changing spark plugs in a specific 4-, 6-, *or* 8-cylinder car and sharpening a box of pencils (assume an infinite supply and an electric sharpener).
8. Make a flow process chart of a person changing a flat tire from the time the person first gets out of the car to check the tire. Follow the person and stop when he gets back in the car (flat tire in the trunk). For the first time, have a person perform the activity as he normally would. Then make a revised flow process chart and show the total time for both. (For each activity, first try to *eliminate the activity*, then try to *combine it* with another, then try to change the sequence for better flow, and finally, try to improve the method of performing the activity.)
9. To test the validity of your allowance for personal delays (8%), you ran a work sampling study. Out of 638 observations, 51 were personal delays. *Assuming* that this is enough observations, should the allowance be changed? If so, to what value?
10. In conducting a work sampling study, you find the following results. Graph the results and from that visually estimate how many more readings will be required.

Number of Observations	*Accumulated %*
50	34
60	26.5
70	32.5
80	28

11. Form into groups of approximately five. Using the Normative Group Technique, determine a list of seven productivity measures that Eskimo Mike's should use. (See Question 11 of this chapter. Comment on the process.)

12. Redo Problem 1 if the performance rating was 90%. Discuss the change and the role of performance rating.

CHAPTER

13 SAFETY ENGINEERING

Major Topics

- Costs of Accidents
- Accidental Deaths in the United States
- Accidents versus Other Causes of Death
- Work Accident Costs and Rates
- Time Lost Because of Work Injuries
- Deaths in Work Accidents
- Work Injuries by Type of Accident
- Death Rates by Industry
- Parts of the Body Injured on the Job
- Chemical Burn Injuries
- Heat Burn Injuries
- Repetitive Strain/Soft Tissue Injuries
- Estimating the Cost of Accidents
- Global Impact of Accidents and Injuries
- Establishing a Safety-First Corporate Culture
- Domino Theory of Accident Causation
- Human Factors Theory of Accident Causation
- Accident/Incident Theory of Accident Causation
- Epidemiological Theory of Accident Causation
- Systems Theory of Accident Causation
- Combination Theory of Accident Causation
- Behavioral Theory of Accident Causation
- Drugs and Accident Causation
- Depression and Accident Causation
- Management Failures and Accident Causation
- Obesity and Accident Causation

There is a long history of debate in this country concerning the effect of **accidents** on industry (the workers and the companies) and the cost of preventing accidents. Historically, the prevailing view was that **accident prevention** programs were too costly. The more contemporary view is that accidents are too costly and that accident prevention makes sense economically. As a result, accident prevention, which had been advocated on a moral basis, is now justified in economic terms.

Accidents are the fourth leading cause of death in this country after heart disease, cancer, and strokes. This ranking is based on all types of accidents, including motor vehicle accidents, drownings, fires, falls, natural disasters, and work-related accidents.

Although deaths from **natural disasters** tend to be more newsworthy than workplace deaths, their actual impact is substantially less. For example, natural disasters in the United States cause fewer than 100 deaths per year on average. **Workplace accidents**, on the other hand, cause more than 10,000 deaths every year in the United States.[1]

This chapter provides prospective and practicing **safety and health professionals** with the information they need to have a full understanding of workplace accidents and their effect on industry in the United States. Such an understanding will help professionals play a more effective role in keeping both management and labor focused appropriately on safety and health in the workplace.

COSTS OF ACCIDENTS

To gain a proper perspective on the economics of workplace accidents, we must view them in the overall context of all accidents. The overall cost of accidents in the United States is approximately $150 billion. These costs include such factors as **lost wages**, **medical expenses**, **insurance administration**, **fire-related losses**, motor vehicle **property damage**, and **indirect costs**.

Figure 13-1 breaks down this overall amount by categories of accidents. Figure 13-2 breaks them down by cost categories. Notice in Figure 13-1 that workplace accidents rank second behind motor vehicle accidents in cost. Figure 13-2 shows that the highest cost category is wages lost by workers who are either injured or killed. The category of indirect losses from work accidents consists of costs associated with responding to accidents (i.e., giving first aid, filling out accident reports, handling production slowdowns).

Clearly, accidents on and off the job cost U.S. industry dearly. Every dollar that is spent responding to accidents is a dollar that could have been reinvested in

FIGURE 13-1 Accident costs by accident type (in billions, in a typical year).

Motor vehicle accidents	$72
Workplace accidents	48
Home accidents	18
Public accidents	12

FIGURE 13-2 Accident costs by categories (in billions, in a typical year).

Wages lost	$37
Medical expenses	24
Insurance administration	29
Property damage (motor vehicle)	27
Fire losses	10
Indirect losses for work accidents	23

modernization, research and development, facility upgrades, and other competitiveness-enhancing activities.

ACCIDENTAL DEATHS IN THE UNITED STATES

Accidental deaths in the United States result from a variety of causes, including motor vehicle accidents, falls, poisoning, drowning, fire-related injuries, **suffocation** (ingested object), firearms, **medical complications**, air transport accidents, interaction with machinery, mechanical suffocation, and the impact of falling objects. The National Safety Council (NSC) periodically computes death totals and death rates in each of these categories. The statistics for a typical year are as follows:

- ***Motor vehicle accidents.*** Motor vehicle accidents are the leading cause of accidental deaths in the United States each year. They include deaths resulting from accidents involving mechanically or electrically powered vehicles (excluding rail vehicles) that occur on or off the road. In a typical year, there are approximately 47,000 such deaths in the United States.
- ***Falls.*** This category includes all deaths from **falls** except those associated with transport vehicles. For example, a person who is killed as the result of falling while boarding a bus or train would not be included in this category. In a typical year, there are approximately 13,000 deaths in the United States from falls.
- ***Poisoning.*** The **poisoning** category is divided into two subcategories: (1) poisoning by solids and liquids and (2) poisoning by gases and vapors. The first category includes deaths that result from the ingestion of drugs, medicine, widely recognized solid and liquid poisons, mushrooms, and shellfish. It does not include poisoning from spoiled food or salmonella. The second category includes deaths caused by incomplete combustion (for example, gas vapors from an oven or unlit pilot light) or from carbon monoxide (for example, exhaust fumes from an automobile). In a typical year, there are approximately 6,000 deaths in the first category and 1,000 in the second.
- ***Drowning.*** This category includes work-related and non-work-related **drownings** but excludes those associated with floods or other natural disasters. In a typical year, there are approximately 5,000 deaths from drowning in the United States.
- ***Fire-related injuries.*** This category includes deaths from burns, asphyxiation, falls, and those that result from falling objects in a fire. In a typical year, there are over 4,000 fire-related deaths in the United States.
- ***Suffocation (ingested object).*** This category includes deaths from the ingestion of an object that blocks the air passages. In many such deaths, the ingested object is food. In a typical year, there are approximately 4,000 suffocation deaths in the United States.
- ***Firearms.*** This category includes deaths that result when recreational activities involving firearms or household accidents involving firearms result in death. For example, a person killed in the home while cleaning a firearm would be included in this category. However, a person killed in combat would not be included. In a typical year, there are approximately 2,000 such deaths in the United States.

FIGURE 13-3 Causes of accidents (ages 25 to 44 years in a typical year).

Accidents	27,500
Cancer	20,300
Motor vehicle	16,500
Heart disease	16,000
Poison (solid, liquid)	2,700
Drowning	1,500
Falls	1,100
Fire related	900

- ***Others.*** This category includes deaths resulting from medical complications arising out of mistakes made by health care professionals, air transport injuries, interaction with machinery, **mechanical suffocation**, and the impact of falling objects. In a typical year, there are over 14,000 deaths in these subcategories.[2]

ACCIDENTS VERSUS OTHER CAUSES OF DEATH

Although there are more deaths every year from **heart disease**, **cancer**, and **strokes** than from accidents, these causes tend to be concentrated among people at or near retirement age.[3] Among people 37 years of age or younger—prime working years—accidents are the number one cause of death.[4] Figure 13-3 summarizes the causes of death for persons from 25 to 44 years of age. Notice that the leading cause is accidents.

Figure 13-3 shows that accidents represent a serious detriment to productivity, quality, and competitiveness in today's workplace. Yet accidents are the one cause of death and injury that companies can most easily control. Although it is true that companies may have some success in decreasing the incidence of heart disease and stroke among their employees through such activities as corporate wellness programs, their impact in this regard is limited. However, employers can have a significant impact on preventing accidents.

WORK ACCIDENT COSTS AND RATES

Workplace accidents cost employers millions every year. Consider the following examples from the recent past. Arco Chemical Company was ordered to pay $3.48 million in fines as a result of failing to protect workers from an explosion at its petrochemical plant in Channelview, Texas. The steel-making division of USX paid a $3.25 million fine to settle numerous health and safety violation citations. BASF Corporation agreed to pay a fine of $1.06 million to settle Occupational Safety and Health Administration (OSHA) citations associated with an explosion at a Cincinnati chemical plant that caused two deaths and 17 injuries.

These examples show the costs of fines only. In addition to fines, these employers incurred costs for safety corrections, medical treatment, survivor benefits, death and burial costs, and a variety of indirect costs. Clearly, work accidents are expensive. However, the news is not all bad. The trend in the rate of accidents is downward.

Work **accident rates** in this century are evidence of the success of the safety movement in the United States. As the amount of attention given to workplace safety and health has increased, the accident rate has decreased.

Many safety and health professionals believe that the major cost of accidents and injuries on the job results from damage to morale. Employee morale is a less tangible factor than documentable factors such as lost time and medical costs. However, it is widely accepted among management professionals that few factors affect productivity more than employee morale. Employees with low morale do not produce up to their maximum potential. This is why so much time and money are spent every year to help supervisors and managers learn different ways to help improve employee morale.

Because few things are as detrimental to employee morale as seeing a fellow worker injured, accidents can have a devastating effect on morale. Whenever an employee is injured, his or her colleagues silently think, "That could have been me," in addition to worrying about the employee. Morale is damaged even more if the injured employee is well-liked and other employees know his or her family.

TIME LOST BECAUSE OF WORK INJURIES

An important consideration when assessing the effect of accidents on industry is the amount of **lost time** due to **work injuries**.[5] According to the NSC, approximately 35 million hours are lost in a typical year as a result of accidents. This is actual time lost from disabling injuries and does not include additional time lost for medical checkups after the injured employee returns to work. Accidents that occurred in previous years often continue to cause lost time in the current year.

DEATHS IN WORK ACCIDENTS

Deaths on the job have decreased markedly over the years. However, they still occur. For example, in a typical year, there are 10,400 work deaths in the United States. The causes of death in the workplace vary. They include those related to motor vehicles, falls, electric current, drowning, fires, air transport, poisoning, water transport, machinery, falling objects, rail transport, and mechanical suffocation.[6] Figure 13-4 gives a complete breakdown of the percentages for the various categories of causes.

WORK INJURIES BY TYPE OF ACCIDENT

Work injuries can be classified by the type of accident from which they resulted. The most common causes of work injuries are

- Overexertion
- Impact accidents
- Falls

FIGURE 13-4 Work deaths by cause for a typical year.

Cause	Percent
Motor vehicle related	37.2%
Falls	12.5
Electric current	3.7
Drowning	3.2
Fire related	3.1
Air transport related	3.0
Poison (solid, liquid)	2.7
Water transport related	1.6
Poison (gas, vapor)	1.4
Other	31.6

- Bodily reaction (to chemicals)
- Compression
- Motor vehicle accidents
- Exposure to radiation or caustics
- Rubbing or abrasions
- Exposure to extreme temperatures

Overexertion, the result of employees working beyond their physical limits, is the leading cause of work injuries. According to the NSC, almost 31 percent of all work injuries are caused by overexertion. **Impact accidents** involve a worker being struck by or against an object. The next most prominent cause of work injuries is falls.[7] The remaining accidents are distributed fairly equally among the other causes listed above.

DEATH RATES BY INDUSTRY

A variety of agencies and organizations including the Bureau of Labor Statistics, the National Center for Health Statistics, and the NSC collect data on **death rates** within industrial categories.[8] Such information can be used in a variety of ways, not the least of which is in assigning workers' compensation rates. The most widely used industrial categories are agriculture, including farming, forestry, and fishing; mining/quarrying, including oil and gas drilling and extraction; construction; manufacturing; transportation/public utilities; trade, both wholesale and retail; services, including finance, insurance, and real estate; and federal, state, and local government.

The rate of fatalities across all occupations fluctuates somewhat from year to year and from industry to industry. However, the rate of 3.5 fatalities per year per 100,000 workers holds fairly constant over time. The greater fluctuation is in which industry sector experiences the most fatalities in a given year. The following order of ranking will change every year, but the industry sectors that comprise the top sectors by annual death rate remain fairly constant. Occupations that typically have the highest rate of fatalities on the job are (not necessarily in order since the order changes every year) as follows:

- Fishing
- Logging
- Mining/quarrying
- Agriculture
- Construction
- Transportation and warehousing
- Manufacturing
- Aircraft pilots, flight engineers, and maintenance
- Refuse and recycling collectors
- Law enforcement
- Professional and business services
- Retail
- Public administration/government

PARTS OF THE BODY INJURED ON THE JOB

In order to develop and maintain an effective safety and health program, it is necessary to know not only the most common causes of death and injury but also the parts of the body most frequently injured. The NSC stated the following:

> Disabling work injuries in the entire nation totaled approximately 1.75 million in 1998. Of these, about 10,400 were fatal and 60,000 resulted in some permanent impairment. Injuries to the back occurred most frequently, followed by thumb and finger injuries and leg injuries.[9]

Typically, the most frequent injuries to specific parts of the body are as follows (from most frequent to least):

1. Back
2. Legs and fingers
3. Arms and multiple parts of the body
4. Trunk
5. Hands
6. Eyes, head, and feet
7. Neck, toes, and body systems

The back is the most frequently injured part of the body. Legs and fingers are injured with approximately the same frequency, as are arms and multiple parts of the body; the hands are next in frequency, followed by the eyes, the head, and feet; and neck, toes, and body systems. This ranking shows that one of the most fundamental components of a safety and health program should be instruction on how to lift without hurting the back.

CHEMICAL BURN INJURIES

Chemical burn injuries are a special category with which prospective and practicing safety professionals should be familiar. The greatest incidence of chemical burns (approximately one-third) occurs in manufacturing.[10] Other high-incidence industries are services, trade, and construction.

The chemicals that most frequently cause chemical burn injuries include acids and alkalies; soaps, detergents, and cleaning compounds; solvents and degreasers; calcium hydroxide (a chemical used in cement and plaster); potassium hydroxide (an ingredient in drain cleaners and other cleaning solutions); and sulfuric acid (battery acid). Almost 46 percent of all chemical burn injuries occur while workers are cleaning equipment, tools, and vehicles.[11]

What is particularly disturbing about chemical burn injuries is that a high percentage of them occur in spite of the use of personal protective equipment, the provision of safety instruction, and the availability of treatment facilities. In some cases, the personal protective equipment is faulty or inadequate. In others, it is not properly used in spite of instructions.

Preventing chemical burn injuries presents a special challenge to safety and health professionals. The following strategies are recommended:

- Familiarize yourself, the workers, and their supervisors with the chemicals that will be used and their inherent dangers.
- Secure the proper personal protection equipment for each type of chemical that will be used.
- Provide instruction on the proper use of personal protection equipment and then make sure that supervisors confirm that the equipment is used properly every time.
- Monitor that workers are wearing personal protection equipment and replace it when it begins to show wear.

HEAT BURN INJURIES

Heat burn injuries present a special challenge to safety and health professionals in the modern workplace. Almost 40 percent of all such injuries occur in manufacturing every year. The most frequent causes are flame (this includes smoke inhalation injuries), molten metal, petroleum asphalts, steam, and water. The most common activities associated with heat burn injuries are welding, cutting with a torch, and handling tar or asphalt.[12]

Following are several factors that contribute to heat burn injuries in the workplace. Safety and health professionals who understand these factors will be in a better position to prevent heat burn injuries.

- Employer has no health and safety policy regarding heat hazards.
- Employer fails to enforce safety procedures and practices.
- Employees are not familiar with the employer's safety policy and procedures concerning heat hazards.
- Employees fail to use or improperly use personal protection equipment.

- Employees have inadequate or worn personal protection equipment.
- Employees work in a limited space.
- Employees attempt to work too fast.
- Employees are careless.
- Employees have poorly maintained tools and equipment.[13]

These factors should be considered carefully by safety and health professionals when developing accident prevention programs. Employees should be familiar with the hazards, should know the appropriate safety precautions, and should have and use the proper personal protection equipment. Safety professionals should monitor to ensure that safety rules are being followed, that personal protection equipment is being used correctly, and that it is in good condition.

REPETITIVE STRAIN/SOFT TISSUE INJURIES

Repetitive strain injury (RSI) is a broad and generic term that encompasses a variety of injuries resulting from cumulative trauma to the soft tissues of the body, including tendons, tendon sheaths, muscles, ligaments, joints, and nerves. Such injuries are typically associated with the soft tissues of the hands, arms, neck, and shoulders.

Carpal tunnel syndrome (CTS) is the most widely known repetitive strain injury. There are also several other RSIs to the body's soft tissues. The carpal tunnel is the area inside the wrist through which the median nerve passes. It is formed by the wrist bones and a ligament. CTS is typically caused by repeated and cumulative stress on the median nerve. Symptoms of CTS include numbness, a tingling sensation, and pain in the fingers, hand, and/or wrist.

Stress placed on the median nerve typically results from repeated motion while the hands and fingers are bent in an unnatural position. However, sometimes the stress results from a single traumatic event such as a sharp blow to the wrist.

Evidence suggests a higher incidence of CTS among women than men. Scientists have found that the incidence rate for CTS among women was 1.96 per 1,000 full-time equivalent (FTE) workers and 1.58 per 1,000 FTE for men. The overall incidence rate for CTS is increasing at a rate of more than 15 percent per year.[14]

A common misconception about RSI is that it is synonymous with CTS. It isn't. In fact, CTS is relatively rare among RSI patients. Following is a list of the broad classifications of RSI and the types commonly associated with each classification:

MUSCLE AND TENDON DISORDERS

- Tendinitis
- Muscle damage
- Tenosynovitis
- Stenosing tenosynovitis

DEQUERVAIN'S DISEASE

- Flexor tenosynovitis (trigger finger)
- Shoulder tendinitis
- Forearm tendinitis
- Cervical radioculopathy

- Epicondylitis
- Ganglion cysts

TUNNEL SYNDROMES

- Carpal tunnel syndrome
- Radial tunnel syndrome

ULNAR NERVE DISORDERS

- Sulcus ulnaris syndrome
- Cubital tunnel syndrome
- Guyon's canal syndrome

NERVE AND CIRCULATION DISORDERS

- Thoracic outlet syndrome
- Raynaud's disease

OTHER ASSOCIATED DISORDERS

- Reflex sympathetic dysfunction
- Focal dystonia (writer's cramp)
- Degenerative joint disorder (osteoarthritis)
- Fibromyalgia
- Dupuytren's contracture[15]

ESTIMATING THE COST OF ACCIDENTS

Even decision-makers who support accident prevention must consider the relative costs of such efforts. Clearly, accidents are expensive. However, to be successful, safety and health professionals must be able to show that accidents are more expensive than prevention. To do this, they must be able to estimate the cost of accidents. The procedure for estimating costs set forth in this section was developed by Professor Rollin H. Simonds of Michigan State College working in conjunction with the Statistics Division of the NSC.

Cost-Estimation Method

Professor Simonds states that in order to have value, a cost estimate must relate directly to the specific company in question. Applying broad industry cost factors will not suffice. To arrive at company-specific figures, Simonds recommends that costs associated with an accident be divided into *insured* and *uninsured* costs.[16]

Determining the insured costs of accidents is a simple matter of examining accounting records. The next step involves calculating the uninsured costs. Simonds recommends that accidents be divided into the following four classes:

- ***Class 1 accidents.*** Lost workdays, permanent partial disabilities, and temporary total disabilities.
- ***Class 2 accidents.*** Treatment by a physician outside the company's facility.
- ***Class 3 accidents.*** Locally provided first aid, property damage of less than $100, or the loss of fewer than eight hours of work time.
- ***Class 4 accidents.*** Injuries that are so minor they do not require the attention of a physician, result in property damage of $100 or more, or cause eight or more work hours to be lost.[17]

FIGURE 13-5 Uninsured costs worksheet.

Class of Accident	Accident Number							
Class 1	**1**	**2**	**3**	**4**	**5**	**6**	**7**	**8**
Cost A	$ 16.00	$ 6.95	$ 15.17	$ 3.26				
Cost B	72.00	103.15	97.06	51.52				
Cost C	26.73	12.62	—	36.94				
Cost D	—	51.36	—	38.76				
Cost E	—	11.17	—	24.95				
Cost F	—	—	—	−13.41				
Cost G	—	—	—	—				
Total	114.73	185.25	112.23	142.02				

Grand Total: $554.23

Average Cost per Accident: $138.56 (Grand Total ÷ Number of Accidents)

Signature: ______________________________ Date: __________

Average uninsured costs for each class of accident can be determined by pulling the records of all accidents that occurred during a specified period and sorting the records according to class. For each accident in each class, record every cost that was not covered by insurance. Compute the total of these costs by class of accident and divide by the total number of accidents in that class to determine an average uninsured cost for each class, specific to the particular company.

Figure 13-5 is an example of how the average cost of a selected sample of Class 1 accidents can be determined. In this example, there were four Class 1 accidents in the pilot test. These four accidents cost the company a total of $554.23 in uninsured costs, or an average of $138.56 per accident. Using this information, accurate cost estimates of an accident can be figured, and accurate predictions can be made.

Other Cost-Estimation Methods

The costs associated with workplace accidents, injuries, and incidents fall into broad categories such as the following:

- Lost work hours
- Medical costs
- Insurance premiums and administration
- Property damage
- Fire losses
- Indirect costs

Calculating the direct costs associated with lost work hours involves compiling the total number of lost hours for the period in question and multiplying the hours times the applicable loaded labor rate. The loaded labor rate is the employee's hourly rate plus benefits. Benefits vary from company to company but typically inflate the hourly wage by 20 to 35 percent. A sample cost-of-lost-hours computation follows:

Employee Hours Lost (4th quarter) × Average Loaded Labor Rate = Cost 386 × 13.48 = $5,203.28

In this example, the company lost 386 hours due to accidents on the job in the fourth quarter of its fiscal year. The employees who actually missed time at work formed a pool of people with an average loaded labor rate of $13.48 per hour ($10.78 average hourly wage plus 20 percent for benefits). The average loaded labor rate multiplied by the 386 lost hours reveals an unproductive cost of $5,203.28 to this company.

By studying records that are readily available in the company, a safety professional can also determine medical costs, insurance premiums, property damage, and fire losses for the time period in question. All these costs taken together result in a subtotal cost. This figure is then increased by a standard percentage to cover indirect costs to determine the total cost of accidents for a specific time period. The percentage used to calculate indirect costs can vary from company to company, but 20 percent is a widely used figure.

Estimating Hidden Costs

Safety professionals often use the *iceberg analogy* when talking about the real costs of accidents. Accident costs are like an iceberg in that their greatest portion is hidden from view.[18] In the case of icebergs, the larger part is hidden beneath the surface of the water. In the case of an accident, the larger part of the actual cost is also hidden beneath the surface.

The supervisor and the accident investigation team probably will need to spend time conducting an investigation, and there will be a lot of time spent on the administration of paper work related to the accident.[19]

There are many different models that can be used for estimating both the direct and indirect costs of accidents. Some of these models are so complex that their usefulness is questionable. The checklist in Figure 13-6 is a simple and straightforward tool that can be used to estimate the hidden costs of accidents.

GLOBAL IMPACT OF ACCIDENTS AND INJURIES

According to the International Labour Organization (ILO) of the United Nations, approximately 2.2 million people die every year of work-related injuries and occupational diseases.[20] Actually, due to poor record keeping and reporting in underdeveloped countries, it is estimated that this figure is low. Rapid development and the pressure of global competition are resulting in increased workplace fatalities in China and the Pacific Rim countries.

What is missing in many of the developing countries that are becoming industrialized is a safety and health infrastructure. Such an infrastructure would include government-enforced safety and health regulations, company-sponsored safety and health training, record-keeping and reporting systems, and management practices that make occupational safety and health a fully integrated component of work processes and the competitive philosophy of organizations. Occupational safety and health must

FIGURE 13-6 Some accident costs that might be overlooked.

Checklist for Estimating the Hidden Costs of Accidents

√	Paid time to the injured employee on the day of the accident.
√	Paid time of any emergency-responder personnel involved (including ambulance driver).
√	Paid time of all employees who were interviewed as part of the accident investigation.
√	Paid time of the safety personnel who conducted the accident investigation.
√	Paid time of the human resources personnel who handled the workers' compensation and medical aspects of the accident.
√	Paid time of the supervisor involved in the accident investigation and accident response.
√	Paid time to employees near the accident working (or slowed down) temporarily as a result of the accident.
√	Paid time to employees who spent time talking about the accident as news of it spread through the company's grapevine.

come to be seen as a strategy for sustaining economic growth and social development in emerging countries.

There is much to be done if developing countries are going to provide safe and healthy working conditions for their citizens. The ILO reports the following:

- Record-keeping and reporting systems in developing countries are deteriorating instead of improving. Consequently, only a fraction of the real toll of workplace accidents and injuries is being reported.
- Men in developing countries tend to die as the result of accidents, lung diseases, and work-related cancers such as those caused by asbestos. Women in developing countries suffer more from musculoskeletal disorders, communicable diseases, and psychosocial problems.
- Occupational injuries in developing countries are more prevalent in such high-risk industries as mining, construction, and agriculture.
- Younger workers in developing countries are more likely to suffer nonfatal injuries, while older workers are more likely to suffer fatal injuries.
- In developing countries, more than half of retirements are taken early to collect pensions based on work-related disabilities rather than normal retirement.

ESTABLISHING A SAFETY-FIRST CORPORATE CULTURE

One could initiate a lively discussion among business experts, safety and health professionals, and academicians by raising the question: What is meant by the term *safety-first corporate culture?* It sometimes appears that there are as many definitions of the concept as there are people who use the term. Safety and health professionals should understand what the concept means. The concept can be defined as follows:

> A safety-first corporate culture exists when the tacit assumptions, beliefs, values, attitudes, expectations, and behaviors that are widely shared and accepted in an organization support the establishment and maintenance of a safe and healthy work environment for all personnel at all levels.

Evidence of an organization's corporate culture includes:

1. ***Its priorities.*** Are safety and health top priorities in the organization?
2. ***How people in the organization succeed.*** Are personnel recognized and rewarded for working safely?
3. ***How decisions are made in the organization.*** Is safety a major consideration when decisions are made?
4. ***Expectations that management has of employees.*** Do executives and management personnel make it clear that safe behavior is the expected behavior in all cases?
5. ***Expectations that employees have of management.*** Are employees encouraged to make their views known about the quality of the work environment?
6. ***Effects of internal peer pressure on safety.*** Does peer pressure among workers support or undermine safety?
7. ***Underwritten rules that are widely accepted.*** Do the organization's underwritten rules support or undermine safety?
8. ***How conflict about safety is handled.*** When conflicts arise between productivity and safety, are they settled in favor of safety?

These questions make the critical point that the cultural elements of safety and health are part of an organization's larger corporate culture, not some separate and distinct component that stands alone. Safety and health should be so fully integrated into an organization's culture that they are seen to be critical elements of the organization's ability to compete in a global marketplace.

Importance of Having a Safety-First Corporate Culture

If asked to summarize in just one word why it is so important for organizations to establish and maintain a safety-first corporate culture, the appropriate word would be competition. The market force behind the need for a safety-first corporate culture is competition. There are also the moral obligations of the employers to provide a safe and healthy work environment for workers and the practical obligations that grow out of regulatory compliance, but competition is the foundational driver behind the need. From a business perspective, competition is the key driver because in order to survive in the global marketplace, businesses must be competitive. A business that fails to provide a work environment that is conducive to peak performance and continual improvement will not—in the long run—be able to compete, and if it cannot compete it will fail.

When the work environment is not safe, organizations begin to experience problems that sap their financial and intellectual strength, making them less able to compete in the global marketplace. These problems include accidents and injuries that erode the organization's talent pool, poor morale among employees, regulatory penalties, negative publicity, increased medical costs, increased workers' compensation costs, and lawsuits.

The competitive value of a safe and healthy workplace is an important concept for modern safety and health professionals to understand and be able to articulate because it is often the pressures of competition that cause organizations to cut corners when it comes to safety and health concerns. Such an approach is shortsighted, ill-advised, and costly in the long run.

Competition has always been an important issue for organizations, but the concept of globalization has magnified the nature and intensity of competition. It has also magnified the need for a safe and healthy work environment. In order to survive and thrive in a globally competitive environment, organizations must be innovative, adaptive, flexible, lean, productive, and able to continually improve their processes, products, people, and services. The undeniable truth is that over the long run, these necessary traits cannot be sustained in an unsafe, unhealthy work environment.

In order to win in the global marketplace, organizations must hire and retain the best workers available, and they must support them in ways that ensure consistent peak performance and continual improvement of performance. A critical aspect of this support is the provision of a safe and healthy work environment—the best type of environment for promoting, encouraging, and allowing peak performance and continual improvement.

What a Safety-First Corporate Culture Looks Like

Part of the process of establishing a safety-first corporate culture is developing an understanding of what such a culture looks like. This is a lot like a person who wants to lose weight taping a picture of a role model to the bathroom mirror. The picture serves not only as a constant reminder of the desired goal, but also as a measurement device that indicates when a goal has been met. If a picture of an organization with a safety-first corporate culture could be taped to an organization's wall for all employees to see, it would have the following characteristics:

- Widely shared agreement among key decision makers that providing a safe and healthy work environment is an essential competitive strategy
- Emphasis on the importance of human resources to the organization and the corresponding need to protect them from hazards
- Ceremonies to celebrate safety- and health-related successes
- Widely shared agreement that the work environment that is most conducive to peak performance and continual improvement is a safe and healthy work environment
- Recognition and rewards are given to high-performing workers and teams for safety- and health-related performance on the job
- Strong customer focus that includes product safety as a critical concern
- Insistence on safety and health as part of supplier relations
- Effective internal network for communicating safety and health information and expectations
- Informal rules of employee behavior that promote safe and healthy work practices

- Strong pro-safety corporate value system as set forth in the strategic plan
- High expectations and standards for performance relating to safety and health

Ten Steps for Establishing a Safety-First Corporate Culture

The process of establishing a safety-first corporate culture in an organization consists of 10 broad steps. These steps are as follows:

1. ***Understand*** the need for safety-first corporate culture.
2. ***Assess*** the current corporate culture as it relates to safety.
3. ***Plan*** for a safety-first corporate culture.
4. ***Expect*** appropriate safety-related behaviors and attitudes.
5. ***Model*** the desired safety-related behaviors and attitudes.
6. ***Orient*** personnel to the desired safety-first corporate culture.
7. ***Mentor*** personnel in the desired safety-related behaviors and attitudes.
8. ***Train*** personnel in the desired safety-related behaviors and attitudes.
9. ***Monitor*** safety-related behaviors and attitudes at all levels.
10. ***Reinforce*** *and maintain* the desired safety-first corporate culture.

THEORIES OF ACCIDENT CAUSATION

Each year, work-related accidents cost the United States almost $50 billion.[21] This figure includes costs associated with lost wages, medical expenses, insurance costs, and indirect costs. The number of persons injured in **industrial place accidents** in a typical year is 7,128,000, or 3 per 100 persons per year.[22] In the workplace, there is one accidental death approximately every 51 minutes and one injury every 19 seconds.[23]

Why do accidents happen? This question has concerned safety and health decision makers for decades because in order to prevent accidents we must know why they happen. Over the years, several theories of accident causation have evolved that attempt to explain why accidents occur. Models based on these theories are used to predict and prevent accidents.

The most widely known theories of accident causation are the domino theory, the human factors theory, the accident/incident theory, the epidemiological theory, the systems theory, the combination theory, and the behavioral theory. This chapter provides practicing and prospective safety professionals with the information they need to understand fully and apply these theories.

DOMINO THEORY OF ACCIDENT CAUSATION

An early pioneer of accident prevention and industrial safety was Herbert W. Heinrich, an official with the Travelers Insurance Company. In the late 1920s, after studying the reports of 75,000 industrial accidents, Heinrich concluded that

- 88 percent of industrial accidents are caused by unsafe acts committed by fellow workers.
- 10 percent of industrial accidents are caused by unsafe conditions.
- 2 percent of industrial accidents are unavoidable.[24]

Heinrich's study laid the foundation for his *Axioms of Industrial Safety* and his theory of accident causation, which came to be known as the **domino theory**. So much of Heinrich's theory has been discounted by more contemporary research that it is now considered outdated. However, because some of today's more widely accepted theories can be traced back to Heinrich's theory, students of industrial safety should be familiar with his work.

Heinrich's Axioms of Industrial Safety

Heinrich summarized what he thought health and safety decision makers should know about industrial accidents in 10 statements he called **Axioms of Industrial Safety**. These axioms can be paraphrased as follows:

1. Injuries result from a completed series of factors, one of which is the accident itself.
2. An accident can occur only as the result of an unsafe act by a person and/or a physical or mechanical hazard.
3. Most accidents are the result of unsafe behavior by people.
4. An unsafe act by a person or an unsafe condition does not always immediately result in an accident/injury.
5. The reasons why people commit unsafe acts can serve as helpful guides in selecting corrective actions.
6. The severity of an accident is largely fortuitous, and the accident that caused it is largely preventable.
7. The best accident prevention techniques are analogous with the best quality and productivity techniques.
8. Management should assume responsibility for safety because it is in the best position to get results.
9. The supervisor is the key person in the prevention of industrial accidents.
10. In addition to the direct costs of an accident (for example, compensation, liability claims, medical costs, and hospital expenses), there are also hidden or indirect costs.[25]

According to Heinrich, these axioms encompass the fundamental body of knowledge that must be understood by decision makers interested in preventing accidents. Any accident prevention program that takes all 10 axioms into account is more likely to be effective than a program that leaves out one or more axioms.

Heinrich's Domino Theory

Perhaps you have stood up a row of dominoes, tipped the first one over, and watched as each successive domino topples the one next to it. This is how Heinrich's theory of accident causation works. According to Heinrich, there are five factors in the sequence of events leading up to an accident. These factors can be summarized as follows:

1. ***Ancestry and social environment.*** Negative character traits that may lead people to behave in an unsafe manner can be inherited (**ancestry**) or acquired as a result of the **social environment** (for example, alcoholism).
2. ***Fault of person.*** Negative character traits, whether inherited or acquired, are why people behave in an unsafe manner and why hazardous conditions exist (for example, carelessness, temper).

3. ***Unsafe act/mechanical or physical hazard.*** **Unsafe acts** committed by people and **mechanical hazards** or **physical hazards** are the direct causes of accidents (failure to follow prescribed safety guidelines).
4. ***Accident.*** Typically, accidents that result in injury are caused by falling or being hit by moving objects.
5. ***Injury.*** Typical injuries resulting from accidents include lacerations and fractures.[26]

Heinrich's theory has two central points: (1) injuries are caused by the action of **preceding factors** and (2) removal of the **central factor** (unsafe act/**hazardous condition**) negates the action of the preceding factors and, in so doing, prevents accidents and injuries.

Heinrich's Theory and Corrective Action

The three "E's" of safety—Engineering, Education, and Enforcement—work well in taking corrective action in conjunction with Heinrich's theory. The engineering component recommends that hazards be controlled through better product design or process improvements. The education component recommends training workers on all aspects of safe work practices and management on their role in providing a safe working environment. The enforcement component recommends that safety professionals, managers, supervisors, and fellow workers act immediately when it is observed that someone is ignoring safe work practices. Safety rules and procedures have no value unless they are followed, and they will not be followed unless they are enforced.

Domino Theory in Practice

Construction Products Company (CPC) is a distributor of lumber, pipe, and concrete products. Its customers are typically small building contractors. CPC's facility consists of an office in which orders are placed and several large warehouses. Contractors place their orders in the office. They then drive their trucks through the appropriate warehouses to be loaded by CPC personnel.

Because the contractors are small operations, most of their orders are also relatively small and can be loaded by hand. Warehouse personnel go to the appropriate bins, pull out the material needed to fill their orders, and load the materials on customers' trucks. Even though most orders are small enough to be loaded by hand, many of the materials purchased are bulky and cumbersome to handle. Because of this, CPC's loaders are required to wear such personal protection gear as hard hats, padded gloves, steel-toed boots, and lower-back-support belts.

For years, CPC's management team had noticed an increase in minor injuries to warehouse personnel during the summer months. Typically, these injuries consisted of nothing worse than minor cuts, scrapes, and bruises. However, this past summer had been different. Two warehouse workers had sustained serious back injuries. These injuries have been costly to CPC both financially and in terms of employee morale.

An investigation of these accidents quickly identified a series of events and a central causal behavior that set up a *domino effect* that, in turn, resulted in the injuries. The investigation revealed that CPC's warehouses became so hot during the summer months that personal protection gear was uncomfortable. As a result, warehouse personnel simply discarded it. Failure to use appropriate personal protection gear in the summer months had always led to an increase in injuries. However, because the injuries were minor in nature, management had never paid much attention to the situation. It was probably inevitable that more serious injuries would occur eventually.

To prevent a recurrence of the summer-injury epidemic, CPC's management team decided to remove the causal factor—failure of warehouse personnel to use their personal protection gear during the summer months. To facilitate the removal of this factor, CPC's management team formed a committee consisting of one executive manager, one warehouse supervisor, and three warehouse employees.

The committee made the following recommendations: (1) provide all warehouse personnel with training on the importance and proper use of personal protection gear; (2) require warehouse supervisors to monitor the use of personal protection gear more closely; (3) establish a company policy that contains specific and progressive disciplinary measures for failure to use required personal protection gear; and (4) implement several heat reduction measures to make warehouses cooler during the summer months.

CPC's management team adopted all the committee's recommendations. In doing so, it removed the central causal factor that had historically led to an increase in injuries during the summer months.

HUMAN FACTORS THEORY OF ACCIDENT CAUSATION

The **human factors theory** of accident causation attributes accidents to a chain of events ultimately caused by **human error**. It consists of the following three broad factors that lead to human error: overload, inappropriate response, and inappropriate activities (see Figure 13-7). These factors are explained in the following paragraphs.

Overload

Overload amounts to an imbalance between a person's capacity at any given time and the load that person is carrying in a given state. A person's capacity is the product of such factors as his or her natural ability, training, state of mind, fatigue, stress, and physical condition. The load that a person is carrying consists of tasks for which he or she is responsible and added burdens resulting from **environmental factors** (noise,

FIGURE 13-7 Factors that cause human errors.

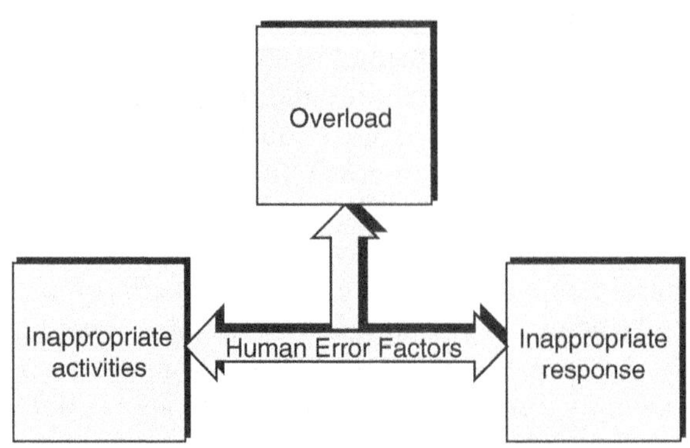

distractions, etc.), **internal factors** (personal problems, emotional stress, and worry), and **situational factors** (level of risk, unclear instructions, etc.). The state in which a person is acting is the product of his or her motivational and arousal levels.

Inappropriate Response and Incompatibility

How a person responds in a given situation can cause or prevent an accident. If a person detects a hazardous condition but does nothing to correct it, he or she has responded inappropriately. If a person removes a safeguard from a machine in an effort to increase output, he or she has responded inappropriately. If a person disregards an established safety procedure, he or she has responded inappropriately. Such responses can lead to accidents. In addition to **inappropriate responses**, this component includes workstation incompatibility. The incompatibility of a person's workstation with regard to size, force, reach, feel, and similar factors can lead to accidents and injuries.

Inappropriate Activities

Human error can be the result of **inappropriate activities**. An example of an inappropriate activity is a person who undertakes a task that he or she doesn't know how to do. Another example is a person who misjudges the degree of risk involved in a given task and proceeds based on that misjudgment. Such inappropriate activities can lead to accidents and injuries. Figure 13-8 summarizes the various components of the human factors theory.[27]

Human Factors Theory in Practice

Kitchenware Manufacturing Incorporated (KMI) produces aluminum kitchenware for commercial settings. After ten years of steady, respectable growth in the U.S. market, KMI suddenly saw its sales triple in less than six months. This rapid growth was the result of KMI's successful entry into European and Asian markets.

FIGURE 13-8 Human factors theory.

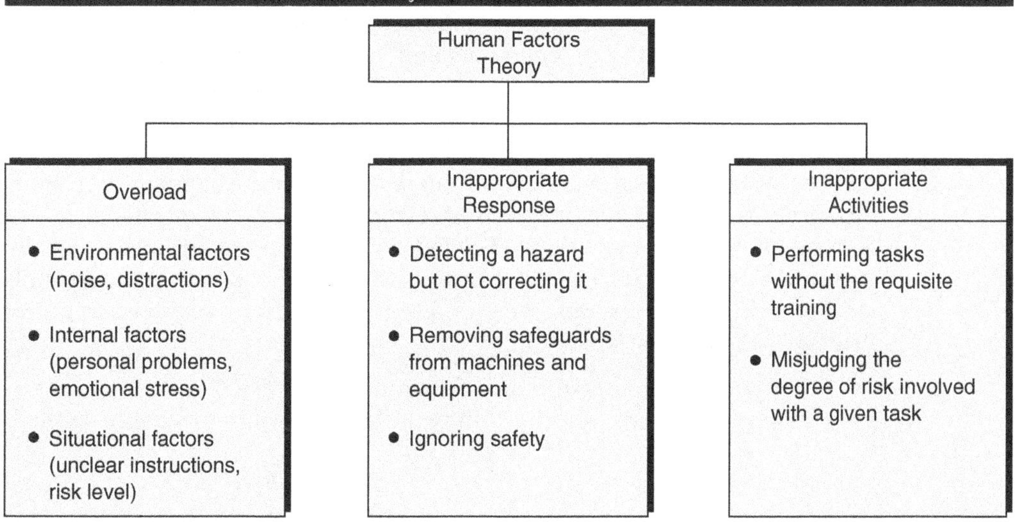

The growth in sales, although welcomed by both management and employees, quickly overloaded and, before long, overwhelmed the company's production facility. KMI responded by adding a second shift of production personnel and approving unlimited overtime for highly skilled personnel. Shortly after the upturn in production, KMI began to experience a disturbing increase in accidents and injuries. During his accident investigations, KMI's safety manager noticed that human error figured prominently in the accidents. He grouped all the human errors identified into three categories: (1) overload, (2) inappropriate response, and (3) inappropriate activities.

In the category of *overload,* he found that the rush to fill orders was pushing production personnel beyond their personal limits in some cases, and beyond their capabilities in others. Stress, insufficient training of new employees, and fatigue all contributed to the overload. In the category of *inappropriate response,* the safety manager determined that many of KMI's production personnel had removed safeguards from their machines in an attempt to speed up production. All the machines involved in accidents had had safeguards removed.

In the category of *inappropriate activities*, the safety manager found that new employees were being assigned duties for which they weren't yet fully trained. As a result, they often misjudged the amount of risk associated with their work tasks.

With enough accident investigations completed to identify a pattern of human error, the safety manager prepared a presentation containing a set of recommendations for corrective measures for KMI's executive management team. His recommendations were designed to prevent human-error-oriented accidents without slowing production.

ACCIDENT/INCIDENT THEORY OF ACCIDENT CAUSATION

The **accident/incident theory** is an extension of the human factors theory. It was developed by Dan Petersen and is sometimes referred to as the Petersen accident/incident theory.[28] Petersen introduced such new elements as **ergonomic traps**, the **decision to err**, and **systems failures**, while retaining much of the human factors theory. A model based on his theory is shown in Figure 13-9.

In this model, overload, ergonomic traps, or a decision to err lead to human error. The decision to err may be conscious and based on logic, or it may be unconscious. A variety of pressures such as deadlines, peer pressure, and budget factors can lead to **unsafe behavior**. Another factor that can influence such a decision is the "It won't happen to me" syndrome.

The systems failure component is an important contribution of Petersen's theory. First, it shows the potential for a **causal relationship** between management decisions or management behavior and safety. Second, it establishes management's role in accident prevention as well as the broader concepts of safety and health in the workplace.

Following are just some of the different ways that systems can fail, according to Petersen's theory:

- Management does not establish a comprehensive safety policy.
- Responsibility and authority with regard to safety are not clearly defined.

FIGURE 13-9 Accident/incident theory.

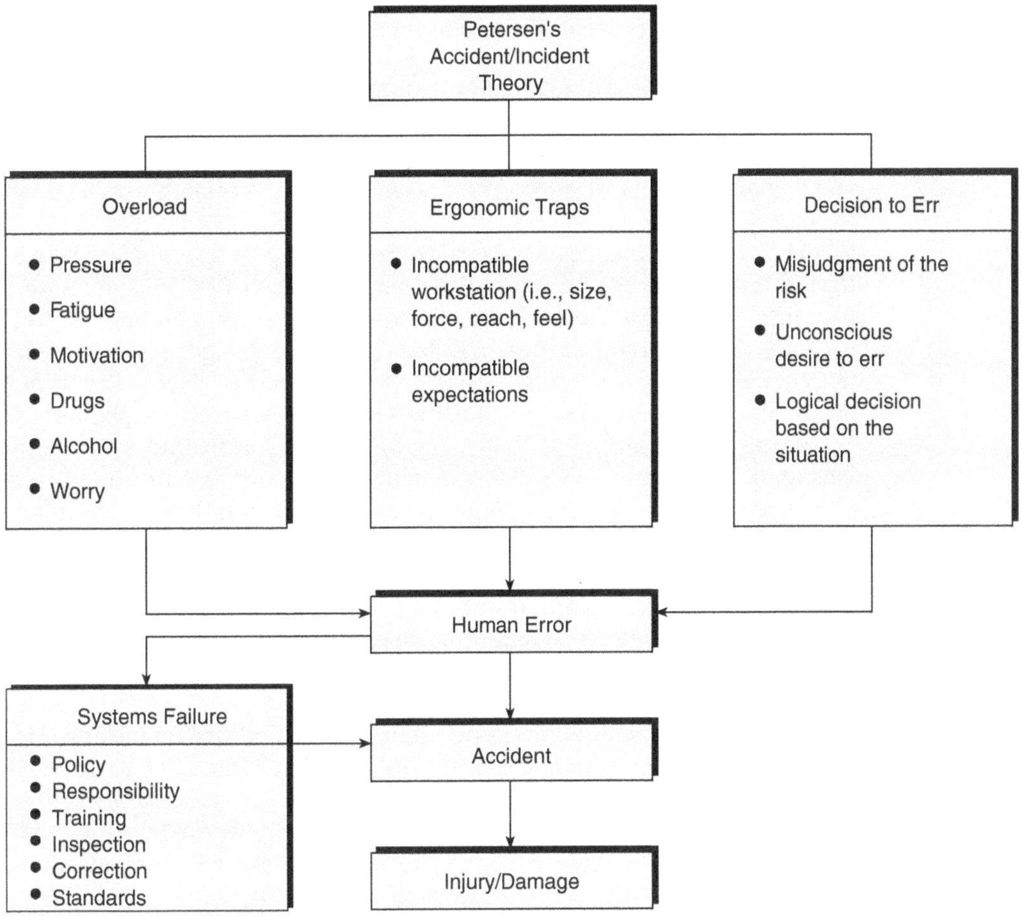

- Safety procedures such as measurement, inspection, correction, and investigation are ignored or given insufficient attention.
- Employees do not receive proper orientation.
- Employees are not given sufficient safety training.

Accident/Incident Theory in Practice

Poultry Processing Corporation (PPC) processes chickens and turkeys for grocery chains. Poultry processing is a labor-intensive enterprise involving a great deal of handwork. A variety of different knives, shears, and cleavers are used. Much of the work is monotonous and repetitive. Selected parts of the overall process must be done in cold conditions.

PPC has gone to great lengths to ensure that workstations are ergonomically sound, that personal protection gear is used as appropriate, and that adequate precautions are taken to prevent illness and injuries. As a result, PPC is an award-winning company in the area of workplace safety and health.

Consequently, the poultry-processing industry was shocked when a class action lawsuit was filed against PPC on behalf of over 50 employees, all of whom claimed to be suffering from carpal tunnel syndrome (CTS). Because of PPC's excellent safety and health record, most observers felt sure that the company would be vindicated in the end.

The company's policies and procedures relating to safety and health were investigated thoroughly by consultants brought in by both PPC and the attorney for the plaintiffs. Over 100 witnesses gave depositions, and several preliminary hearings were held. By the time the trial finally rolled around, both sides had accumulated mountains of paper and filing cabinets full of evidence. Then, suddenly and without advance notice, PPC offered a substantial financial settlement, which the plaintiffs accepted.

It was one of PPC's outside consultants who discovered what had caused the increased incidence of CTS. The company had always used a centralized approach to managing safety and health. Responsibility for such tasks as measurement, inspection, correction, and investigation was assigned to the safety manager, Joe Don Huttle. Huttle had an excellent record during his 20 years in the poultry-processing industry, with the last 5 spent at PPC. In fact, he was so well respected in the industry that his peers had elected him president of a statewide safety organization. This, as it turned out, is where PPC's troubles began.

When Huttle took it over, the safety organization had experienced a three-year decline in membership and was struggling to stay afloat financially. He had been elected as "the man who could save the organization." Intending to do just that, Huttle went right to work. For months at a time he worked seven days a week, often spending as much as two weeks at a time on the road. When he was in his office at PPC, Huttle was either on the telephone or doing paperwork for the safety organization.

Within six months, he had reversed the organization's downhill slide, but not without paying a price at home. During the same six-month period, his duties at PPC were badly neglected. Measurement of individual and group safety performance had come to a standstill. The same was true of inspection, correction, investigation, and reporting.

It was during this time of neglect that the increased incidence of CTS occurred. Safety precautions that Huttle had instituted to guard against this particular problem were no longer observed properly once the workers realized that he had stopped observing and correcting them. Measurement and inspection may also have prevented the injuries had Huttle maintained his normal schedule of these activities.

PPC's consultant, in a confidential report to executive managers, cited the *accident/incident theory* in explaining his view of why the injuries occurred. In this report, the consultant said that Huttle was guilty of applying "it won't happen here" logic when he made a conscious decision to neglect his duties at PPC in favor of his duties with the professional safety organization. Of course, the employees themselves were guilty of not following clearly established procedures. However, because Huttle's neglect was also a major contributing factor, PPC decided to settle out of court.

EPIDEMIOLOGICAL THEORY OF ACCIDENT CAUSATION

Traditionally, safety theories and programs have focused on accidents and the resulting injuries. However, the current trend is toward a broader perspective that also encompasses the issue of industrial hygiene. **Industrial hygiene** concerns environmental factors that can lead to sickness, disease, or other forms of impaired health.

This trend has, in turn, led to the development of an epidemiological theory of accident causation. Epidemiology is the study of causal relationships between environmental factors and disease. The **epidemiological theory** holds that the models used for studying and determining these relationships can also be used to study causal relationships between environmental factors and accidents or diseases.[29]

Figure 13-10 illustrates the epidemiological theory of accident causation. The key components are **predispositional characteristics** and **situational characteristics**. These characteristics, taken together, can either result in or prevent conditions that may result in an accident. For example, if an employee who is particularly susceptible to peer pressure (predispositional characteristic) is pressured by his coworkers (situational characteristic) to speed up his operation, the result will be an increased probability of an accident.

Epidemiological Theory in Practice

Jane Andrews was the newest member of the loading unit for Parcel Delivery Service (PDS). She and the other members of her unit were responsible for loading 50 trucks every morning. It was physically demanding work, and she was the first woman ever

FIGURE 13-10 Epidemiological theory.

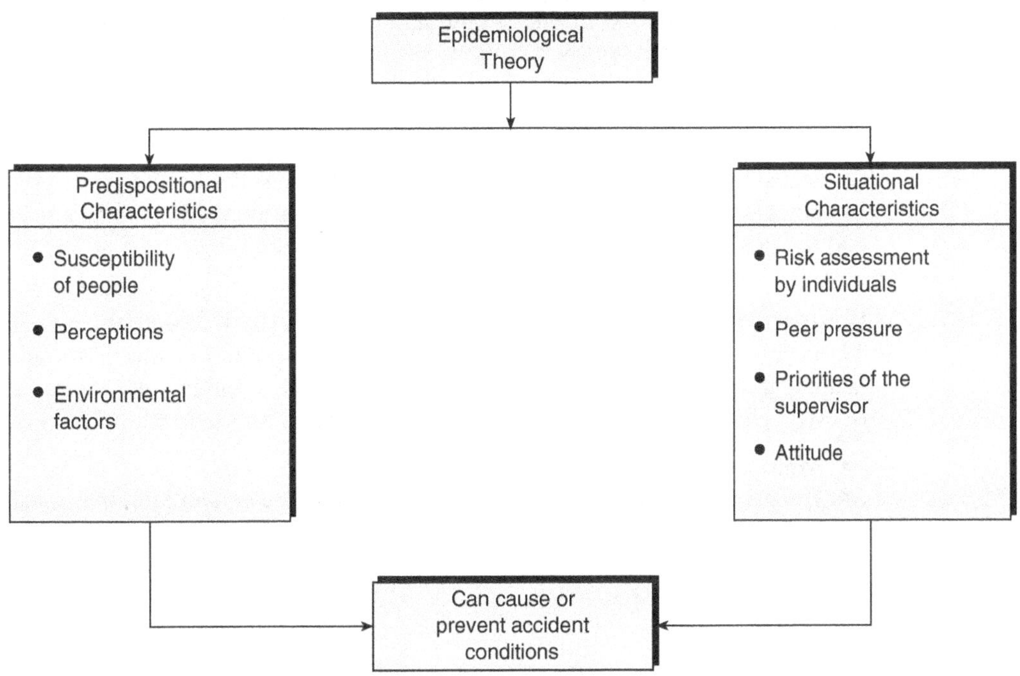

selected by PDS to work in the loading unit. She had gotten the job as part of the company's upward mobility program. She was excited about her new position because within PDS, the loading unit was considered a springboard to advancement. Consequently, she was eager to do well. The responsibility she felt toward other female employees at PDS only intensified her anxiety. Andrews felt that if she failed, other women might not get a chance to try in the future.

Before beginning work in the loading unit, employees must complete two days of training on proper lifting techniques. The use of back-support belts is mandatory for all loading dock personnel. Consequently, Andrews became concerned when the supervisor called her aside on her first day in the unit and told her to forget what she had learned in training. He said, "Jane, nobody wants a back injury, so be careful. But the key to success in this unit is speed. The lifting techniques they teach in that workshop will just slow you down. You've got the job, and I'm glad you're here. But you won't last long if you can't keep up."

Andrews was torn between following safety procedures and making a good impression on her new supervisor. At first, she made an effort to use proper lifting techniques. However, when several of her coworkers complained that she wasn't keeping up, the supervisor told Andrews to "keep up or get out of the way." Feeling the pressure, she started taking the same shortcuts she had seen her coworkers use. Positive results were immediate, and Andrews received several nods of approval from fellow workers and a "good job" from the supervisor. Before long, Andrews had won the approval and respect of her colleagues.

However, after two months of working in the loading unit, she began to experience persistent lower-back pain. Andrews felt sure that her hurried lifting techniques were to blame, but she valued the approval of her supervisor and fellow workers too much to do anything that might slow her down. Finally, one day while loading a truck, Andrews fell to the pavement in pain and could not get up. Her back throbbed with intense pain, and her legs were numb. She had to be rushed to the emergency room of the local hospital. By the time Andrews checked out of the hospital a week later, she had undergone major surgery to repair two ruptured discs.

Jane Andrews's situation can be explained by the *epidemiological theory* of accident causation. The predispositional factor was her susceptibility to peer pressure from her coworkers and supervisor. The applicable situational factors were peer pressure and the priorities of the supervisor. These factors, taken together, caused the accident.

Limitations of Event-Chain Accident Causation Theories

The accident causation theories presented up to this point—Heinrich's domino, human factors, accident/incident, and epidemiological theories—all fall into the broad category of event-chain theories. Event-chain theories explain accident causation in terms of:

- Multiple events that occur in a sequence
- Events that are linked by direct relationships between and among causal factors but ignore indirect relationships
- Events that involve human error, component failure, and/or energy-related factors

All of these event-chain theories have value, but they also have shortcomings—shortcomings that safety and health professionals should be aware of. These shortcomings include the following:

- Neglect broader social and organizational factors.
- Do not adequately account for physical context, social context, personal values, or the dynamics of work processes when assessing human error.
- Neglect human adaptation (the tendency of more experienced workers to behave in ways that involve higher levels of risk).

The shortcomings of event-chain theories of accident causation have led to the development of additional theories that attempt to take a more holistic approach to the analysis of accident causation. These additional theories are explained in the remainder of this chapter.

SYSTEMS THEORY OF ACCIDENT CAUSATION

A *system* is a group of regularly interacting and interrelated components that together form a unified whole. This definition is the basis for the **systems theory** of accident causation. This theory views a situation in which an accident may occur as a system comprised of the following components: person (host), machine (agency), and **environment**.[30] The likelihood of an accident occurring is determined by how these components interact. Changes in the patterns of interaction can increase or reduce the probability of an accident.

For example, an experienced employee who operates a numerically controlled five-axis machining center in a shop environment may take a two-week vacation. Her temporary replacement may be less experienced. This change in one component of the system (person/host) increases the probability of an accident. Such a simple example is easily understood. However, not all changes in patterns of interaction are this simple. Some are so subtle that their analysis may require a team of people, each with a different type of expertise.

The primary components of the systems model are the person/machine/environment, information, decisions, risks, and the task to be performed.[31] Each of the components has a bearing on the probability that an accident will occur. The systems model is illustrated in Figure 13-11.

As this model shows, even as a person interacts with a machine within an environment, three activities take place between the system and the task to be performed. Every time a task must be performed, there is the risk that an accident may occur. Sometimes the risks are great; at other times, they are small. This is where information collection and decision making come in.

Based on the information that has been collected by observing and mentally noting the current circumstances, the person weighs the risks and decides whether to perform the task under existing circumstances. For example, say a machine operator is working on a rush order that is behind schedule. An important safety device has malfunctioned on his machine. Simply taking it off will interrupt work for only five minutes, but it will also increase the probability of an accident. However, replacing it could take up to an hour. Should the operator remove the safety guard and proceed with the task or take the time to replace it? The operator and his supervisor may assess the situation (collect

FIGURE 13-11 Systems theory model.

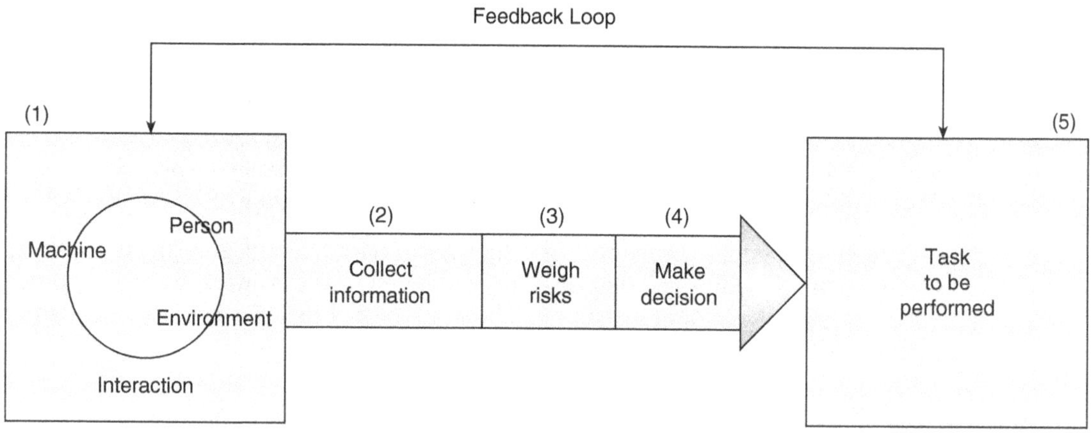

information), weigh the risks, and make a decision to proceed. If their information was right and their assessment of the risks accurate, the task will probably be accomplished without an accident.

However, the environment in which the machine operator is working is unusually hectic, and the pressure to complete an order that is already behind schedule is intense. These factors are **stressors** that can cloud the judgment of those collecting information, weighing risks, and making the decision. When stressors are introduced between points 1 and 3 in Figure 13-11, the likelihood of an accident increases.

For this reason, five factors should be considered before beginning the process of collecting information, weighing risks, and making a decision:

- Job requirements
- The workers' abilities and limitations
- The gain if the task is successfully accomplished
- The loss if the task is attempted but fails
- The loss if the task is not attempted[32]

These factors can help a person achieve the proper perspective before performing the above mentioned tasks. It is particularly important to consider these factors when stressors such as noise, time constraints, or pressure from a supervisor may tend to cloud one's judgment.

Systems Theory in Practice

Precision Tooling Company (PTC) specializes in difficult orders that are produced in small lots, and in making corrections to parts that otherwise would wind up as expensive rejects in the scrap bin. In short, PTC specializes in doing the types of work that other companies cannot, or will not, do. Most of PTC's work comes in the form of subcontracts from larger manufacturing companies. Consequently, living up to its reputation as a high-performance, on-time company is important to PTC.

Because much of its work consists of small batches of parts to be reworked, PTC still uses several manually operated machines. The least experienced machinists operate these machines. This causes two problems. The first problem is that it is difficult for even a master machinist to hold to modern tolerance levels on these old machines. Consequently, apprentice machinists find holding to precise tolerances quite a challenge. The second problem is that the machines are so old that they frequently break down.

Complaints from apprentice machinists about the old machines are frequent. However, their supervisors consider time on the old "ulcer makers" to be one of the rites of passage that upstart machinists must endure. Their attitude is, "We had to do it, so why shouldn't you?" This was where things stood at PTC when the company won the Johnson contract.

PTC had been trying for years to become a preferred supplier for H. R. Johnson Company. PTC's big chance finally came when Johnson's manufacturing division incorrectly produced 10,000 copies of a critical part before noticing the problem. Simply scrapping the part and starting over was an expensive solution. Johnson's vice president for manufacturing decided to give PTC a chance.

PTC's management team was ecstatic! Finally, they had an opportunity to partner with H. R. Johnson Company. If PTC could perform well on this one, even more lucrative contracts were sure to follow. The top managers called a company-wide meeting of all employees. Attendance was mandatory. The CEO explained the situation as follows:

> Ladies and gentlemen, we are faced with a great opportunity. I've just signed a contract with H. R. Johnson Company to rework 10,000 parts that their manufacturing folks produced improperly. The rework tasks are not that complicated, but every part has got to go through several manual operations at the front end of the rework process. This means our manual machining unit is going to have to supply the heroes on this job. I've promised the manufacturing VP at Johnson that we would have his parts ready in 90 days. I know that's a lot to do in so short a period of time, but Johnson is in a real bind here. If we can produce on this one, they won't forget us in the future.

This put PTC's apprentice machinists on the spot. If PTC didn't perform on this contract, it would be their fault. They cursed their old machines and got to work. The CEO had said the rework tasks would not be "that complicated," but, as it turned out, the processes weren't that simple either. The problem was tolerances. Holding to the tolerances specified in the Johnson contract took extra time and a special effort on every single part. Before long, the manual machining unit was behind schedule, and management was getting nervous. The situation was made even worse by the continual breakdowns and equipment failures experienced. The harder the unit supervisor pushed, the more stressed the employees and machines became.

Predictably, it wasn't long before safety procedures were forgotten, and unreasonable risks were being taken. The pressure from management, the inexperience of the apprentice machinists, and the constant equipment failures finally took their toll. In a hurry to get back on schedule, and fearing that his machine would break down again, one machinist got careless and ran his hand into the cutter on his milling machine. By the time the machine had been shut down, his hand was badly mutilated. In the aftershock of this accident, PTC was unable to meet the agreed-upon completion schedule. Unfortunately, PTC did not make the kind of impression on H. R. Johnson's management team that it had hoped.

This accident can be explained by the systems theory. The *person-machine-environment* chain has a direct application in this case. The person involved was relatively inexperienced. The machine involved was old and prone to breakdowns. The environment was especially stressful and pressure packed. These three factors, taken together, resulted in this serious and tragic accident.

COMBINATION THEORY OF ACCIDENT CAUSATION

There is often a degree of difference between any theory of accident causation and reality. The various models presented with their corresponding theories in this chapter attempt to explain why accidents occur. For some accidents, a given model may be very accurate. For others, it may be less so. Often the cause of an accident cannot be adequately explained by just one model or theory. Thus, according to the **combination theory**, the actual cause may combine parts of several different models. Safety personnel should use these theories as appropriate both for accident prevention and accident investigation. However, they should avoid the tendency to try to apply one model to all accidents.

Combination Theory in Practice

Crestview Grain Corporation (CGC) maintains ten large silos for storing corn, rice, wheat, barley, and various other grains. Because stored grain generates fine dust and gases, ventilation of the silos is important. Consequently, all of CGC's silos have several large vents. Each of these vents uses a filter similar to the type used in home air conditioners that must be changed periodically.

There is an element of risk involved in changing the vent filters because of two potential hazards. The first hazard comes from unvented dust and gases that can make breathing difficult, or even dangerous. The second hazard is the grain itself. Each silo has a catwalk that runs around its inside circumference near the top. These catwalks give employees access to the vents that are also near the top of each silo. The catwalks are almost 100 feet above ground level, they are narrow, and the guardrails on them are only knee high. A fall from a catwalk into the grain below would probably be fatal.

Consequently, CGC has well-defined rules that employees are to follow when changing filters. Because these rules are strictly enforced, there had never been an accident in one of CGC's silos—that is, not until the Juan Perez tragedy occurred. Perez was not new to the company. At the time of his accident, he had worked at CGC for over five years. However, he was new to the job of silo maintenance. His inexperience, as it turned out, would prove fatal.

It was time to change the vent filters in silo number 4. Perez had never changed vent filters himself. He hadn't been in the job long enough. However, he had served as the required "second man" when his supervisor, Bao Chu Lai, had changed the filters in silos 1, 2, and 3. Because Chu Lai was at home recuperating from heart surgery and would be out for another four weeks, Perez decided to change the filters himself. Changing the filters was a simple enough task, and Perez had always thought the second-man concept was overdoing it a little. He believed in taking reasonable precautions as much as the next person, but in his opinion, CGC was paranoid about safety.

Perez collected his safety harness, respirator, and four new vent filters. Then he climbed the external ladder to the entrance–exit platform near the top of silo number 4.

Before going in, Perez donned his respirator and strapped on his safety harness. Opening the hatch cover, he stepped inside the silo onto the catwalk. Following procedure, Perez attached a lifeline to his safety harness, picked up the new vent filters, and headed for the first vent. He changed the first two filters without incident. It was while he was changing the third filter that tragedy struck.

The filter in the third vent was wedged in tightly. After several attempts to pull it out, Perez became frustrated and gave the filter a good jerk. When the filter suddenly broke loose, the momentum propelled Perez backward and he toppled off the catwalk. At first it appeared that his lifeline would hold, but without a second person to pull him up or call for help, Perez was suspended by only the lifeline for over 20 minutes. He finally panicked, and in his struggle to pull himself up, knocked open the buckle of his safety harness. The buckle gave way, and Perez fell over 50 feet into the grain below. The impact knocked off his respirator, the grain quickly enveloped him, and Perez was asphyxiated.

The accident investigation that followed revealed that several factors combined to cause the fatal accident—the combination theory. The most critical of these factors were as follows:

- Absence of the supervisor
- Inexperience of Perez
- A conscious decision by Perez to disregard CGC's safety procedures
- A faulty buckling mechanism on the safety harness
- An unsafe design (only a knee-high guardrail on the catwalk)

BEHAVIORAL THEORY OF ACCIDENT CAUSATION

The behavioral theory of accident causation and prevention is often referred to as **behavior-based safety (BBS)**. BBS has both proponents and critics. One of the most prominent proponents of BBS is E. Scott Geller, a senior partner of Safety Performance Solutions Inc. and a professor of psychology. It is appropriate that Geller is a professional psychologist because BBS is the application of behavioral theories from the field of psychology to the field of occupational safety.

According to Geller, there are seven basic principles of BBS: (1) intervention that is focused on employee behavior; (2) identification of external factors that will help understand and improve employee behavior (from the perspective of safety in the workplace); (3) direct behavior with activators or events antecedent to the desired behavior, and motivation of the employee to behave as desired with incentives and rewards that will follow the desired behavior; (4) focus on the positive consequences that will result from the desired behavior as a way to motivate employees; (5) application of the scientific method to improve attempts at behavioral interventions; (6) use of theory to integrate information rather than to limit possibilities; and (7) planned interventions with the feelings and attitudes of the individual employee in mind.[33]

Those who have studied psychology will recognize BBS as an innovative and practical application of standard behavioral theory to the field of occupational safety. These theories are relevant in any situation in which certain types of human behaviors are desired while others are to be avoided. Positive reinforcement in the form of incentives and rewards is used to promote the desired (safe) behaviors and to discourage undesirable (unsafe) behaviors.

Behavioral Theory in Action

Mark Potter is the safety manager for Excello Corporation. Several months ago, he became concerned because employees seemed to have developed a lax attitude toward wearing hard hats. What really troubled Potter was that there was more than the usual potential for head injuries because of the type of work done in Excello's plant, and he had personally witnessed two near misses in less than a week. An advocate of BBS, he decided to apply the ABC model in turning around this unsafe behavior pattern.

His first step was to remove all the old "Hard Hat Area" signs from the plant and replace them with newer, more noticeable signs. Then he scheduled a brief seminar on head injuries and cycled all employees through it over a two-week period. The seminar took an unusual approach. It told a story of two employees. One was in a hospital bed surrounded by family members he did not even recognize. The other was shown enjoying a family outing with happy family members. The clear message of the video was "the difference between these two employees is a hard hat." These two activities were the antecedents to the behavior he hoped to produce (all employees wearing hard hats when in a hard hat area).

The video contained a powerful message and it had the desired effect. Within days, employees were once again disciplining themselves to wear their hard hats (the desired behavior). The consequence was that near misses stopped, and no head injuries have occurred at Excello in months. The outcome of this is that Excello's employees have been able to continue enjoying the fruits of their labor and the company of loved ones.

DRUGS AND ACCIDENT CAUSATION

One of the most pernicious causes of accidents on the job is chemicals—but not the kind industrial hygienists generally concern themselves with.[34] The chemicals alluded to here are the illicit drugs and alcohol used by employees. Drugs and alcohol are the root cause or contributing cause of many accidents on the job every year. Consequently, safety professionals need to be on guard for employees who are drug and alcohol abusers.

This is why so many companies implement drug-free workplace programs. In fact, since 1989 federal contractors have been required to do so. Such programs typically include the following components: drug-free workplace policy, supervisory training, employee education, employee assistance programs, and alcohol and drug testing.

Establishing drug-free workplace programs is typically the responsibility of the human resources department. However, safety and health professionals should be aware of the workplace problems that can be caused by alcohol and drug abuse. Further, if a cross-functional team of representatives from various departments is convened by the human resources department for the purpose of developing a drug-free workplace program, the chief safety and health professional for the organization should be a member of that team.

DEPRESSION AND ACCIDENT CAUSATION

An invisible problem in today's workplace is **clinical depression**. People who suffer from clinical depression are seriously impaired and, as a result, they pose a clear and present safety risk to themselves, fellow workers, and their employer.[35] Mental health professionals estimate that up to 10 percent of the adult population in the United States suffers from clinical depression. This translates to 1 in every 20 people on the job.

The causes of clinical depression are many and varied, but the most common causes are biological (too few or too many of the brain chemicals known as **neurotransmitters**), cognitive (negative thought processes), genetic (family history of depression), and concurring illnesses (strokes, cancer, heart disease, Alzheimer's, and other diseases can increase the incidence of depression).

Warning Signs

Safety and health professionals are not mental health professionals and should not attempt to play that role. However, they should be alert to the warning signs of clinical depression in employees. These signs are as follows:

- Persistent dreary moods (sadness, anxiety, nervousness)
- Signs of too little sleep
- Sleeping on the job or persistent drowsiness
- Sudden weight loss or gain
- General loss of interest, especially in areas of previous interest
- Restlessness, inability to concentrate, or irritability
- Chronic physical problems (headaches, digestive disorders, etc.)
- Forgetfulness or an inability to make simple decisions
- Persistent feelings of guilt
- Feelings of low self-worth
- Focus on death or talk of suicide

Safety and health professionals who recognize any or all of these symptoms in an employee should avoid the natural human tendency to *help the employee deal with the problems*. Rather, the appropriate action is to get the employee into the hands of competent mental health professionals right away. The best way to do this is to approach the employee's supervisor and recommend that he or she refer the employee to the organization's employee assistance program (EAP) or to the human resources department.

Sources of Help

Because clinical depression in employees has become such an all-pervasive problem that increases the risk of accidents and injuries on the job, safety and health professionals need to learn all they can about this problem and keep up-to-date with the latest information concerning it. The following sources may help:

Employee Assistance Professionals Association, 703–522–6272, eapa.org
National Institute of Mental Health, 800–421–4211, nimh.nih.gov
National Mental Health Association, 800–969–NMHA, nmha.org

MANAGEMENT FAILURES AND ACCIDENT CAUSATION

One of the leading causes of accidents in the workplace is the failure of management to do its part to ensure a safe and healthy work environment.[36] Different levels of management have different levels of responsibility. The level of management with the most direct, hands-on, day-to-day responsibility for workplace safety and health is the supervisory level. Supervisors play a critical role in making sure that employees work in a safe and healthy environment.

Role of the Supervisor in Workplace Safety and Health

Safety and health professionals cannot do their jobs effectively without the full cooperation and day-to-day assistance of first-line supervisors. Supervisors and safety professionals must be partners when it comes to providing a safe and healthy workplace for employees. Supervisors should be assigned responsibility for the work environment and for the safety of employees in their units. Safety and health professionals should be readily available to help supervisors fulfill this responsibility.

Key responsibilities of supervisors relating to workplace safety and health include the following:

- Orienting new employees to the safe way to do their jobs
- Ensuring that new and experienced employees receive the safety and health training they need on a continual basis
- Monitoring employee performance and enforcing safety rules and regulations
- Assisting safety and health professionals in conducting accident investigations
- Assisting safety and health professionals in developing accident reports
- Keeping up-to-date on safety issues
- Setting a positive example for employees that says "the safe way is the right way"

Typical Management Failures That Cause Accidents

Management failures represent a major cause of accidents on the job. If management is serious about providing a safe and healthy work environment for employees it must (1) show employees that safe and healthy work practices are expected by including such practices in job descriptions, monitoring employee work practices, and setting an example of safe and healthy work practices; (2) provide training in how to work safely, including orientation training for new employees as well as ongoing updated training for experienced employees; (3) include safe and healthy work practices as criteria in the periodic performance appraisals of employees; and (4) reinforce safe and healthy work practices by rewarding and recognizing employees who use them. Common examples of management failures include the following:

POOR HOUSEKEEPING OR IMPROPER USE OF TOOLS, EQUIPMENT, OR FACILITIES

Management either has not developed the necessary requirements, or has but does not enforce them. The management failure in this case could be lack of safety procedures (failure to let employees know the expectations), lack of training (failure to give employees the knowledge and skills they need to work safely), or failure to properly supervise (failure to monitor employee actions).

PRESSURE TO MEET DEADLINES

Sometimes management has developed a good safety and health policy, established good safety and health procedures, built safety and health expectations into job descriptions and performance appraisals, and provided the necessary training only to put all this aside when a rush order comes in. This may be the most problematic of the many different types of management failures that can occur because it can undermine all of the organization's safety and health efforts. When management allows safety and health procedures to be ignored or, worse yet, encourages them to be ignored to speed up production in the short run, employees soon get the message that safety and health are important only when there is no rush. This is an example of management failing to set the proper example.

OBESITY AND ACCIDENT CAUSATION

Researchers at Ohio State University found that extremely obese people are more likely than normal-weight people to injure themselves.[37] This is bad news for what an Australian study conducted at Queensland University calls "sedentary workplaces"—those that involve a lot of sitting at desks. It is bad news because the Australian study concluded that the more people sit at desks during the workday, the more likely they are to be overweight.

Obesity has long been associated with such chronic diseases as high blood pressure, coronary heart disease, diabetes, and certain types of cancer, but these studies now tie it to workplace injuries too. The ramifications for safety and health professionals are profound. The World Health Organization estimates that there are more than 300 million obese people worldwide. In the industrialized nations of the world—nations such as the United States—the number of people considered obese is growing rapidly. These studies used a standard body mass index (BMI) score of 30 or above to define obesity.

In the study conducted by Ohio State University, researchers collected data on more than 2,500 adults. The data show that 26 percent of extremely obese male subjects reported personal injuries. The percentage for extremely obese women was only slightly lower at approximately 22 percent. Researchers compared these percentages with those for normal-weight people and found a noticeable difference in the percentages of males and females who reported injuries (17 percent for males and 12 percent for women). Researchers used a BMI of 18.5 to 24.9 to define "normal weight."

The most common causes of injuries to obese people were the result of overexertion (35.2 percent) and falls (29.9 percent). Underweight people—BMI of 18.5 or less—reported the fewest number of injuries.

Key Terms and Concepts

- Accident prevention
- Accident rates
- Accidents
- Accident/incident theory
- Ancestry
- Axioms of Industrial Safety
- Behavior-based safety (BBS)
- Cancer
- Carpal tunnel syndrome (CTS)
- Causal relationship
- Central factor
- Chemical burn injuries
- Clinical depression
- Combination theory
- Death rates
- Domino theory
- Drownings
- Environment
- Environmental factors
- Epidemiological theory
- Ergonomic traps
- Falls
- Fault of person
- Fire-related losses
- Hazardous condition
- Heart disease
- Heat burn injuries
- Human error
- Human factors theory
- Impact accidents
- Indirect costs
- Inappropriate activities
- Inappropriate responses
- Industrial hygiene
- Industrial place accidents
- Internal factors
- Insurance administration

- Lost time
- Lost wages
- Mechanical hazards
- Mechanical suffocation
- Medical complications
- Medical expenses
- Natural disasters
- Neurotransmitters
- Obesity
- Overload
- Overexertion
- Physical hazards
- Poisoning
- Preceding factors
- Predispositional characteristics
- Property damage
- Repetitive strain injury (RSI)
- Safety and health professionals
- Situational characteristics
- Situational factors
- Social environment
- Stressors
- Strokes
- Suffocation
- Systems theory
- Unsafe acts
- Unsafe behavior
- Work injuries
- Workplace accidents

Review Questions

1. What are the leading causes of death in the United States?
2. When the overall cost of an accident is calculated, what elements make up the cost?
3. What are the five leading causes of accidental deaths in the United States?
4. What are the leading causes of death in the United States of people between the ages of 25 and 44?
5. Explain how today's rate of accidental work deaths compares with the rate in the early 1900s.
6. What are the five leading causes of work deaths?
7. What are the five leading causes of work injuries by type of accident?
8. When death rates are classified by industry type, what are the leading industry types?
9. Rank the following body parts according to frequency of injury from highest to lowest: neck, fingers, trunk, back, and eyes.
10. Name three chemicals that frequently cause chemical burns in the workplace.
11. Identify three factors that contribute to heat burn injuries in the workplace.
12. Explain the difference between repetitive strain injury and carpal tunnel syndrome.
13. Explain the reasons for high accident rates in developing countries.
14. Explain how to establish a safety-first corporate culture in an organization.
15. Explain the domino theory of accident causation, including its origin and its impact on more modern theories.
16. What were the findings of Herbert W. Heinrich's 1920s study of the causes of industrial accidents?
17. List five of Heinrich's Axioms of Industrial Safety.
18. Explain the following concepts in the domino theory: preceding factor; central factor.
19. What are the three broad factors that lead to human error in the human factors theory? Briefly explain each.
20. Explain the systems failure component of the accident/incident theory.
21. What are the key components of the epidemiological theory? How does their interaction affect accident causation?
22. Explain the limitations of event-chain accident theories.
23. Explain the systems theory of accident causation.
24. What impact do stressors have in the systems theory?
25. List five factors to consider before making workplace decisions that involve risk.
26. Explain the principles of behavior-based safety.
27. What is the role of the safety and health professional with regard to handling employees who might be drug or alcohol abusers?
28. List the warning signs of clinical depression.
29. What must management do if it is serious about providing a safe and healthy work environment for employees?
30. Explain the connection between obesity and injuries.

Endnotes

1. National Safety Council, *Accident Facts* (Chicago: National Safety Council, 2013), 37.
2. Ibid., 4–5.
3. Ibid., 35.
4. Ibid., 36.
5. Ibid., 36.
6. Ibid., 37.
7. Ibid., 38.
8. Ibid., 39.
9. Ibid., 40.
10. Ibid., 41.
11. Ibid., 41.
12. Retrieved from ergooutfitters.com on February 12, 2013.
13. Ibid.
14. National Safety Council, *Accident Prevention Manual for Business and Industry*, 13th ed. (Chicago: National Safety Council, 2009), 158.
15. Ibid., 158.
16. Daniel Corcoran, "The Hidden Value of Safety," *Occupational Health & Safety* 71, no. 6: 20–22.
17. Corcoran, "The Hidden Value of Safety," 22.
18. Occupational Health & Safety Online, "Study: 2.2 Million People Die Worldwide of Work-Related Accidents, Occupational Diseases." Retrieved from ohsonline.com on March 15, 2013, 1–3.
19. Ibid.
20. International Labor Organization of the United Nations. "Safety in Numbers," retrieved from www.ilo.org/public/english/region/eurpro/moscow/areas/safety_in_numbers_en.pdf on October 7, 2013.
21. National Safety Council, *Accident Facts* (Chicago: National Safety Council, 2013), 23.
22. Ibid., 24.
23. Ibid., 26.
24. Industrial Foundation for Accident Prevention. Retrieved from ifap.asn.au on March 11, 2013.
25. Ibid.
26. Ibid.
27. Ibid.
28. Ibid.
29. M. A. Topf, "Chicken/Egg/Chegg!" *Occupational Health Safety* 68, no. 6: 60–66. 2003.
30. D. L. Goetsch, *Implementing Total Safety Management* (Upper Saddle River, NJ: Prentice Hall, 1998), 227.
31. Ibid.
32. Ibid.
33. E. S. Geller, "Behavior-Based Safety: Confusion, Controversy, and Clarification," *Occupational Health & Safety* 68, no. 1: 40–49.
34. Stepher G. Minter, "The Safety Threat from Within," *Occupational Hazards* 64, no. 4: 8.
35. Todd Nighswonger, "Depression: The Unseen Safety Risk," *Occupational Hazards* 64, no. 4: 38–42.
36. Retrieved from online.misu.kodak.edu/19577/AccCautrac.htm on March 12, 2013.
37. Occupational Health & Safety Online, "Obesity Studies Focus on Injuries, Sedentary Workplaces." Retrieved from ohsonline.com/stevens/ohspub.nsf/d3d5b4f938b22b6e8625670c006dbc58/d9 on April 20, 2013.

CHAPTER

14 Energy Systems Engineering

Objectives

After reading this chapter, you will have learned

- What renewable energy is
- How to do a basic energy analysis of photovoltaic solar systems
- How to do a basic energy analysis of wind turbines
- How to do a basic energy analysis of hydropower plants
- How to do a basic energy analysis of geothermal power plants
- How to do a basic energy analysis of marine power plants
- How to do a basic energy analysis of biomass power plants

14.1 INTRODUCTION

Energy has always been crucial to the survival and development of man. Early man used the energy of his own body to walk and run, hunt, build shelters, and transport his belongings, and he used the energy of fire to keep warm and cook his food. As the earth's population grew and societies developed, the energy of man's own body was augmented by the energy of animals that were tamed for livestock. The transition from a hunting to an agricultural society necessitated the use of livestock for planting crops and transporting water and supplies. As cultures became more sophisticated, man learned how to use fire to smelt ore for making copper, iron, and other metals. The energy produced by wind and naturally flowing water was harnessed to grind grain for food, pump water for irrigation, and saw timbers for construction. The steam engine, a development of the industrial revolution, produced mechanical energy for a variety of uses, including transportation, water pumping, and manufacturing. In the modern world, energy is used to operate electrical devices, power transportation systems, and heat and cool living spaces.

Before we discuss renewable energy, we must define the term *energy*. A concise definition of ***energy*** is the *capacity to do work*. If a device or system has the capacity to do work, it possesses at least one form of energy that is available for transformation to another form of energy. A simple illustration is a stretched spring. When a stretched spring is released, its stored (potential) energy is transformed into energy of motion (kinetic energy). If, for example, the spring was connected to a component in a mechanism, work would be done on that component. Another illustration of a system in which an energy transformation occurs is a wind turbine. A wind turbine is basically an electrical generator with a set of blades mounted on its shaft. Wind causes the generator's shaft (the armature of the generator) to rotate. As the wire windings of the armature cut the magnetic field lines of magnets that surround the armature, a voltage is produced. Hence, the kinetic energy of the wind is transformed into electrical energy, which is a form of work. The unit of energy in the SI and English unit systems is the joule (J) and British thermal unit (Btu), respectively.

Renewable energy is *energy that comes from sources that are naturally replenished.* Stated another way, renewable energy comes from sources that are refilled with energy at rates comparable to the rates of energy extraction. Solar energy is one form of renewable energy. The amount of solar energy intercepted by the earth each year is approximately 5.4×10^{24} J. Therefore, the amount of solar energy that is absorbed by the earth and re-radiated plus the solar energy that is reflected from the earth each year is approximately 5.4×10^{24} J. If the energies entering and leaving the earth were not equivalent, the temperature of the earth's surface would increase or decrease, depending on the relative amounts of the incoming and outgoing energies. The average temperature of the earth's surface is fairly stable at about 14°C (57°F). Wind energy is also a form of renewable energy. Winds are caused by pressure differences in the atmosphere induced primarily by uneven heating of the earth's surface by the sun. Hence, wind energy is a form of solar energy.

Fossil fuels—coal, oil, and natural gas—are not renewable energy sources because these materials take an extremely long time to be replenished in the earth. Unlike energy derived from fossil fuels, nuclear energy does not involve the introduction of harmful substances into the atmosphere, but radioactive materials from which nuclear energy is derived are not renewable energy sources either because these materials are finite in the earth and cannot be replaced.

The sources of renewable energy covered in this chapter are solar, wind, hydro, geothermal, marine, and biomass. Brief descriptions of these renewable energy sources are now given.

Solar energy emanates from the sun, the yellow star at the center of our solar system. Like all stars, the sun generates vast amounts of energy by nuclear fusion reactions in its interior. The energy from these reactions is transported to the sun's surface where it radiates into space. The sun has been radiating energy at a reasonably steady rate for several billion years and is predicted to continue to do so for billions of years to come, which is the basis for considering solar energy as a renewable source. The sun radiates energy at a rate of 3.9×10^{26} W. At the outer edge of the earth's atmosphere, the average solar heat flux (solar power intercepted by a plane surface facing directly into the sun) is approximately 1353 W/m^2. This quantity is referred to as the *solar constant.* Due to absorption by atmospheric carbon dioxide and water vapor and interactions with dust and other pollutants, the power density represented by the solar constant does not reach the earth's surface. Thus, the power input for all earth-based solar energy systems is less than the solar constant.

Wind is the movement of atmospheric air across the earth's surface from regions of high pressure to regions of low pressure. The primary cause of winds is the uneven heating of the earth's surface by the sun, which depends on latitude, time of day, and the distribution of land and large bodies of water, particularly the oceans. Another cause of winds is fluid friction between the atmosphere and earth's surface, which allows the earth to drag the atmosphere around producing turbulence. Horizontal components of wind velocities are normally much greater than the vertical velocity components. The kinetic energy of the wind, and therefore the wind's power generating potential, is proportional to the cube of wind velocity. Because winds are primarily caused by uneven heating effects of the sun, wind energy is considered to be a form of solar energy, and is therefore renewable.

Hydro refers to energy derived from flowing or falling water. The terms *hydropower* and *hydroelectric* are commonly used in conjunction with this type of renewable energy source. The kinetic energy of water flowing in a river or stream can be harnessed directly to produce work. More commonly, the flow of water is regulated by means of a reservoir or dam, thereby facilitating a conversion of the water's potential energy to kinetic energy. To a greater extent than wind energy, hydro energy is geographically dependent, being limited to locations where rivers are found. Hydro energy is one of the oldest sources of renewable energy and is a natural consequence of the earth's water cycle, which is driven by the sun.

Geothermal energy originates within the earth. About 80 percent of this internal energy is heat generated by the decay of radioactive substances, and the remainder is residual heat from earth's planetary accretion. The energy from these two sources is transported to the earth's surface where it is manifested in hot ground and surface waters, rocks, and magma. Even though the heat content of the earth is finite, geothermal energy is considered to be renewable because the rate of heat extraction is small compared with the earth's heat content, which is estimated to be 10^{31} J.

Marine energy, often referred to as *ocean* energy, is a broad category that includes energy derived from oceanic tides as well as oceans at locations away from shorelines. *Tidal* energy is derived from the periodic rise and fall of sea levels, which is caused by a combination of gravitational forces exerted by the moon and sun and the rotation of the earth. The potential or kinetic energy of the tides is converted to mechanical energy in a turbine and then electrical energy in a generator. *Ocean* energy is derived from three sources—ocean surface waves, underwater ocean currents, and temperature differences in ocean water. For the first two sources, kinetic energy of the waves and currents is converted to mechanical energy, which is then converted to electrical energy. For the third source, a heat engine utilizes a temperature difference of deep and shallow water to generate mechanical energy in a turbine and then electrical energy in a generator. As long as the earth's oceans remain in the liquid phase, which is equivalent to saying that as long as the sun maintains the earth's surface at suitable temperatures, marine energy will be available. Thus, marine energy, like wind and hydro energy, is a form of solar energy.

Lastly, ***biomass*** is fuel derived from living, or recently living, organisms, which includes plant- and animal-based materials. Plant matter can be burned as a heat source for a heat engine to generate electrical energy. Both plant and animal matter can be converted into *biofuels*. Biomass is a renewable energy source in that plants can be quickly regrown to replace the plants that were consumed. Because biological organisms depend on the sun, solar energy is the ultimate source of energy for biomass.

14.1.1 Environmental Considerations

Through farming, construction, and other activities, man has always changed his surroundings. In the decades following the industrial revolution, the consumption of coal steadily increased, resulting in severe pollution of major cities. In the twentieth century, coal consumption continued to increase, but due to the use of smokeless fuel and tall smokestacks, pollution in populated areas was significantly reduced. However, forests in regions of North America and Europe were dying, and marine life in lakes was dying as well. Scientists discovered that coal combustion was causing *acid rain*, a form of precipitation arising from emissions of sulfur dioxides and nitrogen oxides that react with water in the atmosphere.

Acid rain is not the only harmful effect of coal combustion. The combustion of coal and other fossil fuels releases carbon dioxide into the atmosphere. Carbon dioxide, along with water vapor, methane, nitrogen oxides, and ozone, is a greenhouse gas. In recent decades the surface temperature of the earth has markedly increased. Scientists have proposed that the accelerated use of fossil fuels is at least partly, and perhaps primarily, to blame for this temperature rise. If the combustion of fossil fuels is responsible for global warming, and if we persist in using fossil fuels at current or accelerated rates, the earth's temperature will continue to rise, leading to gradual melting of polar ice caps and subsequent rises in sea level that will inundate coastal areas. Another likely effect of global warming is the increased frequency of extreme weather phenomena such as droughts and heavy rainfall. Global warming could even cause species extinctions due to shifting temperature patterns.

For practical and economic reasons, the consumption of fossil fuels cannot be suspended or even significantly slowed in the coming years. According to the U.S. Energy Information Administration (EIA), approximately 83 percent of the total energy (electrical, transportation, heating, etc.) consumed in the United States in 2010 came from fossil fuels, about 9 percent came from nuclear sources, and approximately 8 percent came from renewable sources. In an attempt to reduce carbon dioxide emissions and become more energy-independent, the United States and other industrialized nations are undergoing a gradual shift from fossil fuels to renewable sources.

In the sections that follow, the fundamental engineering principles of renewable energy sources of solar, wind, hydro, geothermal, marine, and biomass are discussed.

14.2 SOLAR

Solar energy is the most vital of all energy sources, for without it no life on earth could exist. With the exceptions of geothermal and nuclear energy, the sun is the ultimate source of all forms of energy on earth. To begin our discussion of solar energy, let's determine the capacity of the sun to satisfy our energy needs.

Just outside the earth's atmosphere the average solar power density (solar intensity) is 1353 W/m^2. This quantity is known as the solar constant, I_{sc}. Due to reflection, absorption, and re-radiation of solar energy from the earth to space, the percentage of the solar constant that reaches the earth's surface is approximately 47 percent. The average radius, R, of the earth is 6371 km, so the projected area (area of a circle), A, of the earth is

$$\begin{aligned} A &= \pi R^2 \\ &= \pi (6371 \times 10^3\,\text{m})^2 \\ &= 1.275 \times 10^{14}\,\text{m}^2. \end{aligned}$$

Hence, the solar power, $\dot{Q}_{\text{solar}}$, at the earth's surface is

$$\begin{aligned} \dot{Q}_{\text{solar}} &= 0.47 A I_{sc} \\ &= (0.47)(1.275 \times 10^{14}\,\text{m}^2)(1353\ \text{W/m}^2) \\ &= 8.1 \times 10^{16}\ \text{W}. \end{aligned}$$

The global power consumption in 2010 has been estimated to be 1.7×10^{13} W. Dividing this number by the solar power at the earth's surface, we obtain 0.00021. Multiplying this number by 100 to get a percentage, we obtain 0.021 percent. Thus, 0.021 percent of the solar power at the earth's surface could theoretically satisfy the world's power demands.

Let's examine the capacity of solar energy another way. Based on the 2010 global power consumption, how much of the earth's surface, covered with solar panels, is required to satisfy the world's power demands? Using the projected area of the earth in our first calculation, the surface area required is

$$A_{\text{solar panel}} = (0.00021)(1.275 \times 10^{14}\,\text{m}^2)$$
$$= 2.7 \times 10^{10}\,\text{m}^2$$

which is approximately 10,000 square miles, an area about the size of Maryland. Without taking into account energy conversion efficiency, energy storage, and distribution variables, these approximate but revealing calculations indicate that solar energy alone has the capacity to satisfy global energy demands.

14.2.1 Solar Energy Systems

There are basically two types of solar energy systems—***solar thermal*** systems and ***photovoltaic*** systems. The first type of solar thermal system is typically designed to heat water for domestic hot water use, as shown in Figure 14.1, but could also be used to heat air for a living space or provide heat for other purposes. This system incorporates a *collector* that absorbs solar radiation, thereby heating a *working fluid* that is pumped through it. Working fluids in such systems are usually an ethylene-glycol mixture (antifreeze). The warm working fluid passes through a *heat exchanger* in a storage tank where the fluid's heat is transferred to the domestic water. The cool working fluid returns to the collector to be reheated by solar radiation. This type of system is typically found in residential and small commercial applications.

Figure 14.1
Solar water heating system.

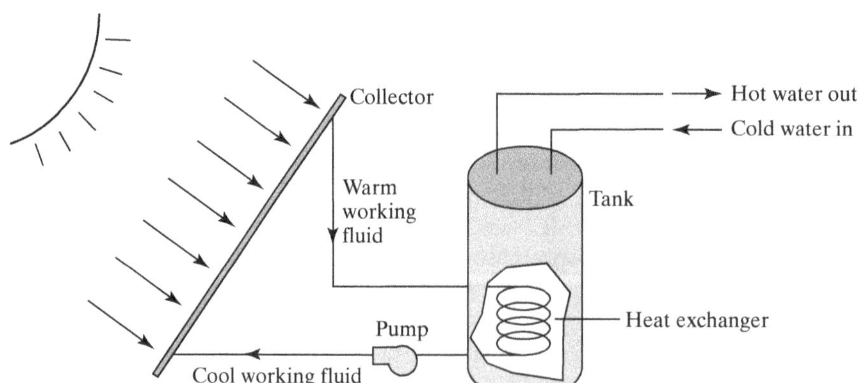

The second type of solar thermal system, commonly called a *concentrating solar power* plant, is designed to generate electrical power on a large scale. This type of system incorporates a *concentrator*, consisting of a mirror or an array of mirrors, that reflects sunlight and focuses it onto a very small area. As shown in Figure 14.2, parabolic dish and heliostat systems focus sunlight at a point, whereas a parabolic trough system focuses sunlight along a line. In order for these concentrators to focus the sunlight properly, the concentrators must be able to track the sun as it

Figure 14.2
Concentrating solar power systems.

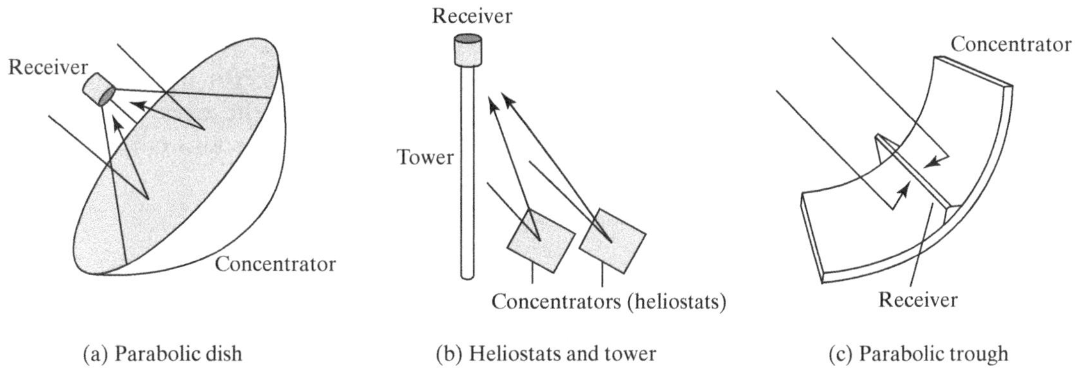

moves across the sky. Dish and heliostat concentrators require dual-axis tracking, whereas a trough concentrator requires single-axis tracking. Located at the focus of the concentrator is a *receiver* that carries a working fluid, typically a molten salt mixture. Because the sunlit area of the receiver is very small, the heat flux (heating rate per unit surface area, measured in W/m^2) at the receiver is high enough to heat the working fluid to temperatures above 500°C. The hot working fluid passes through a heat exchanger where the fluid transfers heat to water, turning the water into steam. The rest of the system resembles a standard power plant in which steam drives a turbine, which in turn drives an electrical generator. Other components in the system include a condenser and pump, as shown in Figure 14.3.

Figure 14.3
Concentrating solar power plant.

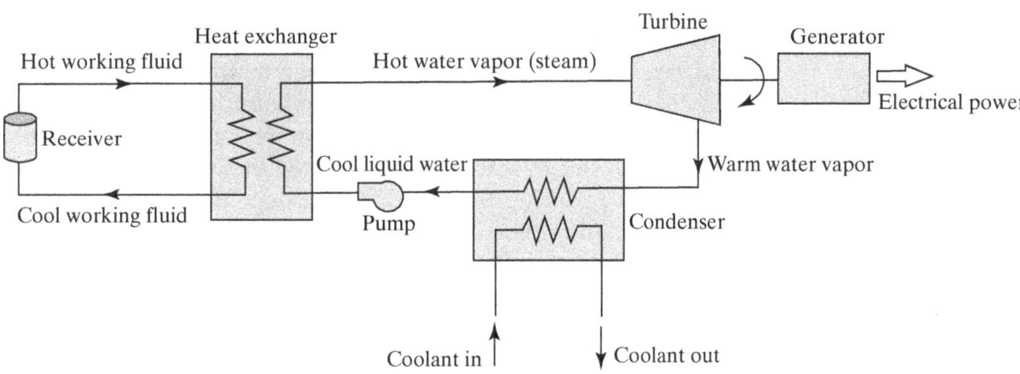

Unlike a concentrating solar plant, a photovoltaic system converts light directly into electricity. The heart of a photovoltaic system is a *solar cell* constructed of silicon or other semiconductor material. The solar cell is arranged to form a *p-n* (positive-negative) junction such that when light strikes the cell, free electrons are created by the photo-electric effect. Under the influence of the junction's electric field, the electrons move through the cell to the cell's surface where they are collected by a metallic grid. The grid is connected to electrical contacts on both sides of the cell. If an external electrical circuit is connected across the contacts, a direct current (DC) electrical current flows through the circuit.

A single solar cell produces an output voltage of approximately 0.5 V, which is insufficient to power most electrical devices. Hence, solar cells are connected in series, forming a *module*, to increase the output voltage. For example, a typical module designed to supply a standard 12 V output may consist of thirty or more cells to ensure reliable operation. To increase the output current, modules are connected in parallel, forming an *array* or *panel.* The desired voltage and current outputs can therefore be achieved by designing the array with the required number of series connections of cells and parallel connections of modules. This photovoltaic hierarchy is illustrated in Figure 14.4.

Figure 14.4
Photovoltaic hierarchy.

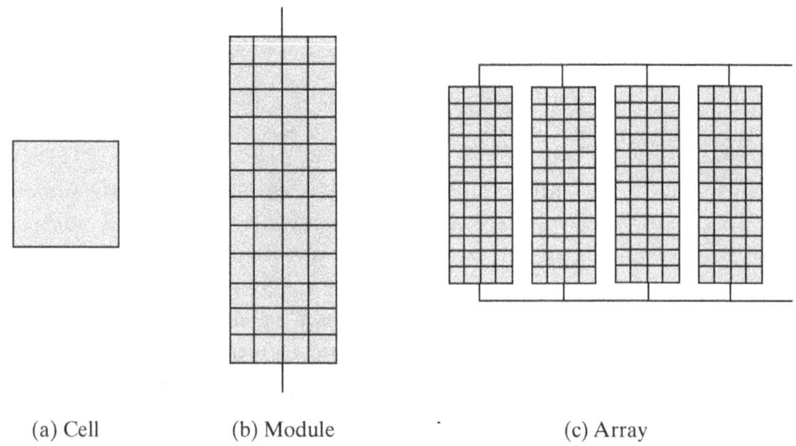

For residential and commercial building applications, photovoltaic systems are typically employed to augment electrical power provided by the local utility. In this *grid-connected* or *grid-tied* system, a solar panel and the power grid function in tandem to supply electrical power to the building, as shown in Figure 14.5. A charge controller prevents storage batteries from being overcharged and eliminates reverse current from the battery bank to the solar panel at night. The battery bank stores electrical energy generated by the solar panel during the day and uses that energy during the day or night, depending on the electrical power demands of the building. The inverter changes direct current to alternating current (AC) to operate lights, appliances, and other electrical devices in the building. If the energy stored in the battery bank is insufficient to meet the entire electrical needs of the building, the power grid supplies the difference. Conversely, if the photovoltaic system generates more electrical energy than the building requires, the excess energy

Figure 14.5
Grid-connected photovoltaic system.

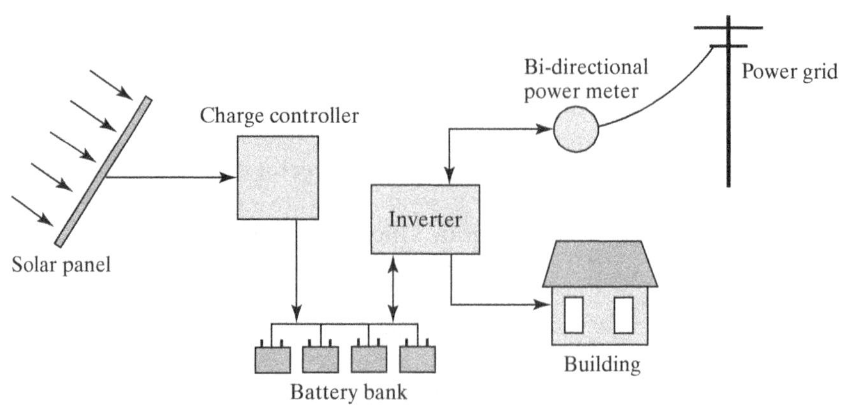

can be delivered to the power grid through the bi-directional power meter, thereby offsetting the electrical power the customer must purchase from the utility company at other times. This process is known as *net metering* and is a straightforward and low-cost way for the consumer to invest in renewable energy.

Concentrating solar systems are more complex than photovoltaic systems and are generally designed for large scale power production rather than residential or small commercial installations. Recent advances in solar cell technology and improvements in conversion efficiencies have broadened the implementation of photovoltaic systems. It is anticipated that photovoltaic systems will proliferate in these small-scale applications in the future, so the fundamental principles of this type of solar energy system are emphasized here. In the next section, the basic engineering principles of photovoltaic systems are presented.

14.2.2 Photovoltaic Systems

Unlike the collector of a concentrating solar plant, the solar panel of a photovoltaic system does not typically track the sun but is fixed in position on a rooftop or other suitable location, as shown in Figure 14.6. This means that throughout the day, and throughout the year, the angle between the sun's rays and the solar panel continually changes. This means that the amount of solar energy collected by the solar panel continually changes also. The total solar energy incident upon a surface is called *total solar irradiance,* I_{tot}, and is expressed in units of W/m^2. As illustrated in Figure 14.7, this quantity consists of three components—*direct* solar radiation, I_D, *scattered* radiation, also called sky radiation, I_S, and reflected radiation, I_R. Total solar irradiance is the sum of these three components, so

$$I_{\text{tot}} = I_D + I_S + I_R. \tag{14.1}$$

Figure 14.6
Photovoltaic solar panels on a rooftop. (Courtesy of Weber State University, Ogden, UT.)

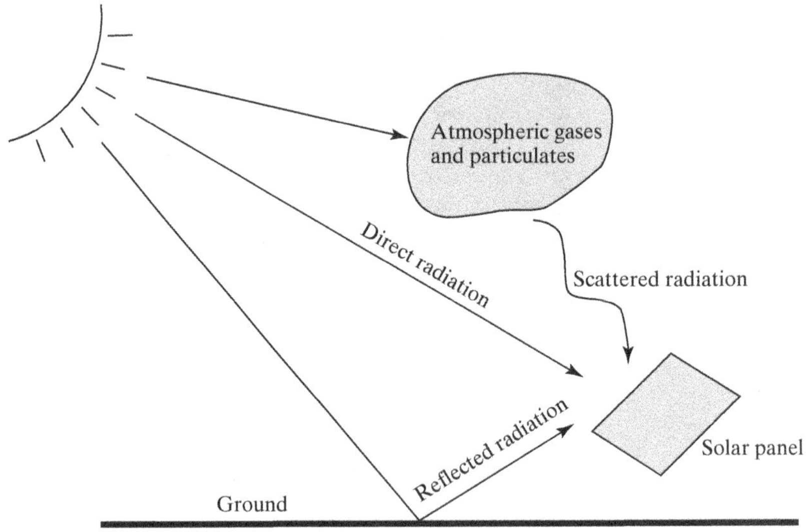

Figure 14.7
Components of solar irradiance.

Total solar power, expressed in units of W, is obtained by multiplying total solar irradiance by surface area.

Total solar irradiance is a function of the sun's position in the sky with respect to a normal (perpendicular) line to the solar panel, atmospheric conditions, and reflectivity of the surface in the foreground of the panel. The position of the sun with respect to the solar panel depends on the latitude and longitude of the solar panel's location, time of day, day of the year, and orientation of the panel with respect to the earth's surface and the north-south compass line. The equations that incorporate all these variables for calculating I_{tot} using Equation (14.1) may be found in most handbooks on solar energy. For the sake of simplicity, total solar irradiance is presented in graphical form in Figures 14.8 through 14.12 for selected U.S. cities as a function of time of day for two days of the year—January 21 and June 21. In all cases the solar panel faces due south, and the reflectivity of the panel's foreground is 0.2, a typical value for the ground or a roof. Results are shown for solar panel tilt angles of 0°, 30°, and 60°. Note the symmetry of the graphs around midday and the effect of latitude, time of year, and panel tilt angle on I_{tot}. The total solar irradiance values represented in these graphs apply to clear atmospheric conditions only; if the sky is overcast or cloudy, the values indicated should be reduced accordingly.

For each total solar irradiance curve in Figures 14.8 through 14.12, the insolation, H, is given. *Insolation*, also referred to as *solar irradiation* or *solar radiant exposure*, is the incident solar energy per unit surface area, found by integrating the total solar irradiance over a specified time period. Thus,

$$H = \int_{t_1}^{t_2} I_{tot}\, dt \tag{14.2}$$

where the time period is one day. This quantity is useful for calculating the daily total energy that can be produced by a solar panel of a given surface area.

Figure 14.8 Total solar irradiance for San Francisco, CA (a) January 21, (b) June 21.

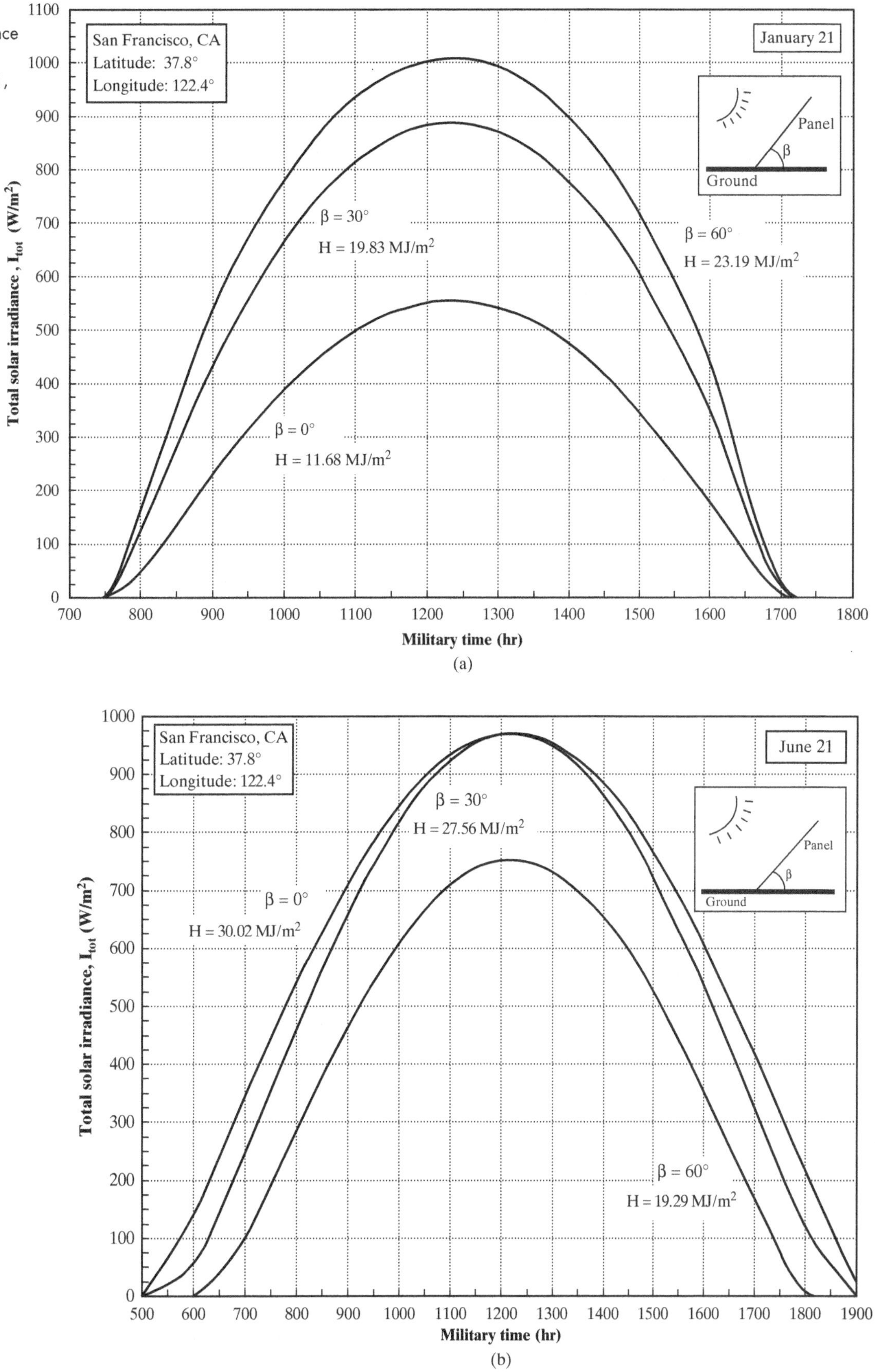

Figure 14.9
Total solar irradiance for San Antonio, TX (a) January 21, (b) June 21.

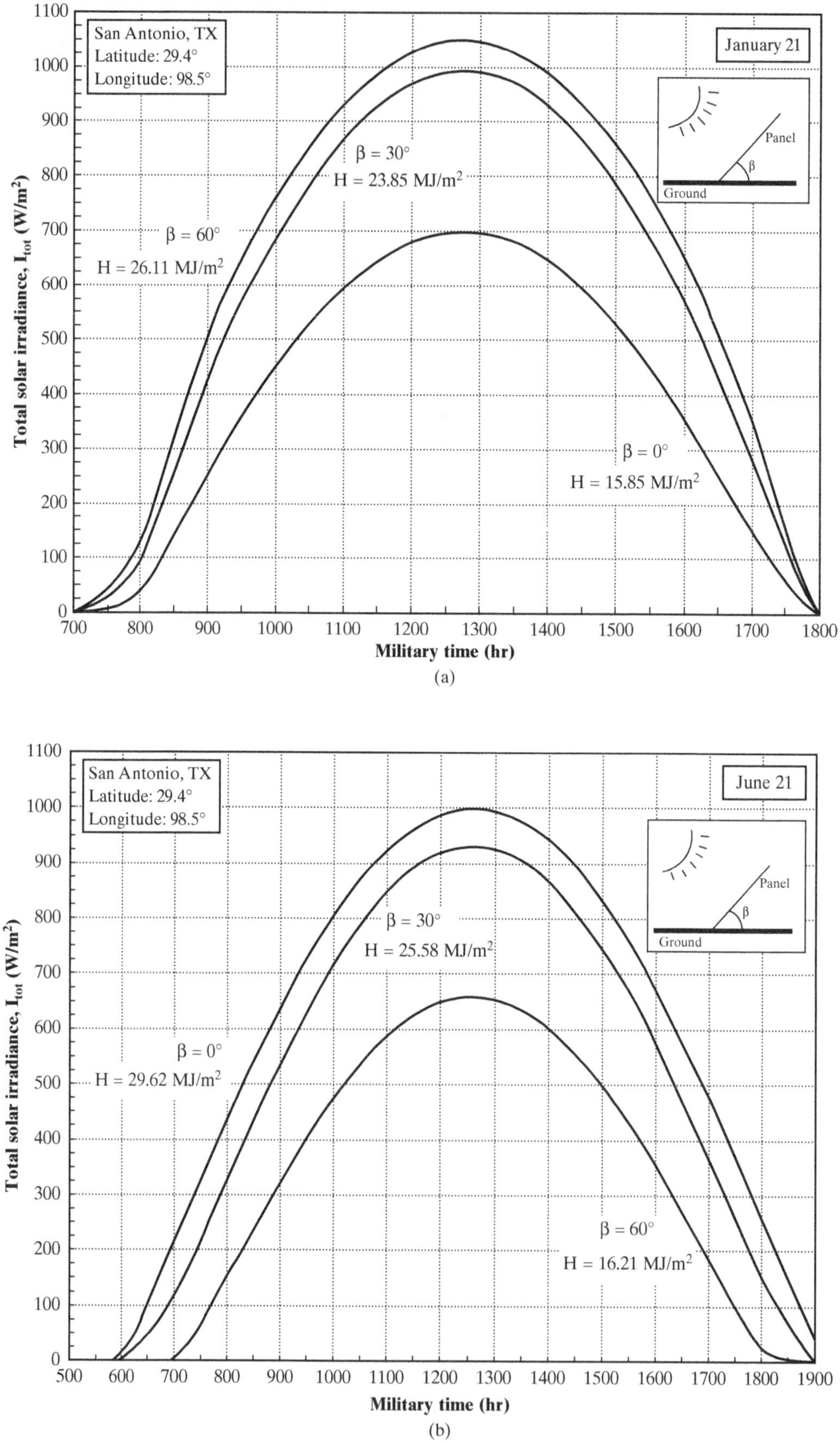

Figure 14.10 Total solar irradiance for Minneapolis, MN (a) January 21, (b) June 21.

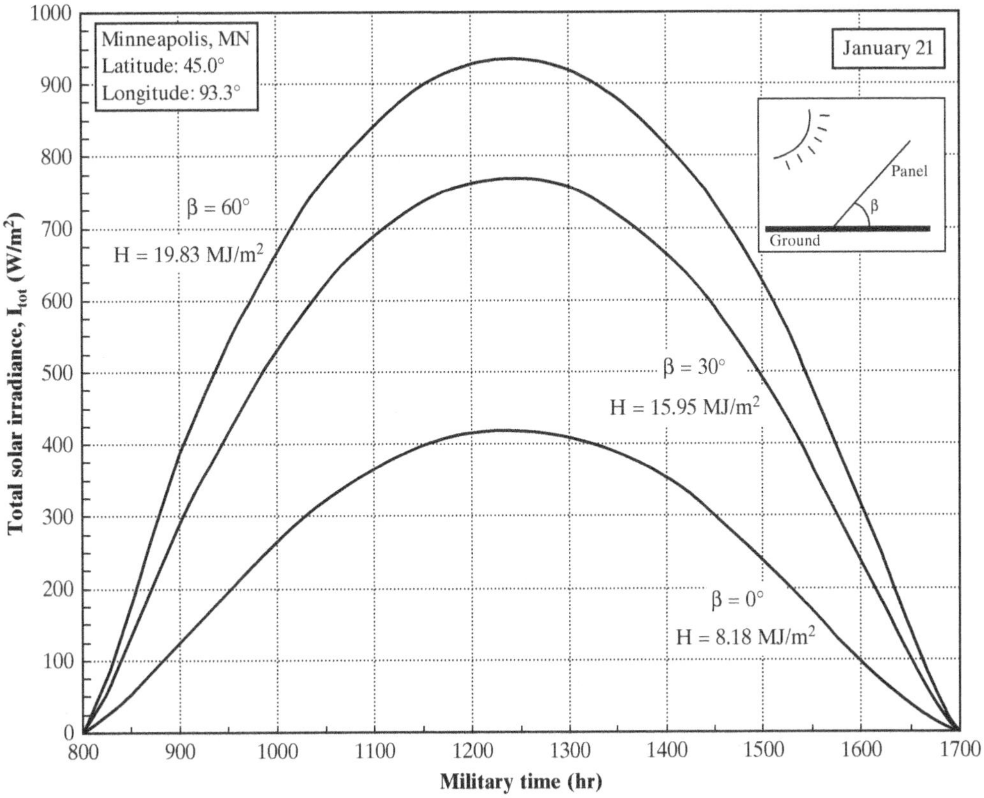

(a)

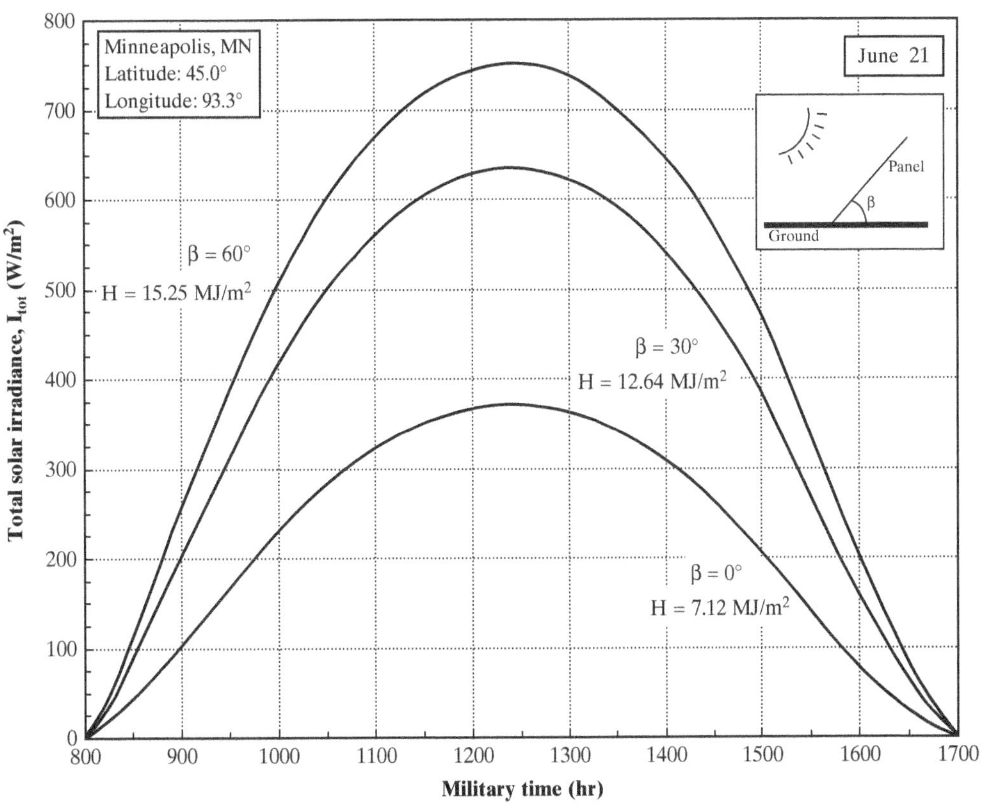

(b)

Figure 14.11
Total solar irradiance for Miami, FL (a) January 21, (b) June 21.

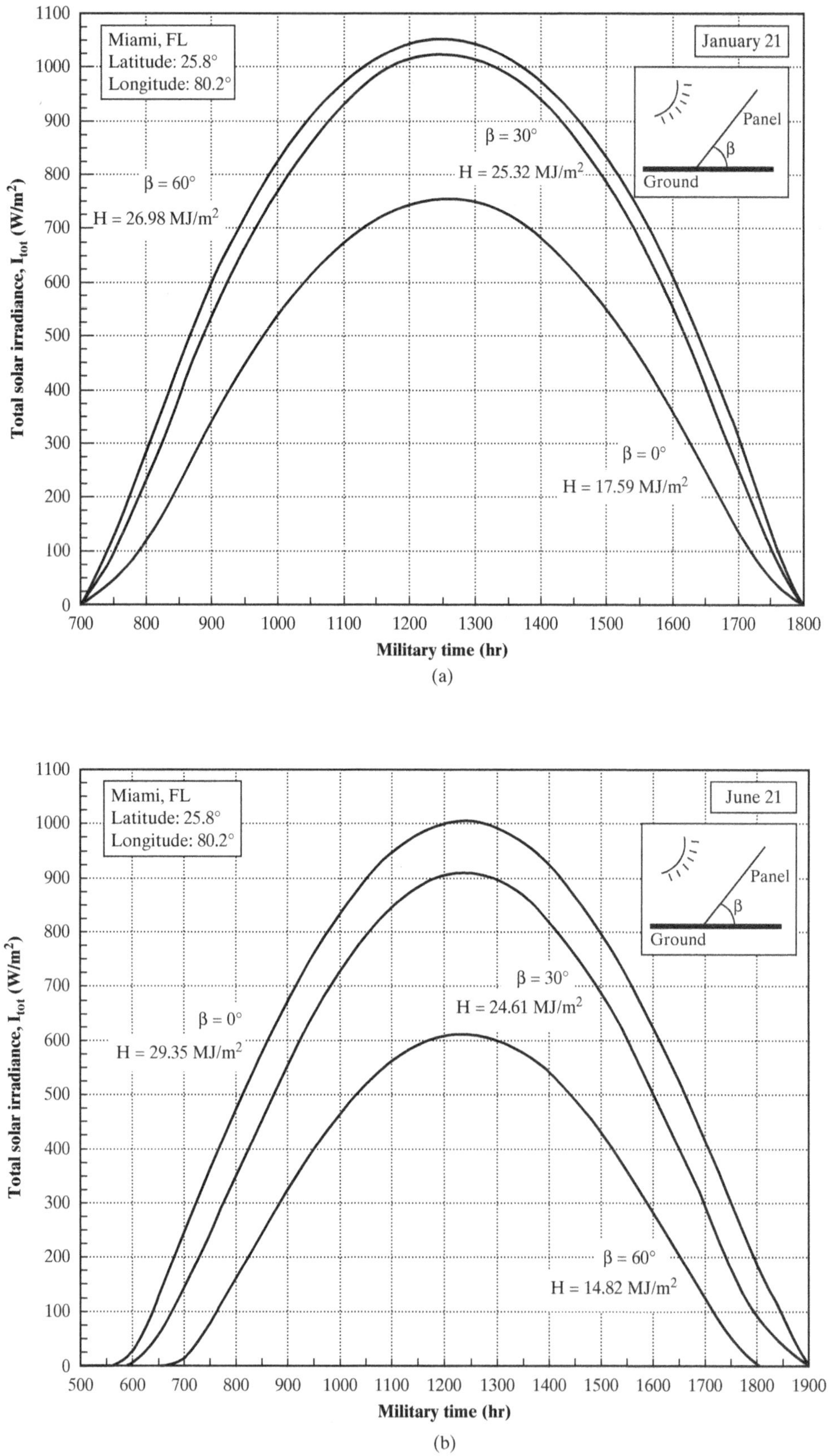

Figure 14.12
Total solar irradiance for Boston, MA
(a) January 21,
(b) June 21.

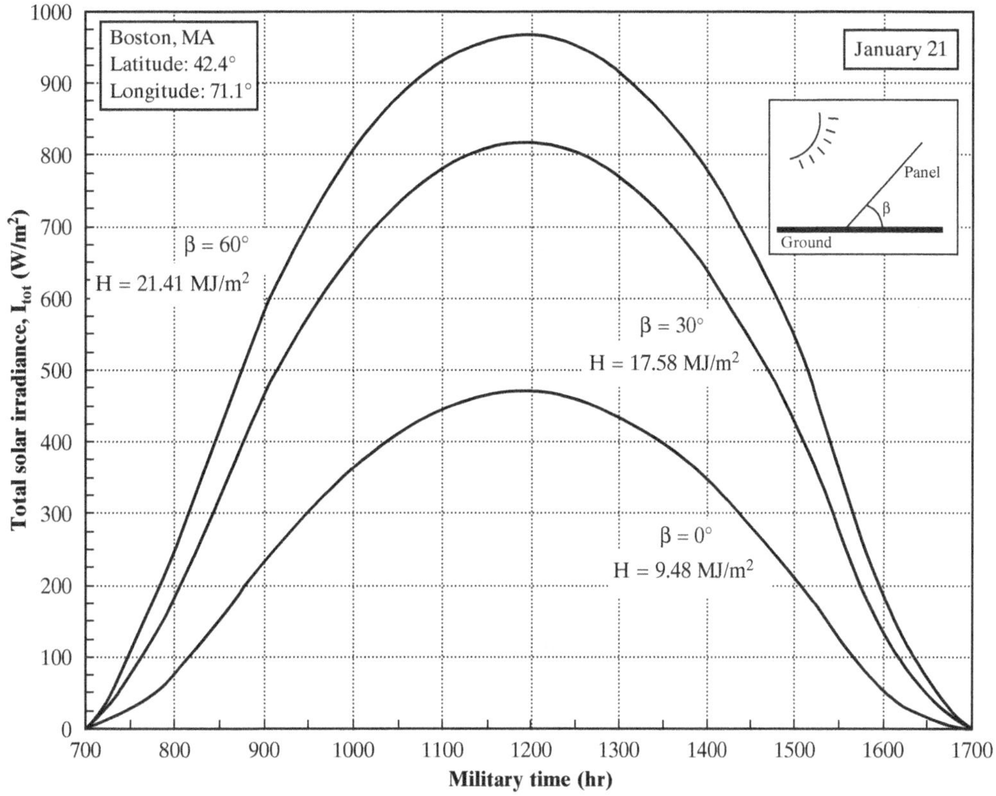

(a)

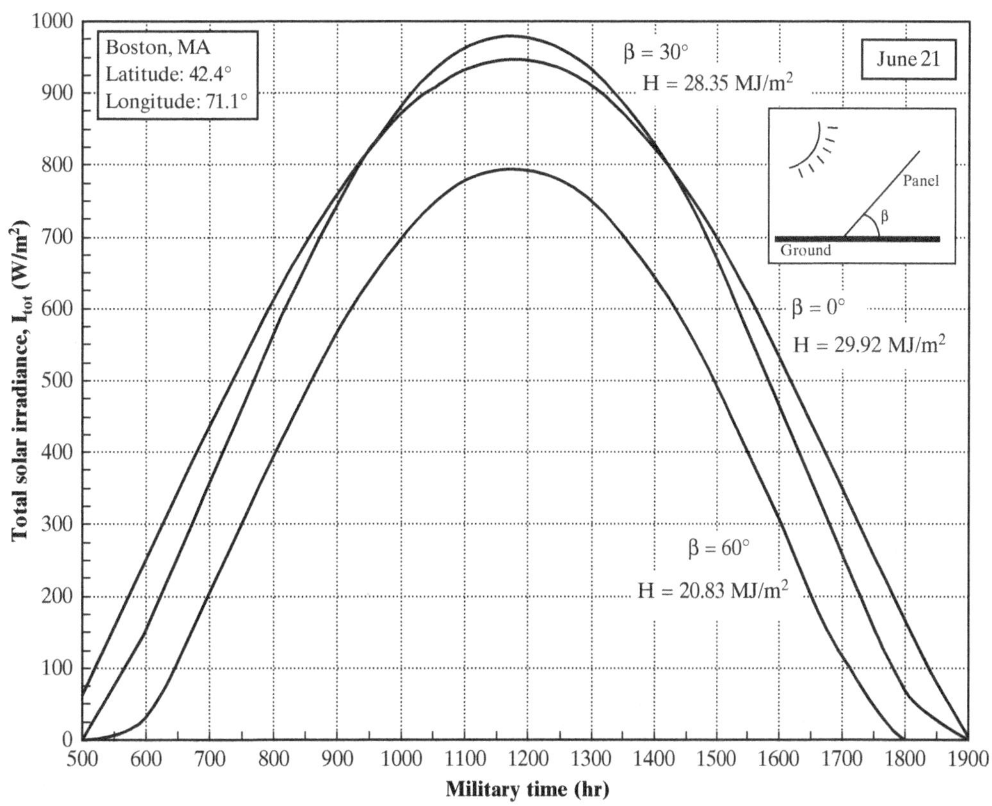

(b)

EXAMPLE 14.1

A photovoltaic solar panel in Boston, MA, measures 3 m × 8 m and has a tilt angle of 60°. What is the total solar power incident on the panel on January 21 at 10:30 am and 3:00 pm? What is the total solar energy incident on the panel during this day? The sky is clear, the panel faces due south, and the foreground reflectivity is 0.2.

Solution

The times 10:30 am and 3:00 pm, expressed in military time, are 1030 h and 1500 h, respectively. From Figure 14.12(a), the total solar irradiances for 1030 h and 1500 h for a panel tilt angle of 60° are approximately 870 W/m^2 and 545 W/m^2, respectively. The total solar power incident on the panel at 10:30 am is

$$\begin{aligned}\dot{Q}_{solar} &= I_{tot} A_{panel}\\ &= (870\ \text{W/m}^2)(24\ \text{m}^2)\\ &= 2.09 \times 10^4\ \text{W} = 20.9\ \text{kW}\end{aligned}$$

and the total solar power incident on the panel at 3:00 pm is

$$\begin{aligned}\dot{Q}_{solar} &= I_{tot} A_{panel}\\ &= (545\ \text{W/m}^2)(24\ \text{m}^2)\\ &= 1.31 \times 10^4\ \text{W} = 13.1\ \text{kW}.\end{aligned}$$

The total solar energy incident on the panel during this day is the insolation multiplied by the surface area of the panel.

$$\begin{aligned}E_{solar} &= HA_{panel}\\ &= (21.41\ \text{MJ/m}^2)(24\ \text{m}^2)\\ &= 513.8\ \text{MJ}.\end{aligned}$$

Photovoltaic solar cells do not convert all of the incident solar power into electrical power. Some of the incident solar radiation is reflected from the front surface of the module or panel. Some of the absorbed radiation does not have the energy required to initiate the photoelectric effect in the semiconductor crystal, and some of the absorbed radiation has more energy than is required, which results in excess energy that is dissipated as heat. Absorption that occurs far from the p-n junctions also results in wasted energy in the form of heat. These and other mechanisms of energy loss contribute to the total energy loss of the solar cell. The ***efficiency*** of a solar cell, η_{cell}, is defined as the ratio of the electrical power produced by the cell to the solar power incident on the cell:

$$\eta_{cell} = \frac{P_{elect}}{\dot{Q}_{solar}}. \tag{14.3}$$

The typical range of η_{cell} is approximately 0.15 to 0.25 (15 to 25 percent), but cell efficiencies of 0.40 and higher have been achieved. Solar cell efficiency can also be expressed in terms of a ratio of electrical energy produced by the cell to the solar energy incident on the cell during a specified time period:

$$\eta_{cell} = \frac{E_{elect}}{E_{solar}}. \tag{14.4}$$

EXAMPLE 14.2

A commercial building in San Antonio, TX, has a roof-mounted photovoltaic solar panel that measures 5 m × 12 m. The tilt angle of the panel is 30°, and the reflectivity of the roof surface is 0.20. If the efficiency of the panel is 0.18, how much electrical energy does the panel produce on January 21?

Solution

As shown in Figure 14.9(a), the insolation for this panel in San Antonio, TX, on January 21 is $H = 23.85\ \text{MJ/m}^2$. The solar energy incident on the panel is the insolation multiplied by the surface area of the panel:

$$\begin{aligned} E_{\text{solar}} &= HA_{\text{panel}} \\ &= (23.85\ \text{MJ/m}^2)(5\ \text{m} \times 12\ \text{m}) \\ &= 1431\ \text{MJ}. \end{aligned}$$

Rearranging Equation (14.4), the electrical energy produced by the panel is

$$\begin{aligned} E_{\text{elect}} &= \eta_{\text{cell}} E_{\text{solar}} \\ &= (0.18)(1431\ \text{MJ}) \\ &= 258\ \text{MJ}. \end{aligned}$$

The commonly used unit for electrical energy is the kilo-watt-hour (kWh). Converting our answer to kWh, we obtain

$$258\ \text{MJ} \times \frac{1\ \text{kWh}}{3.6\ \text{MJ}} = 71.7\ \text{kWh}.$$

Solar energy is considered to be one of the most environmentally friendly types of renewable energy. During operation, neither solar thermal nor photovoltaic systems introduce harmful emissions into the atmosphere. Solar thermal systems generate an insignificant amount of noise, and photovoltaic systems generate none. Despite these attractive features, solar energy is not without environmental impacts. Because the amount of solar radiation that can be intercepted by a solar collector or panel is limited, both types of solar energy systems require a significant amount of space. On a commercial scale, a large area of land is required, and on a residential scale, the space on a rooftop limits the amount of energy that can be gathered. Furthermore, a rooftop installation must share space with other things such as vent pipes, sky lights, and electrical systems. Photovoltaic solar panels incorporate silicon, a semiconductor material that is very energy intensive to manufacture. A significant amount of electricity is required to produce silicon, and since the majority of electricity is generated by fossil-fuel power plants, the production of silicon introduces harmful substances into the atmosphere. Lastly, the manufacture of silicon involves toxic chemicals, such as cadmium, which must be prevented from entering the natural environment.

PRACTICE!

1. Find the solar power intercepted by a 3 m × 5 m sun-facing surface just outside the earth's atmosphere.
Answer: 20.3 kW
2. A horizontal photovoltaic solar panel in Miami, FL, measures 2 m × 6 m. What is the total solar power incident on the panel on January 21 at 12:00 noon and 4:00 pm? What is the total solar energy incident on the panel during this day? The sky is clear, the panel faces due south, and the foreground reflectivity is 0.2.
Answer: 8.94 kW, 4.26 kW, 211 MJ
3. A home in San Francisco, CA, has a roof-mounted photovoltaic solar panel that measures 2 m × 3.5 m. The tilt angle of the panel is 60°, and the reflectivity of the roof surface is 0.20. If the efficiency of the panel is 0.16, how much electrical energy does the panel produce on June 21?
Answer: 6.00 kWh

14.3 WIND

Wind is one of the oldest renewable energy sources. Sail boats have been propelled by the wind for thousands of years, and windmills were traditionally used for grinding grain and spices, sawing wood, making paper materials and paints and dyes. Along with solar and other renewable energy sources, wind energy has been widely developed in recent years. The first wind turbine for generating electrical power was developed in the late 19th century. Wind energy technologies developed gradually during the early to mid 20th century, but during the oil crisis of the 1970s, wind energy research and development expanded rapidly. By the end of the 20th century, wind energy played an integral role in renewable energy development. Wind energy has several advantages such as no pollution, short implementation time and relatively low capital cost. Currently, wind supplies approximately 1.0 percent of the world's electrical energy demand, but it is anticipated that wind energy will supply about 12 percent of the world's electrical energy demand by 2020.

Wind is the movement of atmospheric air from regions of high pressure to regions of low pressure. The main cause of winds is the uneven heating of the earth's surface by the sun, which depends on latitude, time of day and the distribution of land and large bodies of water, particularly the oceans. Another cause of winds is fluid friction between the atmosphere and earth's surface, which allows the earth to drag the atmosphere around producing turbulence. Horizontal components of wind velocities are normally much greater than the vertical velocity components, which is why we normally think of wind as the horizontal movement of air.

Over the earth's land masses, winds generate approximately 6.1×10^{21} J of energy annually. Dividing this energy by the number of seconds in a year, we obtain 1.9×10^{14} W, the global power generated by winds. We noted earlier that the global power consumption in 2010 has been estimated to be 1.7×10^{13} W. We also noted that the solar power at the earth's surface is 8.1×10^{16} W, approximately two orders of magnitude higher than the energy generated by winds. This comparison is somewhat misleading, however, because the quantity given for wind energy generation applies to land masses only, and about 75 percent of the earth's surface is covered by oceans. Nevertheless, we can see that wind, like solar, has the capacity to supply the world's energy demands.

Machines that convert the kinetic energy of the wind to electrical energy are referred to as wind turbines. As illustrated in Figure 14.13, there are two types of wind turbines, each being characterized by the orientation of the axis (shaft) of the turbine. A typical *horizontal axis wind turbine* (HAWT) consists of three blades mounted to a horizontal shaft. The power generating capacity of horizontal axis wind turbines varies widely. Large commercial turbines generate as much as 5 MW of power, whereas turbines used to generate power for a single home or small business generate 100 kW or less. Blade diameter varies widely also, ranging from a few meters to over 100 m. The nacelle of a HAWT is an enclosure that houses the generator, gearbox, drive train, brake system and other mechanical components. The role of the tower is to support the turbine at a height well above the ground to intercept stronger winds and therefore harness more wind energy.

Figure 14.13
Types of wind turbines

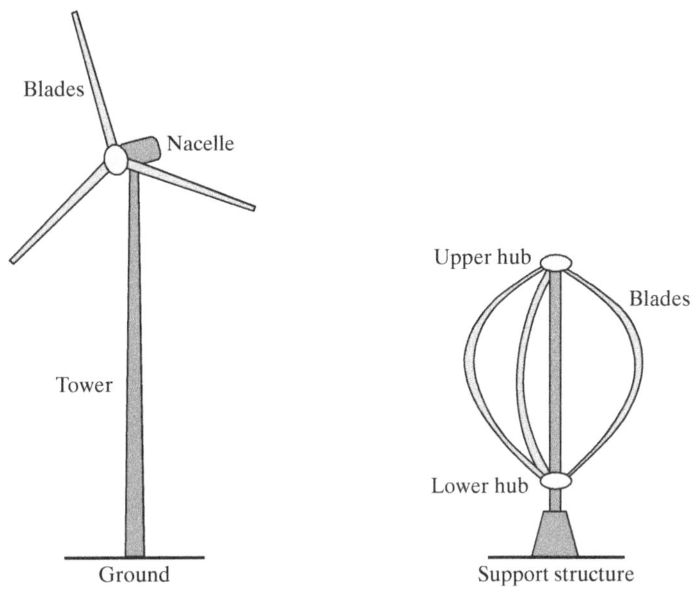

(a) Horizontal axis wind turbine (HAWT) (b) Vertical axis wind turbine (VAWT)

A *vertical axis wind turbine* (VAWT), which resembles an "eggbeater," typically has three blades mounted to a vertical shaft. Historically, the first wind mills were based on the vertical-axis configuration. Vertical axis wind turbines are less common than horizontal axis wind turbines and are primarily used in small-scale applications. This type of wind turbine is normally mounted on a short mast close to the ground or the roof of a building.

Even though HAWT systems are more prevalent than VAWT systems, both configurations have advantages and disadvantages. Advantages of HAWT systems include higher efficiency, lower cost-to-power ratio, and the ability of the blades to self-start the turbine. A disadvantage of HAWT systems is that the generator and gearbox are located at the top of a high tower, thereby restricting maintenance and service. Another disadvantage of a HAWT system is its complexity. A HAWT must be designed with a yaw control system so that the turbine blades always face into the wind. Advantages of VAWT systems include easy maintenance of the generator and gearbox, since they are usually located close to the ground or roof. No yaw control system is needed for a VAWT since the blades will rotate under any wind direction, and a VAWT is less costly to fabricate due to the relative simplicity of its blade design. Disadvantages of VAWT systems include lower efficiency and difficulty in controlling blade over-speed. And, unlike a HAWT, a VAWT is not self-starting, so the generator must run in motor mode to start the blades rotating.

A simplified version of the main components of a HAWT is shown in Figure 14.14. The blades are mounted to a hub that is connected to the main shaft of the gear box. The function of the gear box is to increase the spin rate of the shaft connected to the electrical generator so that the power produced by the generator is maximized. Depending on the size and complexity of the HAWT, various types of braking systems are employed to keep the turbine speed within its allowed limits in high winds and to hold the turbine at rest during maintenance. For large turbines that have blades with controllable pitch (angle of attack), aerodynamic braking is typically used. Braking of small wind turbines can be done by dumping the energy from the generator into a resistor bank, converting the kinetic energy of the shaft into heat. The function of a mechanical brake is to hold the turbine at rest while maintenance is performed. Because the tower produces turbulence behind it, the turbine is positioned upwind of the tower. The anemometer and wind vane measure wind speed and direction, respectively, sending this data to the controller. The controller directs the yaw system, keeping the turbine pointed into the wind, and controls blade pitch and the braking system.

Figure 14.14
Main components of a HAWT.

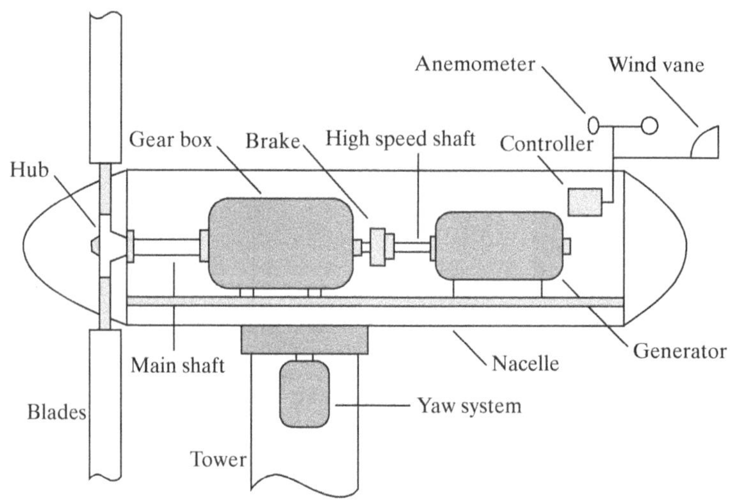

A *single* horizontal axis wind turbine is suitable for residential and small commercial applications. For large-scale power generation applications, however, horizontal axis wind turbines are found in a group called a wind farm or wind park. A *wind farm* consists of a cluster of wind turbines covering an extended area, as shown in Figure 14.15. Wind farms require open spaces with unobstructed access to the wind, making wind farms ideal for agricultural areas, grazing lands, and coastlines. A small wind farm may consist of fewer than twenty turbines covering a few acres, whereas a large wind farm may consist of hundreds of turbines covering hundreds of square miles. Most wind farms are located on land, but wind farms can also be located offshore.

Figure 14.15
A wind farm. (Stephen Bures/Shutterstock)

In a well designed wind farm, the spacing of wind turbines is carefully considered. If the turbines are too close to one another, the power output of downwind turbines will be diminished due to interference of upwind turbines. If the turbines are spaced too far apart, the wind farm site space is underutilized. The optimum spacing between towers in the direction of the wind is between five and nine rotor diameters, and the optimum side-to-side spacing between towers is between three and five rotor diameters. This optimum spacing configuration is illustrated in Figure 14.16.

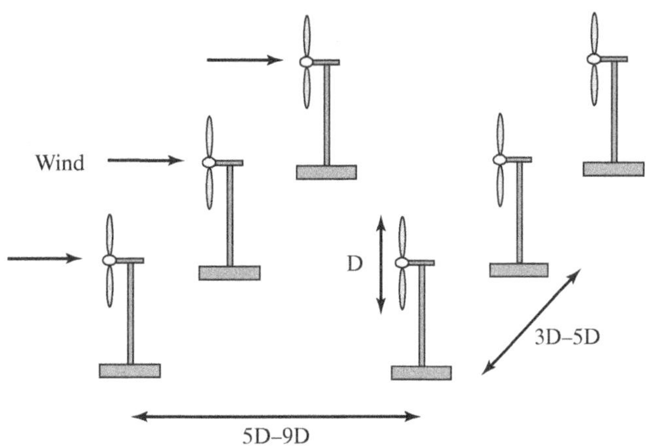

Figure 14.16
Optimum tower spacing in a wind farm.

The generation of electrical power by wind turbines does not release harmful substances into the environment, but wind power impacts the environment in other ways. Wind turbines generate a low to moderate amount of aerodynamic and mechanical noise. Aerodynamic noise comes from the blades moving through the air, and mechanical noise comes from the gear box, generator, and other mechanical components. If wind turbines with metal in their blades are located directly between radio, television, or microwave transmitters and receivers, the blades can distort the received signals. Rotating turbine blades are a threat to birds, particularly in migration corridors. Perhaps the greatest environmental impact of wind turbines is visual. Some people object to having tall structures with rotating blades near their homes or businesses, claiming that the machines detract from the surroundings and are visually distracting. All these environmental impacts can be mitigated by careful site selection and good engineering design.

In the next section, a basic engineering analysis of a horizontal axis wind turbine is given.

14.3.1 Basic Energy Analysis of a Horizontal Axis Wind Turbine

A wind turbine converts the kinetic energy of moving air to electrical energy. Of primary importance to engineers is the maximum amount of energy that a wind turbine can produce for a given wind speed. Consider a HAWT exposed to a wind of velocity, v, whose blades sweep out an area, A, as illustrated in Figure 14.17. From basic physics, kinetic energy is found using the formula

$$\mathrm{KE} = \frac{1}{2}mv^2 \tag{14.5}$$

where m is mass and v is velocity. Dividing both sides of Equation (14.5) by mass, m, we obtain

$$\mathrm{ke} = \frac{1}{2}v^2 \tag{14.6}$$

where ke is kinetic energy per unit mass, expressed in units of J/kg. The mass flow rate, $\dot{m}$, of air flowing through the swept area of the turbine blades is expressed as

$$\dot{m} = \rho A v \tag{14.7}$$

where ρ is the density of air, expressed in units of $\mathrm{kg/m^3}$, and $\dot{m}$ is expressed in units of kg/s. We see that by multiplying both sides of Equation (14.6) by mass flow rate, the left side of the equation now has units of J/s, which is defined as a watt (W), a unit of power. Hence, we have derived a relation for the available power in the wind,

$$P_{\text{wind}} = \frac{1}{2}\rho A v^3. \tag{14.8}$$

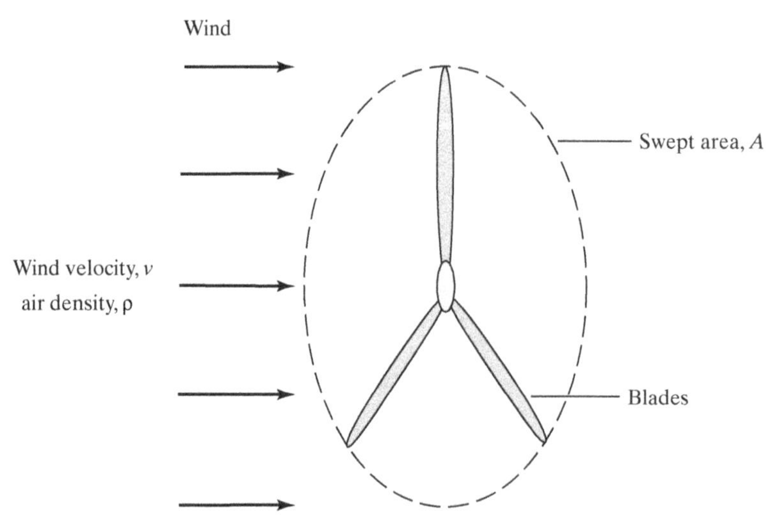

Figure 14.17
Wind turbine diagram for calculating available wind power.

The available power in the wind, found using Equation (14.8), is not the maximum power that the wind turbine can produce. A wind turbine that produces this amount of power implies that all the kinetic energy of the wind is converted to mechanical energy of a rotating shaft, yielding a wind velocity of zero downstream of the turbine blades. The maximum available power is found using the relation

$$P_{\text{wind, max}} = C_p P_{\text{wind}} \tag{14.9}$$

where C_p is the power coefficient, which has a maximum value of 16/27 (0.593). This value is known as the Betz limit. Modern wind turbines have power coefficients in the range of about 0.45 to 0.50. Furthermore, there are energy losses in the gear box, generator, and bearings that further reduce the output power of the turbine. When these losses are also taken into account, only 10 to 30 percent of the power in the wind is converted to electrical power. Thus, the electrical output power of the turbine may be expressed as

$$P_{\text{elect}} = \eta P_{\text{wind, max}} \tag{14.10}$$

where η is the combined efficiency of the gear box, generator, and bearings.

EXAMPLE 14.3

A wind farm consists of 60 horizontal axis wind turbines with a blade diameter of 45 m. The power coefficient of the turbines is 0.48, and the combined efficiency of the gear box, generator, and bearings is 0.36. Assuming that the wind blows steadily at 5 m/s, what is the electrical output power of the wind farm? For the density of air, use 1.16 kg/m^3.

Solution

The area swept out by the turbine blades is

$$\begin{aligned} A &= \frac{\pi D^2}{4} \\ &= \frac{\pi (45\text{ m})^2}{4} = 1590.4\text{ m}^2. \end{aligned}$$

Using Equation (14.8), the available power in the wind is

$$\begin{aligned} P_{\text{wind}} &= \frac{1}{2}\rho A v^3 \\ &= \frac{1}{2}(1.16\text{ kg/m}^3)(1590.4\text{ m}^2)(5\text{ m/s})^3 \\ &= 1.153 \times 10^5\text{ W}. \end{aligned}$$

Combining Equations (14.9) and (14.10), the electrical output power of one turbine is

$$\begin{aligned} P_{\text{elect}} &= C_p \eta P_{\text{wind}} \\ &= (0.48)(0.36)(1.153 \times 10^5\text{ W}) \\ &= 1.992 \times 10^4\text{ W}. \end{aligned}$$

There are 60 turbines in the wind farm, so the total electrical output power of the wind farm is

$$P_{\text{elect, farm}} = (60)(1.992 \times 10^4\,\text{W}) = 1.195 \times 10^6\,\text{W} = 1.195\,\text{MW}.$$

PROFESSIONAL SUCCESS—BEIN' GREEN

Kermit the Frog, a muppet character created by Jim Henson, sang the song, "It's Not Easy Bein' Green." Putting aside Kermit's struggle to cope with blending in with the natural surroundings, this song's title is a legitimate statement to consider in the context of renewable energy. A dictionary definition of *green* is "tending to preserve environmental quality, as by being recyclable, biodegradable, or nonpolluting." The renewable energy resources discussed in this chapter certainly qualify as green because they tend to preserve environmental quality and are, relative to fossil fuels, nonpolluting. But, being green means more than deriving our energy from renewable resources. Being green means turning down the heat in your home by installing a programmable thermostat, unplugging electronic devices that consume "vampire power" while not in use, using mass transit instead of your car, using cold water to wash your clothes, replacing incandescent lights with compact fluorescent lights, using native plants that require fewer pesticides, and, of course, recycling.

Is it easy being green? Ralph Waldo Emerson said, "That which we persist in doing becomes easier to do, not that the nature of the thing has changed, but that our power to do has increased." If it is not easy now, perhaps our persistence to be green will make it easier to be green in the future.

PRACTICE!

1. A house located at the mouth of a windy canyon utilizes a horizontal axis wind turbine to augment electrical power from the grid. The turbine has a blade diameter of 2.7 m. The power coefficient of the turbine is 0.40, and the combined efficiency of the gear box, generator, and bearings is 0.33. If the wind blows steadily at 14 m/s, what is the electrical output power of the turbine? Use 1.16 kg/m^3 for the density of air.
Answer: 1203 W

2. A wind farm operated by a large utility company consists of 125 identical, horizontal axis wind turbines with a blade diameter of 40 m. The power coefficient and combined efficiency of the gear box, generator and bearings are 0.41 and 0.38, respectively. On a given day, the wind blows at an average speed of 4.5 m/s for 5 h. How much electrical energy does the wind farm generate that day? For the density of air, use 1.16 kg/m^3.
Answer: 6.47 MWh

14.4 HYDRO

Of all the renewable energy sources, hydro is the oldest. The use of waterpower dates back to ancient Egypt and Mesopotamia where water wheels powered mills for grinding grain and saws for cutting wood and stone. During the Middle Ages water wheels were used for pumping water from mines and to power machines for forging metals. Other uses for waterpower were textile manufacturing, husking rice, papermaking, and pulping of sugar cane. Even though steam became the major source of energy during the industrial revolution, the use of waterpower continued well into the 19th century.

Today, the power of water is harnessed to generate electricity. The terms *hydropower* and *hydroelectric* are typically used to describe the generation of electrical power from falling or flowing water. The kinetic energy of water flowing in a river or stream can be harnessed directly to produce work. More commonly, the flow of water is regulated by means of a reservoir or dam, thereby facilitating a conversion of the water's potential energy to kinetic energy. Accounting for over 16 percent of global electrical power generation in 2010, hydropower is the most widely used source of renewable energy. In 2010, hydropower accounted for 93 percent of the global renewable energy capacity. While the share of other renewable energy sources, particularly solar and wind, are expected to increase in the future, hydro is projected to remain the largest renewable energy source for many years. The annual global hydropower potential is estimated to be over 16,400 TWh, but only about 19 percent of this potential has been developed.

There are three basic types of hydropower systems. The first and the most common type is a conventional hydropower system that converts the potential energy of water stored in a reservoir or dam to electrical energy. The second type, referred to as a run-of-river system, incorporates little or no water storage. A run-of-river system converts the kinetic energy of flowing water in a river to electrical energy. The third type, called a pumped-storage system, incorporates two reservoirs. During times of low power demand, water is pumped from the lower reservoir to the higher reservoir, where water can be released during times of high power demand. This type of system is not strictly a renewable energy system since electrical power is required to run the pump. Because the conventional hydropower system is the most common, this type will be emphasized here. The largest hydropower plant in the world is shown in Figure 14.18.

Figure 14.18
The Three Gorges Dam, spanning the Yangtze River in China, is the world's largest hydropower plant with a power generation capacity of 22.5 GW. (PRILL/Shutterstock)

A conventional hydropower plant converts the potential energy of stored water to electrical energy. The potential energy of the water is proportional to the depth of the water in the reservoir. As illustrated in Figure 14.19, water enters a downward sloping channel called a penstock. A control gate at the entrance of the penstock is used to control the flow rate of water or close the penstock during periods of maintenance. A screen system at the intake prevents fish and debris from entering the penstock and potentially damaging the turbine. The turbine at the bottom of the penstock is basically a water wheel that converts the kinetic energy of the flowing water to mechanical energy of a rotating shaft. The shaft of the turbine rotates the armature of the electrical generator thereby producing electrical power, which is delivered to the power grid over transmission lines. Water exiting the turbine is carried by a draft tube to the tail water, which is typically a river.

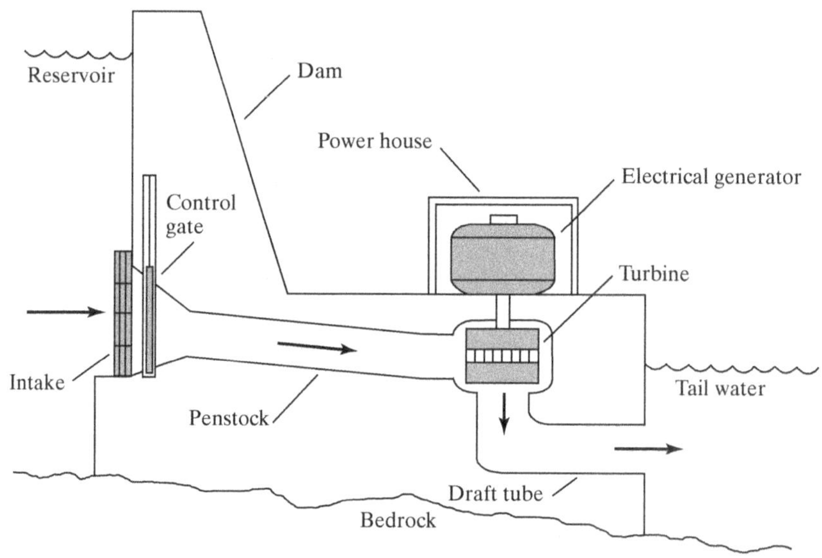

Figure 14.19
Conventional hydropower plant.

Like other renewable energy systems, hydropower has advantages and disadvantages. Hydropower plants do not burn fossil fuels, so they are unaffected by price fluctuations of coal, natural gas and oil. Hydropower plants do not release greenhouse gases into the environment by burning fossil fuels, but greenhouse gases are generated over time by the breakdown of plants that are covered with water when the reservoir is filled. The cost of electricity generated by large hydropower plants, after the building costs have been paid, ranges from about one to four cents per kilowatt-hour, less than any other source of electricity. However, hydropower plants are more costly to build than fossil fuel power plants. Hydropower is a well proven and reliable technology, but a hydropower plant can be built only where a river is located, and the flow of that river depends on precipitation. Reservoirs for hydropower plants may provide recreational uses, but the inundation of land may have displaced people from their homes and disrupted the biological ecosystem. Finally, sediment tends to precipitate onto the reservoir floor, reducing the amount of sediment in the river downstream of the reservoir. This reduction can cause river bank erosion and the disruption of ecostructures that rely on the sediment. Clearly, environmental and cost issues must be carefully considered in the planning of a hydropower plant.

Hydropower systems can be developed on a wide range of scales. The term *small hydro* refers to hydroelectric systems that power a small community or industrial facility. Small hydro systems, which have generating capacities of 10 MW or less, can be connected to the power grid but are sometimes located in isolated regions and therefore power a limited area. Subclassifications of small hydro are referred to as *mini, micro,* and *pico hydro,* which have generating capacity limits of 1000 kW, 100 kW, and 5 kW, respectively. Typically, a mini hydro system can power a few homes, a micro hydro system can power a single home, and a pico hydro system can power some lights and a few small appliances in a home. Small hydro systems are usually of the run-of-river type that use pipes to divert some of the water to a turbine before returning it to the river or stream. Because small hydro systems involve no reservoirs and minimal construction, they have relatively low environmental impact compared to large hydropower systems.

14.4.1 Basic Energy Analysis of a Hydropower Plant

A hydropower plant converts the potential energy of stored water to electrical energy. As with wind turbines, engineers are particularly interested in knowing the maximum amount of energy that a hydropower plant can produce. From basic physics, the potential energy per unit mass of a static liquid (water in the reservoir) is given by the relation

$$\mathrm{pe} = gh \tag{14.11}$$

where g is the acceleration of gravity, and h is the depth of the water. The units of pe are J/kg. The mass flow rate, expressed in units of kg/s, of water flowing through the turbine is

$$\dot{m} = \rho\dot{V} \tag{14.12}$$

where ρ is the density of water expressed in units of kg/m^3, and $\dot{V}$ is volume flow rate expressed in units of m^3/s. We see that by multiplying both sides of Equation (14.11) by mass flow rate, the left side of the equation now has units of J/s, which is defined as a watt (W), a unit of power. Thus, the power potential of the water is

$$P_{\text{water}} = \rho\dot{V}gh. \tag{14.13}$$

The term h, referred to as the *head,* is the elevation difference between the surface of the reservoir and the surface of the tail water. The power potential given by Equation (14.13) is not the maximum power that the hydropower plant can produce. There are energy losses in the turbine and generator that reduce the output power below this value. Accounting for these losses, the electrical output power of the hydropower plant is given by

$$P_{\text{elect}} = \eta P_{\text{water}} \tag{14.14}$$

where η is the combined efficiency of the turbine and generator. Typical values of η range from 0.8 to 0.9.

EXAMPLE 14.4

The head of a hydropower plant is 25 m, and the combined efficiency of the turbine and generator is 0.85. If the volume flow rate of the river downstream of the power plant is 6000 m^3/s, what is the electrical output power of the plant? For the density of water, use 1000 kg/m^3.

Solution

Combining Equations (14.13) and (14.14), the electrical output power of the hydropower plant is

$$\begin{aligned} P_{elect} &= \eta P_{water} = \eta \rho \dot{V} g h \\ &= (0.85)(1000\ kg/m^3)(6000\ m^3/s)(9.81\ m/s^2)(25\ m) \\ &= 1.251 \times 10^9\ W = 1.251\ GW. \end{aligned}$$

Hydropower plants with generating capacities ranging from a few hundred megawatts to more than 10 GW are considered "large" power plants.

PRACTICE!

1. In a pico hydro system, a stream is utilized to generate electrical power by diverting the water into a pipe with an inside diameter of 15.4 cm. If the velocity of the water in the pipe is 4.6 m/s, find the available power in the water. For the density of water, use 1000 kg/m^3. Hint: Adapt Equation (14.8), the equation for available power in the wind, for water flow in a pipe.
Answer: 907 W
2. The electrical power demand of a community near a future hydropower plant site is 750 MW, and the volume flow rate of the river downstream of the reservoir is estimated to be 4500 m^3/s. Assuming a combined efficiency for the turbine and generator of 0.80, what is the minimum reservoir head required? For the density of water, use 1000 kg/m^3.
Answer: 21.2 m

14.5 GEOTHERMAL

Geothermal energy is generated and stored within the earth. Approximately 80 percent of this internal energy is heat generated by radioactive decay, and the remainder is residual heat from the formation of the planet. Heat from these two sources is transported to the earth's surface where it is manifested in hot ground and surface waters, rock, and magma. Even though the heat content of the earth is finite, geothermal energy is considered to be renewable because the rate of heat extraction from the earth is small compared with its heat content.

Like hydro energy, geothermal energy has been utilized by man for centuries. Hot springs from geothermal energy were used for bathing in the Paleolithic period and for space heating in ancient Rome. In modern times, geothermal energy is still used for heating water and living spaces but is also used for electrical power generation. Geothermal accounted for approximately 10,000 MW of global electrical power generation in 2010. A projection suggests that geothermal could provide 1 percent of global electrical power generation by 2050.

There are three primary geothermal resources. The first, and most readily implemented of the three, is hot underground water that either can be tapped by drilling boreholes or is naturally transported to the surface under pressure. The second is rock that is locally heated by natural formations in the earth's crust. This resource is impractical to exploit because rock must be drilled. The third resource is magma, which exists at very high temperatures and pressures, making the exploitation of this resource very challenging.

Heat from a geothermal resource can be converted to useful energy in three principal ways. First, underground hot water can be piped directly to heat living spaces of homes and businesses. A heat exchanger is used to transfer heat from the hot water to air that passes through the living space. This is called *direct geothermal*. The second way utilizes a *heat pump*. Heat pumps take advantage of the year-round constancy of the earth's temperature a few meters below ground. Like a refrigerator, a heat pump incorporates a compressor and two heat exchangers. An underground heat exchanger transfers heat between the refrigerant and the ground, and a heat exchanger in the living space transfers heat between the refrigerant and indoor air. During the heating season, the earth is used as a heat source, whereas during the cooling season, the earth is used as a heat sink. This dual heating/air-conditioning function is accomplished by reversing the direction of the refrigerant. In the third way, steam or a mixture of steam and hot liquid water is used to generate electricity in a *geothermal power plant*.

There are three types of geothermal power plants. The first type is called *direct-steam* or *dry-steam*. If the geothermal resource naturally produces underground dry steam (steam with no liquid water), that steam can be piped directly into a turbine. In this application, the geothermal resource replaces the boiler in a standard fossil-fuel power plant. Geothermal regions do not typically produce dry steam but a mixture of dry steam and hot liquid water. The second type of geothermal power plant, referred to as a *flash steam plant*, converts the hot liquid water into steam, which, when combined with the steam from the borehole, is piped into the turbine. The third type of power plant utilizes relatively low-temperature geothermal water to vaporize a secondary low-boiling point fluid that passes through the turbine in a separate closed loop. Hence, the water from the geothermal resource never comes in contact with the turbine or pump. This type of geothermal power plant, illustrated in Figure 14.20, is called a *binary plant*. Because binary plants can generate electrical power from relatively low-temperature geothermal resources, this type of power plant can exploit more geothermal sites than direct-steam or flash-steam plants. For this reason, the binary plant is the fastest growing type of geothermal power plant.

In a binary plant, hot water from a production well flows through a heat exchanger where the hot water heats a low-boiling point working fluid such as isobutane. The cooled geothermal water is returned to the earth via an injection well, which is located a suitable distance away from the production well so that cooled geothermal water does not affect the temperature of the water at the production well. The hot working fluid is a vapor as it enters the turbine. Energy from the vapor is converted to mechanical energy of a rotating shaft, which is converted to electrical energy by the generator. The working fluid passes through a second heat exchanger, called a condenser, where the vapor condenses to a liquid, exchanging heat with cooling water from a cooling tower. The cooling tower transfers waste heat to the atmosphere by evaporation.

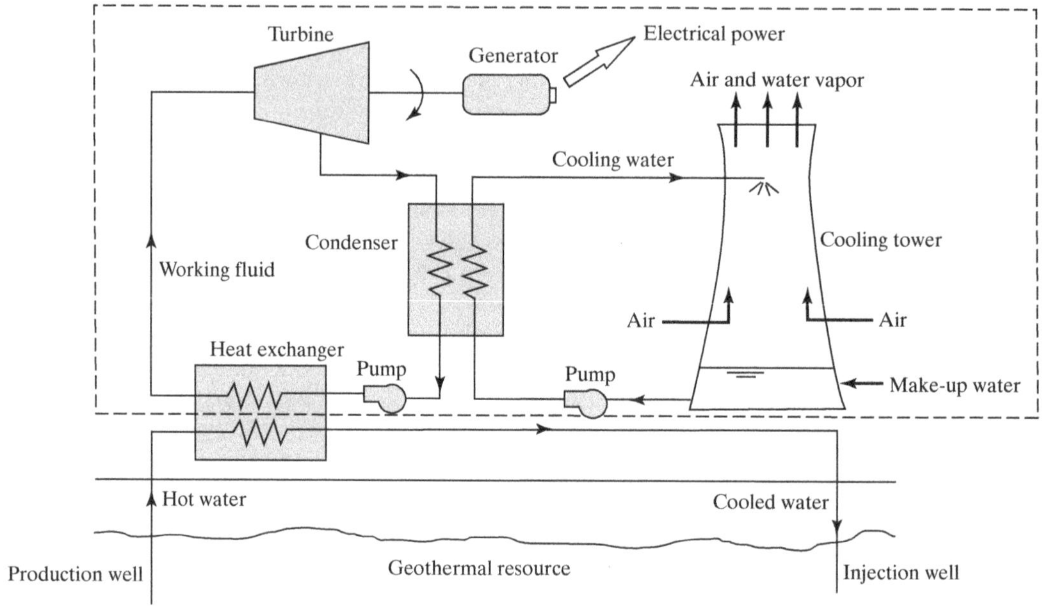

Figure 14.20 Geothermal binary power plant.

In Figure 14.20, the components enclosed by the dashed line are nearly the same as those in a conventional fossil-fuel power plant. The turbine, generator, condenser, pumps, and cooling tower are found in conventional power plants. The exception is that the heat exchanger in a binary plant is replaced by a boiler in a conventional plant. A boiler is a large vessel where the working fluid (water) is heated by a burning fuel (coal, oil or natural gas). The temperatures of the water in conventional plants are much higher than those of the low-boiling point fluid in binary plants, giving conventional plants a higher output power.

14.5.1 Basic Energy Analysis of a Binary Plant

A specific energy analysis of a binary power plant requires a knowledge of engineering topics that is beyond the scope of this book. Hence, a simplified general energy analysis is presented that pertains not only to a geothermal binary plant but a conventional plant as well.

A *heat engine* is a device that converts heat to work, and this is precisely what a power plant does. A power plant converts heat to shaft work of a turbine, which is converted to electrical energy. As illustrated in Figure 14.21, a heat engine receives an amount of heat, Q_{in}, from a high-temperature source and converts a portion of that heat to work, W_{out}. The heat engine rejects the remaining heat, Q_{out}, to a low-temperature sink. In a conventional power plant, Q_{in} is the heat supplied to the boiler from a burning fuel, whereas in a binary power plant, Q_{in} is the heat supplied to the working fluid from the geothermal resource. In both types of plants, Q_{out} is the heat rejected to the atmosphere by a cooling tower or other means. The quantity W_{out} is the shaft work (energy) generated by the turbine.

The first law of thermodynamics states that energy is conserved. Thus, for a heat engine, the heat supplied must equal the output work plus the rejected heat. Expressed mathematically, the first law for the heat engine is written as

$$Q_{in} = Q_{out} + W_{out} \tag{14.15}$$

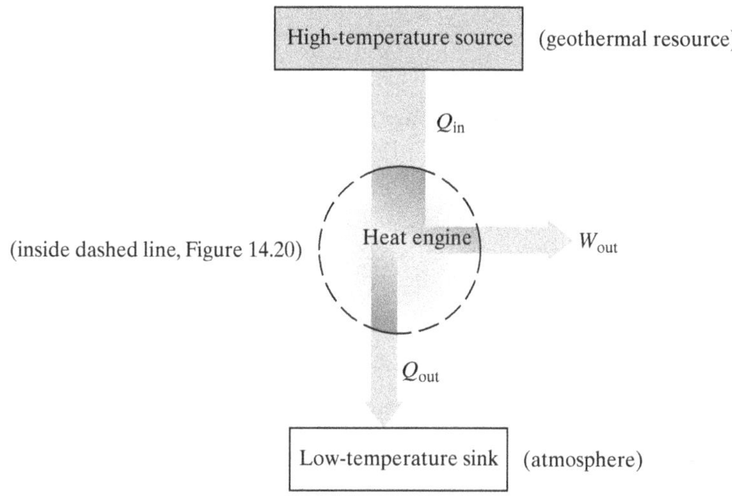

Figure 14.21
A power plant is a type of heat engine, a device that converts heat to work.

where each quantity is expressed in units of J. Stated in terms of power, these quantities can also be expressed in units of W. *Thermal efficiency* of a heat engine, denoted η_{th}, is defined as the shaft work of the turbine divided by the heat input,

$$\eta_{th} = \frac{W_{out}}{Q_{in}}. \tag{14.16}$$

The second law of thermodynamics states that is impossible for a heat engine to convert all the heat it receives from a high-temperature source to work. Hence, the value of η_{th} is limited to values less than 1. A well known theorem from thermodynamics states that for an *ideal* heat engine operating between source and sink temperatures of T_H and T_L, respectively, the maximum thermal efficiency is given by the relation

$$\eta_{th,\ ideal} = 1 - \frac{T_L}{T_H} \tag{14.17}$$

where T_H and T_L are expressed in absolute temperature units of kelvin (K) or rankine (R). The thermal efficiency given by Equation (14.17) is the maximum possible thermal efficiency a heat engine can have, and is referred to as the *Carnot efficiency*, in honor of the French engineer Sadi Carnot.

Turbine shaft work is converted to electrical energy by the generator, which means that the generator also has an efficiency. Generator efficiency is defined as electrical energy divided by shaft work,

$$\eta_{gen} = \frac{W_{elect}}{W_{out}}. \tag{14.18}$$

Generator efficiencies are typically over 0.90. The quantities on the right-hand side of Equation (14.18) can also be expressed in power units of W. The product of the thermal and generator efficiencies given by Equations (14.16) and (14.18) defines the overall efficiency, $\eta_{overall}$,

$$\eta_{overall} = \eta_{th}\eta_{gen} = \frac{W_{out}}{Q_{in}}\frac{W_{elect}}{W_{out}} = \frac{W_{elect}}{Q_{in}} \tag{14.19}$$

where, once again, the quantities can be expressed in power units of W.

The following example illustrates the use of Equations (14.15) through (14.19) for a geothermal binary power plant.

EXAMPLE 14.5

In a binary power plant, 60°C water from the production well supplies 500 kW to an iso-butane working fluid in the heat exchanger. The output power of the turbine shaft is 40 kW, and the efficiency of the generator is 0.93. Find the thermal efficiency, overall efficiency, electrical output power, and rate of heat rejection to the atmosphere. If the temperature of the atmosphere is 22°C, find the Carnot efficiency.

Solution

From Equation (14.18), thermal efficiency is

$$\begin{aligned}\eta_{\text{th}} &= \frac{\dot{W}_{\text{out}}}{\dot{Q}_{\text{in}}}\\ &= \frac{40\ \text{kW}}{500\ \text{kW}}\\ &= 0.080.\end{aligned}$$

From Equation (14.19), overall efficiency is

$$\begin{aligned}\eta_{\text{overall}} &= \eta_{\text{th}}\,\eta_{\text{gen}}\\ &= (0.080)(0.93)\\ &= 0.0744.\end{aligned}$$

From Equation (14.19), the electrical output power is

$$\begin{aligned}P_{\text{elect}} &= \eta_{\text{overall}}\,\dot{Q}_{\text{in}}\\ &= (0.0744)(500\ \text{kW})\\ &= 37.2\ \text{kW}.\end{aligned}$$

Using Equation (14.15), the first law of thermodynamics for a heat engine, the rate of heat rejection to the atmosphere is

$$\begin{aligned}\dot{Q}_{\text{out}} &= \dot{Q}_{\text{in}} - \dot{W}_{\text{out}}\\ &= 500\ \text{kW} - 40\ \text{kW}\\ &= 460\ \text{kW}.\end{aligned}$$

Converting the geothermal water and atmosphere temperatures to units of kelvin, the Carnot efficiency is

$$\begin{aligned}\eta_{\text{th, ideal}} &= 1 - \frac{T_{\text{L}}}{T_{\text{H}}}\\ &= 1 - \frac{(22 + 273)\text{K}}{(60 + 273)\text{K}}\\ &= 0.114.\end{aligned}$$

Thermal efficiencies of binary plants are typically low, so our calculated value of 0.080 is not surprising. Note that the actual thermal efficiency is lower than the Carnot efficiency, which is required by the second law of thermodynamics.

Before leaving geothermal energy, a few remarks about its environmental impacts deserve mentioning. As water is carried from the earth, some dissolved gases are carried with it, notably carbon dioxide (CO_2), methane (CH_4), hydrogen sulfide (H_2S), and ammonia (NH_3). When introduced into the atmosphere, these gases contribute to global warming and acid rain, and, with the exception of CO_2, emit unpleasant odors. As a fraction of generated electrical power, geothermal power plants produce much less of these pollutants than fossil-fuel plants. Geothermal water can also contain trace amounts of toxic elements such as mercury and arsenic. The release of harmful substances into the environment is mitigated by injecting the geothermal fluids back into the earth via the injection well. Continuous operation of a geothermal power plant may result in lowering the local water table, but re-injection of the water mitigates this problem as well. Re-injection can cause a local cooling of the reservoir, but this effect can be minimized by mapping the local reservoir flows and re-injecting the cooled water a suitable distance away from the production well.

Ground subsidence (settling) and seismic activity have been attributed to the long-term operation of geothermal power plants. The magnitude of ground subsidence is typically less than a centimeter, and the link to earthquakes is questionable because geothermal sites are generally prone to earthquakes anyway.

The environmental impact of the construction of geothermal power plants is comparable to that of conventional plants. Unlike conventional plants, however, geothermal plants involve the drilling of wells, which requires great quantities of water from the surroundings. But geothermal drilling is safer than oil or gas drilling because there is no risk of fire, and the potential environmental impact of a drilling accident is relatively minor. Lastly, geothermal power plants require small areas of land, taking up only a few acres per 100 MW of power generation.

PRACTICE!

1. Find the maximum thermal efficiency of a geothermal power plant operating between a ground temperature of 45°C and an ambient air temperature of 18°C.
   Answer: 0.0849

14.6 MARINE

Marine energy is a comprehensive term used to describe energy derived from oceans. There are two main types of marine energy. The first type is *tidal energy*, which is generated by the periodic rise and fall of sea levels. Tides are primarily caused by gravitational forces exerted by the moon and sun but are also caused by the earth's rotation. The second type, called *ocean energy*, consists of three subtypes that exploit different physical phenomena: *ocean currents, ocean waves*, and *ocean thermal energy*. In the first two, energy in ocean currents and waves is converted to mechanical energy, which is subsequently converted to electrical energy. In the third, a heat engine utilizes temperature differences between deep and shallow water to generate mechanical energy in a turbine and then electrical energy in a generator. This type of system is called *ocean thermal energy conversion* (OTEC).

14.6.1 Tidal

Tidal energy is the only renewable energy source that is derived from the earth-moon-sun orbital system. Solar, wind, and hydro energy are directly or indirectly derived from the sun, and geothermal energy is derived from the earth.

There are two principal systems for generating electrical power from tides. The first system is called a *tidal stream generator* (TSG). In a tidal stream generator, kinetic energy of moving water is converted to mechanical energy in a turbine which drives a generator to produce electricity. In principle, this technology is the same as wind power, but the fluid is water instead of air. Tidal stream generators can be secured to the ocean floor by dedicated structures or integrated into existing structures such as bridges.

The second system is called a *tidal barrage*. A tidal barrage, illustrated in Figure 14.22, exploits the potential energy in the water's height difference (elevation head) between high and low tides. As the ocean level rises toward its highest point at high tide, the sluice gates are open to allow the basin to fill. When the ocean reaches high tide, the gates are closed entrapping the water inside the basin, which is typically part of a river estuary. The basin will continue to fill as it is fed by the inland river. When the ocean reaches low tide, the elevation head is maximum, and the gates are opened allowing the stored water to rapidly flow through the turbines back to the ocean. Mechanical energy generated by the turbines is converted to electrical energy in a generator. The turbines continue to operate until the water levels in the basin and ocean are equal, at which point the elevation head is zero. The gates are then closed, and the cycle starts again. A road may be incorporated into the barrage, as depicted in Figure 14.22.

Figure 14.22
Tidal barrage.

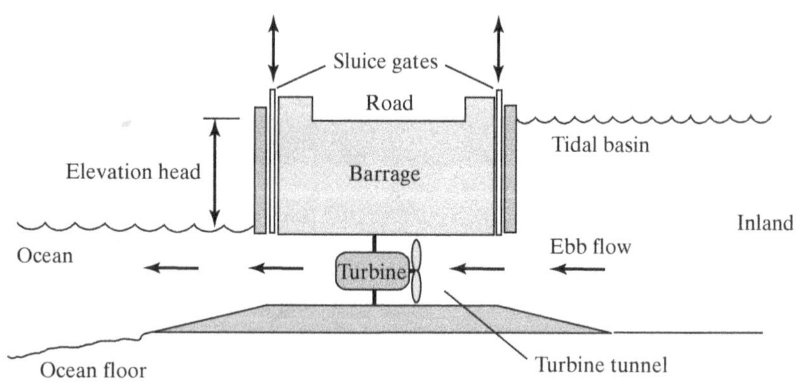

The system just described is a one-way scheme that generates power during ebb tide (when water flows from the basin to the ocean). A tidal barrage may be designed to operate during flood tide (when water flows from the ocean to the basin), but this scheme is less efficient because the flow rate in the turbine is lower due to a slowly filling basin. The barrage may be bi-directional, generating power during the entire tidal cycle. Bi-directional systems are less efficient than one-way systems because the elevation heads are smaller and turbines that are designed to operate in both directions are less efficient than uni-directional units.

Tidal barrages present some environmental challenges. The construction of a barrage across a river alters the sedimentation process, salinity level, and extent of mud flats in the estuary thereby affecting marine and fowl life. Furthermore, fish may pass through the sluice gates and even through the turbines, causing injury or death. Construction of a barrage across an estuary impedes shipping, but the barrage design typically includes ship locks.

14.6.2 Ocean

14.6.2.1 Ocean Currents

Ocean currents are underwater flows caused by tidal motion, temperature gradients, and salinity gradients. The power potential of the earth's ocean currents is estimated to be between 2×10^{12} W and 5×10^{12} W. The technology for converting kinetic energy of ocean currents to electrical energy is fundamentally the same as that for wind. However, for a given velocity and temperature, ocean currents generate more power than winds because the density of seawater is over 800 times the density of air. In Section 14.3 we saw that the available power in the wind is directly proportional to density, as expressed by Equation (14.8). Thus, the power available in ocean currents is

$$P_{\text{ocean current}} = \frac{1}{2}\rho A v^3 \tag{14.20}$$

where A is the cross sectional area through which the seawater flows, and v is velocity. The quantity ρ is density, which for seawater is approximately 1025 kg/m^3. The density of air at standard atmospheric conditions is about 1.2 kg/m^3. As with wind power generation systems, ocean current systems employ either horizontal axis or vertical axis turbines.

EXAMPLE 14.6

A horizontal axis turbine is located where the ocean current has a steady velocity of 0.8 m/s. Find the power available in the ocean current for a cross sectional area of 1 m^2. Compare your result with the power available in the wind for the same velocity and cross sectional area. For the density of seawater and air, use 1025 kg/m^3 and 1.18 kg/m^3, respectively.

Solution

Using Equation (14.20), the power available in the ocean current is

$$\begin{aligned} P_{\text{ocean current}} &= \frac{1}{2}\rho A v^3 \\ &= \frac{1}{2}(1025 \text{ kg/m}^3)(1 \text{ m}^2)(0.8 \text{ m/s})^3 \\ &= 262 \text{ W}. \end{aligned}$$

The power available in the wind is

$$\begin{aligned} P_{\text{wind}} &= \frac{1}{2}\rho A v^3 \\ &= \frac{1}{2}(1.18 \text{ kg/m}^3)(1 \text{ m}^2)(0.8 \text{ m/s})^3 \\ &= 0.30 \text{ W}. \end{aligned}$$

14.6.2.2 Ocean Waves

Ocean waves are caused by winds across the surface of the water. As discussed in Section 14.3, winds are primarily caused by solar heating of the earth, so this form of ocean energy is an indirect form of solar energy. In order to design a system for generating electrical power, engineers need to know the available power in ocean waves. Ocean waves are a complex phenomenon, but a simplified analysis that treats an ocean wave as sinusoidal, as illustrated in Figure 14.23, yields a useful approximation for the available power.

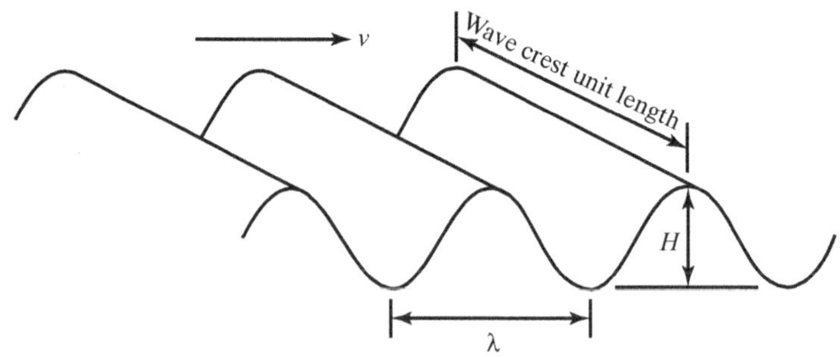

Figure 14.23
Sinusoidal ocean wave.

The quantities H and λ represent the wave peak-to-peak amplitude (height) and wavelength, respectively. The velocity, v, at which the wave moves across the ocean surface can be expressed in terms of wavelength as

$$v = \frac{\lambda}{T} \tag{14.21}$$

where T is the period of the wave, defined as the time it takes for successive peaks or troughs to pass a fixed point. Wave period, T, is the reciprocal of wave frequency, f, so $T = 1/f$. The power of ocean waves is expressed in terms of *energy flux*, defined as the average power per unit length of wave crest, expressed in units of W/m. This important quantity is the average energy per second that passes under one meter of wave crest from the ocean surface to the seabed. Energy flux decreases exponentially with depth such that nearly all of the energy is found near the surface.

For deep ocean waves, it can be shown that the energy flux, J, is given by the expression

$$J = \frac{\rho g^2}{32\pi} TH^2 \tag{14.22}$$

where ρ is seawater density and g is gravitational acceleration. For the mathematical steps leading to Equation (14.22), consult the references at the end of this chapter.

EXAMPLE 14.7

Far off shore, the wave period and height of an ocean wave are 7.5 s and 3.0 m, respectively. Find the energy flux and the total available power of five waves with a wave crest length of 100 m. Use 1025 kg/m^3 for the density of seawater.

Solution

Using Equation (14.22), the energy flux (power per unit length of wave crest) is

$$J = \frac{\rho g^2}{32\pi} TH^2$$
$$= \frac{(1025\ \text{kg/m}^3)(9.81\ \text{m/s}^2)^2(7.5\ \text{s})(3.0\ \text{m})^2}{32\pi}$$
$$= 6.62 \times 10^4\ \text{W/m} = 66.2\ \text{kW/m}.$$

The total available power is the product of the energy flux and wave crest length of five waves,

$$P_{\text{waves}} = JL_{\text{waves}}$$
$$= (66.2\ \text{kW/m})(5)(100\ \text{m})$$
$$= 3.31 \times 10^4\ \text{kW} = 33.1\ \text{MW}.$$

A variety of techniques are employed to convert wave power to electrical power. Referred to as *wave energy converters,* these can be broadly classified into three groups: *wave activated devices, overtopping devices,* and *oscillating water columns.* In a wave activated device, wave motion is directly transferred to mechanical motion of some kind. An overtopping device incorporates a ramp or tapered channel that forces water of incoming waves to spill into a reservoir. In a manner similar to hydropower plants, stored water is directed back to the ocean through a turbine. In an oscillating water column, waves cause the water surface in a partially submerged pipe to oscillate thereby pumping air in the pipe through a turbine. These systems are illustrated in Figure 14.24.

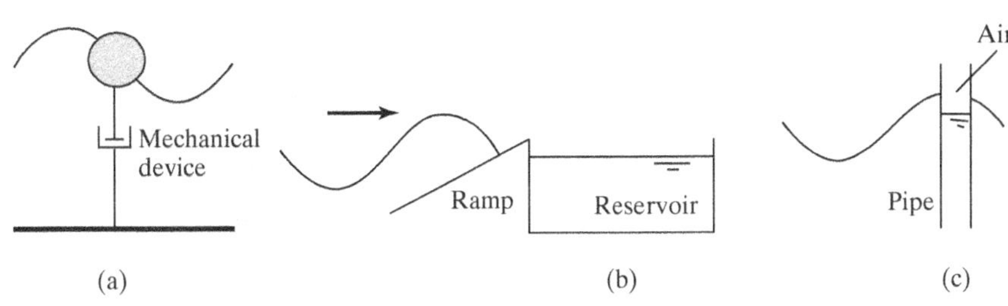

Figure 14.24
Types of wave energy converters: (a) wave activated device, (b) overtopping device, (c) oscillating column.

The environmental impacts of wave energy systems are relatively benign. Pollution from lubricants would occur if the systems are not well sealed. Floating systems, if located in shipping lanes, present a danger to ships. Marine life is not affected, and noise generation is minimal.

14.6.2.3 Ocean Thermal Energy

Ocean thermal energy conversion (OTEC) utilizes temperature differences between shallow and deep seawater. The ocean surface absorbs solar radiation that warms shallow water, but solar radiation cannot reach deep water, so a temperature gradient is established. In deep oceans, the temperature difference between shallow and deep water can be as high as 20°C. This temperature difference can be exploited to run a heat engine. A floating OTEC plant is illustrated in Figure 14.25.

Basic heat engines and their application to geothermal power generation were presented in Section 14.5.1. In principle, a heat engine employed in OTEC and geothermal applications is the same. However, the temperature difference in OTEC systems is relatively small, resulting in very low thermal efficiencies, as pointed out in the following example.

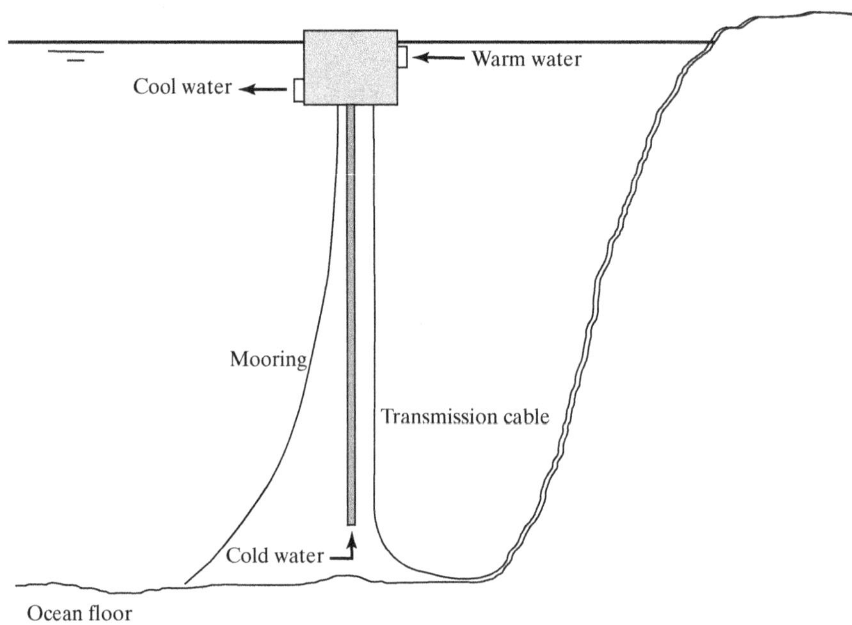

Figure 14.25 Floating ocean thermal energy conversion plant.

EXAMPLE 14.8

An OTEC system operates in a region where the temperatures of the surface and deep ocean waters are 30°C and 10°C, respectively. Find the maximum possible thermal efficiency of a heat engine that is employed in this system.

Solution

Using Equation (14.17), the Carnot efficiency of an OTEC system in which the temperatures of the deep and shallow water, respectively, are 10°C (283 K) and 30°C (303 K), is

$$\begin{aligned}\eta_{\text{th, ideal}} &= 1 - \frac{T_L}{T_H} \\ &= 1 - \frac{283\text{ K}}{303\text{ K}} \\ &= 0.066(6.6\text{ percent}).\end{aligned}$$

Accounting for energy losses, the actual efficiency of a heat engine employed in an OTEC power plant ranges from 2 to 3 percent.

The primary environmental impact of an ocean thermal energy conversion system is the return of cool water to the upper regions of the ocean, which could affect marine life. If released into the ocean, lubricants and the working fluid that is pumped through heat exchangers could likewise disrupt the local marine environment.

PRACTICE!

1. An ocean current power generation plant consists of 12 turbines with a blade diameter of 8.0 m. The turbines are positioned at a depth where the ocean current velocity is 1.7 m/s. Find the available power in the ocean current for all turbines if the density of seawater is 1025 kg/m^3.
   Answer: 1.52 MW
2. An ocean wave has a frequency of 0.20 Hz. If the height of the wave is 2.8 m, find the wave's energy flux. For the density of seawater, use 1025 kg/m^3.
   Answer: 38.5 kW/m
3. If an OTEC power plant has a Carnot efficiency of 3.0 percent and the temperature of the shallow water is 28°C, what is the temperature of the deep water?
   Answer: 19°C

14.7 BIOMASS

Biomass is the earth's living, or recently living, matter found within the biosphere, the thin surface layer near the earth's surface. Energy stored in biomass is naturally recycled through a series of chemical and physical conversion processes in the soil and surrounding atmosphere, which is the reason that energy from biomass is considered renewable. A part of this energy can be captured by intervening at the right time when biomass is available as fuel. If the rate of consumption does not exceed the rate at which biomass is recycled, the combustion of biomass does not generate more heat or carbon dioxide than does natural processes. Biomass can be directly converted to energy by combusting it to heat a living space or to heat water to run a turbine for generating electricity. Alternatively, biomass can first be converted to a secondary substance that is subsequently combusted or combined with other substances prior to combustion.

Man has utilized biomass for heating and cooking for centuries, but the use of biomass for electrical power generation is a relatively new development. Because electrical power generation is the focus of the other renewable energy resources discussed in this chapter, electrical power generation from biomass will be emphasized here.

For power generation, there are two primary types of biomass: wastes and crops. Wastes include various types of human-generated refuse, agricultural and livestock waste, and wood processing waste. Over the long term, wastes alone will not be able to supply enough biomass to meet power demands. Energy crops grown on dedicated plantations will be required to sustain the biomass power generation industry.

The most common technology for converting thermal energy of burning biomass to electrical energy is a *direct-fired power plant*, illustrated in Figure 14.26. This type of power plant is similar to a conventional fossil-fuel power plant in which coal, oil, or gas is burned. Burning biomass transfers heat to a boiler, and a pump circulates water in a closed loop through the boiler, turbine, and condenser. Thermal energy of high-pressure, high-temperature steam is converted to mechanical energy in the turbine, which is subsequently converted to electrical energy in the generator.

Overall efficiencies of biomass power plants are typically below 25 percent, approximately half that of coal-fired plants, due to the lower combustion temperatures of biomass. Efficiencies are higher in a *co-fired power plant* in which a small

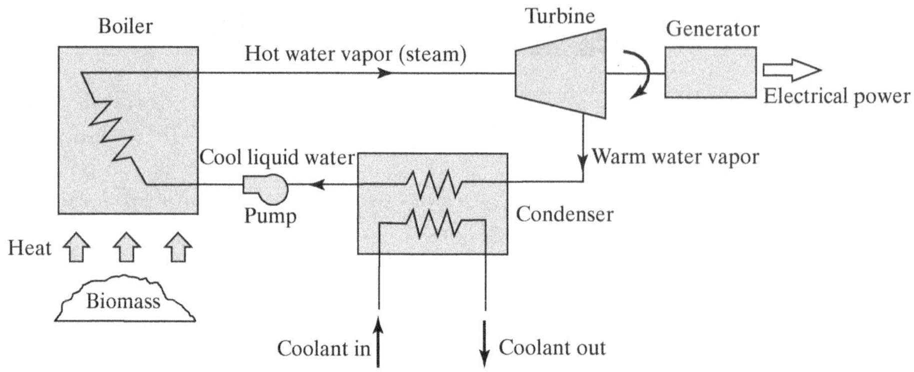

Figure 14.26 Direct-fired biomass power plant.

percentage (less than 2 percent) of biomass is mixed with coal. This mixture can be implemented without making modifications to the coal-fired plant. Another approach to biomass power generation, called *biomass gasification*, involves the partial combustion of biomass to produce the combustible organic compounds carbon monoxide and hydrogen. The heat content of this gas is about half that of natural gas and can be burned in a gas turbine.

Like the burning of fossil fuels, burning of biomass will produce emissions, particularly carbon dioxide. So, why is biomass more desirable than fossil fuels? The reason is that when biomass is burned, the carbon dioxide released is later absorbed by replacement crops during photosynthesis. Hence, over a complete cycle of growth and combustion, no net carbon dioxide is released into the environment. This is not so with coal, oil, and natural gas. When these substances are burned, the carbon dioxide that was once locked in the earth is released into the atmosphere. Like fossil-fuel power plants, biomass power plants emit other harmful gases such as carbon monoxide, so measures have to be taken to remove these substances from the fuel gas.

Because a direct-fired biomass power plant is fundamentally a heat engine, a basic energy analysis parallels that of a conventional power plant or a binary geothermal power plant discussed in Section 14.5.1.

KEY TERMS

biomass
efficiency
energy
fossil fuels
geothermal
hydro
marine
photovoltaic
renewable energy
solar
solar thermal
wind

REFERENCES

Boyle, G. Editor, *Renewable Energy: Power for a Sustainable Future*, New York: Oxford, 1996.

Breeze, P., *Power Generation Technologies*, Burlington, MA: Elsevier, 2005.

Kreith, F. and D.Y. Goswami, Editors, *Handbook of Energy Efficiency and Renewable Energy*, Boca Raton, FL: CRC Press, 2007.

Kruger, P., *Alternative Energy Resources: The Quest for Sustainable Energy*, New York: John Wiley & Sons, 2006.

Sorensen, B., *Renewable Energy: Its Physics, Engineering, Environmental Impacts, Economics and Planning*, 2nd ed., London, UK: Academic Press, 2000.
Zobaa, A.F. and R.C. Bansal, Editors, *Handbook of Renewable Energy Technology*, Hackensack, NJ: World Scientific, 2011.

PROBLEMS

For the following problems, use the general analysis procedure of (1) problem statement, (2) diagram, (3) assumptions, (4) governing equations, (5) calculations, (6) solution check, and (7) discussion.

Solar

14.1 A spacecraft in orbit about the earth has a 4 m × 6 m solar panel that faces directly into the sun for 5 h. Find the solar energy incident on the panel during this time period.

14.2 A horizontal photovoltaic solar panel on the roof of a warehouse in Minneapolis, MN, measures 20 m × 45 m. What is the total solar power incident on the panel on January 21 at 10:00 am, 12:00 noon, and 2:00 pm? What is the total solar energy incident on the panel during this day? The sky is clear, and the foreground reflectivity of the roof and surrounding surfaces is 0.2.

14.3 A home owner in Miami, FL, wishes to power some lights and appliances using a photovoltaic solar panel mounted directly to the roof. The roof has a slope of 30° with respect to the ground, and the daily electrical energy requirement is 8 kWh. Based on the total solar irradiance for January 21, what is the required surface area of the solar panel if its efficiency is 0.16? Assume that the panel faces due south, the sky is clear, and the foreground reflectivity is 0.2.

14.4 Compare the electrical power generation of photovoltaic systems in San Francisco, CA, San Antonio, TX, Minneapolis, MN, and Boston, MA. The solar panel in each location faces due south, has a surface area of 40 m^2, and a tilt angle of 60°. If each panel has an efficiency of 0.15, find the electrical energy generated on June 21. Assume a foreground reflectivity of 0.2 and a clear sky.

14.5 Consider two identical photovoltaic solar panels, one in Miami, FL, and the other in Boston, MA. The tilt angles of both panels are zero. For 12:00 noon on January 21, what are the total solar irradiances for these two locations? Explain why the total solar irradiance in Miami is much greater than that in Boston.

14.6 A photovoltaic solar panel in Minneapolis, MN, has a tilt angle of 60° and a surface area of 15 m^2. The cost of electricity in Minneapolis is \$0.065/kWh. Find the electrical power cost savings for the month of January if the efficiency of the solar panel is 0.17. Assume clear sky solar conditions of January 21 for the entire month (31 days) and a foreground reflectivity of 0.2.

14.7 On January 21 in San Francisco, CA, the sky is clear from sun rise until 2 pm, but the sky is completely overcast for the rest of the day due to an approaching storm. For a 6 m × 8 m photovoltaic solar panel with a tilt angle of 60°, estimate the electrical output power for the day if the panel's efficiency is 0.18. The foreground reflectivity is 0.2, and the solar panel faces due south.

14.8 Compare the performance on June 21 of two photovoltaic solar panels, one in San Antonio, TX, and the other in Minneapolis, MN. The San Antonio panel has a tilt angle of 60° and a surface area of 25 m^2. The Minneapolis solar panel is horizontal. If the electrical output power of the Minneapolis solar panel is to be the same as that of the San Antonio solar panel, what is the required surface area of the Minneapolis solar panel? Both solar panels have the same efficiency. Assume a foreground reflectivity of 0.2 and a clear sky and that both panels face due south.

Wind

14.9 A wind farm consists of 30 horizontal axis wind turbines with a blade diameter of 32 m. The power coefficient of the turbines is 0.46, and the combined efficiency of the gear box, generator, and bearings is 0.33. If the wind blows steadily at 5 m/s, what is the electrical output power of the wind farm? For the density of air, use 1.16 kg/m^3.

14.10 Consider a steady 15 mi/h wind that blows through a cross section of atmosphere measuring 100 ft high and 1.0 mi wide. Assuming an air density of 1.16 kg/m^3, find the available power in this wind in units of MW.

14.11 The desired electrical output power of a horizontal axis wind turbine is 2.75 kW. A site has been selected where the average wind speed is 3.2 m/s. The power coefficient and combined efficiency of the turbine are 0.49 and 0.25, respectively. Using 1.16 kg/m^3 as the density of air, find the required blade diameter of the turbine.

14.12 A power utility company is considering building a wind farm at a site where the wind speed varies from 1.9 m/s to 4.3 m/s. A preliminary design calculation shows that the horizontal axis wind turbines, which have a blade diameter of 40 m, will have a power coefficient of 0.50 and a combined efficiency of 0.31. Find the range for the electrical output power for this wind farm if 75 turbines are to be installed. Assume that the winds are constant, and use 1.16 kg/m^3 for air density.

14.13 A home owner has a single horizontal axis wind turbine that augments electrical power from the local power utility. The wind speed varies throughout the day according to the schedule shown in Table P14.13.

Table P14.13

Time period	Wind speed (m/s)
12 am–7 am	2.1
7 am–9 am	5.5
9 am–1 pm	4.0
1 pm–3 pm	3.2
3 pm–6 pm	3.8
6 pm–9 pm	4.7
9 pm–12 am	3.4

The wind turbine has a blade diameter of 5.0 m, a power coefficient of 0.45, and a combined efficiency of 0.22. Assuming that this daily schedule is repeated throughout a 30-day month, how much electrical energy does the home owner's wind turbine generate per month? For the density of air, use 1.16 kg/m^3.

14.14 An inventor proposes to retrofit utility poles with vertical axis wind turbines, providing some electrical power for street and parking lot lights. The proposed wind turbine design is illustrated in Figure P14.14. The turbine has three blades positioned 120° apart on a rotor that spins around the utility pole. Assuming a constant wind speed of 3.0 m/s and an air density of 1.16 kg/m^3, find the available power in the wind for this turbine. For the area swept out by the turbine blades, use the height times diameter, the "presented area" of the cylinder-shaped turbine. If this turbine can convert 20 percent of the available wind power to electrical power, how much electrical energy can this turbine generate per day? Is this proposal viable?

Figure P14.14

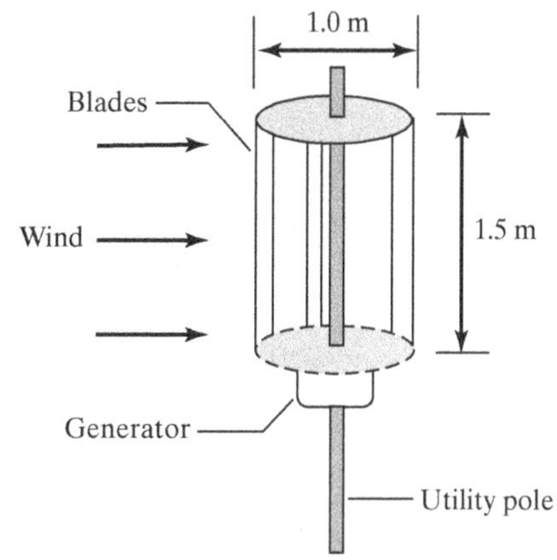

14.15 Compare the electrical output power of two hypothetical renewable energy systems in San Antonio, TX: a photovoltaic solar system and a wind farm. The horizontal photovoltaic solar panels cover 90 percent of a 5-acre square parcel of land (some space between panels is needed for maintenance), and a square array of horizontal axis wind turbines occupies the entire 5-acre parcel. The efficiency of the solar panels is 0.18. The turbines have a blade diameter of 6.1 m, and the power coefficient and combined efficiency of the turbines are 0.48 and 0.16, respectively. Assume a constant wind speed of 6.0 m/s, a clear sky, a foreground reflectivity of 0.2 and an air density of 1.16 kg/m^3. Using a wind turbine tower spacing of five blade diameters in the direction of the wind and three blade diameters in the cross wind direction, how many wind turbines can be installed? On January 21, how much electrical energy is generated by each of these systems?

Hydro

14.16 The head of a hydropower plant is 16 m, and the combined efficiency of the turbine and generator is 0.80. The volume flow rate of the river downstream of the power plant is 4500 m^3/s. What is the electrical output power of the plant? For the density of water, use 1000 kg/m^3.

14.17 The electrical power demand of a town near a future hydropower plant site is 1.75 GW, and the volume flow rate of the river downstream of the reservoir is estimated to be 5000 m^3/s. Assuming a combined efficiency for the turbine and generator of 0.88, what is the minimum reservoir head required? Use 1000 kg/m^3 for the density of water.

14.18 In a micro hydro system, a river is exploited to generate electrical power by diverting the water into a pipe with an inside diameter of 30.3 cm. If the velocity of the water in the pipe is 8.2 m/s, find the available power in the water. For the density of water, use 1000 kg/m^3.

14.19 A hydroelectric power plant has a gate at the spillway of the reservoir to regulate the flow of water downstream. The required range of electrical output power of the plant is 100 to 175 MW. If the head of the reservoir is held constant at 35 m, what is the range of flow rates that the gate must maintain? Assume 1000 kg/m^3 for the density of water and a combined efficiency of 0.83 for the turbine and generator.

Geothermal

14.20 A geothermal power plant extracts thermal energy from 95°C water and rejects thermal energy to a 15°C atmosphere. What is the maximum thermal efficiency of this plant?

14.21 In a binary power plant, 55°C water from the production well supplies 400 kW to an iso-butane working fluid in the heat exchanger. The output power of the turbine shaft is 35 kW, and the efficiency of the generator is 0.91. Find the thermal efficiency, overall efficiency, electrical output power, and rate of heat rejection to the atmosphere. If the temperature of the atmosphere is 20°C, find the Carnot efficiency.

14.22 A person proposes a power plant that utilizes hot rock as the thermal energy source and the atmosphere as the thermal energy sink. The temperature of the source and sink are 60°C and 18°C, respectively. The person states that the geothermal power plant extracts 120 kW from the rock and rejects 95 kW to the atmosphere. Evaluate this proposal.

14.23 Consider the design of a geothermal power plant that extracts thermal energy from 750°C magma and rejects thermal energy to 15°C seawater. If the rate of heat extraction from the magma is 1.2 MW, what is the maximum possible electrical output power?

14.24 A geothermal power plant rejects its waste heat to a 10°C atmosphere. The rate of heat extraction from underground water is 125 kW. If the rate at which mechanical work is done by the turbine is 22 kW, what is the minimum required temperature of the underground water?

Marine

14.25 Six horizontal axis turbines are attached to a bridge spanning an ocean inlet where the current has a steady velocity of 0.6 m/s. Find the available power in this current for a turbine blade diameter of 8.0 m. For the density of seawater, use 1025 kg/m^3.

14.26 For the system described in Problem 14.25, find the electrical output power for an overall efficiency of 0.65.

14.27 In deep ocean water, the wave period and height of a wave are 5.5 s and 2.8 m, respectively. Find the energy flux and the total available power of one wave with a wave crest length of 250 m. Use 1025 kg/m^3 for the density of seawater.

14.28 An OTEC power plant extracts thermal energy from 30°C shallow water and rejects thermal energy to 9°C deep water. If the rate of energy extraction from the shallow water is 15 kW, what is the maximum possible output power of the turbine shaft? If the combined efficiency of the turbine and generator is 0.75, what is the maximum possible electrical output power?

Biomass

14.29 The combustion temperature of wood chips is approximately 350°C. If a biomass power plant uses wood chips as fuel, and the plant rejects its waste heat to a 15°C lake, what is the maximum possible thermal efficiency of the power plant?

14.30 In a biomass power plant that burns livestock and human waste, water passing through the boiler receives 500 MJ of energy. The power plant has an overall efficiency of 20 percent. Find the electrical output power.

15

Engineering Ethics

SECTIONS

OBJECTIVES

After reading this chapter, you will be able to:

- explain why engineers should be ethical;
- list the canons in the NSPE Code of Ethics;
- identify the correct ethical choices in engineering applications.

15.1 INTRODUCTION

Engineering is a profession. This means in part that engineers must practice their profession in accordance with high ethical standards. In this chapter, you will explore *why* engineers must act ethically and examine the basic codes of engineering ethics. Examples will be used to illustrate the ethical conflicts that occur on the job.

Ethics can be studied on many levels. Please understand that the discussions in this chapter are on a very practical level. The philosophy of ethics (sometimes called *formal ethics*) is a beautiful and interesting field of study. This is not a chapter on formal ethics. Rather, the purpose of this chapter is to provide you with guidance on how to live your life as an engineer.*

15.2 PROFESSIONAL ISSUES

Ethics refers to a system of moral principles. All professions have standards of ethics to which their members are bound. In fact, a code of ethics allowing practitioners to "police themselves" is really a prerequisite for a profession.

Ethical issues in the medical, legal, and political arenas appear almost daily in the newspaper. Engineers also must follow ethical standards.

*A student of moral philosophy might say that this chapter is about *normative ethics* (i.e., the principles that guide how we should live our lives), rather than *metaethics* (i.e., the study of what is good).

PONDER THIS

Why should engineering follow ethical standards?

ethics: a system of moral principles

Recall that engineering ethics is important because (1) engineering affects public safety and health, and (2) public trust in engineers must be preserved, since engineering work is not easy to understand.

15.3 CODES OF ETHICS

15.3.1 Introduction

Many engineering societies have their own codes of ethical behavior. Engineering societies having codes of ethics (and other ethics documents) are listed in Table 15.1. The differences between the codes are fairly minor. Most of the ethical standards are modeled after the code of ethics of the National Society of Professional Engineers (NSPE). To understand the concepts underlying the various ethics doctrines, the NSPE Code of Ethics will be examined in more detail.

15.3.2 NSPE Code of Ethics

NSPE Fundamental Canons of Ethics: the six basic principles outlining the professional responsibilities of engineers

Board of Ethical Review: an NSPE committee that offers commentary on ethics cases

NSPE has established six principles called the ***Fundamental Canons of Ethics***. The principles refer to the responsibilities of engineers when conducting their work and when approving documents. When a PE approves a document, he or she is *personally certifying* that the work meets professional standards. An NSPE committee called the ***Board of Ethical Review*** (BER) offers commentary on ethics cases to show engineers how the fundamental canons have been interpreted in the past. Each of the fundamental canons will be presented with examples on their interpretation. (*Note*: The entire NSPE Code of Ethics is given at the end of the chapter. More examples of the interpretation of the fundamental canons can be found in the Rules of Practice and Professional Obligations sections of the NSPE Code of Ethics.)

Key idea: Engineers shall hold paramount the safety, health, and welfare of the public.

Canon 1: Engineers shall hold paramount the safety, health, and welfare of the public. This is the most important of the fundamental canons. It states that nothing—not profit, not inconvenience, not personal gain—comes before the safety, health, and welfare of the public. Remember, safety, health, and welfare are paramount. If you face a contradiction in your work life between canons, Canon 1 prevails.

Canon 1 has three important interpretations. First, the canon has been interpreted to mean that individual engineers must notify their employer, client, or the proper authorities when life or property is endangered. Notification must occur *even if the client requests that data or any other engineering work be held back*. For example, if you discover a potentially dangerous situation in the course of your professional work, you **must** report it even if the paying client *demands* that you squelch the information.

TABLE 15.1 Engineering Societies with Codes of Ethics

Organization	Last Revision	Other Ethics Documents
American Consulting Engineers Council (ACEC)	1980	none
American Institute of Chemical Engineers (AIChE)	1989	none
American Society of Civil Engineers (ASCE)	1993	ASCE's Guidelines on Practice
American Society of Mechanical Engineers (ASME)	1991	ASME Criteria for Interpretation of the Canons
Institute of Electrical and Electronics Engineers (IEEE)	1990	Employment Guidelines

Safety outweighs all other considerations.

A second implication of the first canon is that engineers must approve only documents that conform to applicable standards. Approving plans that you know are not in compliance with standards is an ethical violation and grounds for loss of your engineering license.

A third implication of the first canon is that engineers have a responsibility to report any violations of the code of ethics. Reporting known violations is a form of *whistle-blowing*. Professionals who come public with mistakes made by themselves, their firm, or their colleagues risk being fired or ostracized. However, because of the overriding importance of the safety, health, and welfare of the public, engineers have a moral duty to report ethical violations committed by themselves or others.

Key idea: Engineers shall perform services only in the areas of their competence.

Canon 2: Engineers shall perform services only in the areas of their competence. At present, engineers usually are licensed in a particular discipline. This canon makes it clear that engineers must stay in their areas of expertise. Thus, a chemical engineer cannot approve structural engineering plans.

One ramification of this canon is that you must approve only documents prepared under your supervision. For example, a mechanical engineer cannot approve heating, ventilation, and air conditioning (HVAC) plans drawn up by people being supervised by another engineer. As a practical matter, the second canon means that each technical portion (e.g., electrical, mechanical, and structural portions) of a set of plans usually must be approved individually.

Key idea: Engineers shall issue public statements only in an objective and truthful manner.

Canon 3: Engineers shall issue public statements only in an objective and truthful manner. As stated in Section 15.2, the engineering profession thrives because of the public trust. This canon states that you must be honest in your dealings with the public. For example, you must acknowledge if you are being paid to issue a public statement about an engineering issue. Suppose you are being paid by a developer to lay out a new neighborhood. If you speak before a city council meeting in favor of the development, then you must identify yourself as an engineer paid by the developer.

Key idea: Engineers shall act for each employer or client as faithful agents or trustees.

Canon 4: Engineers shall act for each employer or client as faithful agents or trustees. The practice of engineering depends on the trust between the engineer and the public. Engineering also is dependent on the trust between the engineer and the client. This canon says in part that the client has the right to expect that the engineer will use his or her best engineering judgment in solving the client's problems.

This canon also is interpreted to mean that engineers must disclose to the client all known or *potential* conflicts of interest. Identification of potential conflicts of interest can be difficult. Clearly, the engineer cannot represent two clients who may come into conflict. For example, you could not represent both a potentially polluting industry and the town owning the wastewater treatment facility or both a developer and a city where the developer does work.

Key idea: Engineers shall avoid deceptive acts.

Canon 5: Engineers shall avoid deceptive acts. This canon has implications in the procurement of work. For example, it is a violation of the code of ethics to falsify your qualifications. In addition, it is considered unethical to offer or give a contribution to influence a public authority's decision about who should be awarded a contract.

The difference between ethical and unethical behavior can be very small. Clearly, it is improper to offer a bribe to the head of a housing authority to get a contract. What about taking the town engineer out to lunch right before a contract is awarded? What about giving the daughter of the mayor extra playing time on the soccer team you coach?

Key idea: Engineers shall conduct themselves honorably, responsibly, ethically, and lawfully so as to enhance the honor, reputation, and usefulness of the profession.

Canon 6: Engineers shall conduct themselves honorably, responsibly, ethically, and lawfully so as to enhance the honor, reputation, and usefulness of the profession. This canon greatly affects the lives of engineers. It means that engineers must advise their clients if the engineers believe a project will *not* be successful. In addition, engineers are prohibited by this canon from accepting free material from suppliers or contractors in return for specifying their products or services.

Language similar to that in the sixth canon is used by the Accreditation Board for Engineering and Technology (ABET) in the fundamental principles section of their code of ethics. (One responsibility of ABET is to certify undergraduate engineering programs.) The ABET fundamental principles state that engineers can "uphold and advance the integrity, honor, and dignity of the engineering profession by"

I. Using their knowledge and skill for the enhancement of human welfare;
II. Being honest and impartial, and serving with fidelity the public, their employers, and clients;
III. Striving to increase the competence and prestige of the engineering profession; and
IV. Supporting the professional and technical societies of their disciplines (Wright, 1994).

Although the NSPE canons remain silent on the issue, both the ABET and ASCE codes of ethics include ethical obligations for professional development. In other words, engineers are ethically obligated to continue their technical training and education throughout their career. Thus, even though some states do not require continuing education for relicensing as a professional engineer, continued technical training is part of your ethical obligation as an engineer.

15.4 EXAMPLES OF ENGINEERING ETHICS

Engineering ethics can become very complex. Sometimes, one ethical canon contradicts another. Two examples will be presented to illustrate the applications of the fundamental canons. These examples are taken from case studies developed by NSPE's Board of Ethical Review. Take a moment to consider your responses before reading the comments that follow the discussion questions. For examples of ethics that crop up in business practice, see the *Focus on Ethics: Workplace Ethics*.

15.4.1 Not Reporting Violations

Case: A civil engineer is hired to assess the structural integrity of a 60-year-old apartment building. The structure of the building is determined to be sound. However, during the inspection, mechanical and electrical problems are noted. The problems are severe enough that they may lead to safety concerns. The engineer tells the client of the problems, but does not include them in the report. The client reminds the engineer of his obligations regarding client confidentiality and then moves to sell the building quickly without repairs. The engineer decides not to report the violations to the authorities.

Discussion: This example represents a conflict between the responsibility to protect public safety (Canon 1) and client confidentiality (Canon 4). Does Canon 2 (not working outside your area of expertise) play in this case?

PONDER THIS

Which factor is more important? Could the engineer have avoided the conflict by including the problems in his original report?

Board of Ethical Review comments: The Board concluded that the engineer had an ethical obligation to report the problems to the authorities because of the paramount importance of public safety (Canon 1). The engineer was correct not to include the problems in his report, since the problems were outside of the engineer's area of expertise (Canon 2).

15.4.2 Whistle-Blowing

Case: The city engineer/director of public works for a medium-sized city is the only licensed professional engineer in a position of responsibility within the city government. The city has several large food processing plants that discharge large amounts of waste into the sewage system during canning season. The engineer is responsible for the wastewater treatment plant. She reports to her supervisor about the inadequate capacity of the treatment plant to handle potential overflow during the rainy season and offers several possible solutions. The engineer also privately notifies other city officials about the plant problem, but her supervisor removes the responsibility for the wastewater treatment plant from her. She also is placed on probation and warned not to discuss the matter further or she will be fired.

Discussion: This example explores what engineers must do to meet their ethical obligations.

PONDER THIS

Has the engineer discharged her ethical responsibilities by notifying city officials of the potential problem?

Board of Ethical Review comments: The Board concluded that removal of responsibility for the treatment plant did not terminate the engineer's ethical obligations, even when threatened with loss of employment. Again, public health and safety are of highest importance (Canon 1). The Board thought that the engineer should have reported the potential problem to higher authorities in the state or federal government.

15.5 SUMMARY

Engineers are professionals and must meet high ethical standards. Engineering ethics is required for two reasons. First, engineering directly influences public safety, health, and welfare. Second, engineering work is not easy to understand, so the trust between the public and engineers must not be diminished by unethical behavior.

Several codes of engineering ethics exist. Most are similar to the six NSPE Fundamental Canons of Ethics. The six canons require that engineers shall (1) hold paramount

FOCUS ON ETHICS: WORKPLACE ETHICS

Engineering ethics often is taught with the high-stakes ethics cases discussed in Problems 15.4 through 15.6. While interesting and instructive, the large cases do not provide you as a new engineer with tools you can use to address everyday ethics questions.

In this section, you will be presented with a series of ethics questions and possible solutions. You may wish to discuss the questions in a group, so that you can exchange ideas. One of the lessons is that ethics questions often do not have one correct answer. Responses to ethical dilemmas in the workplace can be ranked from better to worse. (Of course, some responses are just plain wrong.)

The questions below are from an ethics game called *Gray Matters*. The game was devised by George Sammet, Jr., to teach business ethics to the employees of Martin Marietta (now Lockheed Martin). Sammet was Vice President of International Ethics and Business Conduct for Lockheed Martin and now works for Grainer and Associates in Ottawa. The questions presented here were selected for their relevance to the engineering workplace and interest to entry-level engineers.

For each question below, a Web site is given to allow you to select from potential responses. The questions are quoted from the onlineethics.org Web site.

Enjoy the game and your discussion of the answers.

GrayMatters Case #8
Appropriating Office Supplies for Personal Use

"Two of your subordinates routinely provide their children with school supplies from the office. How do you handle this situation?"

Web site: http://onlineethics.org/corp/graymatters/case8.html

GrayMatters Case #64
Instructed to Distort the Truth?

"You are on a proposal-writing team. In the orientation briefing, the head of the team gives the following guidance: 'We really have to win this one. I want you to be really optimistic in what you write.' How do you interpret her advice?"

Web site: http://onlineethics.org/corp/graymatters/case64.html

GrayMatters Case #87
Supplier Offers You a Discount

"You are in Production Control. Planning on adding a porch onto your house, you visit a lumberyard to get ideas and a price. During the discussion, the sales manager says, 'Oh, you work for XYZ Company. They buy a lot from us, so I'm going to give you a special discount.' What do you do?"

Web site: http://onlineethics.org/corp/graymatters/case87.html

GrayMatters Case #24
HIV Positive Employee

"A female employee tells you, her manager, that a fellow employee is HIV positive. What do you do?"

Web site: http://onlineethics.org/corp/graymatters/case24.html

GrayMatters Case #72
He Calls All the Women "Sweetie"

"When a male supervisor talks to any female employee, he always addresses her as 'Sweetie.' You have overheard him use this term several times. As the supervisor's manager, should you do anything?"

Web site: http://onlineethics.org/corp/graymatters/case72.html

How did you do in your group? The maximum and minimum possible scores for the five scenarios above are +50 and −55, respectively.

the safety, health, and welfare of the public; (2) perform services only in the areas of their competence; (3) issue public statements only in an objective and truthful manner; (4) act for each employer or client as faithful agents; (5) avoid deceptive acts; and (6) conduct themselves honorably, responsibly, ethically, and lawfully to enhance the honor, reputation, and usefulness of the profession.

SUMMARY OF KEY IDEAS

- Engineers shall hold paramount the safety, health, and welfare of the public.
- Engineers shall perform services only in the areas of their competence.
- Engineers shall issue public statements only in an objective and truthful manner.
- Engineers shall act for each employer or client as faithful agents or trustees.
- Engineers shall avoid deceptive acts.
- Engineers shall conduct themselves honorably, responsibly, ethically, and lawfully so as to enhance the honor, reputation, and usefulness of the profession.

Problems

15.1. What are the six canons in the NSPE Code of Ethics?

15.2. Is there a hierarchy among the canons in the NSPE Code of Ethics?

15.3. Review the Professional Obligations section of the NSPE Code of Ethics at the end of the chapter. Discuss the ethical implications of engineering consulting firms making campaign contributions.

15.4. Discuss the ethical behavior of engineers in the Space Shuttle *Challenger* disaster. For background information, see the World Wide Web Ethics Center for Engineering and Science, hosted by Case Western Reserve University at http://www.onlineethics.org. This site contains a fine discussion of engineering ethics, including detailed examples of whistle-blowing in engineering.

15.5. Discuss the ethical behavior of engineers in the walkway collapse at the Kansas City Hyatt Regency Hotel. For background information, see Pfrang and Marshall (1982).

15.6. Discuss the ethical behavior of engineers with regard to the structural problems in New York City's Citicorp Towers. For background information, see the World Wide Web Ethics Center for Engineering and Science, hosted by Case Western Reserve University at http://www.onlineethics.org.

15.7. Pick a field of engineering and discuss how an ethical violation would adversely affect public health, safety, or welfare.

Problems 15.8 through 15.10 are taken from actual ethics cases that came before NSPE's Board of Ethical Review. For each scenario, discuss the applicable parts of the NSPE Code of Ethics and decide whether the behavior was ethical.

15.8. Due to potential dangers during construction, an engineer recommends to a client before a project begins that a full-time person should be hired for on-site monitoring of the project. The client rejects the request for on-site monitoring, stating that the monitor would increase project costs to an unreasonable level. The engineer starts work on the project anyway. Was it ethical for the engineer to begin the project knowing that the client would not agree to hire an on-site monitor?

15.9. A company was advised by a state agency to get permission to discharge waste into a river. The company hires a consulting engineer to help them respond to the state's request. The engineer finds that the waste will degrade the quality of the river below established standards and that treating the waste will be expensive. The engineer tells the company of these findings orally, before the final report is written. The company then pays the engineer, terminates the engineering service agreement, and tells the engineer not to write a report. The engineer hears later that the company reported to the state agency that the waste discharge will not degrade the quality of the river. Does the engineer have an ethical obligation to report the findings to the state agency?

15.10. A state agency hires an engineer to conduct a feasibility study on a proposed highway spur. The spur will go through an area near where the engineer owns property. The engineer tells the state agency of the potential conflict of interest, but the agency does not object to the engineer working on the project. The engineer completes the study and the highway spur is built. Did the engineer act ethically by performing the study, even though the engineer's property may be affected?

NSPE CODE OF ETHICS FOR ENGINEERS

Preamble

Engineering is an important and learned profession. As members of this profession, engineers are expected to exhibit the highest standards of honesty and integrity. Engineering has a direct and vital impact on the quality of life for all people. Accordingly, the services provided by engineers require honesty, impartiality, fairness, and equity, and must be dedicated to the protection of the public health, safety, and welfare. Engineers must perform under a standard of professional behavior that requires adherence to the highest principles of ethical conduct.

I. Fundamental Canons

Engineers, in the fulfillment of their professional duties, shall:

1. Hold paramount the safety, health and welfare of the public.
2. Perform services only in areas of their competence.
3. Issue public statements only in an objective and truthful manner.
4. Act for each employer or client as faithful agents or trustees.
5. Avoid deceptive acts.
6. Conduct themselves honorably, responsibly, ethically, and lawfully so as to enhance the honor, reputation, and usefulness of the profession.

II. Rules of Practice

1. Engineers shall hold paramount the safety, health and welfare of the public.
    a. If engineers' judgment is overruled under circumstances that endanger life or property, they shall notify their employer or client and such other authority as may be appropriate.
    b. Engineers shall approve only those engineering documents that are in conformity with applicable standards.

c. Engineers shall not reveal facts, data, or information without the prior consent of the client or employer except as authorized or required by law or this Code.

d. Engineers shall not permit the use of their name or associate in business ventures with any person or firm that they believe are engaged in fraudulent or dishonest enterprise.

e. Engineers shall not aid or abet the unlawful practice of engineering by a person or firm.

f. Engineers having knowledge of any alleged violation of this Code shall report thereon to appropriate professional bodies and, when relevant, also to public authorities, and cooperate with the proper authorities in furnishing such information or assistance as may be required.

2. Engineers shall perform services only in the areas of their competence.

   a. Engineers shall undertake assignments only when qualified by education or experience in the specific technical fields involved.

   b. Engineers shall not affix their signatures to any plans or documents dealing with subject matter in which they lack competence, nor to any plan or document not prepared under their direction and control.

   c. Engineers may accept assignments and assume responsibility for coordination of an entire project and sign and seal the engineering documents for the entire project, provided that each technical segment is signed and sealed only by the qualified engineers who prepared the segment.

3. Engineers shall issue public statements only in an objective and truthful manner.

   a. Engineers shall be objective and truthful in professional reports, statements, or testimony. They shall include all relevant and pertinent information in such reports, statements, or testimony, which should bear the date indicating when it was current.

   b. Engineers may express publicly technical opinions that are founded upon knowledge of the facts and competence in the subject matter.

   c. Engineers shall issue no statements, criticisms, or arguments on technical matters that are inspired or paid for by interested parties, unless they have prefaced their comments by explicitly identifying the interested parties on whose behalf they are speaking, and by revealing the existence of any interest the engineers may have in the matters.

4. Engineers shall act for each employer or client as faithful agents or trustees.

   a. Engineers shall disclose all known or potential conflicts of interest that could influence or appear to influence their judgment or the quality of their services.

   b. Engineers shall not accept compensation, financial or otherwise, from more than one party for services on the same project, or for services pertaining to the same project, unless the circumstances are fully disclosed and agreed to by all interested parties.

c. Engineers shall not solicit or accept financial or other valuable consideration, directly or indirectly, from outside agents in connection with the work for which they are responsible.

d. Engineers in public service as members, advisors, or employees of a governmental or quasi-governmental body or department shall not participate in decisions with respect to services solicited or provided by them or their organizations in private or public engineering practice.

e. Engineers shall not solicit or accept a contract from a governmental body on which a principal or officer of their organization serves as a member.

5. Engineers shall avoid deceptive acts.

a. Engineers shall not falsify their qualifications or permit misrepresentation of their or their associates' qualifications. They shall not misrepresent or exaggerate their responsibility in or for the subject matter of prior assignments. Brochures or other presentations incident to the solicitation of employment shall not misrepresent pertinent facts concerning employers, employees, associates, joint venturers, or past accomplishments.

b. Engineers shall not offer, give, solicit or receive, either directly or indirectly, any contribution to influence the award of a contract by public authority, or which may be reasonably construed by the public as having the effect of intent to influencing the awarding of a contract. They shall not offer any gift or other valuable consideration in order to secure work. They shall not pay a commission, percentage, or brokerage fee in order to secure work, except to a bona fide employee or bona fide established commercial or marketing agencies retained by them.

III. Professional Obligations

1. Engineers shall be guided in all their relations by the highest standards of honesty and integrity.

a. Engineers shall acknowledge their errors and shall not distort or alter the facts.

b. Engineers shall advise their clients or employers when they believe a project will not be successful.

c. Engineers shall not accept outside employment to the detriment of their regular work or interest. Before accepting any outside engineering employment they will notify their employers.

d. Engineers shall not attempt to attract an engineer from another employer by false or misleading pretenses.

e. Engineers shall not promote their own interest at the expense of the dignity and integrity of the profession.

2. Engineers shall at all times strive to serve the public interest.

a. Engineers shall seek opportunities to participate in civic affairs; career guidance for youths; and work for the advancement of the safety, health, and well-being of their community.

b. Engineers shall not complete, sign, or seal plans and/or specifications that are not in conformity with applicable engineering standards. If the client or employer insists on such unprofessional conduct, they shall notify the proper authorities and withdraw from further service on the project.

c. Engineers shall endeavor to extend public knowledge and appreciation of engineering and its achievements.

3. Engineers shall avoid all conduct or practice that deceives the public.

a. Engineers shall avoid the use of statements containing a material misrepresentation of fact or omitting a material fact.

b. Consistent with the foregoing, engineers may advertise for recruitment of personnel.

c. Consistent with the foregoing, engineers may prepare articles for the lay or technical press, but such articles shall not imply credit to the author for work performed by others.

4. Engineers shall not disclose, without consent, confidential information concerning the business affairs or technical processes of any present or former client or employer, or public body on which they serve.

a. Engineers shall not, without the consent of all interested parties, promote or arrange for new employment or practice in connection with a specific project for which the engineer has gained particular and specialized knowledge.

b. Engineers shall not, without the consent of all interested parties, participate in or represent an adversary interest in connection with a specific project or proceeding in which the engineer has gained particular specialized knowledge on behalf of a former client or employer.

5. Engineers shall not be influenced in their professional duties by conflicting interests.

a. Engineers shall not accept financial or other considerations, including free engineering designs, from material or equipment suppliers for specifying their product.

b. Engineers shall not accept commissions or allowances, directly or indirectly, from contractors or other parties dealing with clients or employers of the engineer in connection with work for which the engineer is responsible.

6. Engineers shall not attempt to obtain employment or advancement or professional engagements by untruthfully criticizing other engineers, or by other improper or questionable methods.

a. Engineers shall not request, propose, or accept a commission on a contingent basis under circumstances in which their judgment may be compromised.

b. Engineers in salaried positions shall accept part-time engineering work only to the extent consistent with policies of the employer and in accordance with ethical considerations.

c. Engineers shall not, without consent, use equipment, supplies, laboratory, or office facilities of an employer to carry on outside private practice.

7. Engineers shall not attempt to injure, maliciously or falsely, directly or indirectly, the professional reputation, prospects, practice, or employment of other engineers. Engineers who believe others are guilty of unethical or illegal practice shall present such information to the proper authority for action.

   a. Engineers in private practice shall not review the work of another engineer for the same client, except with the knowledge of such engineer, or unless the connection of such engineer with the work has been terminated.

   b. Engineers in governmental, industrial, or educational employ are entitled to review and evaluate the work of other engineers when so required by their employment duties.

   c. Engineers in sales or industrial employ are entitled to make engineering comparisons of represented products with products of other suppliers.

8. Engineers shall accept personal responsibility for their professional activities, provided, however, that engineers may seek indemnification for services arising out of their practice for other than gross negligence, where the engineer's interests cannot otherwise be protected.

   a. Engineers shall conform with state registration laws in the practice of engineering.

   b. Engineers shall not use association with a nonengineer, a corporation, or partnership as a "cloak" for unethical acts.

9. Engineers shall give credit for engineering work to those to whom credit is due, and will recognize the proprietary interests of others.

   a. Engineers shall, whenever possible, name the person or persons who may be individually responsible for designs, inventions, writings, or other accomplishments.

   b. Engineers using designs supplied by a client recognize that the designs remain the property of the client and may not be duplicated by the engineer for others without express permission.

   c. Engineers, before undertaking work for others in connection with which the engineer may make improvements, plans, designs, inventions, or other records that may justify copyrights or patents, should enter into a positive agreement regarding ownership.

   d. Engineers' designs, data, records, and notes referring exclusively to an employer's work are the employer's property. The employer should indemnify the engineer for use of the information for any purpose other than the original purpose.

   e. Engineers shall continue their professional development throughout their careers and should keep current in their specialty fields by engaging in professional practice, participating in continuing education courses, reading in the technical literature, and attending professional meetings and seminars.

   —As revised January 2003

16

The Engineering Profession and Professional Registration

SECTIONS

OBJECTIVES

After reading this chapter, you will be able to:

- explain why engineering is a profession;
- explain the benefits of becoming a licensed professional engineer;
- list the steps in the registration process.

16.1 INTRODUCTION

In this part of the text, you will explore engineering as a profession. To do so, you must consider two questions. First, what does *profession* mean? If you are getting paid, is that enough to qualify a job as a profession? This area will be explored in Section 16.2.

Second, does engineering qualify as a profession? To answer this question, the characteristics that make engineering unique and important in today's society will be discussed.

16.2 PROFESSIONAL ISSUES

16.2.1 What Is a Profession?

A profession is more than just a job requiring specialized knowledge and intensive study.

Common elements include (Martin and Schinzinger, 1989)

- Compensation
- Professional practices that result in public good
- Need for formal education
- Requirement for judgment, discretion, and skill in performing your job
- Requirement to be admitted to the profession
- Self-policed conduct

Do these concepts make sense? You can apply them against known professions, such as medicine and law. Both medicine and law require all these elements. The fact that doctors and lawyers satisfy the first four criteria is obvious: they are paid, they do good, they are trained formally, and their jobs require judgment, discretion, and skill.

PONDER THIS

What other elements come to mind when you think of the word *profession*?

Key idea: Common elements of a profession include compensation; performing a public good; a need for formal education; a requirement for judgment, discretion, and skill; a requirement to be admitted to the profession; and self-policing.

Key idea: Engineers must exercise professional judgment in the practice of their profession.

The fifth criterion also is satisfied, since both jobs require admission to the profession (passing medical board examinations or the bar examination). The admission to both professions is regulated by each state. In other words, doctors are licensed and lawyers are admitted to the bar by the state in which they practice. Finally, practicing professionals sit on the boards regulating professional conduct (for example, medical and legal ethics panels). Thus, the last criterion is satisfied.

16.2.2 Engineering as a Profession

For engineering to be a profession, it must satisfy the elements listed in Section 16.2.1. That engineers are compensated is obvious. Further, it is sincerely hoped that you are convinced by now that engineering practice results in public good. What about formal education? You are at the beginning of your formal education now.

What about the requirements for judgment and discretion, admission to the profession, and self-policing? In fact, these elements are at the heart of engineering as a profession.

16.2.3 Judgment and Discretion in Engineering

Engineering is all about generating alternative solutions to a problem and choosing from among alternatives. Thus, judgment is a critical part of engineering decision-making. Professional judgment is at the core of "thinking like an engineer."

What about discretion in engineering?

PONDER THIS

Why should engineers be thoughtful and discreet?

Key idea: Discretion in engineering is important because (1) engineering affects public safety and health, and (2) public trust in engineers must be preserved, since engineering work is not easy to understand.

There are two reasons why discretion and thoughtfulness are imperative for engineers. First, engineering has a significant impact on public safety and health. A poorly designed automobile or bridge or electric transmission line could result in terrible suffering and loss of human life. Any profession that affects *every* citizen *every* day (as is the case with engineering) should be held to the highest standards of ethical behavior.

Second, most engineering work is difficult for the average person to comprehend. As a result, the public must trust the word of engineers. If an engineer declares that the probability of the collapse of a high-rise building is extremely small, then the public is reassured. Reassurance comes *without* an understanding of the details of the materials employed or the construction techniques or even elementary statics. To *preserve the public trust*, engineers must be held accountable to a very strict professional code.

16.2.4 Admission to the Profession

Engineering also satisfies the criteria that you must be formally accepted into the profession. To use the title "engineer" to make money, you must be *admitted* to the profession. This process is called *registration*. The steps required for professional registration are discussed in more detail in Section 16.4.

16.2.5 Self-Policing

Engineers police themselves. Self-policing is accomplished by having engineers sit on state review boards that oversee licensing and the removal of licenses.

For professionals to police themselves, the profession needs a list of guidelines. Engineering has such guidelines, called *professional ethics*.

Thus, engineering fulfills all the requirements normally associated with being a profession. This analysis does not capture the awesome potential you have for influencing society or the serious responsibilities you will carry with pride as an engineer. As a source of inspiration, read the *Focus on Professionalism: Standing on the Shoulders of Giants*.

FOCUS ON PROFESSIONALISM: STANDING ON THE SHOULDERS OF GIANTS

The formal analysis of whether engineering is a profession misses the most important point: *you* are joining this profession, this band of brothers and sisters devoted to changing the world in a positive way. You will reap the satisfaction of helping others and you will make part of your being the obligations of excellence.

You are not alone. The contributions you will make build on the advances made by engineers before you. As Isaac Newton (1642–1727) said, "If I have seen further, it is by standing on the shoulders of giants." By standing on such a tall perch, you should feel that your potential to help others is unlimited.

In your everyday world of studying and working, it is easy to lose sight of the traditions of engineering, of the pride and obligations implicit in the profession. Presented here are two statements of engineering pride and obligations. The first statement, the "Obligation of an Engineer," is recited as part of the ceremony of joining the American association called the Order of the Engineer. The "Obligation of an Engineer" was written as an engineering version of the Hippocratic Oath taken by physicians. The Order of the Engineer was formed in 1970.

Obligation of an Engineer

(from the Order of the Engineer)

I am an Engineer, in my profession I take deep pride. To it I owe solemn obligations.

Since the Stone Age, human progress has been spurred by the engineering genius. Engineers have made usable Nature's vast resources of material and energy for Humanity's benefit. Engineers have vitalized and turned to practical use the principles of science and the means of technology. Were it not for this heritage of accumulated experience, my efforts would be feeble.

As an Engineer, I pledge to practice integrity and fair dealing, tolerance and respect, and to uphold devotion to the standards and the dignity of my profession, conscious always that my skill carries with it the obligation to serve humanity by making the best use of Earth's precious wealth.

As an Engineer I shall participate in none but honest enterprises. When needed, my skill and knowledge shall be given without reservation for the public good. In the performance of duty and in fidelity to my profession, I shall give the utmost.

The American ceremony was based on the Canadian "Ritual of the Calling of an Engineer," which dates back to 1926. The Canadian engineer's oath, called "The Obligation," was written (in three weeks!) by Nobel Laureate Rudyard Kipling.

The Obligation

(from the Iron Ring Ceremony in the Canadian Ritual of the Calling of an Engineer)

I (your name), in the presence of these my betters and my equals in my Calling, bind myself upon my Honour and Cold Iron, that, to the best of my knowledge and power, I will not henceforward suffer or pass, or be privy to the passing of, Bad Workmanship or Faulty Material in aught that concerns my works before mankind as an engineer, or in my dealings with my own Soul before my Maker.

My Time I will not refuse; my Thought I will not grudge; my Care I will not deny towards the honour, use, stability and perfection of any works to which I may be called to set my hand.

My Fair Wages for that work I will openly take. My Reputation in my Calling I will honourably guard; but I will in no way go about to compass or wrest judgement or gratification from any one with whom I may deal. And further, I will early and warily strive my uttermost against professional jealousy and the belittling of my working-colleagues in any field of their labour.

For my assured failures and derelictions I ask pardon beforehand of my betters and my equals in my Calling here assembled, praying that in the hour of my temptations, weakness and weariness, the memory of this my Obligation and of the company before whom it was entered into, may return to me to aid, comfort, and restrain.

Upon Honour and Cold Iron, God helping me, these things I purpose to abide.

When times are tough—when you have two tests on Tuesday and two homework assignments due tomorrow and you wonder why you ever wanted to be an engineer—reread the "Obligation of an Engineer" and "The Obligation." Remind yourself of your potential and your obligations to help.

16.3 PROFESSIONAL ENGINEERS

16.3.1 Introduction

registration: the act of being admitted to the engineering profession

professional engineer (PE): a licensed (registered) engineer

Engineering is a profession in part because engineers are admitted to practice. Every state regulates the practice of all professions. Examples include medicine (e.g., physicians, dentists, optometrists, and nurses), law, and engineering. In engineering, admission to the profession is called ***registration*** (or licensing). A registered engineer is called a ***professional engineer*** or PE. In this chapter, the registration process will be described.

16.3.2 Why Become a Professional Engineer?

Key idea: People become licensed to show their competence and commitment, to perform duties unavailable by law to unlicensed personnel, and to gain personal benefits.

The registration process (described in Section 16.4) is long, typically requiring at least seven years. Why go to the trouble of becoming licensed? The most important reason to become licensed is that it demonstrates your extra commitment to the profession. In addition, it is a badge of honor that shows your increased level of competence in the field.

Another reason to become licensed is that a PE may conduct professional activities that unlicensed personnel are forbidden to do. For example, only a licensed engineer may prepare, approve, and submit engineering plans and drawings to a public authority. Only a licensed engineer may approve engineering work for public and private clients. Consulting engineers in positions of authority over work performed in their office must, by law, be licensed. In general, you cannot use the term *engineer* in your title unless you are licensed. State engineering boards are successfully using the courts to impose civil penalties on individuals using the term *engineer* without an engineering license. As a example, penalties have been imposed on people without a PE doing business as *software engineers*.

In addition, PEs reap personal benefits from registration. Surveys have shown that professional engineers can expect salaries 15% to 25% higher than unlicensed colleagues. Some jobs are available only to professional engineers. This is particularly true in some government positions, where obtaining a PE license often is required for promotion. PEs also may find themselves more desirable to another firm in the event of downsizing. For advice about registration from professional engineers, see the *Focus on Registration: PE or Not PE?*

In spite of these arguments, not all engineers become licensed. Registration is more common in some disciplines than others. In particular, engineers working primarily in private practice (e.g., civil and environmental engineers) are more likely to become licensed than engineers working primarily in industry.

16.4 THE REGISTRATION PROCESS

16.4.1 Overview

Key idea: The steps in the registration process are as follows: obtain a baccalaureate degree from an accredited engineering department, pass the Fundamentals of Engineering Exam, acquire the requisite experience, and pass the Principles and Practice Exam.

Engineer-in-Training: a prelicensure certificate given to individuals completing the first steps in the registration process

The registration procedure for engineers is a four-step process. First, would-be engineers must obtain a baccalaureate degree from an accredited engineering department. Second, after earning the degree (or in the final stages of the program), potential engineers must take and pass a written test called the Fundamentals of Engineering Examination (or FE Exam). At this point, individuals can earn a prelicensure certificate. This level of licensing generally is called the ***Engineer-in-Training*** (or EIT). Third, EITs must acquire experience under the supervision of a licensed engineer. Finally, the would-be engineer must take and pass the Principles and Practice Examination in a specific engineering discipline. At this point (after the requisite licensing fees are paid), a person may be called an engineer and have the title *professional engineer*. You indicate the title by the initials *PE* after your name. Each of these steps will be discussed in more detail.

16.4.2 The Accredited Degree

The registration process starts at a college or university.* In accordance with the idea that professions require extensive training, potential engineers must obtain an undergraduate degree through an accredited engineering program.

What does *accredited* mean? To ensure the high technical level of the profession, engineering programs must request accreditation through the Accreditation Board for Engineering and Technology (ABET). Engineering programs must be reaccredited periodically, typically every six years. ABET's Engineering Accreditation Commission conducts site visits of each department seeking accreditation.

In 2005, there were over 1,600 accredited engineering programs in the United States. For a list of accredited programs, see the ABET Web site (www.abet.org).

The accreditation process is the beginning of the self-policing feature of the engineering profession. To be eligible for professional registration later, it is important for you to seek a degree from an accredited engineering program. Accreditation is awarded to each program, not the college or university in general.

16.4.3 Fundamentals of Engineering Examination

Fundamentals of Engineering (FE) Exam: an eight-hour exam covering general science, general engineering, and discipline-specific engineering topics

The ***Fundamentals of Engineering Examination*** (FE Exam) generally is taken in the last semester of the undergraduate program. The FE Exam[†] is administered by the National Council of Examiners for Engineering and Surveying (NCEES). The FE Exam consists of two parts. The morning portion deals with general science and engineering concepts and is common to all disciplines. The morning portion consists of 120 one-point questions to be answered in four hours. Topics include (with the percentage of questions in parentheses)

- Chemistry (9%)
- Computers (6%)

*In most states, the requirement for a baccalaureate degree can be satisfied by many years of experience. However, the most common pathway for professional registration includes an undergraduate degree from an accredited engineering program.

[†]You may hear the FE Exam referred to by its previous name: the EIT Exam.

- Dynamics (7%)
- Electrical circuits (10%)
- Engineering economics (4%)
- Ethics (4%)
- Fluid mechanics (7%)
- Materials science/structure of matter (7%)
- Mathematics (20%)
- Mechanics of materials (7%)
- Statics (10%)
- Thermodynamics (9%)

The afternoon session is discipline-specific. It consists of 60 two-point questions to be answered in four hours. Afternoon sessions are offered in chemical, civil, electrical, environmental, industrial, and mechanical engineering. A general engineering section also is available.

Following successful completion of the FE Exam, the potential engineer is eligible for certification as an Engineer-in-Training (or EIT; in some states, it is also called an Intern Engineer or Engineer Intern). The EIT receives a registration number from the state.

16.4.4 Experience

Key idea: For registration purposes, your work experience must be under the direction of a PE.

For an EIT to become a PE, you first must acquire engineering experience. Most states require four years of experience under the supervision of a PE. This distinction is important. If a newly minted EIT takes his or her first job at a firm that has no PEs on staff, it would be difficult to acquire experience under the direction of a PE. Thus, the first job after becoming an EIT is very important for future licensure. A portion of the time spent obtaining a graduate degree in engineering typically may be applied towards the experience requirement.

16.4.5 Principles and Practice Examination

Principles and Practice (PP) Exam: an eight-hour exam of a specific engineering discipline

After acquiring the necessary experience, potential engineers can apply to take the ***Principles and Practice Examination*** (PP Exam).[°] Like the afternoon session of the FE Exam, the PP Exam is discipline-specific. Exams are offered in 17 areas. Not all areas are offered in every state.

The PP Exam is eight hours long and may be all essay questions, all multiple-choice, or a mixture of essay and multiple-choice questions. The exam, administered by NCEES, is the same nationwide, but each state sets its own passing grade.

16.5 AFTER REGISTRATION

To maintain registration, an engineer must follow the ethical guidelines of his or her field and maintain competence in the field. Some states require PEs to engage in continuing education activities. Even without such requirements, maintaining state-of-the-art knowledge of the field is mandated by the ethical codes. In addition, fees must be paid periodically to the state to maintain the license.

reciprocity: the recognition of a license in one state by another state

There is general agreement between the states on the licensing process. Most states recognize licenses obtained in another state (a process known as ***reciprocity***). Thus, it is often possible for a engineer licensed in one state to gain licensure in another state relatively easily.

[°]The PP Exam is sometimes referred to by its previous name: the PE Exam.

FOCUS ON REGISTRATION: PE OR NOT PE?

In this section, advice will be presented from registered professional engineers. The professional engineers offer sound comments on why to become licensed and suggestions for the licensing process. The quotations are from interviews published on one of the Web sites run by the National Council of Examiners for Engineering and Surveying (www.engineeringlicense.com).

Kathy Caldwell, PE; civil engineer

President, JEA Construction Engineering Services

. . . (T)ake the FE exam just as soon as you possibly can before you leave the college environment. It's very difficult to go back and take the fundamentals exam once you have been practicing in your major area of practice. A lot of fundamentals slip away from you and it becomes much more difficult to go back and relearn the fundamentals later.

Jim Parrish, PE; electrical engineer

General Manager, Duke Energy, Energy Delivery Services International

Achieving a PE also means that I have more career choices and can benefit from higher financial rewards that are commensurate with industry-recognized professional status. In my [unit of Duke Energy], engineers without a PE license can be promoted to only certain levels of engineering-related management jobs.

Brett Pielstick, PE; civil engineer

Vice President, PTG Construction Services

I think people know that as an engineer with a PE I will provide an honest day's work, an honest answer, and that they can trust in what we're going to do for them.

Deborah Grubbe, PE; chemical engineer

Corporate Director of Safety and Health, The DuPont Company

Ethics is an important aspect of professional licensure. Obtaining a PE (license) registration not only targets technical competence, but also requires a clear understanding of both ethics and professionalism. Ethics is essential to one's personal reputation, and once compromised, can almost never be reclaimed. Personal credibility becomes even more important when working outside the United States.

16.6 SUMMARY

In all aspects of the word, engineering qualifies as a profession. First, engineers are compensated for their work. Second, the general public benefits significantly from engineering practice. Third, engineers require a formal education prior to practicing their craft. Fourth, engineering practice requires judgment, discretion, and skill. Fifth, to legally use the title *engineer*, a person must be admitted to the engineering profession. The process of admission to the profession in engineering is called registration. Sixth, the professional conduct of engineers is self-policed through strong written codes of professional ethics.

Professional registration as a professional engineer (PE) is a demonstration of competence and commitment to the engineering profession. Registration also allows engineers to perform duties unavailable by law to unlicensed personnel and to gain personal benefits. Registration is a four-step process: obtain a baccalaureate degree from an accredited engineering department, pass the FE Exam, acquire the requisite experience (generally four years under the supervision of a PE), and pass the PP Exam. To maintain registration, PEs must follow the ethical guidelines of their field and maintain professional competence in the field.

SUMMARY OF KEY IDEAS

- Common elements of a profession include compensation; performing a public good; a need for formal education; a requirement for judgment, discretion, and skill; a requirement to be admitted to the profession; and self-policing.
- Engineers must exercise professional judgment in the practice of their profession.
- Discretion in engineering is important because (1) engineering affects public safety and health, and (2) public trust in engineers must be preserved, since engineering work is not easy to understand.
- People become licensed to show their competence and commitment, to perform duties unavailable by law to unlicensed personnel, and to gain personal benefits.
- The steps in the registration process are as follows: obtain a baccalaureate degree from an accredited engineering department, pass the Fundamentals of Engineering Exam, acquire the requisite experience, and pass the Principles and Practice Exam.
- For registration purposes, your work experience must be under the direction of a PE.

Problems

16.1. Write a short essay to justify why engineering is a profession.

16.2. List the steps in the professional registration process. Make a timeline to show when you plan to complete each step.

16.3. From the ABET Web site, make a list of accredited engineering programs in your discipline of interest in your state.

16.4. Can a person without a PE license use the title "Web design engineer" in a business? Why or why not?

16.5. Using data from an interview with a practicing engineer or from the Internet, determine the impact of earning a PE on salary in your area.

16.6. Find and report the distribution of questions on the disciple-specific portion of the Fundamentals of Engineering Examination for an engineering discipline of interest to you.

16.7. Find and report the distribution of questions on the disciple-specific portion of the Principles and Practice Examination for an engineering discipline of interest to you.

16.8. Some states have continuing education requirements for professional engineers. Report on the continuing education requirements, if any, for professional engineers in your state.

16.9. Some states do not offer reciprocity without further examination. Research and report on any additional examination for obtaining a PE license in civil engineering by reciprocity in California.

16.10. Using the Web site of your state licensing office, find a case of an engineer who lost his or her license (for reasons other than nonpayment of fees). What did the engineer do to lose it?

17

Succeeding in the Engineering Classroom

Sections

17.1 INTRODUCTION

As an engineering student, an important goal is to succeed in the classes needed to attain an engineering degree. How this is done depends to a great deal on the individual student and the style, temperament and strengths of that student. This chapter presents strategies proven to be effective in helping students succeed. As with any tips on success, you need to look at the techniques being outlined and decide if they can be applied to you and your own individual style. If you are unsure about some of the suggestions, give them a try. You may find yourself in the same situation as one freshman engineering student after being given a book on study skills.

The student enters a professor's office very excited and asks for a short conference with the professor:

Student: Professor! It worked!
Professor: That's great. What worked?
Student: The book!
Professor: That's great. What book?
Student: That study skills book.
Professor: So you read it?
Student: Yes, I read it like you suggested. It sure sounded like a lot of unworkable stuff at first. But then I decided, what the heck. I'll try it for a week. Couldn't hurt, I figured. Well, it really did work! It even worked for physics lab. I got an A on the last lab. I still can't believe it.
Professor: That's great. I am glad it helped.
Student: Could I get another copy of that book? I have some friends who want to read it now...

Just like that student, we are often reluctant to try new things. We are comfortable in what we do. You may have been very successful in high school with your study methods and wonder why you need to develop different study methods and techniques. The answer is that your environment has changed. College is very different from high school. The courses are structured differently and require significantly more preparation outside of class. The expectations of the instructors are different and the grading is different. Modifying your habits to become more effective is a key to succeeding in engineering studies. Learning to adapt to meet new challenges is a skill that will serve you well not only in college, but for the rest of your life.

The methods for excelling in the classroom outlined in this chapter have proven effective for numerous students. We suggest that you consider all of these ideas and adapt the ones that fit best with your own personality. If you are unsure whether they will work, try some out for a trial period and see. If those don't work, move on to others until you find something that fits with your own style.

The three components to succeeding in your academics are your ability, your attitude and your effort. There is no book that can address the first item. While there is unquestionably a level of ability that is needed to succeed in engineering, there are numerous students who had the ability but failed at engineering because of poor attitudes or poor or inefficient effort. There are also many students who lacked natural ability but because of their positive attitude and extraordinary effort are now excellent engineers. This chapter focuses on techniques to improve these two components for success that are totally under your control.

17.2 ATTITUDE

"Ability is what you're capable of doing. Motivation determines what you do. Attitude determines how well you do it."

—Lou Holtz, South Carolina football coach

Your attitude is the first thing that you have control over when beginning any challenge, including studying engineering. You can expect success or expect failure. You can look for the positives or dwell on the negatives.

Approaching your classes, professors and teaching assistants with a positive attitude is the first key to succeeding in your engineering studies. Each class should be looked at as an opportunity to succeed. If there are difficulties along the way, learn to deal with those and move ahead. Many students will decide that a certain class is too difficult or that the professor is a poor teacher and will expect to do poorly. Many fulfill their own prophecy and actually fall. Other students look at the same situation and overcome the hurdles and excel.

It is a distinct possibility that you may have professors you don't like or who you think teach poorly. You may even have an experience like the following example.

Example 17.1: A physics professor of mine discovered the overhead projector during a semester and moved his lecture from the blackboard to overheads. The problem was that he was very tall and stood so that he blocked the overhead. It shined on his coat and tie but we couldn't see what he was writing. When we told him in class that we couldn't see, he moved out of the way and looked at the overhead. He then adjusted the focus and asked if we could see now. After we responded yes, he moved back to write more and blocked the overhead again. When another student tried to explain that he was blocking his view of the overhead, the professor explained that no matter where he stood, he would block someone's view. He suggested that the student move to the other side of the room. A third student from the other side of the room chimed in that he couldn't see from that side either. Unfortunately he was in the back of the lecture hall, so the professor thought that was the reason. He moved out of the way of the projector and asked one of the students in the front row if she could see. She said, "Yes, but. . ." The professor cut her off before she could finish and asked the guy in the back of the room if he had glasses. When he said no, the professor recommended he get his eyes checked because the woman in front said she could see. At this point we all gave up taking notes for the rest of the lecture.

Later, I complained to an advisor who was the director of the university honors program. Rather than a strategy for communicating with the professor, I received a story about a professor he had had. It seems that he had an English professor who was about 108 years old (or *seemed* that old). He couldn't stand up without leaning on something, so he used the blackboard. His shoulder made contact with the blackboard, which wasn't a problem in itself except that he wrote at shoulder level. The result is that he erased what he wrote as he moved along the board. So if you weren't in a seat where you could see the three or four words he wrote before he erased them with his shoulder, you couldn't take notes.

The point he was making was that there are challenges in learning as well as later in life. What we have to do is to find ways to overcome obstacles, not to dwell on them. If your attitude is that it is a professor's fault that you won't learn, you won't. If you keep a positive attitude you can turn challenges into opportunities. He shared that his class became much closer and studied more together because they had to piece the lecture together from their collective notes. They learned a lot, and the positive outcome was the closeness the class developed. He challenged us to go and talk to the professor outside of class, during his office hours, and try to explain our difficulty with the lecture. Instead of just complaining, take action with a positive attitude to solve our problem. It was a real lesson for life.

I am still not sure what lesson I was supposed to learn from that story. It struck me, however, that even with professors like that, the advisor earned his Ph.D., became a professor and eventually ran the university's honors program. He had developed strategies to cope and kept a good attitude. Attitude is totally under your control.

17.3 GOALS

"Do you feel the world is treating you well?

If your attitude toward the world is excellent, you will receive excellent results. If you feel so-so about the world, your response from that world will be average. Feel badly about your world and you will seem to have only negative results from life."

—Dr. John Maxwell, founder of INJOY, Inc.

As freshmen, there were a number of us who were in an honors program. One in the group set a goal to be a Rhodes Scholar from the first day of his freshman year. The rest of us were too busy adjusting to college and taking it all in. In our senior year, when we were interviewing for jobs and deciding whether to go to graduate school, the friend who had set the goal was being toured around the state as the newest Rhodes Scholar from our university. The difference over our college careers was astounding, simply because he had established clear, elevated goals. The rest of us did okay, but he excelled, in large part because he had defined a standard and kept striving for it.

The lesson that I took away from that friendship in college was that setting clearly defined goals early is a key to success in school, in an engineering career or in life as a whole.

In setting goals, there are two essential elements— "height" and time. You might ask, "How high do I set my goals?" This will vary from individual to individual. Goals should be set high, but attainable. A common management practice is for a team to decide on a reasonable goal to accomplish. Then the bar is raised and a stretch goal is defined that encourages the team to produce more. This stretch goal is designed to push the group farther than they would have gone with only the original goal. Sometimes this stretch goal looks reasonable and sometimes it looks totally unreachable. What often happens, though, is that the stretch goal is met, realistic or not. As an individual, setting stretch goals helps us grow. When setting goals for yourself, look at what you think you can do and set base goals. Then establish stretch goals a little higher than the base goals. You may just find yourself meeting those higher goals.

The second key element in goal setting is time. When will the goal be accomplished? It is important to have long-term goals and short-term goals. Long-term goals help guide where you are headed. For example, what do you want to do with your engineering degree? An appropriate long-term goal as a freshman could involve determining the kind of entry-level job you desire. This will help answer the question "Why am I doing this?" when your classes get tough.

Short-term or intermediate goals should be stepping-stones to the achievement of your long-term goals. An advantage of an academic setting is that there are natural breaks, semesters or quarters, for the establishment of goals within specific time frames. Goals can be set for the academic year, and for each semester or quarter. These provide intermediate goals.

Short-term goals need to be set and rewarded. Rewarding short-term goals is essential to keeping motivation high in the quest for long-term goals. Many students lose sight of their long-term goals because they didn't set short-term, "rewardable" goals. Short-term goals may be an "A" on an upcoming quiz or test, or it may be completing a homework set before the weekend.

Example 17.2:

Long-term goal	Job as a design engineer for a major manufacturer of microprocessors
Intermediate goals	3.7 GPA after freshman year 3.5 GPA at graduation Hold an office in the local chapter of IEEE Internships after sophomore and junior years
Short-term goals	'B' on first calculus test 'A' in calculus 3.7 GPA after first semester Join IEEE and attend meetings Complete chemistry lab by Friday

In the list of short-term goals above, there are opportunities for rewards. Perhaps there's a compact disk you want. Set a goal of a certain grade on that calculus test. If you meet the goal, you can go buy the CD.

Clearly define your goals by writing them down. It is much more powerful to write down your goals than to simply think about them. Some people post them where they can see them daily. Others put them in a place where they can retrieve and review them regularly. Either method is effective. For short-term goals, establish weekly goals. Select a day, probably Sunday or Monday, and establish the goals for the week. This is a great time to establish the rewards for your weekly goals. Plan something fun to do on the weekend if you meet those goals.

At the end of each semester, examine your progress toward your long-term goals. This is an advantage to being in school. Every semester, there is an opportunity to assess your progress and start fresh in new classes. Take advantage of this. Examine your long-term goals and make any changes if necessary.

17.3.1 Determining Goals

It can be intimidating or difficult for students early in their careers to decide on appropriate goals. Fortunately, there are numerous resources on campus to help. Academic advisors can provide criteria for grade-point goals. Things to consider include:

1. What is the minimum GPA to stay enrolled in school?
2. What grades do you need to continue in engineering or to enter engineering?
3. What grades are needed to be eligible for co-op or intern positions?
4. What GPA is required to make the dean's list or honor list?
5. What GPA is needed to be eligible for the engineering honorary organizations?
6. What grades are needed for admittance to graduate school?
7. What grades are needed for scholarships?

There are also people on campus who work with the placement of graduates. They are excellent resources to get information on what employers are looking for. What credentials do you need to be able to get the kind of job you desire? If you know this as a freshman, you can work toward it. You can also go directly to the employers and ask these same questions of them. If companies come to your campus, talk to them about what credentials are needed to get a job. Job fairs are great times to do this. Approaching company representatives about establishing goals is also a great way to show initiative and distinguish yourself from your classmates.

Goals are important for outside the classroom too. It is important to have goals in all areas of your life. Later in this chapter we will discuss such goals.

17.4 KEYS TO EFFECTIVENESS

Once goals have been set regarding grades, the next step is to achieve those grades. Effort and effectiveness may be the most important components of your success as a student. There are numerous cases of well-equipped students who end up failing. There are also students who come to the university poorly prepared or start off poorly but eventually succeed. The difference is in their effort and their effectiveness. The following suggestions contain strategies to improve personal effectiveness in your studies.

A general rule is that it takes a minimum of two hours of study outside of class for each hour of class lecture.

17.4.1 Take Time to Study

One of the strongest correlations with a student's performance is the time he or she spends studying.

According to the reminder above, if you are taking 15 hours, that means that you need to spend 30 effective hours a week studying outside of class. In a typical high school, the time requirement is much less than this. Developing the study habits and discipline to spend the needed time studying is a prime factor in separating successful and unsuccessful students.

17.4.2 Go to Class

There is also a high correlation with class attendance and performance. This may seem obvious, but you will run into numerous students who will swear that it doesn't matter if you go to class. Many instructors in college do not take attendance, so no one but you knows if you were there, which is different from high school. The classroom is where important information will be presented or discussed. If you skip class, you may miss this information. Also, by skipping class, you miss an opportunity to be exposed to the material you will need to know, and you will have to make up that time on your own. Missing class only delays the expenditure of time you will need to master the material. Most often, it may take much more than an hour of independent study to make up an hour's worth of missed class time. It is possible for you to never miss a class in your undergraduate years, except for illness or injury, serious family problems or trips required by groups to which you belong. Never missing a class could be one of your goals.

Also, you may want to schedule classes at 8 a.m. and 4 p.m. This will get you up and going and keep you there! You will seldom, if ever, study at 8 a.m. and at 4 p.m. Do your studying in the middle of the day. This will help you to use your time effectively.

"After class, reread the assigned sections in your textbook. Concentrate on the sections highlighted by the lecture. By doing so, you will not waste time trying to understand parts of the book that are not critical to the course."

17.4.3 Make Class Effective

The first component of making class effective is sitting where you can get involved. If the classroom is large, can you see and hear in the back? If not, move up front. Do you need to feel involved in the class? Sit in the front row. Do you fall asleep in class? Identify why. Ask yourself, "Am I sleeping enough at night?" If the answer is no, get more sleep.

The second step for making class time effective is to prepare for class. Learning is a process of reinforcing ideas and concepts. As such, use the class time to reinforce the course material by reviewing material for the lecture beforehand. Most classes will have a textbook and assigned reading for each lecture. Take time before the lecture to skim over the relevant material. Skimming means reading it through but not taking time to understand it in depth. This will make the class time more interesting and more understandable, and it will make note taking easier.

By following the above advice, you will have had three exposures to the material, which helps you remember and understand the material better.

17.4.4 Keep Up with the Class

Class is most effective if you keep up. An excellent short-term goal is to master the lecture material of each course before the next lecture. A course is structured in such a way that you master one concept and then move on to the next during the next lecture. Often in science and engineering classes the concepts build on each other. The problem with falling behind is that you will be trying to master concepts that depend on previous lectures, and you are going to hear lectures that you will not understand.

There is a very practical reason that a three-credit class meets three times per week. During a 15-week semester, there are only 45 hours of class meeting time. That amount of class time could be held in a week of consecutive nine-hour days. Arranged that way, a whole semester could be taken in a month. Then it would only take a year to get a bachelor's degree. So why don't we do it that way?

The reason is that the current educational model is designed to introduce a concept and give you, the learner, time to digest it. It is an educationally sound model. In engineering study, digesting means reviewing notes, reading the text and working several problems related to the topic. If you are still confused about an aspect of the previous lecture, the next class provides a great opportunity to bring it up with the professor. Get your questions answered before going on to the next topic.

A note of caution should be given here. Friends of yours may be studying subjects where they can study from test to test rather than from class to class. They may insist that you don't need to keep up so diligently because they don't. Some majors cover material that is more conceptually based. That is, once you understand the concept, you have it. Engineering, math and science courses are not this way. To be mastered they require extensive study and preparation over an extended period of time. In some courses there are only a few basic concepts, yet the entire course is dedicated to the application of these few concepts. An example is Statics, a sophomore-level course, in which there are only two fundamental equations germane to the course: the sum of forces equals zero, and the sum of moments equals zero. The entire course involves applying these two equations to various situations, and using the information in engineering applications.

17.4.5 Take Effective Notes

A main component in keeping up with your classes is taking effective notes for each lecture. Effective notes capture the key points of the lecture in a way that allows you to understand them when you are reviewing for the final exam three months later. Suggested note-taking strategies follow:

1. Skim the assigned reading prior to class to help identify key points.
2. Take enough notes to capture key points but don't write so much that you fall to adequately listen to the presentation in class.

3. Review your notes after class to annotate them, filling in gaps so they will still make sense later in the semester.
4. Review your notes with other students to ensure that you captured the key points.
5. Review your notes early in the semester with your professor to be certain you are capturing the key points.

In class, your job is to record enough information to allow you to annotate your notes later. You don't have to write everything down. Getting together with classmates after class is a terrific way to annotate your notes. That way you have different perspectives on what the main points were. If you really want to make sure you captured the key points, go and ask your professor. Doing so early in the semester will not only set you up for successful note taking throughout the semester, it will also allow you to get to know your professor.

17.4.6 Work Lots of Problems

Because the applications of the concepts are the core of most of your classes, the more problems you can work, the better prepared you will be. In math, science or engineering courses, doing the assigned homework problems is a minimum. Search for additional problems to work. These may be from the text or from old exams. Your professor should be able to steer you to appropriate problems to supplement the homework.

17.4.7 Caution Using Solution Manuals and Files

For some classes, there are homework files or solution manuals available. While these tools may help get the homework done quicker, they very often adversely affect exam performance. The problem is that most professors don't allow the solutions to be used during the test. Using them becomes a crutch and can impair your ability to really learn the material and excel on exams. If they are used at all, use them only after exhausting all other possibilities of working out the answer yourself. If it takes a lot of work to figure out a homework problem, that concept will be learned well and you will have a higher probability of demonstrating your knowledge on the next test. Taking the easy way out on homework has a very consistent way of showing up on test results. Minimal learning results from copying down a solution.

17.4.8 Group Studying

A better model for studying than using solutions is to study in a group. Numerous studies have shown that more learning takes place in groups than when students study by themselves. Retention is higher if a subject is discussed, rather than just listened to or read. If you find yourself doing most of the explaining of the ideas to your study partners, take heart! The most effective way to learn a subject is to teach it. Anyone who has taught can confirm this anecdotally; they really learned the subject the first time they taught it.

> *"The most important single ingredient to the formula of success is knowing how to get along with people."*
>
> —Theodore Roosevelt

If you aren't totally convinced of the academic benefits of group studying, consider it a part of your engineering education. Engineers today work in groups more than ever before. Being able to work with others effectively is essential to being an effective engineer. If you spend your college years studying in groups, working in groups will be second nature when you enter the workforce.

Studying with others will also make it more bearable, and even fun, to study. There are numerous stories about students preparing for exams and spending the whole day on a subject. It is difficult to stick to one subject for an entire day by yourself. With study partners you will find yourself sticking to it and maybe even enjoying it.

Studying with others makes it easier to maintain your study commitments. Something we all struggle with is discipline. It is much easier to keep a commitment to study a certain subject if there are others who are depending on you. They will notice if you are not there and studying.

> *"I will pay more for the ability to deal with people than any other ability under the sun."*
>
> —John D. Rockefeller

Group studying is more efficient if properly used. Chances are that not everyone in the group will be stuck on the same problems. So right away, you will have someone who can explain the solution. The problem areas common to everyone before the group convenes can be tackled with the collective knowledge and perspectives of the group. Solutions are found more quickly if they can be discussed. More efficient study time will make more time for other areas of your life!

Choosing a group or partner may be hard at first. Try different study partners and different size groups. Ray Landis, Dean of engineering at California State University, Los Angeles and a leading expert on student success, suggests studying in pairs. "That way each gets to be the teacher about half the time." Larger groups also can be effective. The trick is to have a group that can work efficiently. Too large a group will degenerate into a social gathering and not be productive in studying. Whichever group size you choose, here are some basic tips for group studying.

1. Prepare individually before getting together.
2. Set expectations for how much preparation should be done before getting together.
3. Set expectations on what will be done during each group meeting.
4. Find a good place to convene the group that will allow for good discussion without too many distractions.
5. Hold each other accountable. If a group member is not carrying his or her weight, discuss it with that person and try to get them to comply with the group's rules.
6. If a member continues to fail to carry their own weight or comply with the group's rules, remove that person from the group.

17.4.9 Select a Good Study Spot

In a subsequent section, differences in learning styles will be discussed. Depending on your own learning style and personality, you may need a certain kind of environment to study efficiently. Some resources attempt to describe the "perfect study environment." In reality, there is no one perfect study environment or method of studying. What is crucial is that you find what you need and find a place that meets your needs. For some, it involves total quiet, sitting at a desk. Others may prefer some noise, such as background music, and prefer to spread out on the floor. Whatever you decide is the best environment, pick one where you can be effective.

Cynthia Tobias describes the conflict she has with her husband's view of a proper study environment in her book *The Way They Learn:*

> *I have always favored working on the floor, both as a student and as an adult. Even if I'm dressed in a business suit, I close my office door and spread out on the floor before commencing my work. At home, my husband will often find me hunched over books and papers on the floor, lost in thought. He is concerned.*
>
> *"The light is terrible in here!" he exclaims. "And you're going to ruin your back sitting on the floor like that. Here, here! We have a perfectly clean and wonderful rolltop desk." He sweeps up my papers and neatly places them on the desk, helps me into the chair, turns on the high-intensity lamp, and pats my shoulder.*
>
> *"Now, isn't that better?" he asks.*
>
> *I nod and wait until he is down the hall and out of sight. Then, I gather all my papers and go back down on the floor... It does not occur to him that anyone in their right mind could actually work better on the floor than at a desk, or concentrate better in 10-minute spurts with music or noise in the background than in a silent 60-minute block of time.*

Different people have different ways in which they can be most efficient. You need to discover yours so that you can be efficient. If you are unsure of what works best for you, test some options. Try different study environments and keep track of how much you get done. Stick with the one that works best for you.

If you and your roommate have different styles and needs, then you will need to negotiate to decide how the room will be used. Both of you will have to be considerate and compromise, because it is a shared space.

One item that is a distraction for almost everyone is television. Because it is visual, it is almost impossible to concentrate on homework with a TV on, although many students swear they are still "productive." A few may be, but most are not. Watching TV is one of the biggest time-wasters for students. It is hard enough for first-year students to adjust to college life and the rigors of pre-engineering classes. We strongly recommend not having a television set in your residence hall room or apartment for at least your first year. It is not even good for study breaks, which should be shorter than 30- or 60-minute programs. This may be one temporary sacrifice you have to make to be successful in your studies.

17.5 TEST-TAKING

Most courses use written, timed exams as the main method of evaluating your performance. To excel in a course, you need to know the material well and be able to apply it in a test situation. The early part of this chapter provided tips on how to improve your understanding of the course material. The second part is test-taking skills. Ask any student who has been away from school for a few years. Taking tests is a skill and needs to be practiced. Just like a basketball player who will spend hours practicing lay-ups, it is important to practice taking tests. A great way to do this is to obtain past exams. A couple of days before the test, block out an hour and sit down and "take" one of the old tests. Use only the materials allowed during the actual test. For instance, if it is a closed-book test, don't use your book. In this way, you are doing two things. The first is assessing your preparation. A good performance on the practice test indicates that you are on the right track. The second thing you are doing is practicing the mechanics of taking the test. Again, with the basketball analogy, players will scrim mage in practice to simulate game conditions. You should simulate test conditions. What do you do when you reach a problem you can't do? Most people will start to panic, at least a little bit. It is critical that you stay calm and reason your way beyond whatever is blocking you from doing the problem. This may mean skipping the problem and coming back to it, or looking at it in another way. In any case, these are the types of things you have to do on tests, and by practicing them under test conditions, you will do better on the actual tests.

Before the first test in the course, visit your professor and ask him or her how to assemble a simulated test. It may be that old exams are the way to go, or possibly certain types of problems will be suggested. Professors don't like the question "What is going to be on the test?" but are much more receptive to requests for guidance in your preparation. The professor's suggestions will also help you focus your studying, to make it more efficient.

17.5.1 Taking the Test

There are some general guidelines for taking any test. The first one is to come prepared to take the test. Do you have extra pencils just in case? Can you use your calculator efficiently? Breaking your only pencil or having your calculator fail during the test can be very stressful and prevent you from doing your best, even if you are given a replacement. Proper planning for the test is the first step toward succeeding.

The second step is to skim over the entire test before beginning to work the problems. This is partially to make sure that you have the complete test. It also gives you an overview of what the test is going to be like so you can plan to attack it efficiently. Take note of the weighting of the points for each problem and the relative difficulty.

The next step is to look for an easy problem or one that you know you can do. Do this problem first. It will get you into the flow of doing problems and give you confidence for the rest of the test. Also, doing the problems you know you can do will ensure that you have made time for the problems for which you should get full credit.

Always keep track of the time during a test. Tests are timed, so you need to use the time efficiently. Many students will say they "just ran out of time." While this may prevent you from finishing the test, it should not prevent you from showing what you know. Look at the number of points on the test and the time you have to take the test. This will tell you how long to spend on each problem. If there are 100 points possible on the test and 50 minutes to complete it, each point is allotted 30 seconds. So a 10-point problem should only take 5 minutes. A common mistake is wasting too much time struggling through a difficult problem and not even getting to others on the test. Pace yourself so you can get to every problem even if you don't complete them. Most instructors will give partial credit. If they do, write down how you would have finished the problem and go on to the next when the allotted time for that one is up. If there is time at the end, you can go back to the unfinished problem. But if not, the instructor can see that you understood the material and can award partial credit.

Remember that the goal of a test is to get as many points as possible. A professor can't give you any credit if you don't write anything down. Make sure you write down what you know and don't leave questions blank. Another common mistake occurs when students realize they have made a mistake and erase much of their work. Often, the mistake they made was minor and partial credit could have been given, but none is awarded because everything was erased.

Concentrate on the problems that will produce the most points. These are either problems you can do relatively quickly and get full credit for, or ones worth a large portion of the total points. For instance, if there are four problems, three worth 20 points and one worth 40 points, concentrate on the 40-point problem.

Leave time at the end, if possible, to review your answers. If you find a mistake and there is not enough time to fix it, write down that you found it and would have fixed it if you had time. If you are taking a multiple-choice test, be careful about changing your answers. Many studies find that students will change right answers as often as they change wrong answers. If you make a correction on a multiple-choice test, make certain you know that it was wrong. Otherwise, your initial answer may have been more likely to have been correct.

Think. This may sound basic, but it is important. What is the question asking? What does the professor want to see you do? When you get an answer, ask yourself if it makes sense. If you calculate an unrealistic answer, comment on it. Often a simple math error produces a ridiculous answer. Show that you know it is ridiculous but that you just ran out of time to find the error.

17.5.2 After the Test

After you get your test back, look it over. It is a great idea to correct the problems you missed. It is much easier to fill in the holes in what you missed as you go along in the semester than to wait until you are studying for the final exam. Most final exams cover the material from all the semester tests. If you can't correct the problems yourself, go and see the professor and ask for help. Remember, you ***will*** see the material again!

17.6 MAKING THE MOST OF YOUR PROFESSORS

One of the most under-utilized resources at a university is the faculty. Many students do not take full advantage of the faculty that they are paying to educate them. The professor is the one who decides on course material, what is important and what is not, and how you will be evaluated. Yet few students take the time to get to know their professors. Here are some reasons to get to know your professors:

1. They are professionals and have experience in the fields you are studying and, therefore, have perspectives that are valuable to you.
2. Every student will need references (for scholarships, job applications or graduate school applications) and professors can provide them.
3. They are the ones in charge of the course and can help you focus your studying to be more effective.
4. They are the experts in the field and can answer the hard questions that you don't understand.
5. They assign your grades, and at some point in your college career, you may be on the borderline between grades. If they know you and know that you are working hard, they may be more likely to give you the higher grade.
6. They are likely to know employers and can provide job leads for full-time positions or internships.

7. They may be aware of scholarship opportunities or other sources of money for which you could apply.
8. They can provide opportunities for undergraduates to work in labs, which is great experience. You may even get paid!

It is interesting to examine the history of the university. In centuries past, a student did not go to a university to get "trained" to do a job. They went to be mentored by great scholars (the faculty). In modern universities, many students come looking primarily for the training needed to get a good job, and getting mentored by faculty is low among their priorities. Consider this: a school could send you the course materials at home and have you show up for one day to take your exams. Yet students still come to a university to study. Why? A main reason is for the interaction with the people at the university—faculty, staff, and other students.

If there are so many reasons to get to know professors, why don't most students do so? Many professors don't encourage students to come see them, or don't present a welcoming appearance. Professors may remind students of their parents, people who seem to be out of touch with college students. It may also be that many students don't really understand the advantages of getting to know their professors.

Here are some things to keep in mind when getting to know professors. The first is that they have spent their careers studying the material you are covering in your courses and, therefore, they enjoy it. So it would be a very bad idea to start by telling your calculus professor that you find math disgusting. This is insulting and suggests he has wasted his career.

The second is that professors teach by choice, especially in engineering. Industry salaries are higher than academic salaries for engineering Ph.D.'s, so professors teach because they want to. They also probably think that they are good at it. If you are having a problem with an instructor's teaching style, approaching him or her with a problem solving strategy such as the following can be very constructive: "I am having trouble understanding the concepts in class. I find that if I can associate the concepts with applications or examples I understand them better. Can you help me with identifying applications?"

As a group, professors also love to talk, especially about themselves and their area of expertise. They have valuable experience that you can benefit from. Ask them why they chose the field they did. Ask them what lessons they learned as students. You will get some valuable insights and also get to know someone who can help you succeed in your class, and possibly in your career. Steven Douglass, in his book *How to Get Better Grades and Have More Fun* suggests asking each professor, "What is the main objective in the course?"

Faculty members are very busy. They have many demands on their time (research, securing grants, writing, committee assignments, etc.) in addition to your class. These other demands are what keep the university running. So respect their time. It is okay to get to know them but don't keep popping in just to shoot the breeze. Also, when you come with questions, show them the work you have been doing on your own; you only need to get past a hurdle. A busy faculty member won't mind helping a hard-working student clear up an idea or concept. However, showing up and asking how to do the homework can get interpreted as "Do my homework for me." This would not make the positive impression you want.

17.7 LEARNING STYLES

In the last several decades scientists have made many discoveries regarding the complexities of the human brain. While the brain is an important organ in each person's body, we now know that each person's brain is unique to him or her. Even before birth our genetic code, as well as our environment after birth, shapes our developing brain. Proper nutrition, stress, drugs and alcohol are some of the factors that can affect a developing brain.

Each person is born with all the brain cells, or neurons, that they will ever have. (Estimates reach as high as 180 billion neurons.) Neurons differ from other cells in that they do not reproduce. However, they do form connections with other neurons. As they are used, neurons can grow tentacle-like protrusions called dendrites which can receive signals from other neurons. In the first five years of life, with normal stimulation, a multitude of connections are created as the brain learns to move the body, speak, analyze, create, and control. However, people can continue to create more neurological connections as they continue to use their brain and learn new material. With the exception of diseases, like Alzheimer's, or physical injury, the brain is capable of creating new connections throughout our lifetime.

This is great news! None of us is ever too old or too dumb to learn something new. Another benefit of brain research is that it has identified several different ways that people think and memorize. Memorizing refers to how people assimilate, or add, new material to existing knowledge and experience. It also includes how we accommodate, or change our previous way of organizing material. Thinking refers to how we see the world, approach problems and use the different parts of our brain. One method does not fit all when it comes to how we learn best, so time spent identifying your own strengths and weaknesses will prove very beneficial. Once you identify the aspects of your own preferred learning style, information can be manipulated so that you learn more information in less time with higher rates of recall.

17.7.1 Memory Languages

Place a check mark by all the statements that strongly describe your preferences.

Auditory

______ I enjoy listening to tapes of lectures.
______ I often need to talk through a problem aloud in order to solve it.
______ I memorize best by repeating information aloud or explaining it to others.
______ I use music and/or jingles to memorize.
______ 1 remember best when information fits into a rhythmic pattern.
______ I would rather listen to the recording of a book than read it.
______ I remember names of people easily.

Visual

______ I follow pictures or diagrams when assembling something.
______ I am drawn to flashy, colorful, visually stimulating objects.
______ I prefer books that include pictures or illustrations with the text.
______ I create mental pictures to help me remember.
______ I usually remember better when I can see the person talking.
______ I remember faces of people easily.
______ I prefer to follow a map rather than ask for directions.

Kinesthetic

_______ I can memorize well while exercising.
_______ I usually learn best by physically participating in a task.
_______ I almost always have some part of my body in motion.
_______ I prefer to read books or hear stories that are full of action.
_______ I remember best when I can do something with the information.
_______ I solve problems best when I can act them out.

You probably checked some in each of the three areas; however, most people have a preferred modality for learning new material. See which area you checked most often and develop ways to study within that area. For example, if you are an auditory learner, buy a small tape recorder and record your lectures to listen to over and over again—perhaps while going to and from class. In lecture, sit where you can hear the professor well. Try focusing on what is being said during the class and taking notes on the material from your tape recording later. Ask the professor questions during class and office hours. Read your assignments out loud to yourself, or into your tape recorder. Keep visual distractions in your study place to a minimum so you can focus on active listening, to your own voice or to a tape. Create jingles, songs, poems, or stories to help you remember. When studying with others, have them drill you on what you're learning, or spend time explaining it to them.

If you are a visual learner, you will benefit from a different set of strategies. During class make sure you sit where you can see the professor and the board or screen clearly. Write notes during lecture with plenty of pictures and meaningful doodles. Rewrite notes later in a more organized fashion, highlighting the main ideas emphasized in class. Create diagrams, lists, charts, and pictures to help clarify material. Write out questions that you have to ask the professor, either in class or during office hours. Highlight main ideas in your books as you read, and don't be afraid to write notes in the margins. Use more than one color to emphasize different things. Use mental images to help you remember how to solve problems. When studying with others, have them give you written practice problems and tests. Use mental maps to illustrate how ideas fit together or to plan upcoming papers or projects. Use video material to reinforce concepts and events.

Example 17.3: I sat down next to a Ph.D. student for a lecture. As we got ready for class, she prepared a little differently. Instead of taking out one writing instrument, she had four—she held four different-colored pens in her left hand. She would take one at a time and use them with her right hand to write. Different ideas had different colors. As I watched, she drew diagrams between points with the different colors. Her notes looked like a work of art. She was one of the most visual learners I have ever seen. Early in her academic career she struggled until she developed methods for succeeding. And she has succeeded, all the way through her Ph.D. program in engineering.

Kinesthetic learners may have the biggest disadvantage, because most formal college learning environments lack kinesthetic activities. The exception to this would be classes with labs. Take advantage of these as often as possible, since doing something with the information you are learning is easiest for you. During lecture, make connections between what is being said and similar things you have done in the past. Talk with your professor during office hours about ways to be more hands-on with the material. Perhaps you could volunteer in your professor's lab, or use models and experiments on your own at home. Try memorizing while exercising, especially repetitive activities like jogging, running stairs, jumping rope, or swimming laps. When studying with others, try acting out the material. Shorter, more frequent study times may be more productive than a single marathon session.

17.7.2 Thinking Skills

The way the brain works is much more complex than simply by adding new material to old. Thinking refers to how we see the world, approach problems and use the different parts of our brain. As science has rapidly improved its ability to measure the brain's activity, our understanding of brain functions has dramatically changed. As research continues, our understanding of this wondrously complex organ is continuing to evolve.

Models have been developed to try to categorize how different individuals think. Early models focused on the specialization of the right and left hemispheres of the brain. Further research has unlocked other complexities and required the development of more complex models. A common current model was developed by Ned Herrmann and shows the brain as four quadrants. Originally, these were four physiological components of the brain. As research continued, it was determined that brain function was too complex to be precisely represented in this format. However, the model has proven effective as a representation of our preferred thinking style and the four quadrants are currently used as a metaphorical model of how we prefer to process information. Currently, the four quadrants are used as a metaphorical model.

Figure 17.1 shows the evolution of the quadrant model from the hemispherical models. Quadrants A and B represent the "left brain" and quadrants C and D represent the "right brain." Herrmann's complete model is detailed in his book *The Creative Brain*. These quadrants illustrate personal preferences and a dominant thinking style. However, most people exhibit characteristics of each quadrant to some degree. Successful people are aware of their strong areas and can capitalize on them while compensating for their weaker areas, and possibly building them up.

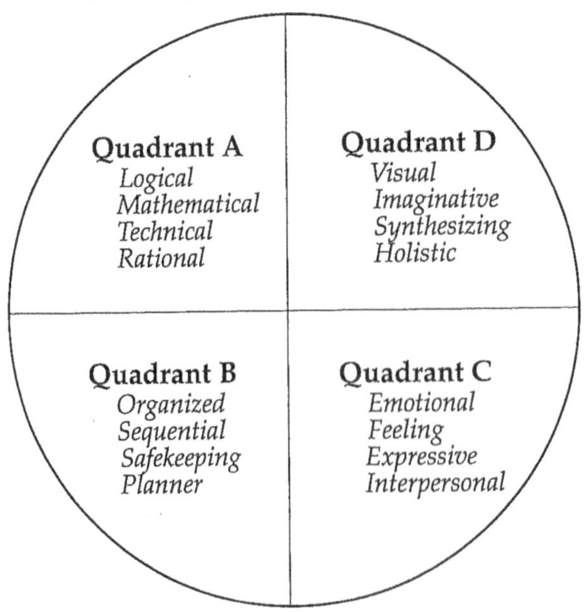

Figure 17.1. Herrmann's "whole brain model" describing four quadrants of the brain.

(Copyright Herrmann International.)

Tables 17.1–17.4 highlight examples of the various thinking skills, preferred learning activities and practice activities for each quadrant. Examine the lists and try to place yourself in one or more quadrants.

The most successful students pick and choose a variety of strategies that work best for their learning style and specific classes. Observe what other successful students are doing and learn from their example. Remember, your one-of-a-kind brain is very capable of learning new material. Spend some time learning how you learn best and you've paved the way for successful learning.

The model can also be used to help understand professors who have a different thinking style. Most people teach like they would like to be taught. If you are in a class where your professor has a very different style than you do, use the model to problem-solve ways to get past this hurdle. It is great experience for later in life. In your professional career, you will have to deal with all different styles effectively to be successful. Being aware of the differences and being able to work with them will be a tremendous asset.

If you would like to know which quadrants apply to you, you can take an assessment profile called the Herrmann Brain Dominance Instrument. An academic advisor or testing center on campus should be able to provide you with information about the HBDI. You may also contact Herrmann International at 794 Buffalo Creek Road, Lake Lure, North Carolina 28746, or at their Web site at *www.hbdi.com*.

TABLE 17.1 Quadrant A

Thinking	Preferred Learning Activities	Practice Activities
Factual Analytical Quantitative Technical Logical Rational Critical	Collecting data and information Listening to lectures Reading textbooks Studying example Problems and solutions Using the scientific method to do research Doing case studies Dealing with things rather than people	Develop graphs, flow charts and outlines from info and data Do a library search on a topic of interest Join an investment club Do logic puzzles or games Play "devil's advocate" in a group decision process Break down a machine and identify the parts and their functions

TABLE 17.2 Quadrant B

Thinking	Preferred Learning Activities	Practice Activities
Organized Sequential Controlled Planned Conservative Structured Detailed Disciplined Persistent	Following directions Doing repetitive, detailed homework problems Doing step-by-step lab work Using programmed learning and tutoring Listening to detailed lectures Making up a detailed budget Practicing new skills through repetition Making time management schedules Writing a "how to" manual	Cook a new dish following a complicated recipe Organize a desk drawer or closet Prepare a family tree Be exactly on time all day Assemble a model kit by instructions Develop a personal budget Learn a new habit through self-discipline Set up a filing system for your paper work

TABLE 17.3 Quadrant C

Thinking	Preferred Learning Activities	Practice Activities
Sensory	Sharing ideas with others	Study in a group
Kinesthetic	Learning by teaching others	Share your feelings with a friend
Emotional	Using group study opportunities	Get involved in a play or musical
Interpersonal	Enjoying sensory experiences	Volunteer in your community
Symbolic	Taking people-oriented field trips	Explore your spirituality
	Respecting other's rights and views	Become a penpal to someone from another culture
	Study with background music	

TABLE 17.4 Quadrant D

Thinking	Preferred Learning Activities	Practice Activities
Visual	Getting actively involved	Daydream
Holistic	Using visual aids in lectures	Play with clay, Legos, Skill Sticks
Innovative	Leading a brain storming session	Focus on big picture, not the details of the problem
Metaphorical	Experimenting	
Creative	Doing problems with many answers	Create a logo
Imaginative	Relying on intuition, not facts	Invent a new recipe and prepare it
Conceptual	Thinking about the future	Imagine yourself 10, 20, 50 years in the future
Flexible		Use analogies and metaphors Intuitive in writing

17.8 WELL-ROUNDED EQUALS EFFECTIVE

Being an effective person goes beyond being a good student. Developing the habits that make you truly effective during your college years will set you up for success in life. Part of being effective is functioning at full capacity. To do this, you must have the various dimensions of your life in order. One analogy that is frequently used is to look at a person as a wheel. A wheel will not roll if it is flat on one side. Similarly, people cannot function optimally if there is a problem area in their life. Five key areas are graphically represented in Figure 17.2. It is important to maintain each area to become as effective as you can be.

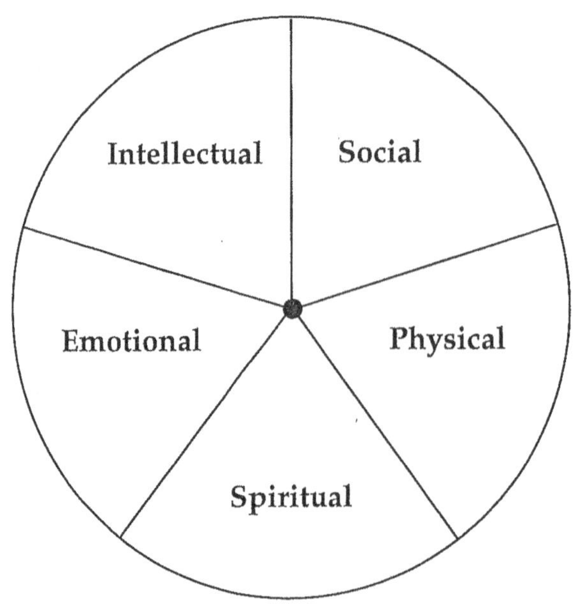

Figure 17.2. Areas of wellness for students

17.8.1 Intellectual

One of the main purposes of going to college is to expand the intellectual dimension of your life. Take full advantage of the opportunities at your institution. Besides engineering courses, schools will require general education courses (languages, humanities or social science classes). These non-technical classes are a great way to broaden yourself. Many students find these classes a refreshing break from their engineering classes.

In addition to course work, colleges have a wide range of activities to enhance your intellectual development. These range from student organizations to seminars to special events. Planning some non-engineering intellectual activities will help to keep you motivated and feeling fresh for your engineering classes. Some students find that reading a novel or two during the semester also provides a refreshing, intellectual break from engineering studies.

17.8.2 Social

The social aspect of college is one that most students seem to master. Planning appropriate social activities will make your college experience more fun. There is a strong correlation between enjoying your college experience and being successful. Also, establishing a group of friends to socialize with will provide a support structure to help you when times get tough. College friends also provide a network that can be very beneficial later in your professional life.

If you have trouble identifying appropriate social activities, there are campus resources to help you. The first place to go is your residence hall staff. Also, most campuses have a central office for student activities. This office can provide you with a list of student organizations and activities.

17.8.3 Physical

All the studying and preparation is a waste of time if you are not physically able to perform when the tests or quizzes come. Also, you can waste time if you are not able to study efficiently. There are three physical areas that are essential to becoming an effective student. They are fitness, sleep and nutrition.

People are more productive when they are physically fit, and students are no exception. Exercise is an activity which will not only help you study better but will allow you to live a longer, more productive life. College is a great time to develop a fitness habit. Fitness activities can include aerobics, jogging, biking, walking or participating in intramural sports. On every campus there are people who can help you develop a fitness plan that is right for you.

Sleep is an area that many students abuse. Finally they are away from their parents and can go to bed any time they want. The problem is that classes still come at the same time regardless of when you fall asleep. Sleep deprivation reduces a person's productivity, which includes studying. Each person needs a different amount of sleep. Typically, most adults need six to eight hours of sleep. It is important to schedule enough time for sleep or you won't be able to be effective in the classroom. Studies have shown that student performance is actually reduced by staying up very late the night before a test to study. You cover more material the night before the test but are so tired at test time that you can't retrieve the material. Honestly evaluate how much sleep you need. Do you fall asleep in class? If so, you probably aren't getting enough rest. Another test for sleep deprivation is to not set an alarm one morning and see when you wake up. If you are on a schedule, most people will wake up at the same time without the alarm. If you sleep several hours past when the alarm would have gone off, you very possibly are sleep-deprived and need more rest.

The third physical area is diet. Again, Mom and Dad are not there to make you eat that broccoli, so you don't. Our bodies are designed to work properly with the right input. While you don't have to become a diet fanatic, eating a sensible diet will enhance your ability to succeed in your academics.

A final note on the physical dimension pertains to non-dietary substances, such as alcohol. The number one problem on most college campuses is alcohol abuse. Every year, thousands of very capable students fail due to alcohol. If you choose to indulge, make honest assessments of the impact on your studying. Don't let alcohol become the barrier to your success.

17.8.4 Spiritual

Few would disagree that humans are spiritual beings and function much better and more efficiently with a balanced spiritual component in their lives. Stephen Covey, in his book *The Seven Habits of Highly Effective People*, puts it this way:

> *The spiritual dimension is your core, your center, your commitment to your value system. It's a very private area of life and a supremely important one. It draws upon the sources that inspire and uplift you and tie you to the timeless truths of all humanity. And people do it very, very differently.*
>
> *I find renewal in daily prayerful meditation on the Scriptures because they represent my value system. As I read and meditate, I feel renewed, strengthened, centered and recommitted to serve.*

Search for the spiritual activities that keep you renewed. This may be through a small group study, attending campus worship services, meditating or immersing yourself in good music for a time of personal reflection. If you aren't sure where to start, try something which is close to what you were used to before college. This will probably be the most comfortable for you. If it doesn't seem to be the right thing for you, branch out from there.

17.8.5 Emotional

Bill Hybels, pastor of a suburban Chicago church, explained the need to attend to this dimension when he addressed a gathering in 1996 in Detroit. He is an avid runner, eats well, and is physically sound. He has many casual friends and is socially active. He has written a number of books and articles and is sound intellectually. Yet he reached an emotional crisis that prevented him from being effective in his job and with his family.

Hybels describes emotional reserves like a gas tank. We need to put reserves into our tank so we have the ability to take it out when we need it. There will be times in your college career, and beyond, when you need those reserves. The lesson Hybels shared is the need to watch the gas gauge and not to let it run out. Students are no different.

A positive step to take to keep the emotional dimension in check is to set up a support network. This network is what you use to make deposits into your emotional tank. Friends are an integral part of the network. Do you have a person or people whom you can talk honestly with about important issues in your life? This might be a close friend, a parent, a relative, an advisor, a professor or a counselor. Cultivate at least one friendship where you feel the freedom to share honestly.

The second aspect to maintaining emotional reserves is to schedule time that is emotionally neutral or energizing. Many students get themselves into a lot of activities that are emotionally draining and don't schedule time for themselves for recovery. Eventually this catches up with them and adversely impacts their effectiveness.

A final note on this area is that on each campus there are advisors and counselors who are available to talk to you. Don't feel that you have to have a severe psychological problem to simply talk with them. These people are trained to help you and will just listen if that is what you need. They are a great resource for maintaining the emotional dimension and thereby helping to keep your academic effectiveness high.

17.9 YOUR EFFECTIVE USE OF TIME

You may be wondering how to find time to maintain all these areas every day: studying two hours for each hour spent in class, keeping a balanced lifestyle, etc. And what happened to all that fun time you've heard about? Aren't the college years supposed to be the best years of your life?

The answer is effective time management. Time management is a skill that will help you succeed as a student and will serve you well throughout your entire career. Mastering time management will set you ahead of your peers. It is interesting that when engineers and managers are surveyed and asked why they don't take the time to manage their time, the most common response is inevitably that they don't have time. Students have the same feelings.

If we consider the academic requirements with a 15-credit load, we will have 45 hours per week of class and study time. That could be done Monday through Friday from 8 a.m. until 6 p.m., assuming you take an hour for lunch every day. Think about it. That would give you every evening and every weekend free to have fun or to do other activities.

Many students don't believe they can do this, but the hours add up. The key to making such a schedule work is to make every minute count. Treat school like a full-time job, working similar hours. This will not only help you to effectively manage your life during your academic years, but it also will provide excellent preparation for your future work life.

Let's assume that you have a schedule like the one in Table 17.5. This student has Chemistry, Math, English, Computer Science and an introduction to Engineering course. On Monday, the day starts at 8 a.m. with two classes back to back. Then there is an hour break before Computer Science. If you are on the job, you need to be working, and

TABLE 17.5 Sample Class Schedule

	Mon	Tues	Wed	Thurs	Fri
8:00	English		English		English
9:00	Chem 101		Chem 101		Chem 101
10:00		Chem Lab		CS Lab	
11:00	CS 101	Chem Lab	CS 101	CS Lab	
12:00		Chem Lab			
1:00				Chem Rec	
2:00	Math	Math	Math		Math
3:00		Engr101		Engr101	
4:00					
5:00					
6:00					
7:00					
8:00					
9:00					

we'll define working as attending class or studying. You need to find a place to study during this hour break. This is where many students go astray. They will waste an hour like this between classes. Similarly, there is a two-hour time block before Math class. This is time to grab lunch and get some more studying done.

It will take some planning to make your day effective because you will need to find suitable places to study. You can use these times to do your individual studying or meet with a study partner or group. It is much more efficient to decide in advance what you will study and when. It can be a waste of precious time to try decide which subject you will study during breaks between classes. Plan enough time for each subject so you can study from class to class and not fall behind in any way. For example, you can't do all your studying for Chemistry on Monday, Math on Tuesday and Computers on Wednesday or you will fail behind the lectures.

Another thing that scares many students away from effective time management is that it sounds too rigid. One of the reasons for going to college is to have fun and meet people. How can you "go with the flow" if everything is scheduled? The answer is responsible flexibility. If you have your schedule planned out, you can decide when it's a good time to interrupt your "work time" to have some fun. A criterion for making this decision is recovery time. In your schedule, is there enough non-work time to convert to work time before deadlines arrive in your classes? For instance, taking a road trip the day before an exam would be irresponsible. However, right after you have midterms, there might be a window of time to take such a break and still recover.

Scheduling two hours of study per class period is usually enough to keep up during your first year. However, there are classes or times in a semester when this won't be enough. Once you fall behind, it is hard to catch up. To avoid this, a weekly make-up time is an excellent idea. This is like an overtime period at work. At an aerospace company where I worked, Saturday morning from 6 to 10 was my favorite time to schedule this overtime work. This didn't interfere with anything else in my normal schedule and I could get the work done and still have a full Saturday to enjoy. For some reason, 6 a.m. on Saturday isn't a very popular time for students. However, an evening or two or Saturday afternoon will work, too. These extra time blocks will allow you to keep up and give you the flexibility to take time off with a friend for something fun but still stay on your weekly schedule.

17.9.1 Scheduling Work and Other Activities

No matter how good you get at squeezing in study time between classes, there are still only 24 hours in each day. As a result, there is a limit to what anyone can do in one week. Planning a schedule complete with classes, study time (including make-up time), eating, sleeping and other activities is essential to deciding how much is too much. If you need to work while you're in school, schedule it into your week. If you join a student organization, plan those hours into your schedule. You need to see where it all fits. A common pitfall with students is that they over-commit themselves with classes, work and activities. Something has to give. It is easier to drop an activity than to have to repeat a class.

Ray Landis, from California State University at Los Angeles, describes a 60-hour rule to help students decide how much to work. Fifteen credit-hours should equate to 45 hours of effort. That leaves 15 hours for work, totaling 60 hours. Students may be able to work more than this, but each individual has a limit. You need to learn for yourself what your limit is.

If you need to work to make it through college financially, you may need to look at the number of credits you can handle effectively each semester. A schedule can help guide you in determining what is manageable. Before deciding to take fewer credits so you can work more hours, examine the value of the money you can make at your job compared with the starting salary you would expect by graduating a semester earlier. In some cases, it will make sense to work more while taking fewer classes. In others, it will make financial sense to take out a loan, graduate sooner, and pay the loan back with your higher salary.

17.9.2 Calendars

It is important to have a daily and weekly calendar. Another critical item in effective time management is a long-range schedule. For students, this is a semester calendar. This will help you organize for the crunch times, like when all five of your classes have midterms within two days, or you have three semester-projects due on the same day at the end of the semester. It is a fact of student life that tests and projects tend to be bunched together. That is because there are educationally sound times to schedule these things, and most professors follow the same model. Given this fact, you can plan your work schedule so that you spread out your work in a manageable way. This is the part you have control over.

Once, while I was in school, we complained to a professor about this on the first day of class. He offered to give the final exam in the next class period and have the semester project due the second week of the semester to free us up for our other classes. We declined his offer and decided it was best to learn the material first.

17.9.3 Organizers: Paper or Electronic?

Calendars and task lists can be kept in numerous forms, and you will need to determine which is most effective for you. There has been a proliferation of electronic tools and gadgets that are designed to supplement or replace paper calendars and organizers. Among the most common electronic options are PDA's (Personal Digital Assistants) which can function like a personal organizer as well as a cellular phone, fax sender, and Internet portal. PDA's began with pen-based input but now incorporate handwriting recognition features as well as keyboard inputs. PDA's are also called palmtops, handheld computers, and pocket computers. Most of the traditional time management companies, such as Franklin Covey, have created electronic versions of their products to be used with PDA's.

Among my colleagues, there are mixed feelings about the use of electronic tools to replace paper tools. Some, including me, have found them useful. Others, however, prefer the traditional calendars or a combination of the two. In a "race," my colleagues who continue to use paper organizers and calendars typically can locate an open date for a meeting more quickly than those who use an electronic organizer.

The media you use to organize your schedule is a personal choice that you have to make. For any system to be effective, it must fit your own needs and style. Our recommendation is that you explore the options before settling on what is best for you, and stay flexible. As new tools are developed, there may be options that enhance your effectiveness and are worth exploring.

17.9.4 Organizing Tasks

A common method for organizing one's day is to make a "To Do" list each morning (or the evening before). This is a good way to ensure that everything needing attention gets it. Things not done on one day's list get transferred over to the next.

The one drawback to making a list is determining what to do first. You can approach the tasks in the order you write them down, but this ignores each task's priority. For instance, if you have a project due the following day for a class and it gets listed last, will it get done? A popular planning tool is the Franklin Planner produced by the Franklin Covey Co. (Material here used courtesy of Franklin Covey.) They stress the importance of prioritizing your life' activities according to your values, personal mission statement, and goals. They suggest using a two-tier approach to prioritizing activities. The first is to use A, B and C to categorize items as follows:

A = Vital (needs attention today)
B = Important (should be taken care of soon)
C = Optional (no one will notice if it is not done today)

Keep your long-term and intermediate goals in mind when assigning priority. And be sure to write simple statements for each task on your list, beginning each item with an action verb.

Once you have categorized the items, look at all the A priorities. Select the item to do first and give it a 1. This item now becomes A1. The next becomes A2, etc. Now you can attack the items on your list starting with A1 and working through the A's. Once you have finished with the A's, you move on to B1 and work through the B's. Often, you will not get to the C priorities, but you have already decided that it is okay if they don't get done that day.

Table 17.6 shows a sample list for a typical student's day. If you are starting a time management routine, include your list. This keeps you reminded of it, and it gives you something to check off your list right off the bat!

Some lessons can be learned by looking at how this student has scheduled activities. One is that paying the phone bill is not necessarily a top priority, but it requires little time to address. A helpful approach is to look at the quickly done A items and get them out of the way first. Then move on to the more time-consuming items. Also note that a workout was planned into the schedule, right after the chemistry lab. The workout is a priority that comes after some extensive studying. This can help motivate the student to finish a difficult task and then move on to something enjoyable.

TABLE 17.6 Sample Prioritized List of Tasks

Priority	Task
A3	Finish chemistry lab (due tomorrow)
A5	Read Math assignment
B3	Do math homework problems
B4	Outline English paper
A4	Workout at Intramural Building
B2	Debug computer program
A6	Annotate Math notes
B1	Annotate chemistry notes
A7	Spiritual meditation
C1	Call Stacy
C2	Write parents
A2	Mail phone bill
C3	Shop for CD's
A1	Plan Day

Looking at this schedule, you may think that poor Stacy is never going to get called or the letter to the parents written. It often happens that the more fun things are, the more likely they end up a C priority. But it is okay to put relationship-building or fun activities high in priority. People need to stay balanced. Maybe today Stacy is a C, but tomorrow might have that call as an A priority.

The Franklin Covey people provide additional help for identifying priorities. They rate tasks by two criteria, urgency and importance, as shown in Table 17.7. Franklin Covey stresses that most people spend their time in Quadrant I and III. To achieve your long-term goals, Quadrant II is the most important. By spending time on activities that are in II, you increase your capacity to accomplish what matters most to you. The urgency is something that is usually imposed by external forces, whereas importance is something you assign.

You may well want to use a schedule that shows a whole week at a time and has your goals and priorities listed. A general example is shown in Table 17.8.

A weekly schedule allows you to see where each day fits. It also keeps your priorities in full view as you plan. Franklin Covey offers many versions of their Planner, including the Collegiate Edition. Contact them at 1-800-654-1776 or visit their website at www.franklincovey.com.

TABLE 17.7 Franklin Covey's 'Time Matrix' for Daily Tasks (copyright 1999 Franklin Covey Co.)

	Urgent	**Not Urgent**
Important	I • Crises • Pressing deadlines • Deadline-driven projects, meetings, preparations	II • Preparation • Prevention •Values clarification • Planning • Relationship building • Recreation • Empowerment
Not Important	• Interruptions, some phone calls • Some mail, some reports • Some meetings • Many pressing matters • Many Popular activities III	• Trivia, busywork •Some phone calls • Time-wasters • 'Escape' activities • Irrelevant mail • Excessive TV IV

17.9.5 Stay Effective

There are two final notes on time management that can help you stay effective. The first is to plan to tackle important activities when you are at your best. We are all born with different biological clocks. Some people wake up first thing in the morning, bright-eyed and ready to work. These people are often not very effective in the late evening. Others would be dangerous if they ever had to operate heavy equipment first thing in the morning, especially before that first cup of coffee. However, once the sun goes down, they can be on a tear and get a lot done. Know yourself well enough to plan most of your studying when you are at your best. It is a common tendency among students to leave studying until late in the evening. This is often because they haven't done it earlier in the day. There are always stories about how the best papers are written at 3:00 a.m. Some are, but most are not. Know yourself well enough to plan effectively.

TABLE 17.8 Sample Generic Weekly Schedule and Planner

Weekly Schedule			Sunday	Monday	Tuesday	Wednesday	Thursday	Friday	Saturday
Musts	**Goals**					**Day's Priorities**			
Academics:				Run laps		Run laps		Finish CS	Run laps
Math	Finish Prob							Finish Math	
Chem lab	Anal, data							Movie	
Comp Sci	Debug Prog								
English	Essay								
						Appointments/Commitments			
		8:00			English		English		English
Hall Dance		9:00	Chapel Synagogue		Chem		Chem		Chem
		10:00				Chem Lab		CS Lab	
		11:00			CS		CS		
		12:00							
		1:00						Chem Rec	
		2:00			Calculus	Calculus	Calculus		Calculus
		3:00			Meet prof		Engr101		Engr101
		4:00							
		5:00							
Physical Mental Spiritual Social Emotional	Run and Ride Read a novel Devotions Hall dance Lunch w/ Dave	Evening							

And finally, pace yourself. You are entering a marathon, not a sprint. Plan your study time in manageable time blocks and plan built-in breaks. A general suggestion is to take a 5-10 minute break for every hour spent studying. This will help keep you fresh and effective.

17.10 ACCOUNTABILITY

The single most important concept to being successful is to establish accountability for your goals and intentions. The most effective way to do this is with an accountability partner. This is someone whom you trust and from whom you feel comfortable receiving honest feedback. Ideally, you will meet regularly with this person and share how things are going. Regularly means about once per week. Throughout a semester you can fall behind quickly, so weekly checks are a good idea. The person you choose can be a peer, a mentor, a parent, a relative, or any one you feel comfortable with, respect and trust.

Your accountability could also be to a group of people. Bill Hybels describes the change in his life when he entered into an accountability group with three other men. These men held each other accountable in all areas of their lives—professional, familial and spiritual. You can do the same as a student.

The first step is to give your accountability partners the standard by which you want to be held. An easy way to do this is to share your goals with them. Let them see and hear from you where you want to be in five years, in ten years and beyond that. Let them also see the short-term goals you have set to achieve those longer-term objectives. Then, weekly, share with them how you are doing in meeting your goals. You will find that it is much easier to stay on course with the help of at least one close friend. Success will come much easier.

17.11 OVERCOMING CHALLENGES

We all at some point in our lives encounter adversity. As a student, adversity may take the form of a bad grade on a test or for an entire course, a professor you feel is treating you unfairly, not getting into the academic program of your choice, or any of a myriad other college-related obstacles. Since everyone encounters adversity and failure at some point in their life, your personal success is not always dependent on how well you avoid failing, but rather on how you learn from it, move on and improve. This is especially true as a student.

Almost every student will run into a course or a test where they struggle. The key is to avoid focusing on the negative and continuing to be productive. Every academic advisor has stories about good students who tripped up on a test or in one class and let that affect their performance in other classes.

17.11.1 Focus on "Controllables"

If something bad happens, figure out what can be done. This may mean writing down a list of actions you can take. Don't focus on things you can't change. If you receive a low grade on a test, you can't change the grade. You may, however, be able to talk to the professor about the test. You can study the material you missed and be ready for the next one.

17.11.2 Keep It in Perspective

Examine the real impact of the situation. What are the consequences of the situation? One class can affect your graduation GPA by about 0.05. This will not produce life-altering consequences, unless you would have graduated with a 4.0! If you are unsure about the consequences, see your professor or an academic advisor and have them help you put it in perspective.

17.11.3 Move to the Positive

It is very important if you hit a bump in the road to move beyond it quickly and remain effective in your studies. It is necessary to examine the problem to learn from it. Perhaps you failed a test. This is a great opportunity to examine how you are studying and take corrective action to improve your effectiveness. You can affect your present and future, but not your past.

17.11.4 Remember, Everyone Fails

Again, keep things in perspective. If you run into difficulty, you are not alone. There are numerous examples of failures.

Henry Ford neglected to put a reverse gear in his first automobile. But he recovered and went on to be successful.

Abraham Lincoln, likely our greatest president, provided a tremendous example of overcoming adversity. Here is a brief list of some of his accomplishments and setbacks;

1831 – failed in business
1832 – defeated for legislature
1833 – failed in business again
1834 – elected to legislature
1835 – fiancée died
1838 – defeated for Speaker
1840 – defeated for Elector
1843 – defeated for Congress
1846 – elected to Congress
1848 – defeated for Congress
1855 – defeated for Senate
1856 – defeated for Vice-President
1858 – defeated for Senate
1860 – elected President

Lincoln never stopped achieving just because he ran into a setback. You don't have to either.

Another example comes from a grandfather who relayed a story about someone in his family failing. "We were visiting your grandmother's brother and his family. It was the time of year when college was in session, so I was surprised when my nephew, Hughey, opened the door and greeted us. He was in his first year of college at the state university. I jokingly said, "What happened to you, Hughey? Flunk out already?" I sure felt stupid when he replied that in fact he had. After that I felt obligated to try to help him out. That evening, he, his father and I talked about what he was going to do. I encouraged him to enroll in the other big state university. I had heard a lot of good things about it even though their science and engineering programs were not as prestigious. He enrolled and did very well. Not only did he graduate, he even got his master's degree before taking a job with a top company."

As a footnote to the story, Hughey continued to do well. The job he took was with a Fortune 100 company. He married and had four children and five grandchildren at last count. He stayed with the same company for his entire career and retired a senior vice president, one step below CEO. Failing in his first attempt at college didn't stop him. He learned from it, moved on and did very, very well. You can do the same with the right attitude and effort.

17.12 REFERENCES

Douglass, S. and A. Janssen, *How to Get Better Grades and Have More Fun*, Success Factors, Bloomington, Indiana, 1997.

Maxwell, J. C., *The Winning Attitude, Your Key to Personal Success*, Thomas Nelson Publishers, Nashville, Tennessee, 1993.

Landis, R., *Studying Engineering, A Road Map to a Rewarding Career*, Discovery Press, Bur- bank, California, 1995.

Tobias, C.U., *The Way They Learn, How to Discover and Teach to Your Child's Strengths*, Focus on the Family, Colorado Springs, Colorado, 1994.

Lumsdaine, E., and M. Lumsdalne, *Creative Problem Solving, Thinking Skills for a Changing World*, 2nd ed., McGraw-Hill, New York, 1993
Herrmann, N., *The Creative Brain*, Brain Books, Lake Lure, North Carolina, 1988.
Covey, S.R., *The 7 Habits of Highly Effective People*, Simon and Schuster, New York, 1989.

17.13 EXERCISES AND ACTIVITIES

17.1. Identify three things you can do to keep a positive attitude toward your studies.

17.2. What could you have done if you were in the Physics class as described in Section 17.2?

17.3. Actually writing goals as described is an important element in defining your goals.
- a) Write five goals to achieve by graduation.
- b) Write five goals to reach within ten years.
- c) Write ten goals to reach during your lifetime.

17.4. Write some specific goals:
- a) Set a grade goal for each class this semester.
- b) Calculate your expected semester GPA.
- c) Define a stretch goal for your semester GPA.

17.5. Find out what the following grade point averages are:
- a) Minimum GPA to stay off academic probation
- b) Minimum GPA to be admitted to Tau Beta Pi
- c) Minimum GPA to be admitted to the engineering honor society in your engineering discipline (e.g., Pi Tau Sigma for mechanical engineering, Eta Kappa Nu for electrical engineering)
- d) Minimum GPA to be on the Dean's List or Honors List
- e) Minimum GPA to graduate with honors

17.6. Identify a particular employer and an entry level position with that employer that you would like to obtain following graduation. Find out what is required to be considered for that position.

17.7. List the barriers you see to achieving your graduation goals.
- a) Categorize them as those within your control and those beyond your control.
- b) Develop a plan to address those within your control.
- c) Develop strategies to cope with those beyond your control.

17.8. For each class you are currently taking:
- a) Map out where you sit in class.
- b) Rate your learning effectiveness (A, B, C, D or F) for each class.
- c) For each class not rated an A, develop a plan to increase your learning effectiveness.

17.9. For one week, make a log of everything you do. Break it up into categories of attending class, studying, working, playing, eating, sleeping, physical activity, intellectual pursuits, emotional health, spiritual well-being, and other.
- a) Total your time spent on each area. Are you spending the time needed on your studies (2 hours for each hour in class)?
- b) Is this the kind of schedule that will make you successful? If not, what will you change?

17.10. Review your current class schedule and identify times and places you can study between classes.

17.11. Develop a "work" schedule. Identify the times you are "on the job," that is, studying and taking classes. Identify a make-up time for weekly catch-up.

17.12. List all your current courses. Put a "+" next to each one in which you are caught up. Put a "–" next to each one in which you are not caught up right now. ("Caught up" is defined as prepared for the next lecture with all assignments completed and all assigned reading done.)

17.13. Ask one of your professors the main objective of his or her course and report the response. Does this information help you in organizing your studying?

17.14. Annotate your notes from a lecture in one of your courses, then take those notes in to your professor to get his or her comments on your effectiveness. Report the professor's comments.

17.15. Make a list of the students you already know who are pursuing the same major as you. Then make a list of the students you know in your science and math classes. Rate each one according to their suitability as a study partner.

17.16. Identify one or more people with whom you might form a study group. Try studying together on a regular basis until the next test. Report on your experience.

17.17. List the group study rules you used in your group studying. Report on the things that went well and those that did not.

17.18. Make a list of problems you can work, in addition to the assigned homework problems, in one of your math, science or engineering courses. Report on the method you used to identify these problems.

17.19. Visit one of your professors and report three personal things about that professor. (For example: Where did she go to school? Why did he choose the field he is in? What does she like best about her job? What one piece of advice does he have for freshmen that he didn't share in class?)

17.20. After completing Exercise 19 above, did this change your feelings about your professor? Does he or she seem more approachable now? Did you learn anything else during your visit that could help you succeed in that class?

17.21. Identify at least one opportunity for you as an undergraduate to become involved in engineering research. What advantages could you gain from this experience?

17.22. Identify the type of environment in which you study the best. Describe it and include a sketch of the location.

17.23. Test your conclusion in Exercise 22 with an experiment. Do one assignment in one environment (perhaps in your room, alone and totally quiet) and another in a different environment (perhaps in a more public place, with background noise). Report your findings.

17.24. At the halfway point in the semester, calculate your grades in each class. Compare these to the goals set in Exercise 4. Are you on track? If not, identify the changes you will make to reach these goals.

17.25. Define at least one long-term goal in each of the five wellness areas discussed in this chapter. Next, define short-term goals to achieve these long-term goals.

17.26. List at least two needs you have in each of the five wellness areas. List at least two activities you can plan to meet these needs.

17.27. Rate your roundness in the wellness wheel (on next page), using 5 for very well, 4 for well, 3 for okay, 2 for some improvement needed, and 1 for much improvement needed. Do you have a well functioning wheel? If not, identify steps to correct the problem areas. List at least two new activities for each problem area.

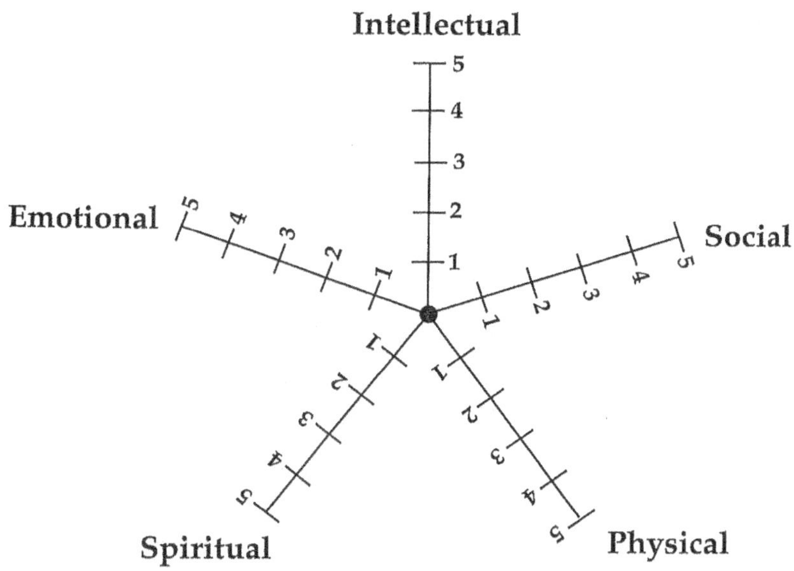

Figure for Excercise 17.27. Wellness assessment.

17.28. Write a paragraph on how each wellness area can help you succeed:
   a) as an engineering student.
   b) as a practicing engineer.

17.29. Make a weekly schedule with two hours of studying for every hour spent in class. Identify activities which will help develop the different areas in your life. Is this a sustainable schedule? If the answer is no, identify things which can be dropped.

17.30. Identify on your weekly schedule the times of each day when you study most efficiently. Are these the times you are actually scheduled to study? If not, explain why you've scheduled other items instead of studying at these times.

17.31. Adopt one of the weekly schedule instruments discussed in this chapter. Try using it for two weeks and report the results. Was this the right instrument for you?

17.32. Make a "To Do" list. Prioritize the list using A, B and C categories, and then number the items in each category.

17.33. Place the items of Exercise 17.32 into the four quadrants of Covey's matrix shown in Table 17.1. How many A priorities are in quadrant II? Based on this ranking of your "To Do" list, would you change any of your priorities?

17.34. List the top three things you are failing to do that would most help you succeed in your career. Briefly explain why you are currently failing to do each item.

17.35. List three people you would consider for an accountability partner.

17.36. Make a list of at least five questions you want your accountability partner to ask you at each check-up.

17.37. Meet with an accountability partner for one month. Report back on your experience. Did it help? Is this something you will continue?

17.38. Call home and tell your parents that you are developing a success strategy for college. Explain the strategy to them.

17.39. Assume that your grades will be lower than expected at the end of the semester, and write out a strategy for recovering from these low grades.

17.40. Make a list of at least three university people who can help you deal with a problem that is preventing you from succeeding. List each by name and position.

17.41. Make a list of your activities over the last 24 hours. Compare the list and the hours spent on each with your planned schedule.

17.42. List your long-term goals, including every area of your life. Fill out a schedule of what you actually did during the last seven days. Put each activity into categories corresponding to one of your long-term goals, or an extra category titled "not related." Grade yourself (using A, B, C, D or F), indicating how well you did the past week working toward your long-term goals.

17.43. Take the inventory test discussed in Section 17.7 on memory languages. Which is your dominant mode? List three actions you can take to improve your learning based on this dominant mode.

17.44. Which of Herrmann's quadrants best describes your dominant thinking style? Which one is the weakest?

17.45. Which of Herrmann's four quadrants is best suited for engineering? Are there any dominant modes of thinking that are not ideal for engineering? Explain your answers.

17.46. List all your current professors and identify their probable Herrmann quadrant. Compare this with your own style. For the professors with styles that differ from yours, develop an action plan to deal with these differences.

17.47. Explore the availability of electronic organizers and prepare a brief report on the options for students.

17.48. Compare and contrast electronic organizers versus paper ones. What are the benefits of each and which would you recommend to your classmates?

17.49. Interview two professors and report what they use to organize and prioritize their time.

APPENDIX A

Solutions to Try Your Skills

Chapter 10

10-A Other information needed includes total number of miles driven each year and the gas consumption (miles per gallon) of the average delivery vehicle.

10-B Some nonmonetary factors (attributes) that might be important are as follows:

- Safety
- Reliability (from the viewpoint of user service)
- Quality in terms of consumer expectations
- Aesthetics (how it looks, and so on)
- Patent considerations

APPENDIX

B Answers to Selected Problems

CHAPTER 14

14.2 239 kW, 378 kW, 315 kW, 7.36 GJ

14.4 38.6 kWh, 43.5 kWh, 33.1 kWh, 35.7 kWh

14.6 $28.30

14.8 56.9 m^2

14.10 8.58 MW

14.12 58.1 kW to 674 kW

14.14 0.113 kWh

14.16 565 MW

14.18 19.9 kW

14.20 0.217

14.22 $\eta_{th,\,idea} = 0.126$, $\eta_{actual} = 0.208$. Proposed geothermal power plant is impossible.

14.24 70°C

14.26 21.7 kW

14.28 1.04 kW, 0.780 kW

14.30 100 MW

References

Chapter 9

1. O. Mayr, *The Origins of Feedback Control*, MIT Press, Cambridge, Mass., 1970.
2. O. Mayr, "The Origins of Feedback Control," *Scientific American*, Vol. 223, No. 4, October 1970, pp. 110–118.
3. O. Mayr, *Feedback Mechanisms in the Historical Collections of the National Museum of History and Technology*, Smithsonian Institution Press, Washington, D. C., 1971.
4. E. P. Popov, *The Dynamics of Automatic Control Systems*, Gostekhizdat, Moscow, 1956; Addison-Wesley, Reading, Mass., 1962.
5. J. C. Maxwell, "On Governors," *Proc. of the Royal Society of London*, 16, 1868, in *Selected Papers on Mathematical Trends in Control Theory,* Dover, New York, 1964, pp. 270–283.
6. I. A. Vyshnegradskii, "On Controllers of Direct Action," *Izv. SPB Tekhnolog. Inst.*, 1877.
7. H. W. Bode, "Feedback—The History of an Idea," in *Selected Papers on Mathematical Trends in Control Theory,* Dover, New York, 1964, pp. 106–123.
8. H. S. Black, "Inventing the Negative Feedback Amplifier," *IEEE Spectrum*, December 1977, pp. 55–60.
9. J. E. Brittain, *Turning Points in American Electrical History*, IEEE Press, New York, 1977, Sect. II-E.
10. W. S. Levine, *The Control Handbook*, CRC Press, Boca Raton, Fla., 1996.
11. G. Newton, L. Gould, and J. Kaiser, *Analytical Design of Linear Feedback Controls*, John Wiley & Sons, New York, 1957.
12. M. D. Fagen, *A History of Engineering and Science on the Bell Systems*, Bell Telephone Laboratories, 1978, Chapter 3.
13. G. Zorpette, "Parkinson's Gun Director," *IEEE Spectrum*, April 1989, p. 43.
14. R. C. Dorf and A. Kusiak, *Handbook of Automation and Manufacturing*, John Wiley & Sons, New York, 1994.
15. W. D. Rasmussen, "The Mechanization of Agriculture," *Scientific American*, September 1982, pp. 26–37.
16. M. M. Gupta, *Intelligent Control*, IEEE Press, Piscataway, N. J., 1995.
17. A. G. Ulsoy, "Control of Machining Processes," *Journal of Dynamic Systems*, ASME, June 1993, pp. 301–307.
18. M. P. Groover, *Fundamentals of Modern Manufacturing*, Prentice Hall, Englewood Cliffs, N. J., 1996.
19. R. T. O'Brien, "Vertical Lateral Control for Automated Highway Systems," *IEEE Transactions on Control Systems Technology*, May 1996, pp. 266–273.
20. J. G. Kassakian, "Automotive Electrical Systems circa 2005," *IEEE Spectrum*, August 1996, pp. 22–27.
21. P. M. Moretti and L. V. Divone, "Modern Windmills," *Scientific American*, June 1986, pp. 110–118.
22. B. Preising and T. C. Hsia, "Robots in Medicine," *IEEE Engineering in Medicine and Biology*, June 1991, pp. 13–22.
23. R. C. Dorf and J. Unmack, "A Time-Domain Model of the Heart Rate Control System," *Proceedings of the San Diego Symposium for Biomedical Engineering*, 1965, pp. 43–47.
24. R. C. Dorf, *Electrical Engineering Handbook*, 2nd ed., CRC Press, Boca Raton, Fla., 1998.
25. R. C. Dorf, *Introduction to Computers and Computer Science*, 3rd ed., Boyd and Fraser, San Francisco, 1982, Chapters 13, 14.
26. K. Sutton, "Productivity," in *Encyclopedia of Engineering*, McGraw-Hill, New York, pp. 947–948.
27. R. C. Dorf, *The Encyclopedia of Robotics*, John Wiley & Sons, New York, 1988.
28. R. C. Dorf, *Robotics and Automated Manufacturing*, Reston Publishing, Reston, Va., 1983.
29. S. S. Hacisalihzade, "Control Engineering and Therapeutic Drug Delivery," *IEEE Control Systems*, June 1989, pp. 44–46.
30. E. R. Carson and T. Deutsch, "A Spectrum of Approaches for Controlling Diabetes," *IEEE Control Systems*, December 1992, pp. 25–30.
31. J. R. Sankey and H. Kaufman, "Robust Considerations of a Drug Infusion System," *Proceedings of the American Control Conference*, San Francisco, Calif., June 1993, pp. 1689–1695.
32. W. S. Levine, *The Control Handbook*, CRC Press, Boca Raton, Fla., 1996.

33. D. Auslander and C. J. Kempf, *Mechatronics*, Prentice Hall, Englewood Cliffs, N. J., 1996.
34. "Things that Go Bump in Your Flight," *The Economist*, July 3, 1999, pp. 69–70.
35. P. J. Brancazio, "Science and the Game of Baseball," *Science Digest*, July 1984, pp. 66–70.
36. C. Klomp, et al., "Development of an Autonomous Cow-Milking Robot Control System," *IEEE Control Systems*, October 1990, pp. 11–19.
37. M. B. Tischler et al., "Flight Control of a Helicopter," *IEEE Control Systems*, August 1999, pp. 22–32.
38. G. B. Gordon and J. C. Roark, "ORCA: An Optimized Robot for Chemical Analysis," *Hewlett-Packard Journal*, June 1993, pp. 6–19.
39. K. Hollenback, "Destabilizing Effects of Muscular Contraction in Human-Machine Interaction," *Proceedings of the American Control Conference*, San Francisco, Calif., 1993, pp. 736–740.
40. L. Scivicco and B. Siciliano, *Modeling and Control of Robot Manipulators*, McGraw-Hill, New York, 1996.
41. O. Mayr, "Adam Smith and the Concept of the Feedback System," *Technology and Culture*, Vol. 12, No. 1, January 1971, pp. 1–22.
42. A. Goldsmith, "Autofocus Cameras," *Popular Science*, March 1988, pp. 70–72.
43. R. Johansson, *System Modeling and Identification*, Prentice Hall, Englewood Cliffs, N. J., 1993.
44. M. DiChristina, "Telescope Tune-Up," *Popular Science*, September 1999, pp. 66–68.
45. K. Capek, *Rossum's Universal Robots*, English version by P. Selver and N. Playfair, Doubleday, Page, New York, 1923.
46. D. Hancock, "Prototyping the Hubble Fix," *IEEE Spectrum*, October 1993, pp. 34–39.
47. A. K. Naj, "Engineers Want New Buildings to Behave Like Human Beings," *Wall Street Journal*, January 20, 1994, p. B1.
48. E. H. Maslen, et al., "Feedback Control Applications in Artifical Hearts," *IEEE Control Systems*, December 1998, pp. 26–30.
49. M. DiChristina, "What's Next for Hubble?" *Popular Science*, March 1998, pp. 56–59.
50. K. G. Ashar, *Magnetic Disk Drive Technology*, IEEE Press, Piscataway, N. J., 1997.
51. T. Coughlin, "2010 Could Be Great Year for Hard Disk Drives," Analysis published at quicktake.morningstar.com, http://www.glgroup.com/News/2010-could-be-great-year-for-hard-disk-drives-46098.html.
52. R. Stone, "Putting a Human Face on a New Breed of Robot," *Science*, October 11, 1996, p. 182.
53. P. I. Ro, "Nanometric Motion Control of a Traction Drive," *ASME Dynamic Systems and Control*, Vol. 55.2, 1994, pp. 879–883.
54. K. C. Cheok, "A Smart Automatic Windshield Wiper," *IEEE Control Systems Magazine*, December 1996, pp. 28–34.
55. D. Dooling, "Transportation," *IEEE Spectrum*, January 1996, pp. 82–86.
56. G. Norris, "Boeing's Seventh Wonder," *IEEE Spectrum*, October 1995, pp. 20–23.
57. D. Hughes, "Fly-by-Wire 777 Keeps Traditional Cockpit," *Aviation Week & Space Technology*, McGraw-Hill Publication, May 1, 1995, pp. 42–48.
58. R. Reck, "Design Engineering," *Aerospace America*, American Institute of Aeronautics and Astronautics, December 1994, p. 74.
59. C. Rist, "Angling for Momentum," *Discover*, September 1999, p. 37.
60. S. J. Elliott, "Down With Noise," *IEEE Spectrum*, June 1999, pp. 54–62.
61. W. Ailor, "Controlling Space Traffic," *AIAA Aerospace America*, November 1999, pp. 34–38.
62 "Seagate Outlines the Future of Storage," An interview with Jim Chirico, 2006, http://www.hardwarezone.com/articles/view.php?cid=1&id=1805&pg=2.
63. G. F. Hughes, "Wise Drives," *IEEE Spectrum*, August 2002, pp. 37–41.
64. R. H. Bishop, *The Mechatronics Handbook*, 2nd ed., CRC Press, Inc., Boca Raton, Fla., 2007.
65. N. Kyura and H. Oho, "Mechatronics—An Industrial Perspective," *IEEE/ASME Transactions on Mechatronics*, Vol. 1, No. 1, 1996, pp. 10–15.
66. T. Mori, "Mecha-tronics," *Yasakawa Internal Trademark Application Memo 21.131.01*, July 12, 1969.
67. F. Harshama, M. Tomizuka, and T. Fukuda, "Mechatronics—What is it, Why, and How?—An Editorial," *IEEE/ASME Transactions on Mechatronics*, Vol. 1, No. 1, 1996, pp. 1–4.
68. D. M. Auslander and C. J. Kempf, *Mechatronics: Mechanical System Interfacing*, Prentice Hall, Upper Saddle River, N. J., 1996.
69. D. Shetty and R. A. Kolk, *Mechatronic System Design*, PWS Publishing Company, Boston, Mass., 1997.
70. W. Bolton, *Mechatronics: Electrical Control Systems in Mechanical and Electrical Engineering*, 2nd ed., Addison Wesley Longman, Harlow, England, 1999.

71. D. Tomkinson and J. Horne, *Mechatronics Engineering*, McGraw-Hill, New York, 1996.
72. H. Kobayashi, Guest Editorial, *IEEE/ASME Transactions on Mechatronics*, Vol. 2, No. 4, 1997, p. 217.
73. D. S. Bernstein, "What Makes Some Control Problems Hard?" *IEEE Control Systems Magazine*, August 2002, pp. 8–19.
74. S. Douglas, "No Pilot Required," *Popular Science*, June 2001, pp. 40–44.
75. O. Zerbinati, "A Direct Methanol Fuel Cell," *Journal of Chemical Education*, Vol. 79, No. 7, July 2002, p. 829.
76. D. Basmadjian, *Mathematical Modeling of Physical Systems: An Introduction*, Oxford University Press, New York, N.Y., 2003.
77. D. W. Boyd, *Systems Analysis and Modeling: A Macro-to-Micro Approach with Multidisciplinary Applications*, Academic Press, San Diego, CA, 2001.
78. F. Bullo and A. D. Lewis, *Geometric Control of Mechanical Systems: Modeling, Analysis, and Design for Simple Mechanical Control Systems*, Springer Verlag, New York, N.Y., 2004.
79. P. D. Cha, J. J. Rosenberg, and C. L. Dym, *Fundamentals of Modeling and Analyzing Engineering Systems*, Cambridge University Press, Cambridge, United Kingdom, 2000.
80. P. H. Zipfel, *Modeling and Simulation of Aerospace Vehicle Dynamics*, AIAA Education Series, American Institute of Aeronautics & Astronautics, Inc., Reston, Virginia, 2001.
81. D. Hristu-Varsakelis and W. S. Levin, eds., *Handbook of Networked and Embedded Control Systems*, Series: Control Engineering Series, Birkhäuser, Boston, MA, 2005.
82. See the website http://www.gps.gov.
83. B. W. Parkinson and J. J. Spilker, eds., *Global Positioning System: Theory & Applications,* Vol. 1 & 2, Progress in Astronautics and Aeronautics, AIAA, 1996.
84. E. D. Kaplan and C. Hegarty, eds., *Understanding GPS: Principles and Applications*, 2nd ed., Artech House Publishers, Norwood, Mass., 2005.
85. B. Hofmann-Wellenhof, H. Lichtenegger, and E. Wasle, *GNSS-Global navigation Satellite Systems*, Springer-Verlag, Vienna, Austria, 2008.
86. M. A. Abraham and N. Nguyen, "Green Engineering: Defining the Principles," *Environmental Progress*, Vol. 22, No. 4, American Institute of Chemical Engineers, 2003, pp. 233–236.
87. "National Electric Delivery Technologies Roadmap: Transforming the Grid to Revolutionize Electric Power in North America," U.S. Department of Energy, Office of Electric Transmission and Distribution, 2004.
88. D. T. Allen and D. R. Shonnard, *Green Engineering: Environmentally Conscious Design of Chemical Processes,* Prentice Hall, N. J., 2001.
89. "The Modern Grid Strategy: Moving Towards the Smart Grid," U.S. Department of Energy, Office of Electricity Delivery and Energy Reliability, http://www.netl.doe.gov/moderngrid/.
90. "Smart Grid System Report," U.S. Department of Energy, July 2009, http://www.oe.energy.gov/ DocumentsandMedia/SGSRMain_090707_lowres.pdf.
91. Climate Group, "SMART 2020: Enabling the Low Carbon Economy in the Information Age," A report by the Climate Group on behalf of the Global eSustainability Initiative, published by Creative Commons, 2008.
92. J. Machowski, J. Bialek, and J. Bumby, *Power System Dynamics: Stability and Control*, 2nd ed., John Wiley & Sons, Ltd, West Sussex, United Kingdom, 2008.
93. R. H. Bishop, ed., *Mechatronics Handbook*, 2nd ed., CRC Press, 2007.
94. See http://www.burjdubai.com/
95. R. Roberts, "Control of high-rise high-speed elevators," *Proceedings of the American Control Conference*, Philadelphia, Pa., 1998, pp. 3440–3444.
96. N. L. Doh, C. Kim, and W. K. Chung, "A Practical Path Planner for the Robotic Vacuum Cleaner in Rectilinear Environments," *IEEE Transactions on Consumer Electronics*, Vol. 53, No. 2, 2007, pp. 519–527.
97. S. C. Lin and C. C. Tsai, "Development of a Self-Balancing Human Transportation Vehicle for the Teaching of Feedback Control," *IEEE Transactions on Education*, Vol. 52, No. 1, 2009, pp. 157–168.
98. K. Li, E. B. Kosmatopoulos, P. A. Ioannou, and H. Ryaciotaki-Boussalis, "Large Segmented Telescopes: Centralized, Decentralized and Overlapping Control Designs," *IEEE Control Systems Magazine*, October 2000.
99. A. Cavalcanti, "Assembly Automation with Evolutionary Nanorobots and Sensor-Based Control Applied to Nanomedicine," *IEEE Transactions on Nanotechnology*, Vol. 2, No. 2, 2003, pp. 82–87.